T0263874

The *Chlamydomonas* Sourcebook

SECOND EDITION

Volume 1: Introduction to *Chlamydomonas* and Its Laboratory Use

The *Chlamydomonas* Sourcebook

SECOND EDITION

Volume 1: Introduction to *Chlamydomonas* and Its Laboratory Use

Elizabeth H. Harris, Ph.D.
Department of Biology
Duke University
Durham, North Carolina, USA

AMSTERDAM • BOSTON • HEIDELBERG • LONDON • NEW YORK
OXFORD • PARIS • SAN DIEGO • SAN FRANCISCO • SYDNEY • TOKYO

Academic Press is an imprint of Elsevier

Cover image: A cluster of post-mitotic *Chlamydomonas* daughter cells from a mother cell that underwent two rounds of division. Courtesy of Su-chiung Fang and James Umen.

Academic Press is an imprint of Elsevier

The Boulevard, Langford Lane, Kidlington, Oxford OX5 8GB
30 Corporate Drive, Suite 400, Burlington, MA 01803, USA
525 B Street, Suite 1900, San Diego, CA 92101-4495, USA

First edition 1989
Second edition 2009

Copyright 2009 © Elsevier Inc. All rights reserved

No part of this publication may be reproduced, stored in a retrieval system or transmitted in any form or by any means electronic, mechanical, photocopying, recording or otherwise without the prior written permission of the publisher.

Permissions may be sought directly from Elsevier's Science & Technology Rights Department in Oxford, UK: phone (+44) (0) 1865 843830; fax (+44) (0) 1865 853333; email: permissions@elsevier.com. Alternatively you can submit your request online by visiting the Elsevier web site at http://elsevier.com/locate/permissions, and selecting *Obtaining permission to use Elsevier material*.

Notice
No responsibility is assumed by the publisher for any injury and/or damage to persons or property as a matter of products liability, negligence or otherwise, or from any use or operation of any methods, products, instructions or ideas contained in the material herein. Because of rapid advances in the medical sciences, in particular, independent verification of diagnoses and drug dosages should be made.

Set ISBN: 978-0-12-370873-1

Volume 1: ISBN: 978-0-12-370874-8
Volume 2: ISBN: 978-0-12-370875-5
Volume 3: ISBN: 978-0-12-370876-2

For information on all Academic Press publications
visit our website at www.elsevierdirect.com

Typeset by Charon Tec Ltd., A Macmillan Company. (www.macmillansolutions.com)

Printed and bound in the United Kingdom
Transferred to Digital Printing, 2011

Working together to grow
libraries in developing countries

www.elsevier.com | www.bookaid.org | www.sabre.org

ELSEVIER   BOOK AID International   Sabre Foundation

Contents of Volume 1

Preface

When the first edition of *The Chlamydomonas Sourcebook* went to press in 1988, our most urgent issue was to develop methods for transforming *Chlamydomonas* cells with exogenous DNA. Within two years, the transformation problem had been solved through the joint efforts of many members of the Chlamy community (thereby rendering the last part of Chapter 10 obsolete shortly after publication), and *Chlamydomonas* joined the ranks of model organisms that could be manipulated with the tools of molecular biology. Over the past 20 years new lines of research have emerged, new investigators have joined us, a draft genome sequence has been completed, and more than 4000 papers have been published. Clearly it is time for a new edition.

In visiting colleagues over the years I have been pleased to see copies of the first edition of the book on laboratory shelves – and even more pleased when their bindings are disintegrating and pages are stained with green, an indication that I've fulfilled my goal of providing a comprehensive and useful reference tool for the investigator. In thinking about a second edition, this was still a primary objective, but I also realized that one person could no longer cover everything adequately, and that we needed expert reviews of the many aspects of current *Chlamydomonas* research. The three-volume format was developed to meet these requirements.

Volume 1 provides an introduction to *Chlamydomonas* and the essential information for working with it in the laboratory. My target audience is the newcomer to Chlamy, perhaps a graduate student or postdoc, or a more experienced investigator from another discipline who realizes its advantages for a particular study. I hope, however, that old hands will also find interesting and useful information here. Like the first edition, this volume provides historical background that is often neglected, even in review articles, because of page limitations. Some text and many classic figures from the first edition have been retained, but I've tried to bring the overview of research topics up to date.

Volume 2, edited by David Stern, covers the "green side" of Chlamy – chloroplasts, photosynthesis, and related processes – and also various aspects of intermediary metabolism, respiration, and other biochemical functions. Volume 3, edited by George Witman, provides similarly detailed reviews of topics related to flagellar structure and assembly, motility, and behavior. We hope that the whole set will find homes on more laboratory shelves, and that these volumes will acquire their own green stains as *Chlamydomonas* research goes forward.

Elizabeth H. Harris

Acknowledgments

I chose David Stern and George Witman as editors for Volumes 2 and 3, respectively, not only because of their expertise on the topics to be covered, but also because I knew from past experience that both wrote well and were attentive to detail. Although they may never forgive me for persuading them to take on this project, I am immensely grateful to both of them for accepting the challenge, and for the many long hours they have put into producing first-rate books. Thank you for a job well done!

I also want to convey my gratitude to the authors of all the chapters in those volumes. They covered their topics thoroughly and well, and without exception they were quick to respond and never complained about our many requests for changes in the interest of consistency.

Erik Hom, a postdoctoral fellow at Harvard, was my "guinea pig" reader for Volume 1. He reminded me of things that weren't obvious to a Chlamy newcomer, asked good questions, and made many excellent suggestions. I also thank Ursula Goodenough, Thomas Pröschold, and George Witman who read and critiqued chapters. I am grateful to Ron Hoham and Wayne Armstrong for information on red snow algae, and to Judy Acreman, Jerry Brand, Ken Hoober, Bill Inwood, Moira Lawson, Pete Lefebvre, Tasios Melis, Sabeeha Merchant, Saul Purton, Bill Snell, and Jim Umen, who answered questions or sent material on various topics.

Many of the figures used in Volume 1 were carried over from the first edition, and once again I thank everyone who contributed them. For new figures, I thank Annette Coleman, Patricia Daniel, Carol Dieckmann, Susan Dutcher, Mariano García-Blanco, Ursula Goodenough, Ulrich Kück, Matt Laudon, Jac-Hycok Lee, Telsa Mittelmeier, Jon Nield, Yoshiki Nishimura, Thomas Pröschold, Bill Snell, Jim Umen (who also provided the cover illustration), James Uniacke, Karen VanWinkle-Swift, Sabine Waffenschmidt, Qian Wang, George Witman, and Bill Zerges.

I also want to acknowledge Françoise Bouazzat, Judy Edwards, Maike Lorenz, and Claude Yéprémian for sending photographs that ultimately I didn't use. On the advice of an editor at Elsevier, the decision was made to eliminate color figures from this volume to reduce costs, and as a result I abandoned a plan to include portraits of some notable persons in *Chlamydomonas* research. I hope to be able to post some of these photos on a web site eventually.

Finally, I thank my husband Albert Harris for his companionship and ironic sense of humor that have sustained me through the whole process. For more than 40 years, he's always been there when I needed him. His book will be our next project.

Conventions Used

We assume that the reader is familiar with the basic vocabulary of cell and molecular biology. Abbreviations and acronyms specific to particular areas of research or to *Chlamydomonas* are defined in the list that follows. *Chlamydomonas* is assumed to refer to *C. reinhardtii* unless otherwise specified. The reference genome sequence is the DOE Joint Genome Institute (JGI) version 3.0 (http://www.genome.jgi-psf.org/Chlre3/Chlre3.home.html; Merchant et al., 2007). Supplementary material can be found on the Elsevier companion web site, http://www.elsevierdirect.com/companions/9780123708731.

Nomenclature for genes and proteins follows the guidelines, contributed by S.K. Dutcher and E.H. Harris, in the *Genetic Nomenclature Guide* published by *Trends in Genetics* in 1998, and also summarized at http://www.chlamy.org/nomenclature.html. In brief, genes encoded in the nucleus are designated by uppercase italic letters, often followed by an arabic number to distinguish different loci with the same name (*ARG7*). Three-letter names are preferred. Unless otherwise named, proteins are designated by the relevant uppercase gene symbol, but not italic (e.g., ARG7). Mutant alleles are designated in lowercase italics; different mutant alleles at the same locus are distinguished by a number separated from the gene symbol by a hyphen (e.g., *arg7-1*).

Nomenclature for organelle-encoded genes and their products is based on the system used for cyanobacteria and for plants: gene symbols consist of three letters in lowercase followed by one uppercase letter or number, in italics (*petG*, *rps12*). When a protein has no more familiar name, it may be expressed as the same letters in mixed case, not italics (PetG).

Abbreviations

ATCC	American Type Culture Collection, Rockville, MD, USA
BP	biparental, referring to inheritance of organelle genes from both parents in a cross
CC-	prefix for strains in the Chlamydomonas Resource Center collection
CCAP	Culture Centre of Algae and Protozoa, Oban, Scotland, UK
CCM	carbon concentrating mechanism
CV	contractile vacuole
DAPI	4,6-diamidino-2-phenylindole, a DNA-binding fluorochrome
DCMU	the herbicide 3-(3,4-dichlorophenyl)-1,1-dimethylurea, which inhibits photosynthesis at the level of the PS II reaction center
EM	electron microscopy
EMS	the mutagen ethyl methanesulfonate
IAM	Institute of Applied Microbiology, Tokyo, Japan
IFT	intraflagellar transport
ITS1, ITS2	internal transcribed spacer sequences in the nuclear genes encoding cytosolic ribosomal RNAs
LHCI, LHCII	light-harvesting antennae of photosystems I and II respectively
MID	"minus-dominance," a gene in the *MT* locus of *minus* cells
MMS	the mutagen methyl methanesulfonate
MNNG	the mutagen N-methyl-N'-nitro-N-nitrosoguanidine
MT	mating type locus
ORF	open reading frame
PAR	photosynthetically active radiation
PS I, PS II	photosystem I, photosystem II
QFC	quadriflagellate cell, formed by fusion of *Chlamydomonas plus* and *minus* gametes
Rubisco	ribulose-1,5-bisphosphate carboxylase/oxygenase
SAG	Sammlung von Algenkulturen, Göttingen, Germany
SDR	short dispersed repeat sequences in chloroplast DNA
TAP	Tris-acetate-phosphate culture medium
UP	uniparental, referring to inheritance of organelle genes from only one parent in a cross
UTEX	University of Texas Algal Collection, Austin TX, USA

List of Tables

List of Figures

Contents of Volume 2

edited by David B. Stern

Contents of Volume 3

edited by George B. Witman

The Genus *Chlamydomonas*

I. INTRODUCTION

This chapter reviews the history of research on species of the chlorophyte algal genus *Chlamydomonas*, with emphasis on the genetically important species *C. reinhardtii*. The origins of the principal laboratory strains of this species are given in detail (insofar as the historical records permit), as questions of strain identity may be important in modern experimental work. The chapter concludes with a discussion of other *Chlamydomonas* species that have received more than passing attention in laboratory studies.

II. DESCRIPTION OF THE GENUS

The genus *Chlamydomonas* (Greek: *chlamys*, a cloak or mantle; *monas*, solitary, now used as a generic term for certain unicellular flagellates) was named by C.G. Ehrenberg (1833, 1838), and probably corresponds to the flagellate *Monas* described in 1786 (O.F. Müller, cited by Gerloff, 1940; Ettl, 1976a). The species described by Ehrenberg is uncertain; Ettl (1976a) suggested it may have been *C. pulvisculus*, but since the published description and illustration could apply to several of the species recognized today, Ettl considered the type genus to be *C. reinhardtii*, which was not described until 1888 but by the 1960s had become the most widely used species in laboratory work. Pröschold et al. (2001; Pröschold and Silva, 2007) have proposed that the genus be redefined based on the laboratory wild-type strain of *C. reinhardtii* isolated in 1945 in Massachusetts. The essential features of the genus are two anterior flagella of equal length, whose points of emergence from the cell body are not widely separated; a cell wall; and a single chloroplast or chromatophore containing one or more pyrenoids (Figure 1.1).

Dill (1895) listed 15 species of *Chlamydomonas*, of which six were new descriptions. By 1927 the list had grown to 146 species found in central Europe. Pascher (1927) delineated six subgenera based on chloroplast shape and number and position of the pyrenoid(s), and Gerloff (1940) provided a new key to these and described additional species, bringing the total to 321. Ettl's comprehensive monograph *Die Gattung Chlamydomonas*

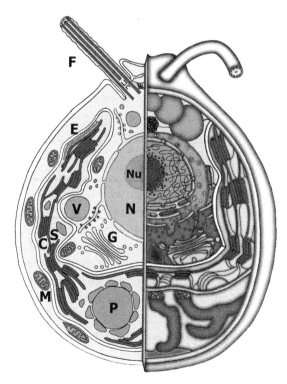

FIGURE 1.1 *Cell structure of Chlamydomonas reinhardtii, showing the central nucleus (N) with the nucleolus (Nu), the two isoform flagella (F), the cup-shaped chloroplast (C) with the eyespot (E) and the starch-containing pyrenoid (P) and the mitochondria (M). In addition, one may distinguish the Golgi vesicle (G), starch grains (S), and vacuoles (V). From Nickelsen and Kück (2000).*

(Ettl, 1976a) summarized the literature on 459 species, elevated the previous subgenus *Chloromonas* to a separate genus, and divided the remaining species into nine groups which he preferred to call Hauptgruppen rather than subgenera, implying no formal taxonomic rank (Figure 1.2). About a third of the species discussed by Ettl are available from one or more of the major algal collections (see http://www.chlamy.org/species.html).

Apart from his assignment of snow- and ice-dwelling forms to one group (Sphaerella), Ettl considered neither habitat nor mode of reproduction in making these major divisions. Within most of the nine groups, Ettl further separated species with similar chloroplast morphology. Species were distinguished from one another by several traits, including presence or absence of a pronounced apical papilla, number and position of contractile vacuoles, overall body shape, thickness of the cell wall, shape and position of the eyespot, and whether a gelatinous sheath surrounded the cell.

Ettl continued refining his *Chlamydomonas* taxonomy until his death in 1997, and collaborated with U.G. Schlösser in reexamination of cultures from the Sammlung von Algenkulturen at Göttingen. Schlösser (1984) had previously separated *Chlamydomonas* species into 15 groups on the basis of their response to vegetative cell autolysins, lytic enzymes that dissolve the mother cell wall shortly after cytokinesis to release newly formed cells, and Ettl and Schlösser (1992) revised several species designations to take this analysis into account. Thomas Pröschold and his collaborators are continuing this work.

The order Volvocales includes both unicellular and colonial algae. Within this order, the family Chlamydomonadaceae consists of approximately 30 genera of unicellular algae with cell walls and with either two or four flagella (Smith, 1938; Bourrelly, 1966; Jakubiec, 1984). DNA sequence analysis clearly demonstrates, however, that this family comprises multiple phylogenetic lineages that do not correspond to the morphologically defined genera (Jupe et al., 1988; Simard et al., 1988; Buchheim et al., 1990, 1996, 1997a, b; Coleman and Mai, 1997), and that some species of *Chlamydomonas*, including *C. reinhardtii*, are much more closely related to colonial Volvocacean algae than to some other unicellular species traditionally included among the Chlamydomonadaceae (see below). The sequence data do suggest that the structural considerations underlying the Hauptgruppen and the specificity of the lytic enzymes have some phylogenetic validity, but the correspondence is imperfect. Coleman and Mai (1997) pointed out, for example, that species in Ettl's group Chlorogoniella belong to at least two different clades, one of which is more closely allied with species in the Chlamydella group.

Pröschold et al. (2001) began a comprehensive revision of the genus *Chlamydomonas* by defining 18 monophyletic lineages within the class Chlorophyceae, based on 18S rRNA sequences, of which 8 clades include species traditionally assigned to the genera *Chlamydomonas* and/or

1b. Snow- or ice-dwelling species, chloroplast structure more difficult to discern	Sphaerella	
2a. Cells with only one pyrenoid	3	
2b. Cells with two or more pyrenoids	7	
3a. Chloroplast cup-shaped or derived from this form	4	
3b. Chloroplast not cup-shaped	5	
4a. Pyrenoid basal	Euchlamydomonas	
4b. Pyrenoid lateral	Chlamydella	
5a. Chloroplast appressed to wall on one side, pyrenoid lateral	Chlorogoniella	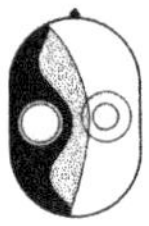
5b. Chloroplast tubular, with cross-bridges in which the pyrenoid lies embedded (H-shaped in longitudinal section)	6	
6a. Nucleus in lumen in front of the pyrenoid	Pseudagloë	
6b. Nucleus in lumen behind the pyrenoid	Agloë	
7a. Only two pyrenoids present	8	
7b. Several to numerous pyrenoids	Pleiochloris	
8a. Chloroplast cup-shaped, pyrenoids lying laterally and opposite	Bicocca	
8b. Chloroplast tubular, with two cross-bridges, before and behind the nucleus, in each of which lies a pyrenoid (in the long axis one behind the other)	Amphichloris	

FIGURE 1.2 *Ettl's taxonomic key to the major groups of Chlamydomonas species. The groups have provided a useful framework for comparison of species, but have never had formal taxonomic rank. Adapted from Ettl (1976a).*

Chloromonas (Figure 1.3). Species in one of these clades were redescribed as two new genera, and some members of other clades were reclassified at the species level. Eventually the genus name *Chlamydomonas* is to be restricted to a single clade, of which *C. reinhardtii* will be the type species,

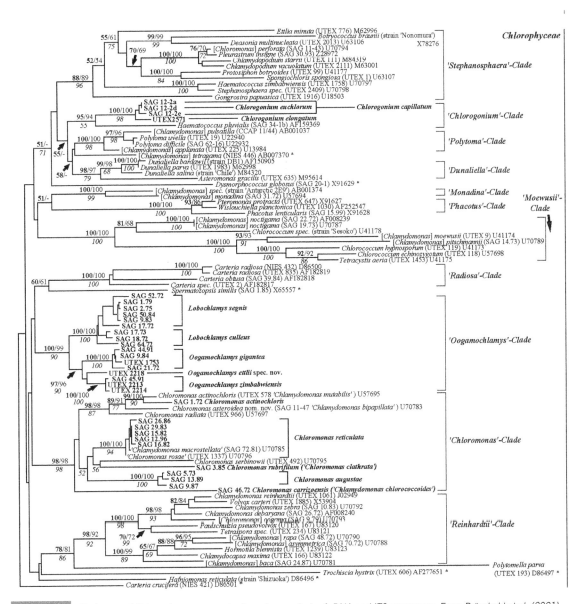

FIGURE 1.3 *Phylogeny of Chlamydomonas species based on analysis of rRNA and ITS sequences. From Pröschold et al. (2001).*

but within the context of the present book, we will consider the genus in its broader traditional sense.

Chlamydomonas was historically considered by some phycologists to include the genus *Chloromonas*, whose cells are similar in overall architecture but lack pyrenoids. Pascher (1927) combined these two genera in his comprehensive treatment of the Volvocales, but subsequent works (for example, Gerloff, 1962; Ettl, 1976a) usually separated them. However, pyrenoids can become disrupted or disappear under some environmental conditions or in different phases of the cell cycle, and molecular analysis shows that some *Chloromonas* and *Chlamydomonas* species form a monophyletic group (Buchheim et al., 1997a; Morita et al., 1999; Pröschold et al., 2001; Hoham et al., 2002). Pröschold et al. (2001) emended the definition of *Chloromonas* to include some strains with pyrenoids that were formerly considered to be *Chlamydomonas* species, and Hoham et al. (2002) revised the taxonomy of several snow-dwelling isolates from both genera. Hoham further suggested that pyrenoids may have been gained and lost several times in the evolution of this group of algae, and that the distinction between *Chloromonas* and *Chlamydomonas* may be as simple as the expression level of the *CCM1* gene which regulates the carbon-concentrating mechanism (see Chapter 6).

There is also argument whether *Gloeomonas* should be regarded as a separate genus within the Chlamydomonadaceae, the principal distinguishing feature of this group being a slightly wider separation of the flagellar origins compared to those of most *Chlamydomonas* species (Ettl, 1965a, c; Fott, 1974). Another major genus of the same family is *Carteria*, whose cells have four rather than two flagella but otherwise look very much like those of *Chlamydomonas*. As in the case of *Chlamydomonas*, molecular studies of *Carteria* species indicate that this is a polyphyletic genus. Buchheim and Chapman (1992) determined based on rRNA sequences that the genus comprises at least two monophyletic groups, and Nozaki et al. (1997) proposed four groups based on sequences of the chloroplast *rbcL* gene. At least one "*Carteria*" species has been demonstrated to be a long-lived quadriflagellate product of *Chlamydomonas* mating (Behlau, 1939; see Chapter 5). *Sphaerellopsis* and *Smithsonimonas* have a wide, gelatinous sheath that differs from the shape of the protoplast, in contrast to *Chlamydomonas* and *Chloromonas*, in which the sheath, if any, conforms closely to the protoplast shape.

Polytoma is a genus of non-photosynthetic flagellates that closely resemble *Chlamydomonas* in body structure and appear to retain some vestige of chloroplast nucleic acids and ribosomes (Pringsheim, 1963; Siu et al., 1976a–c; Rumpf et al., 1996). The genus *Polytomella* comprises another group of colorless species that differ from *Polytoma* in lacking cell walls. The absence of chloroplasts makes these algae especially useful subjects for study of mitochondrial proteins and their function, and for comparison of these properties with those of *Chlamydomonas* (Antaramian et al., 1996; van Lis et al., 2005).

Dunaliella and related genera form an analogous group of wall-less, halotolerant green flagellates which in many respects appear to be very closely allied to the Chlamydomonadaceae, although sequence analysis suggests they form distinctly different clades (Coleman and Mai, 1997; González et al., 2001). Placing them in a class distinct from the walled Volvocales, as proposed by Ettl (1981), is probably not warranted, however. *Dunaliella* species are grown commercially for production of β-carotene and glycerol, and have been the subject of substantial research over the past century (see Oren, 2005, for review; also Raja et al., 2007 and http://www.dunaliella.org/). *Haematococcus pluvialis*, another wall-less alga, is used for commercial production of astaxanthin and has been the subject of many studies on carotenoid biosynthesis (see for example Lorenz and Cysewski, 2000; Yoshimura et al., 2006; Eom et al., 2006).

The evolutionary relationships of *Chlamydomonas* to other genera, and particularly the position of the Volvocales as a side branch in the development of higher plants from green algae, were explored at the morphological level by investigators in several laboratories. The composition and organization of the cell wall, the morphology of the flagellar root system, and the structures involved in cytokinesis are the most significant features contributing to structure-based evolutionary schemes within the Chlorophyta (Pickett-Heaps, 1975; Ettl, 1981; McCourt, 1995; Lewis and McCourt, 2004). All *Chlamydomonas* and *Chloromonas* species belong to a subgroup of the class Chlorophyceae in which the basal bodies are displaced in a clockwise orientation (O'Kelly and Floyd, 1983–1984; Nakayama et al., 1996), as do *Dunaliella, Polytoma, Haematococcus, Chlorococcum,* and several other genera (Pröschold et al., 2001).

Molecular studies have provided additional, and sometimes surprising, insights into algal phylogeny. The highly conserved 18S and 26S rRNAs have proved especially useful in providing a broad overview of algal phylogeny (Buchheim and Chapman, 1991; Chapman and Buchheim, 1992), and the more rapidly evolving internal transcribed sequences (ITS1 and ITS2) within the ribosomal DNA cistrons have been used to separate lineages within the Volvocales (Coleman and Mai, 1997; Mai and Coleman, 1997; Coleman et al., 1998), as have intron sequences within other genes (Liss et al., 1997). Sequences of chloroplast genes, such as *rbcL* and *atpB*, have also been used for phylogenetic studies among the Volvocales (Morita et al., 1999; Nozaki et al., 1995, 2000, 2003). Fawley and Buchheim (1995) have suggested that presence or absence of the pigment loroxanthin might also be a useful taxonomic character, since it can be quickly assayed and seems to have some correlation with phylogenies determined from sequence analysis. Of particular interest for *Chlamydomonas* research is the discovery that *C. reinhardtii*, by far the best-investigated species, is much more closely related to some species of colonial genera such as *Volvox* and *Gonium* than to many other algae previously identified as *Chlamydomonas* (Buchheim et al., 1990; Larson et al., 1992; Woessner and Goodenough, 1994; Nozaki et al., 1999; Pröschold et al., 2001).

Although cell body shape and size vary among *Chlamydomonas* species (as defined on morphological criteria), the overall polar structure, with paired apical flagella and basal chloroplast surrounding one or more pyrenoids, is constant. Cells are usually free-swimming in liquid media but on solid substrata may be non-flagellated, and are often seen in gelatinous masses similar to those of the algae *Palmella* or *Gloeocystis* in the order Tetrasporales. This condition has been referred to as a "palmelloid" state (Fott and Novakova, 1971; see also Chapter 2). There has even been some discussion that *Gloeocystis* may comprise palmelloid *Chlamydomonas* species for which no motile stage has been identified (Badour et al., 1973). The *Chlamydomonas* wall is distinct, with some variation in thickness among species. Some species also secrete a mucilaginous polysaccharide coating. Most species have a prominent eyespot, usually red or orange, and two or more contractile vacuoles. Asexual division occurs by lengthwise division of the protoplast. Usually two successive divisions occur to form four daughter cells, which are then released from the mother cell wall (Chapter 3). The forms of sexual reproduction among the *Chlamydomonas* species cataloged by Ettl (1976a) range from isogamy (fusion of morphologically similar gametes, the most prevalent form) to oogoniogamy or oogamy (formation of clearly differentiated egg and sperm cells; see Wiese et al., 1979; Kirk, 2006, also Chapter 5). Ettl did not consider sexual reproduction to be a diagnostic character for the genus. However, Pröschold et al. (2001) have defined a new genus *Oogamochlamys* comprising three oogamous taxa previously considered to be *Chlamydomonas* species, which form a monophyletic lineage based on their 18S rRNA sequences. Sexual fusion of whatever style leads to formation of a zygospore with a hard, thick wall, which is resistant to adverse environmental conditions. Some species may also form asexual resting spores, or akinetes, in which the original vegetative cell wall becomes much thicker and carotenoids, starch, and lipids may accumulate (see Coleman, 1983, for review).

The genus *Chlamydomonas* is of worldwide distribution and is found in a diversity of habitats. Although most of the described species were collected in central Europe, this bias undoubtedly reflects the distribution of phycologists, not of Chlamydomonads. Collection sites include temperate, tropical, and polar regions. *Chlamydomonas* species have been isolated from freshwater ponds and lakes, sewage ponds, marine and brackish waters, snow, garden and agricultural soil, forests, deserts, peat bogs, damp walls, sap on a wounded elm tree, an artificial pond on a volcanic island, mattress dust in the Netherlands, roof tiles in India, and a Nicaraguan hog wallow. A petri plate exposed for 1 minute from an airplane flying at 1100 m altitude produced *Chlamydomonas* among other algae (Brown et al., 1964). Symbiotic species have been found associated with foraminifera (*C. hedleyi* and *C. provasolii*, Lee et al., 1974; Lee and McEnery, 1983; Pawlowski et al., 2001).

III. *CHLAMYDOMONAS* GENETICS: 1830–1960

Descriptive studies in the 19th century led to comprehension of the life cycle of *Chlamydomonas* and to its early recognition as an organism with possibilities for genetic analysis. Sexual reproduction was first described by J.N. Goroschankin in 1875 and further studied by L. Reinhardt, P.A. Dangeard, W. Schmidle, O. Dill, and G. Klebs in the period 1876–1900 (see Ettl, 1976a). Desroche (1912) in a remarkable monograph extolled the virtues of *Chlamydomonas* for studies of motility and response to light, temperature, compression, and gravity, but it was many years before much further work was done in these areas. Pascher (1918) reported segregation of genetic differences in crosses of two *Chlamydomonas* strains differing in several morphological characteristics. Although the identities of the species used by Pascher are uncertain, it is noteworthy that the traits in which they differed included body shape, thickness of the cell wall, presence or absence of the apical papilla, lateral vs. basal position of the chloroplast, and shape of the eyespot, all of which were later used as criteria separating species. The chloroplast position was in fact one of the principal features defining Ettl's Hauptgruppen (see Figure 1.2).

Since Pascher actually observed mating in progress, including nuclear fusion, and obtained recombinant progeny, one suspects that the genetic distances between some species are not nearly so great as the taxonomic keys suggest. Pascher apparently did not pursue these genetic studies, although he continued to work with *Chlamydomonas* and described several new species. Kater (1929) published an extensive cytological study, including descriptions of mitosis and the flagellar apparatus, in *Chlamydomonas nasuta*, and Strehlow (1929) described mating in a homothallic species, *C. monoica*, and in certain heterothallic algae that are now considered to be *Chloromonas* species, but neither of these investigators carried out any genetic analysis.

Chlamydomonas gained prominence, and eventually notoriety, with the publications of Franz Moewus on relative sexuality, Dauermodifikationen, mating substances, and other topics. These studies were important insofar as they demonstrated the potential utility of *Chlamydomonas* for genetic analysis. However, as Lewin (1976) wrote, Moewus "seems to have strayed from the path of strict veracity," and his experimental work therefore will not be discussed in detail here. The reader is referred instead to C.S. Gowans' excellent summary of Moewus's publications (Gowans, 1976b), to Sapp's essay (1987) and book (1990), and to other papers that evaluate particular aspects of Moewus's data (Patau, 1941; Smith, 1946; Sonneborn, 1951; Raper, 1952; Ryan, 1955; Renner, 1958; see also Lewin, 2000).

Although publications by Moewus dominated the *Chlamydomonas* literature in the 1930s, by the next decade his reports were being challenged, and investigations were under way in other laboratories that would lay the foundations for the great progress made in *Chlamydomonas* genetics and cell biology in later years. Apart from his work, the first attempts to preserve sexually competent strains in culture and to investigate the physiological basis of sexuality in *Chlamydomonas* were made by G.M. Smith, Luigi Provasoli, and Ralph Lewin in the 1940s (Smith, 1946, 1950; Lewin, 1949; Smith and Regnery, 1950). By seeking zygospores rather than vegetative cells in natural material, and then germinating these, Smith was able to obtain mating pairs of 15 heterothallic strains, most of which survive in the University of Texas Algal Collection (UTEX) and other collections. Homothallic strains were also identified. In contrast to Moewus, Smith found that all the strains he tested were capable of gametogenesis and mating in darkness and was unable to detect the sexual substances (crocetins) described by Moewus. Provasoli isolated a pair of *C. moewusii* strains which were subsequently used by Lewin for induction of mutations and further studies in genetics. These were first described in an abstract (Lewin, 1949) prophetically entitled, "Genetics of *Chlamydomonas* – paving the way."

Apart from Pascher's early observations, which were not followed up, and Moewus's dubious results, no systematic isolation and analysis of mutants had been done on any *Chlamydomonas* species prior to about 1950. Lewin (1949–1954; see 1953a paper for summary) published a series of papers on the genetics of *C. moewusii*, with paralyzed, vitamin-requiring, and several other phenotypes being used as markers. Linkage was found between two pairs of loci. Gowans (1960) extended this genetic analysis to auxotrophic and other mutants of the related alga *C. eugametos* and determined many gene–centromere distances but was unable to construct a complete linkage map. Although work with these species continued in several laboratories, *C. reinhardtii* has become the species of choice for genetic studies, largely as a result of early work by Ruth Sager, Bill Ebersold, and Paul Levine.

Wishing to study maternal inheritance, Sager was advised by C.B. van Niel, in consideration of Smith's work, to use *C. reinhardtii* because its life cycle was known and it would grow in the dark on an organic carbon source, whereas *C. eugametos* and *C. moewusii*, at that time the better-known species, would not. Sager's early studies on the control of the sexual cycle by nitrogen availability in *C. reinhardtii* (Sager and Granick, 1953, 1954) and on pigment-deficient and antibiotic-resistant mutants (Sager and Palade, 1954; Sager 1955) began a long series of papers. The discovery of non-Mendelian (uniparental) inheritance of certain antibiotic resistance mutations (Sager, 1954) opened the field of experimental organelle genetics, for which *Chlamydomonas* has remained one of the best model systems.

Contemporary studies by Ebersold, Levine, and their collaborators (Ebersold and Levine, 1959; Ebersold et al., 1962) led to construction of the first nuclear genetic maps for *C. reinhardtii* and to the use of *Chlamydomonas* mutants for diverse studies in cell biology, plant physiology, and other disciplines. All these topics will be discussed at length in the chapters that follow. For a brief review of the entire body of early experimental work using *Chlamydomonas* as a model system, see Trainor and Cain (1986).

IV. ORIGINS OF THE MAJOR LABORATORY STRAINS OF *C. REINHARDTII*

The species *C. reinhardtii* was described in 1888 by Dangeard, and named for Ludwig Reinhard(t), a Ukrainian botanist who in 1876 had published a description of copulation in a species he identified as *C. pulvisculus*. Since details of sexual reproduction in this isolate were different in several respects from those typical of *C. pulvisculus* as described by Goroschankin (Ettl, 1976a), Dangeard described Reinhardt's isolate as a new species, *C. reinhardti*. This name was later cited by Goroschankin (1891) and Gerloff (1940) as *C. reinhardi* and by Pascher (1927) as *C. reinhardii*. Reinhardt's name appears in bibliographies (Gerloff, 1940; Ettl, 1976a) with the t, but in an obituary (1922) and in some reference materials as Reinhard, suggesting that this may have been how he spelled it himself. Nevertheless, the presence of the t in the original species description is binding for taxonomic purposes. The use of two i's rather than one is dictated by rules of botanical Latin nomenclature: when the epithet is for the discoverer of a plant, the specific name is in the genitive singular; to form this when the name ends in a consonant, the letters ii are added. Thus the correct spelling is *reinhardtii* (Ettl, 1976a).

Three principal strains of *C. reinhardtii* have been widely used for genetic and biochemical analyses (Figure 1.4). All have been identified in the literature as descendants of a mating pair (*plus* and *minus*) from the collection of G.M. Smith, supposedly derived from a single zygospore in a soil sample taken from a potato field near Amherst, Massachusetts, in 1945 and designated by Smith as isolate 137c, referring to the third (c) zygote colony recovered from soil sample number 137 (see Table 1.1). Sequence analysis supports the presumption of a common origin, but is inconsistent with the hypothesis that all three lines derive from clonal propagation of individual progeny of a single zygote (Kubo et al., 2002, Pröschold et al., 2005). Probably they are descendants from crosses made among F1 or subsequent progeny of a single zygospore, but we cannot exclude the possibility that they derive from different zygospores collected from the same site, especially in view of the fact that several of the early references to these strains are to 137+ and 137−, rather than to 137c specifically. In any case, the three lines have been separate since at least the early 1950s.

Sager obtained strains of both mating types from Smith, either directly or perhaps by way of S.H. Hutner or L. Provasoli, prior to 1953. The strains

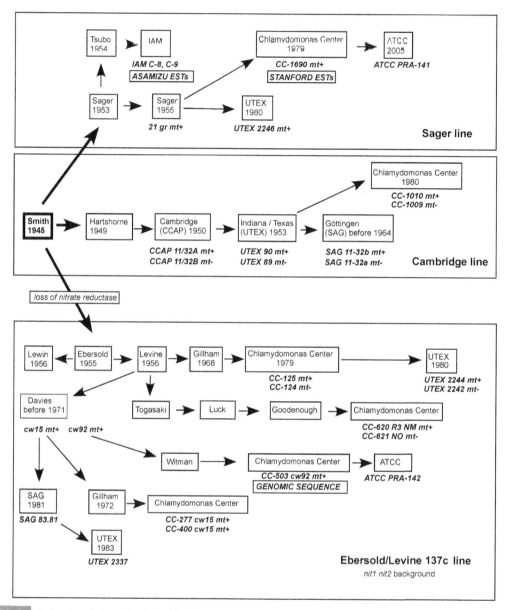

FIGURE 1.4 *Laboratory strains of C. reinhardtii derived from the 1945 isolate of G.M. Smith. Based on Pröschold et al. (2005), in which information on the molecular analysis can be found.*

C-8 and C-9 of the IAM algal collection in Tokyo are descended from stocks sent to Yoshihiro Tsubo by Sager in 1954. In 1955, Sager described a clone of her *plus* strain that grew well in the dark and designated this as strain 21 gr. A direct descendant of this strain has been used for production of cDNA libraries for EST sequences at Stanford University (Shrager et al., 2003).

She also found a spontaneous mutant in her *minus* culture that was yellow in the dark and named this strain 4Y (later known as the *y1* mutant). Many of her subsequent stocks were developed from crosses between these two strains.

The second line is derived from cultures given by Smith to Hartshorne (1953, 1955), who gave samples to the British culture collection, then at Cambridge, in 1950. The *plus* and *minus* isolates of this strain were designated CCAP 11/32a and 11/32b respectively. In 1953 cultures of these stocks were sent from CCAP to the Indiana University collection of algae (now the University of Texas collection, UTEX), where they were numbered 89 and 90, with 89 supposedly being *plus* and 90 being *minus*. Subsequently it was discovered that UTEX 89 is in fact *minus* and UTEX 90 is *plus*, suggesting that a mix-up occurred either in Cambridge or Indiana at the time this transfer was made (I. Friedmann, personal communication to R.C. Starr, confirmed by the *Chlamydomonas* Genetics Center at Duke in 1980). The stocks 11-32b and 11-32a of the Sammlung von Algenkulturen in Göttingen (SAG), were obtained by W. Koch prior to 1969 from Indiana, not from CCAP. Thus SAG 11-32b is equivalent to UTEX 90 and to CCAP 11/32A, and SAG 11-32a is conversely the same mating type as UTEX 89 and CCAP 11/32B. These differences should be borne in mind when reading the early literature on gametogenesis and mating, in which behaviors specific to one mating type are described.

Strains of the third line, which was obtained from Smith by Ebersold in the early 1950s, are distinguished from other two lines by their inability to utilize nitrate as their sole nitrogen source. This distinction results from the presence in the Ebersold stocks of two unlinked mutations, *nit1* and *nit2*, either of which is sufficient to prevent nitrate utilization (see Chapter 6). This strain is the one that has consistently been designated 137c in publications. In 1955–1956 Ebersold went to Harvard to work with Levine, and his wild-type strain became the ancestral stock of the many non-photosynthetic and other mutants isolated and characterized in Levine's laboratory. All these mutant stocks appear to carry both the *nit1* and *nit2* mutations, indicating that both these mutations were present very early in the history of the Ebersold/Levine line. Furthermore, isolates obtained from Ebersold by Lewin in 1956 were observed at the time to be unable to grow on nitrate medium (Lewin, personal communication), indicating that at least one of the *nit* mutations must have arisen prior to 1956. DNA for genomic sequencing was prepared from CC-503 *cw92 mt*$^+$, a cell wall-deficient strain from the Ebersold/Levine lineage. The close congruence overall of the cDNA and genomic sequences supports the common origin of the Sager and Ebersold/Levine strains, despite the difference in nitrate utilization.

It would appear therefore that although these three strains are closely related, the designation 137c may correctly apply only to one of them

Table 1.1 Equivalence of isolates of *Chlamydomonas reinhardtii* and its close relatives in the Chlamydomonas Resource Center and other collections

Isolate	SAG	UTEX	CCAP	IAM	ATCC	Source	Comments
Standard laboratory strain							
CC-1690 mt^+ (Sager 21 gr)						Near Amherst, Massachusetts, 1945	See text and Figure 1.4, also Pröschold et al. (2005)
CC-407 mt^+ (Tsubo C8)	37.89			C-562	PRA-141		Sager (1955)
CC-3348 mt^+	73.72			[C-8]			May be the same strain as CC-407
CC-408 mt^- (Tsubo C9)	38.89			C-9			
CC-1010 mt^+	11-32b	90	11/32A				
CC-1009 mt^-	11-32a	89	11/32B	C-238			
CC-125 mt^+ (Ebersold/Levine 137c)				C-541			
CC-620 mt^+ (Goodenough R3 NM)							High efficiency mating strain, subclone of equivalent to CC-125
CC-124 mt^- (Ebersold/Levine 137c)							
CC-621 mt^- (Goodenough NO)							High efficiency mating strain, subclone of equivalent to CC-124
CC-410 mt^-	11-32c						Formerly listed as "Caroline's and Lewin;" see Ferris (1989)
CC-1418 mt^-	18.79						Formerly listed as "Florida red tide, Provasoli;" see Ferris (1989)
CC-1374 mt?	77.81						Formerly listed as "G. Paris, France;" see Ferris (1989); does not mate

Strains from other locations

Strain			Location	Reference
CC-1373 *C. smithii mt⁺*	54.72	1062	South Deerfield, Massachusetts, 1945	Hoshaw and Ettl (1966)
CC-2290 *mt⁻* (Gross S1 D2)	32.89	PRA-143	Plymouth, Minnesota, 1987	Gross et al. (1988)
CC-2342 *mt⁻* (Jarvik #6)			Pittsburgh, Pennsylvania, 1988	Spanier et al. (1992)
CC-2344 *mt⁻* (Jarvik #356)			Ralston, Pennsylvania, 1988	Spanier et al. (1992)
CC-2343 *mt⁺* (Jarvik #124)			Melbourne, Florida, 1988	Spanier et al. (1992)
CC-2931 *mt⁻* (Harris #6) CC-2932 *mt⁺* (Harris #10)			Durham, North Carolina, 1991	Harris (1998)
CC-2935 *mt⁻* (LEE-1) CC-2936 *mt⁺* (LEE-2) CC-2937 *mt⁺* (LEE-3) CC-2938 *mt⁻* (LEE-4)			Farnham, Quebec, 1993	Sack et al. (1994)
Not in CC collection		18798	Japan	Listed by ATCC as *C. reinhardtii*, but identity is questionable
Not in CC collection	11-31		Unknown	Sensitive to same lytic enzyme as *C. reinhardtii*; see Table 2.4
CC-3871 *C. incerta mt⁻*	7.73		Probably Cuba	Closely related based on molecular data, but not interfertile with *C. reinhardtii* standard strains
Not in CC collection	81.72		Uncertain	"*C. globosa*," ostensibly from the Netherlands, but may = *C. incerta*

CC- numbers correspond to accessions in the *Chlamydomonas* Resource Center (http://www.chlamycollection.org)
SAG: Sammlung von Algenkulturen (http://www.epsag.uni-goettingen.de/html/sag.html)
UTEX: University of Texas Algal Collection (http://www.utex.org)
CCAP: Culture Collection of Algae and Protozoa (http://www.ccap.ac.uk/)
IAM: Institute of Applied Microbiology, Tokyo; cultures now at NIES (http://www.nies.go.jp/biology/mcc/home.htm)
ATCC: American Type Culture Collection (http://www.atcc.org/)

(see Harris, 1989, for a more detailed discussion of the historical identification of these strains.) Following Pröschold et al. (2005), we will refer to all these strains and their derivatives collectively as the "standard laboratory strains," in contrast to isolates from other locations. Pröschold and Silva (2007) have proposed designating a cryopreserved sample of CCAP 11/32A (=UTEX 90, SAG 11-32b) as the new epitype of the genus *Chlamydomonas*.

Three additional isolates originally from the SAG collection appear to be misidentified representatives of the standard strains: SAG 11-32c, "Caroline Islands, Lewin;" SAG 77.81, "France, G. Paris;" and SAG 18.79, "Florida red tide, Provasoli." These strains have the Gulliver transposon in positions typical of the standard strains (Ferris, 1989), and ribosomal ITS sequences corresponding to those strains (Pröschold et al., 2005).

"Strain 2137" was created by Spreitzer (1980) by crossing Sager's 21 gr strain to the *minus* strain of Ebersold-Levine 137c. A product was selected for ability to grow as single cells in minimal medium, for negative phototaxis (*agg1*, derived from the Ebersold-Levine strain equivalent to CC-124; see Chapter 4), and for green color when grown in the dark (a property of the 21 gr strain used by Spreitzer; both *plus* and *minus* isolates of the Ebersold-Levine wild type in Spreitzer's collection were yellow in the dark and were confirmed to carry a *y1* allele). Many non-photosynthetic and herbicide-resistant mutants have been isolated in this background.

V. OTHER ISOLATES OF *C. REINHARDTII*

After G.M. Smith's death in 1959, his collection was reviewed by R.W. Hoshaw for the Indiana University collection (now UTEX). Smith had previously identified four stocks as *C. reinhardtii*: a mating pair, 137c *plus* and *minus*, from Amherst, Massachusetts, 1945, which Hoshaw assumed were the isolates already in the Indiana collection as strains 89 and 90; 136f, from South Deerfield, Massachusetts, 1945; and 684c, from Santa Cruz, California, 1946. A catalog from Smith's notes of "Heterothallic species of *Chlamydomonas* on hand, 9/1/51," communicated by Robert Page to Richard Starr at Indiana in 1959, lists three additional numbers in a group with *C. reinhardtii*: 358 from Bluefields, Nicaragua, originally collected in 1940; 375 from Mayville, Florida, 1946; and 413 from Livermore, California, 1949. By the time these were given to the Indiana collection (Hoshaw, 1965), 358 had been identified as *C. elliptica* var. *britannica* (now UTEX 1059 and 1060), and 375 as *C. frankii* (now UTEX 1057 and 1058). Both these pairs of strains were designated *C. culleus* by Starr and Zeikus in their 1987 revision of the UTEX collection catalog. Based on DNA sequence analysis, Pröschold et al. (2001) redescribed the SAG strains corresponding to UTEX 1057 and UTEX 1059 as *Lobochlamys culleus*, together with a strain collected in Czechoslovakia in 1969. Smith's number 413, which had been a pair of stocks in 1951 but was now a single isolate and had been moved by Smith to a group by itself, was not further identified and appears to have been discarded. It would appear that none of these latter strains was in fact closely allied with *C. reinhardtii*.

On examination of Smith's strains, Hoshaw in collaboration with Ettl concluded that strains 136f and 684c differed sufficiently from Indiana strains 89 and 90 in body shape and chromatophore (chloroplast) structure to warrant description as a new species, which they named *C. smithii* (Hoshaw and Ettl, 1966). The Massachusetts (136f) strain was taken as the type species. Because Hoshaw tested the *C. smithii* strains for mating with UTEX 89 and 90, whose mating types were incorrectly labeled at the time, his descriptions of the C. *smithii* strains likewise have their mating types reversed. The 136f strain (UTEX 1062, SAG 54.72, CC-1373) is *plus* with respect to the standard strains, and mates well with them, giving many viable progeny. It should therefore probably be regarded as a true *C. reinhardtii* isolate. In contrast, Smith's 684c strain appears to be a totally different species, similar to *Chlamydomonas culleus* and now redescribed as *Lobochlamys culleus* (Pröschold et al., 2001, 2005). Previous reports that this strain could mate with standard strains of *C. reinhardtii* (Hoshaw and Ettl, 1966; Bell and Cain, 1983; Harris, 1989) may be explained by its being homothallic (Coleman and Mai, 1997). We therefore can exclude this strain from the catalog of valid *C. reinhardtii* isolates.

Additional field isolates interfertile with the standard *C. reinhardtii* strains have been obtained from Minnesota (Gross et al., 1988), Pennsylvania (Spanier et al., 1992), Florida (Spanier et al., 1992), North Carolina (Harris, 1998), and Quebec (Sack et al., 1994). The Minnesota isolate S1 D2 has been used to identify single nucleotide polymorphisms with respect to the standard strains (Shrager et al., 2003; Merchant et al., 2007).

Isolate 7.73 in the Sammlung von Algenkulturen (SAG) was formerly known as *Chlamydomonas incerta*, but now appears in their catalog as *C. reinhardtii* based on morphological criteria and its ability to be lysed by the same vegetative cell autolysin as the laboratory strains (Schlösser, 1984; Schlösser and Ettl, unpublished), although it does not mate with them. Ferris et al. (1997) confirmed its close relationship to the standard *C. reinhardtii* strains in a comparison of genes within the mating type region. This isolate is genetically mating type *minus* based on the presence of the *MID* gene (see Chapter 5). Efforts to obtain a *plus* partner have been unsuccessful. Additional sequence comparisons have been made by Liss et al. (1997), Coleman and Mai (1997), Pröschold et al. (2005), Kramzar et al. (2006), and by Popescu et al. (2006) who prepared a cDNA library from SAG 7.73.

SAG 81.72, formerly *C. globosa*, appears to be identical with SAG 7.73 with respect to chloroplast DNA restriction digest patterns and ITS sequences, suggesting that one of these cultures may have replaced the other in the SAG collection at some time in the past (Harris et al., 1991;

Pröschold et al., 2005). An isolate from the Norwegian (NIVA) algal collection is very similar to *C. incerta* SAG 7.73. The next closest relatives to *C. reinhardtii* based on ribosomal ITS2 sequences are a group of isolates from Kenya (Pröschold et al., 2005). No mating has been observed among these strains or with the *C. reinhardtii* laboratory strains. Work on these strains is continuing and will probably result in redescription and name changes (Pröschold, personal communication). In the present book, *C. incerta* will be retained as the working name for the SAG 7.73 isolate.

VI. OTHER *CHLAMYDOMONAS* SPECIES USED EXPERIMENTALLY

A. *C. eugametos* and *C. moewusii*

C. eugametos and *C. moewusii* are the names commonly used for two pairs of strains that based on DNA analysis are only very distantly related to *C. reinhardtii* (Buchheim et al., 1990; Lemieux et al., 1985c). These strains are of considerable historical importance, but the spectrum of mutants that can be isolated is limited by their obligate photoautotrophy, and their use as research organisms has declined substantially in recent years. Genetic mapping was abandoned at a very early stage, and only a few genes have been sequenced.

Their taxonomic history is complicated. Franz Moewus conducted his studies primarily on a group of 16 natural and 10 derived strains assigned to the species *C. eugametos*, which he described in 1931, and on isolates of several additional species that he found to be interfertile with these, although by the usual taxonomic criteria they would appear to be widely separated (Smith, 1946). Isolates obtained from Moewus and supposedly equivalent to his type species of *C. eugametos* were found by Czurda (1935) and Gerloff (1940) to differ significantly from Moewus's description, and Gerloff redescribed one of these isolates as the new species *C. moewusii* (Gowans, 1963, 1976a). The species described by Moewus resembles *C. sphagnophila* Pascher (Ettl, 1976a). Gowans (1963) suggested that the name *C. eugametos* Moewus be retained for the laboratory strains, even though they did not confirm to the description given by Moewus in 1931. After comparison of the "*C. eugametos*" strains in the Cambridge (CCAP) collection, Farooqui (1974) suggested that Moewus had used the same species name for two distinctly different isolates, one (CCAP 11/5) corresponding to Czurda's emendation of Moewus' original description of *C. eugametos*, and the other (CCAP 11/6) resembling *C. hydra* (Ettl, 1965b). Farooqui recommended that *C. eugametos sensu* Czurda be retained as a valid species name to include both the CCAP 11/5 series of *C. eugametos* and the laboratory strains of *C. moewusii*, but Ettl (1976a) designated all these strains as *C. moewusii* Gerloff, thereby discarding *C. eugametos* as a species name. Nontaxonomists have continued to use *C. eugametos* for the strains obtained from Moewus and *C. moewusii* for all subsequent isolates, and this terminology will be followed in the present book.

Early work by Gowans (prior to 1954 for mating type *plus* and prior to 1959 for *minus*) was done with *C. eugametos* strains obtained from Moewus by Smith, but when these stocks were lost in a culture chamber failure Gowans continued his work with the UTEX stocks 9 and 10, obtained from Moewus by H.C. Bold in 1951 (R.C. Starr, personal communication, cited in Cain, 1979). Gowans isolated many auxotrophic mutants in this background, including auxotrophs for nicotinamide, *p*-aminobenzoic acid, and thiamine, and mutants resistant to various antibiotics and metabolic inhibitors (Gowans, 1960; Nakamura and Gowans, 1964, 1972; Gowans, 1976a). The same strains were also used by van den Ende and colleagues to study gametogenesis and mating (see van den Ende, 1994 for review, also Chapter 5).

Gerloff's type material for *C. moewusii* was apparently lost, and the principal cultures for this species used for research (UTEX 96 and 97) are those isolated in New York by Provasoli in 1948 (Lewin, 1949). Lewin obtained many mutants in this background, primarily strains with impaired motility or with abnormalities in cell division, although a few auxotrophs were also described.

The UTEX stocks 9 and 10 are interfertile with stocks 96 and 97, albeit with high lethality among meiotic products in some combinations (Gowans, 1963; Cain, 1979; Lemieux et al., 1981). Lemieux, Turmel, and colleagues published an extensive series of papers on genetic and molecular analysis of chloroplast mutations in both these species (Lemieux et al., 1988 and references cited therein; see Chapter 7).

B. Homothallic species: *C. monoica*, *C. geitleri*, and *C. noctigama*

Chlamydomonas monoica (now included in the *C. noctigama* group) is a large homothallic species initially described by Strehlow (1929). Like *C. eugametos* and *C. moewusii*, it was placed by Ettl in his Chlamydella group. It has been used by VanWinkle-Swift and colleagues to investigate the genetic control of sexuality (VanWinkle-Swift and Thuerauf, 1991; van den Ende and VanWinkle-Swift, 1994) and formation and germination of zygospores (VanWinkle-Swift et al., 1998; Malmberg and VanWinkle-Swift, 2001), and to examine the structure of the zygospore wall (VanWinkle-Swift and Rickoll, 1997; Blokker et al., 1999; see Figure 5.22).

Chlamydomonas geitleri, described as a new species by Ettl (1969) but later included with several other isolates as *C. noctigama*, is another homothallic species which was studied intensively by Nečas and colleagues (1981–1986c; Žárský et al., 1985; Vyhnalek, 1990; Tetík and Sulek,

1994) with respect to its life cycle, gametogenesis, and mating. François and Robinson (1988) studied its response to herbicides, and Kalina and Stefanova (1992) described the fine structure of the eyespot in this species. Sommaruga et al. (1997) included it as a representative freshwater alga in a study of thymidine and leucine incorporation.

Chlamydomonas monadina is a homothallic species that differentiates into gametes of unequal sizes (Rosowski and Hoshaw, 1988). Other homothallic strains were described by Heimke and Starr (1979) and by Deason and Ratnasabapathy (1976).

C. Snow algae

Chlamydomonas nivalis is the name usually given to the principal organism found in red snow in arctic and alpine regions of the northern hemisphere (Viala, 1967; Thomas, 1972; Fjerdingstad et al., 1974; Mosser et al., 1977), although microscopic reexamination of some red snow isolates and molecular analysis (Hoham et al., 2002) indicate that other species share the same habitat, and that some isolates originally listed as *C. nivalis* were probably incorrectly identified. Another common red snow alga is *Chloromonas rosae*, which is closely related to *Chlamydomonas augustae* (Buchheim et al., 1997a; Hoham et al., 2002; Nozaki et al., 2002). A species known as *Chlamydomonas sanguinea* was described from red snow deposits in Europe and in the southern hemisphere, and additional species have been reported from Australia, New Zealand, and the Antarctic (Marchant, 1982; Novis, 2002). Snow algae typically appear in late spring as air temperatures rise above freezing, but can tolerate temperatures below zero (Hoham, 1975). The red snow phenomenon is well known to mountain hikers and skiers, and nontechnical articles with colorful photographs can be found on-line and in print (Hardy and Curl, 1972; Armstrong, 1987; see also Wikipedia). Some wilderness guides warn that ingestion of red snow can cause diarrhea, but Fiore et al. (1997) reported no adverse effects in seven volunteers who consumed red snow collected in the Sierra Nevada.

The red color developed by these species results from accumulation of carotenoids, identified by Viala (1966) for *C. nivalis* as astaxanthin esters, in cytoplasmic granules around the cell periphery (Weiss, 1983b; Remias et al., 2005). These pigments are produced in cells grown at high light intensity and have been assumed by most authors to serve a photoprotective function (Bidigare et al., 1993; Gorton et al., 2001; Gorton and Vogelmann, 2003; but see also Fan et al., 1998). Czygan (1970) reported that astaxanthins also accumulated under conditions of nitrogen deficiency, with a concomitant decrease in chlorophyll content. Exposure to light in the UV-C range induces synthesis of phenolic compounds with antioxidant properties (Duval et al., 1999). Snow algae may also accumulate electron-dense mineral particles on their extracellular surface and within vacuoles (Lütz-Meindl and Lütz, 2006). The red cells of *C. nivalis*, which are non-flagellated apla-nospores (Kawecka, 1981), are also described in some publications as "rest-ing cells" although they are metabolically active (Williams et al., 2003). The elevated lipid content of these cells is correlated with increased tolerance of temperatures below freezing (Bidigare et al., 1993). Similar accumulations of pigment are seen in other Volvocacean algae living in conditions of high light intensity, for example *Dunaliella* species growing in salt lakes in desert regions (Ben-Amotz and Avron, 1983) and *Haematococcus* (Boussiba and Vonshak, 1991; Fan et al., 1998).

"Green snow" species of *Chlamydomonas* are also known in frozen habitats with lower light intensity. *C. balleniana*, from the Antarctic, and *C. yellowstonensis*, described from Yellowstone National Park but also reported in the Caucasus, thrive and produce motile cells at temperatures below freezing (Kol, 1941; Kol and Flint, 1968). A species formerly classified as *C. subcaudata*, but now identified by sequence analysis as *C. raudensis* (Pocock et al., 2004) was collected from the permanently ice-covered Lake Bonney in Antarctica, and has been compared in physiological and molecu-lar studies with *C. reinhardtii* and other temperate-zone species (Morgan et al., 1998; Morgan-Kiss et al., 2002a, b; 2005; Gudynaite-Savitch et al., 2006; Szyszka et al., 2007). Nitrate reductase, argininosuccinate lyase, and ribulose bisphosphate carboxylase were compared between *C. reinhardtii* and an Antarctic *Chloromonas* isolate designated ANT1 (Loppes et al., 1996; Devos et al., 1998). Sequence analysis of the chloroplast *rbcL* gene suggests this strain is closely related to *Chlamydomonas pulsatilla* (Hoham et al., 2002). *Chlamydomonas* sp. ICE-L and ICE-W are isolates collected from floating ice in the Antarctic that have 18S rDNA and ITS1 sequences similar to *C. monadina* (Liu et al., 2006; see also Ding et al., 2005). *Chlamydomonas altera* is a cryotolerant species collected from lakes in the American upper midwest that is currently under investigation.

D. Acidophilic species

C. acidophila was described by Negoro in 1944 (cited in Ettl, 1976a) based on samples collected in Japan from extremely acid environments. Several additional isolates have subsequently been assigned to this species, repre-senting samples from Argentina, Canada, the Czech Republic, Germany, Italy, and the United States (see Gerloff-Elias et al., 2005, for DNA sequence analysis of some of these isolates). *C. sphagnophila* is another acidophilic species that has received some attention (Cassin, 1974) and some strains of *C. applanata* are also acid tolerant (Visviki and Santikul, 2000). Fott and McCarthy (1964) characterized a *C. acidophila* strain and also a *Carteria* species from acid water near a spa in Czechoslovakia. Fott and McCarthy, and also Ettl (1976a), made the point that a Danish isolate identified as *C. acidophila* by Nygaard is not the same species as *C. acidophila* Negoro.

Pollio et al. (2005) have reported an isolate resembling *C. pitschmannii* from volcanic soil at an acidic hot spring in Italy.

Typical habitats for these species include acid bogs, lakes in volcanic craters, sulfur springs, and contaminated water in mining regions, and interest in these strains has increased with concerns about environmental acidification (Twiss, 1990; Visviki and Palladino, 2001). These algae are able to maintain an internal pH near neutrality while growing in media in the range of pH 1.7 to 2.5, apparently by a combination of low membrane permeability to H^+ and ATP-dependent transport of H^+ from the cytosol (Messerli et al., 2005). Plasma membrane H^+ ATPases have been characterized from acidophilic *Dunaliella* species (Weiss and Pick, 1996), but not yet from an acidophilic *Chlamydomonas*. Variations in membrane lipid composition may also facilitate growth at very low pH (Tatsuzawa et al., 1996).

Some of these strains are able to grow at neutral pH but others grow only in acid medium (Cassin, 1974). There is also variation among these strains in ability to use acetate or other compounds for growth (Cassin, 1974; Nishikawa and Tominaga, 2001; see also Chapter 6). Spijkerman (2005) characterized the carbon-concentrating mechanism in a *C. acidophila* strain from an acid lake. Ultrastructural studies of *C. acidophila* have been made by Visviki and Palladino (2001), who worked with a moderately acid-tolerant strain collected from a copper-contaminated site in Ontario, and by Nishikawa and colleagues (Nishikawa and Tominaga, 2001; Nishikawa et al., 2003), using an isolate from an acid lake in Japan. Nishikawa et al. (2006) also studied the effects of phosphate stress on their isolate. Additional studies on these species are cited in Chapter 6.

E. Halotolerant species

Ettl and Green (1973) described *C. reginae*, a new marine species, but very little seems to have been done with it subsequently. Cann and Pennick (1982) presented an ultrastructural study of another marine isolate, *C. bullosa*, which has also been compared to *Dunaliella salina* in terms of metal toxicity (Visviki and Rachlin, 1994a, b). Physiological and ecological studies have been made of various other marine isolates of *Chlamydomonas*, including *C. pulsatilla* (Hellebust and Le Gresley, 1985; Hellebust and Lin, 1989; Hellebust et al., 1989), *C. angulosa* (Hellebust et al., 1985 and references cited therein), *C. parkeae* (Sasa et al., 1992; Kim et al., 1994; see also Ettl, 1967, for the original description of this species), *C. palla* (Antia et al., 1975, 1977), *C. provasolii* (Saks, 1982), *Chlamydomonas* strain MGA161 (Miura et al., 1992; Maeda et al., 1994; Miyasaka et al., 2000), and several strains identified only as *Chlamydomonas sp.* (Paul and Cooksey, 1979; Turner, 1979).

Chlamydomonas strain W80 (Miyasaka et al., 1998) and strain HS-5 (Miyasaka and Ikeda, 1997) are halotolerant isolates that have been investigated more extensively in terms of gene expression, especially under conditions of osmotic and other metabolic stresses (Takeda et al., 2000, 2003; Tamoi et al., 2001, 2005; Tanaka et al., 2004), and from which numerous genes have been sequenced. Both were collected from marine environments on the Japanese coast. Hydrogen production by marine species has been studied by Greenbaum et al. (1983a,b) and by Miura et al. (1992).

F. Other species

Species that produce large amounts of extracellular polysaccharide have found practical use as soil conditioners. Among these are *C. mexicana*, *C. ulvaensis*, and *C. sajao* (see for example Lewin, 1977, 1984; Kroen, 1984). *Chlamydomonas mexicana* in particular has been grown commercially for agricultural purposes (Barclay and Lewin, 1985; Metting, 1986, 1987; http://www.soilspray.com/), but its utility may be limited by the need for thorough irrigation (Metting and Rayburn, 1983).

Chlamydomonas segnis (now *Lobochlamys segnis*; Pröschold et al., 2001) is a freshwater species, homothallic and anisogamous, which forms mucilaginous palmelloid colonies and is noteworthy for having a pigment-deficient eyespot and for accumulation of unusual crystalline protein bodies in the chloroplast (Ettl, 1965b, 1976a). Badour and colleagues published a series of physiological studies of this species, concentrating especially on photosynthesis (Badour et al., 1973; Badour and Irvine, 1990; Weinberger et al., 1987, and earlier papers cited therein). *C. segnis* has also been used to investigate pesticide effects on freshwater algae (Caunter and Weinberger, 1988; Kent and Weinberger, 1991; Dechacin et al., 1991; Kent and Caux, 1995; Kent and Currie, 1995).

Chlamydomonas chlamydogama is a vitamin B_{12}-requiring species that received a modest amount of experimental attention prior to 1980 (Bold, 1949a, b; Bell, 1955; Wiese et al., 1979; see also Chapter 5). *Chlamydomonas pallens*, a partially chlorophyll-deficient, acetate-requiring species, is also a natural B_{12} auxotroph and was proposed by Pringsheim (1962, 1963) as an assay organism for this vitamin.

Chlamydomonas gymnogama (Deason, 1967), a homothallic species that sheds its cell walls early in mating, was used by Miller et al. (1974) as a source of wall material uncontaminated by cytoplasmic debris, and in one study of synchronization of cell division (Carroll et al., 1970) but otherwise has received little attention. Based on DNA analysis the original *C. gymnogama* isolate is now considered to be part of the *C. segnis* group.

Although *C. snowiae* was one of the earliest described species, it has been studied only in regard to phototaxis (Mayer and Poljakoff-Mayber, 1959; Stahl and Mayer, 1963; Chorin-Kirsh and Mayer, 1964a, b) and in one investigation of alternative respiration pathways (Myroniuk and Kurinna, 2000). It does not appear in the catalogs of any of the major culture collections.

Chlamydomonas dysosmos (=*C. sphagnophila* var. *dysosmos* or *C. applanata*; Ettl, 1976a) was used for a few cytological and physiological studies, and some mutants were isolated (Lewin, 1954b; Neilson et al., 1972; Silverberg, 1974; Silverberg and Sawa, 1974). Albertano et al. (1997) used this species as a test system for soft X-ray contact microscopy, but did not explain why they chose it in preference to *C. reinhardtii* or another better-known species.

Ettl (1985) chose several strains of *C. debaryana* for a study of structural variability over the life cycle. The ultrastructure of *C. pseudopertusa* was described by Boldina (2000). Sedova (1998–2002) published a series of studies of mitosis and chromosome cytology of various *Chlamydomonas* species using light and electron microscopy. *C. allensworthii*, which produces a sexual pheromone (Coleman et al., 2001), will be discussed in Chapter 5.

CHAPTER 2

Cell Architecture

CHAPTER CONTENTS

I. INTRODUCTION

As discussed in Chapter 1, all species of *Chlamydomonas* share a basic body plan. Cell shape varies greatly among species, however, as does the shape and relative position of the chloroplast. Most species are ellipsoid or ovate in shape, and all have a clearly polar structure, with two anterior flagella and a single basal chloroplast that may partially surround the nucleus. One or more pyrenoids are found within the chloroplast, either at the basal region or along the side of the cell, and starch bodies are often seen surrounding the pyrenoids. All

25

species have a distinct cell wall or hull, usually closely appressed to the plasma membrane, but this varies greatly in thickness within the genus. A mucilaginous coat is formed external to the wall in some species. The nucleus is usually centrally located, with a prominent nucleolus. All but a few species have a colored eyespot, and most also have one or more, most often two, contractile vacuoles. These features are all apparent by light microscopy under good conditions, and they have been the basis for species recognition. The early cytological studies and descriptions of individual species were thoroughly reviewed by Ettl (1976a). Further descriptions here will be restricted primarily to *C. reinhardtii*, the best-known laboratory species (Figure 2.1). The cell wall and the eyespot will both be covered in detail, since these have been extensively studied and are not reviewed elsewhere in this series. Basal bodies and the flagellar root structures are discussed in Volume 3, Chapter 2.

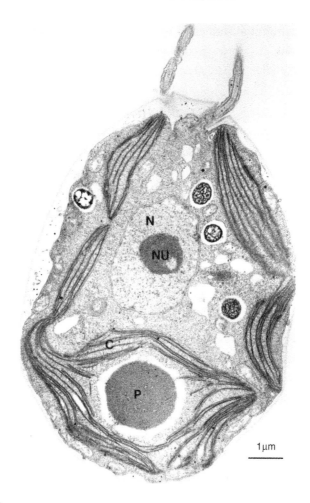

FIGURE 2.1 *Median section through mixotrophically grown wild-type cell of C. reinhardtii, showing prominent nucleus (N) and nucleolus (NU), chloroplast (C) pyrenoid (P), and other cellular features which are depicted schematically in Figure. 1.1. Courtesy of J.E. Boynton.*

II. EARLY ULTRASTRUCTURAL STUDIES

Sager and Palade (1954) published electron micrographs of wild-type *C. reinhardtii* (strain 21 gr) and the *y-1* (now *y1*) mutant, which does not form chlorophyll in the dark. Their sections showed chloroplast lamellae, the pyrenoid, eyespot, and mitochondria, and demonstrated the absence of a developed chloroplast membrane system in dark-grown *y1* cells. The time courses of chloroplast disintegration and formation in *y1* cells shifted from light to darkness and back were subsequently studied in greater detail by Ohad et al. (1967a, b; see also Chapter 7). In 1957 Sager and Palade published a true atlas of ultrastructure in wild-type *C. reinhardtii*, with particular attention paid to the chloroplast. Ringo (1967a, b) presented the first detailed study of the flagella and their cellular roots. Johnson and Porter (1968) in a study primarily of cell division also described the principal ultrastructural features of the non-dividing *C. reinhardtii* cell, as did Triemer and Brown (1974) for *C. moewusii*. Bray et al. (1974) published freeze-etch studies of *C. eugametos*. The endomembrane system of *C. reinhardtii*, comprising the Golgi apparatus, vacuoles, and cytoplasmic vesicles, was studied by Gruber and Rosario (1979). A three-dimensional model of a *C. reinhardtii* cell (Figure 2.2) was constructed based on electron micrographs of serial sections by Schötz et al. (1972; see also Schötz, 1972). Calculations of volume occupied by the various cell components for two typical cells are shown in Table 2.1. Estimates for chloroplast and mitochondrial area were also made by Boynton et al. (1972) and were in good agreement with the values obtained by Schötz.

Together these classic papers set forth the basic features of *Chlamydomonas* that are revealed by electron microscopy, and they have been the foundation for subsequent studies of specific structures and organelles.

III. THE CELL WALL

A. Composition

Early reports often referred to *Chlamydomonas* cell walls as cellulosic (Lewin, 1952c; Sager and Palade, 1957), and this erroneous statement is still occasionally seen in textbooks and encyclopedias. Later studies leave no question that the major wall constituents are hydroxyproline-rich glycoproteins, with arabinose, mannose, galactose, and glucose being the predominant sugars, at least in *C. reinhardtii*. These glycoproteins are somewhat similar to the hydroxyproline-rich extensins of higher plants, but the sugar composition and molecular organization are distinctive (Miller et al., 1972; O'Neill and Roberts, 1981; Woessner and Goodenough, 1992, Ferris et al., 2001). No evidence for cellulose fibrils has been found, nor are wall components digestible with cellulase (Horne et al., 1971; Adair and Snell, 1990). This conclusion is probably also valid for other algae in the Volvocales. One possible exception is a report by Hagen et al. (2002) that there is cellulose in the Volvocacean alga *Haematococcus pluvialis*, based on a colorimetric test for carbohydrate

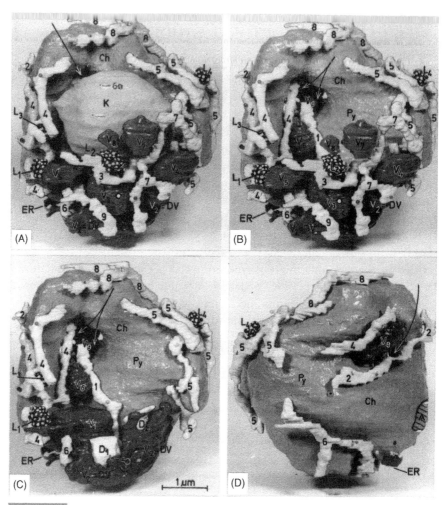

FIGURE 2.2 *Scale model of a C. reinhardtii cell (plus gamete) portrayed from the flagellar side (A–C) and from the chloroplast side (D). The model consists of 77 serial sections. (A) Entire cell. (B) Cell after removal of the nucleus, in order to provide an unobstructed view of the cell interior. (C) In addition to the nucleus, the mitochondrial complex situated at the flagellar base (mitochondria numbers 3,7,9) and a portion of the vacuoles (V3-6 and V7-8) have been removed, to permit recognition of the dictyosome (Golgi).*

The mitochondria are identified by numbers (1–9). The flagellar insertion is located in front of V3 at ○. V2 = DV is the inflated vacuolar peridictyosomal cisterna of the endoplasmic reticulum. The construction includes only the narrowest region of the dictyosomes, represented in each case as compact bodies, and only a few small characteristic cisternae of the endoplasmic reticulum. The arrows show perforation of the chloroplast with an embedded vacuole (V9); the perforation is traversed by mitochondrion number 4. The marker in (C) near D1 points to the point of connection between V2 and V3. The black stripes on the cell nucleus and white ones on the chloroplast (only partly visible) show the distance separating every 10 sections. From Schötz et al. (1972).

after acid extraction. These authors presented no corroborating evidence from microscopy or other assays, however, and their electron micrographs show a tripartite wall structure similar to that of *C. reinhardtii* (as described below).

Jiang and Barber (1975) reported that an alkali-extractable polysaccharide fraction from *C. reinhardtii* had a molecular weight of 41.5 kD and

| Table 2.1 | Volumes of cell components of two typical cells examined by electron microscopy in serial sections |

Cell component	Cell#1		Cell#2	
	Volume (μm³)	% of cell volume	Volume (μm³)	% of cell volume
Whole cell, including cell wall	56.08		80.38	
Cell inside the plasma membrane	44.08	100.00	59.07	100.00
Nucleus	4.42	10.04	4.33	7.31
Chloroplast	17.41	39.46	5.17	42.66
Total mitochondria (individual mitochondria 0.1–0.5 μm³)	1.36	3.09	2.01	3.39
Total Golgi (two dictyosomes in each cell)	0.40	0.90	0.31	0.51
Total vacuoles (individual vacuoles 0.01–1.5 μm³)	2.85	6.46	5.38	9.14
Total lipid bodies (individual bodies 0.01–0.08 μm³)	0.09	0.22	0.28	0.47
Total cytoplasm (including small vesicles as well as endoplasmic reticulum)	17.55	39.83	21.59	36.52

From Schötz et al. (1972).

contained galactose, arabinose, mannose, and glucose in molar ratios of 3.5, 3.3, 1.0, and 0.8 respectively. D- and L-galactose enantiomers were present in a ratio of 84:16. A water-soluble polysaccharide fraction contained mannose, galactose, arabinose, rhamnose, and glucose in molar ratios 3.4, 2.3, 1.4, 1.0, and 1.0. Miller et al. (1974) analyzed cell walls of *C. gymnogama*, which sheds its walls intact during mating. A complex mixture of oligosaccharides was found. As in *C. reinhardtii*, arabinose and galactose accounted for the greatest fraction of the sugar residues, but some differences were found in specific oligosaccharide sequences. Other *Chlamydomonas* species surveyed by Schlösser (1976) were found to have walls similar in composition to that of *C. reinhardtii*, again with galactose and arabinose the predominant sugars. Xylose, fucose, rhamnose, and uronic acids were present in smaller quantities and in varying proportions in the species examined. One species tested, *C. ulvaensis*, was divergent in having a high content of glucose and xylose and relatively little galactose or arabinose.

B. Ultrastructure

Roberts et al. (1972) defined seven wall layers in electron micrographs of vegetative cells of *C. reinhardtii*, and the structure of these layers was examined by Goodenough and Heuser (1985b) using quick-freeze, deep-etch techniques (Figures 2.3 and 2.4). The innermost layer, W1, varies in thickness

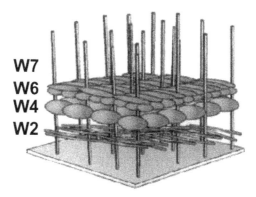

FIGURE 2.3 *Diagram of the cell wall of C. reinhardtii. Adapted from an original color figure courtesy of Sabine Waffenschmidt.*

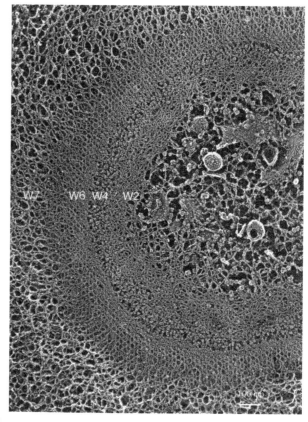

FIGURE 2.4 *Cell wall of detergent-treated, wild-type C. reinhardtii gamete after tangential fracture of quick-freeze, deep-etch specimen. Layers are as defined in Figure 2.3. Bar = 0.5 μm. Courtesy of U.W. Goodenough and J.E. Heuser.*

from 30 to 200 nm, and is insoluble in chaotropic agents such as sodium perchlorate, lithium chloride, or urea. It appears to contain anastomosing fibers of variable size that radiate outward from the cell membrane in a trabecular network. Goodenough and Heuser (1985b) found 15–20 nm granules associated with these fibers. Layers W2 and W6 are arrays of glycoprotein surrounding a granular layer W4, which was described by Roberts et al. as showing a periodic "beaded" structure. Roberts and colleagues regarded the W2 and W6 layers as identical, whereas Goodenough and Heuser were able to distinguish them. In deep-etch micrographs, the W2 layer appears to consist of a fibrous network, more closely connected than that of W1 and with thicker fibers lying mostly parallel to the cell surface. W6 comprises a densely woven crystalline mat (W6A) of thick fibers interconnected by thin crossfibrils and an open-weave lattice (W6B).

Layers W3 and W5 are electron-transparent regions that are probably spaces rather than true wall components (Goodenough and Heuser, 1985b). An outer amorphous layer, W7, is absent under some growth conditions and was presumed by Roberts et al. (1972) to consist of adsorbed material from the culture medium. Roberts et al. (1985a, b) revised this interpretation to suggest that the W7 layer represents nonstructural glycoproteins in a transitional stage prior to release into the culture medium. Goodenough and Heuser described branching fibers in the W7 layer similar to those in W1. Layers W2, W4, and W6, referred to as the "central triplet," are very constant in appearance and size regardless of growth conditions or fixation procedure; they will be described in greater detail below.

Optical diffraction analysis of the lattice structure of *Chlamydomonas* cell walls was made by Horne et al. (1971) and Hills et al. (1973). Roberts (1974; Roberts et al., 1982) classified cell walls of a number of species of green algae based on their crystalline structure. *Chlamydomonas* species were represented in each of the groups as originally defined, although by far the greatest number were in class I, which included *C. reinhardtii* together with all the multicellular species examined (*Volvox, Pandorina, Eudorina*), and class II, which included *C. eugametos* and its relatives as well as many other species (Shaw and Hills, 1984). Class III included *C. asymmetrica*; class V, *C. incisa. C. angulosa*, included in class IV in Roberts (1974), was moved to class I in Roberts et al. (1981). Goodenough and Heuser (1988a, b) noted especially the differences in organization of the W6 layer in class I vs. class II walls, and pointed out that the class II W6 layer was very similar in structure to the crystalline structure of the flagellar collar in *C. reinhardtii* and in *Volvox carteri* (see below, Figure 2.7). They proposed that class I walls evolved from class II in one line of Volvocales, and that in this branch the class II structure was retained specifically in the flagellar collar. Woessner and Goodenough (1994; also Woessner et al., 1994) and Ferris et al. (2001) continued this line of evolutionary investigation with comparison of the major glycoprotein components by electron microscopy, antibody binding, and amino acid sequences. Consistent with data from other types of investigations (Chapter 1), their studies support a close evolutionary relationship between *C. reinhardtii* and specific colonial algae, including *V. carteri*.

C. Fractionation of wall components

The outer layers of the central triplet (W4–W6) are readily solubilized in 1–2 M NaClO$_4$ or in LiCl, leaving the W2 inner layer in an insoluble fraction together with the W1 and W7 fibers (Hills et al., 1975; Goodenough and Heuser, 1985b; Voigt, 1986). Goodenough and Heuser (1985b) reported that the remaining W1 and W2 layers could be dissociated by the gamete lytic enzyme (see below) into fibrillar "fishbone" or "bottlebrush" structures. The proteins of the insoluble fraction can also be released by chemical deglycosylation (Vogeler et al., 1990). The W2 layer of *V. carteri* is indistinguishable in micrographs from that of *C. reinhardtii*. Similar W2 fishbone units are also released on lysis of walls of *C. eugametos* (Goodenough and Heuser, 1988a; Woessner and Goodenough, 1994). Hills (1973; Hills et al. 1975; Catt et al., 1978) and Goodenough and Heuser (1988a) found that the proteins of the perchlorate-soluble fraction could reassemble on dialysis into crystalline fragments resembling the outer cell wall. Dialysis in the presence of the insoluble fraction produced complete wall structures. Walls can even be reconstituted from mixtures of wall fractions from *C. reinhardtii* and *V. carteri*, both of which are in Roberts' class I, but not from *C. reinhardtii* and *C. eugametos* (Adair et al., 1987). Separation of the salt-soluble fraction on Sepharose 2B yielded two glycoprotein fractions, 2BI and 2BII (Catt et al., 1976), which were eventually equated with the W4 and W6 layers (Goodenough et al., 1986; Adair et al., 1987). The 2BI fraction contains a glycine-rich glycoprotein, GP1.5 that is apparently equivalent to the W4 granules. The 2BII fraction contains the three major hydroxyproline-rich glycoproteins of the W6 layer. The W6B layer consists of GP1, a fibrous, 100-nm protein with a globular 7-nm head and, typically, two bends or kinks along its shaft (Ferris et al, 2001). The W6A layer is formed from two more globular proteins, GP2 and GP3, which are distinguishable from one another both in structure and in amino acid composition (Goodenough et al., 1986; Adair and Apt, 1990; Grief and Shaw, 1990; Voigt et al., 2007).

Studies in the 1980s with polyclonal and monoclonal antibodies (Smith et al. 1984; see also Roberts et al., 1985b), surface labeling experiments (Monk et al., 1983), and electrophoretic separation of proteins in wall material recovered from the culture medium of mating gametes (Imam et al., 1985) all suggested that walls comprise something on the order of 20–30 distinct proteins, some of which are also seen in flagellar preparations. Cooper et al. (1983) and Adair (1985) noted the striking similarities between cell wall proteins and the mating-type-specific flagellar agglutinins (Adair et al, 1983), which are also hydroxyproline-rich, fibrous glycoproteins with a high arabinose content, and proposed that these molecules are evolutionarily related. Woessner and Goodenough (1992) characterized several proteins with (SerPro)$_x$ repeat sequences, some of which were specific to the zygospore rather than the vegetative cell wall. Similar proteins were also found in *C. eugametos* and in *V. carteri* (Woessner et al., 1994). Ferris et al. (2001) showed that the GP1 protein has a domain containing this motif and another Ser–Pro-rich domain with a repeating (ProProSerProX)$_x$ sequence and correlated the amino acid sequence with the shaft and kink appearance of this protein in electron micrographs. Kurvari (1997) sequenced a cDNA encoding a 46 kD protein, Mrp47, whose expression was upregulated in cells recovering from wall removal by gamete lytic enzyme. Several other structural wall proteins were characterized by Waffenschmidt et al. (1993).

Pherophorins are a class of glycoproteins with a central hydroxyproline-rich domain and globular ends that have been identified as important components of the complex extracellular matrix of *V. carteri* (Ender et al., 2002). Genes encoding pherophorins have been found in the simpler colonial algae *Gonium pectorale* and *Pandorina morum*, and have been annotated in the *C. reinhardtii* genome (Hallmann, 2006; Table 2.2) but identification of specific pherophorins with cell wall or other cellular fractions is as yet incomplete. Voigt and Frank (2003) identified a 14-3-3 protein in the insoluble wall fraction which they postulate is involved in cross-linking of soluble glycoprotein precursors to form the wall matrix. Prolyl hydroxylases, one of which appears to be specifically involved in wall assembly, have been characterized by Keskiaho et al. (2007). Proteomic studies of *Chlamydomonas* cell wall fractions are in progress (Kilz et al., 2000, 2002). A proteomic study of cell walls of *Haematococcus* has been published (Wang et al., 2004). Table 2.2 summarizes the likely genes encoding wall proteins identified in the *C. reinhardtii* genome to date.

D. Synthesis and assembly

Biochemical and structural aspects of cell wall biogenesis were first studied by Lang, Roberts, Voigt, and their respective colleagues. In synchronous cultures on a 12:12 light:dark regime, Lang and Chrispeels (1976) found that cell wall protein was synthesized throughout the cycle, but peaks of activity (up to 15% of total cellular protein synthesis) were seen at the end of the light period and in the second half of the dark period, after daughter cells separated following division. Voigt (1986) used pulse-labeling with [^{3}H]proline and [^{35}S]methionine to demonstrate synthesis of daughter cell walls *de novo* during cytokinesis and turnover of wall components during cell enlargement. Roberts et al. (1985b) found that the degree of fluorescent antibody binding to cell wall polypeptides was also cell cycle dependent. Walls of young cells, particularly mitotic daughter cells still within the

Table 2.2	Proteins associated with the vegetative and zygospore cell walls	
Gene/Protein		**References**
Proteins originally identified from fractionation of walls		
GP1	Chaotrope-soluble protein	Goodenough et al. (1986); Adair and Apt (1990); Ferris et al. (2001)
GP1.5	Glycine-rich protein of W6 layer	Goodenough et al. (1986)
GP2	Chaotrope-soluble protein	Goodenough et al. (1986); Adair and Apt (1990); Voigt et al. (2007)
GP3	Chaotrope-soluble protein	Goodenough et al. (1986)
Proteins identified from gene expression and genome sequencing		
ECP88	Extracellular protein	Takahashi et al. (2001)
ECP76	Extracellular protein	Takahashi et al. (2001)
	14-3-3 protein of insoluble wall fraction	Voigt and Frank (2003)
GAS28	HRGP upregulated by wall removal or gametogenesis	Hoffmann and Beck (2005)
GAS30		Hoffmann and Beck (2005)
GAS31	Pherophorin family protein	Hoffmann and Beck (2005)
HRP2-1	Protein with HR domain with PPSPX-repeats	Identified in genome sequence
HRP2-2	Protein with HR domain with PPSPX-repeats	Identified in genome sequence
HRP3	HR domain with $(SP)_n$ repeats	Identified in genome sequence
HRP4	HR domain with $(SP)_n$ repeats	Identified in genome sequence
ISG-C2	Similar to *Volvox* ISG and *C. reinhardtii* VSP3 (see below)	Identified in genome sequence
ISG-C3	Similar to *C. incerta* hydroxyproline-rich glycoprotein VSP-3	Identified in genome sequence
ISG-C4	Similar to *Volvox* ISG and *C. reinhardtii* VSP3, also known as flagella associated protein FAP137	Identified in genome sequence
MG1	HRGP identified from gametic cells	Adair and Apt (1990)
MMP1	Gamete lytic enzyme; *note that other matrix metalloproteinases have been identified from genome sequencing but have not been confirmed to be associated with the cell wall*	Kubo et al. (2001)

(Continued)

Table 2.2	*Continued*	
Gene/Protein		**References**
Mrp47		Kurvari (1997)
PHC1 through PHC24	Pherophorin family proteins	Hallmann (2006)
V-LYSIN	Vegetative cell lysin	Spessert and Waffenschmidt (1990); Matsuda et al. (1995)
VSP1	Tyrosine-rich HRGP	Waffenschmidt et al. (1993)
VSP3	Glycoprotein with HR domain, similar to *Volvox* ISG	Woessner et al. (1994)
VSP4	HRGP with $(SP)_n$-repeats	Waffenschmidt et al. (1993)
VSP6	HRGP with $(SP)_n$-repeats, similar to VSP4	Waffenschmidt et al. (1993)
VSP7	Similar to VSP4 and VSP6, but function unknown	Identified in genome sequence
ZSP2-1	Zygote-specific lectin-like protein	Suzuki et al. (2000)
ZSP2-2	Zygote-specific lectin-like protein	Suzuki et al. (2000)

mother cell wall, were highly fluorescent, whereas walls of older single cells and mother cell walls were much more weakly labeled.

In later studies, Voigt et al. (1996b) showed that polyclonal antibodies to the 150 kD GP3B protein reacted most strongly with walls of mature sporangia just prior to release of the daughter cells, and concluded that this protein becomes exposed at the cell surface as the mother cell wall structure begins to change in preparation for lysis. From work with antibodies tagged with ferritin or colloidal gold and protein A, Roberts et al. (1985b) hypothesized that wall polypeptides are transferred from the Golgi apparatus to cytoplasmic vesicles, then to vacuoles, from which they are extruded to diffuse around the cell between the existing wall and the plasma membrane before insertion into the wall. Also using gold-tagged antibodies to mature and deglycosylated proteins, Grief and Shaw (1987) demonstrated the presence of a major wall protein in the Golgi apparatus and endoplasmic reticulum. Zhang et al. (1989) showed that hydroxyproline arabinosyl and galactosyl transferases were associated with membrane fractions from the endoplasmic reticulum and Golgi, implying that these are the sites where wall proteins are glycosylated prior to assembly. In *Gloeomonas kupfferi*, a unicellular alga with a somewhat more complex wall than *C. reinhardtii*, Domozych and Dairman (1993) reported that wall deposition begins during cytokinesis with appearance of fibrillar materials on the surface of each daughter cell. This is followed by deposition of a crystalline median layer, beneath which the inner wall then begins to condense. As in Grief and Shaw's 1987 study of *C. reinhardtii*, wall materials appear to be released from a vacuole.

Using polyclonal antibodies reactive with glycoprotein epitopes and with deglycosylated wall proteins, together with pulse-chase experiments with [³H]proline, Voigt et al. (1991, 1996a) identified several salt-soluble glyco proteins as probable precursors of the inner, insoluble portion of the wall. Among these was a prominent 150 kD protein equated with the GP3B protein described by Goodenough et al. (1986). Deposition of the insoluble wall appears to involve formation of isodityrosine cross-bridges (Waffenschmidt et al., 1993). As discussed above, a specific 14-3-3 protein has been postulated to have a role in the cross-linking process (Voigt and Frank, 2003), as does a specific transglutaminase (Waffenschmidt et al., 1999).

Cell wall formation can also be conveniently studied by treating vegetative cells with the gamete lytic enzyme (see below) and allowing them to regenerate walls on transfer to fresh medium. Using *C. smithii* cells (SAG 54.72; see Table 1.1), Robinson and Schlösser (1978) saw the first evidence of wall regeneration 40–60 minutes after lysin treatment, with appearance first of globular material, then of a 6-nm fringe external to the plasma membrane. The W1 layer of *C. smithii* appeared to be less well defined than that of *C. reinhardtii*, and Robinson and Schlösser could not ascertain whether the fringe layer they saw consisted of W1, a W2 precursor, or both. Within the next hour the central triplet structure developed, with the inner W2 layer appearing before the outer W6 later. Cycloheximide completely inhibited cell wall regeneration. Concanavalin A delayed regeneration and resulted in an abnormal, amorphous wall that was not readily digestible with gamete lytic enzyme. The effects of both these agents were reversible on transfer to fresh culture medium. EDTA and 2-deoxyglucose (which inhibits cell wall regeneration in yeast) had little effect. Su et al. (1990) used a similar regeneration system to follow accumulation of poly(A)+ RNA in *C. reinhardtii*. Cytosine-rich mRNAs were prominent, and were presumed to correspond to the transcripts for hydroxyproline-rich glycoproteins.

E. Zygospore walls

Walls of *C. reinhardtii* zygospores resemble vegetative cell walls in consisting of hydroxyproline-rich glycoproteins, but differ in their structural organization and in composition (Woessner and Goodenough, 1989, 1992; Suzuki et al., 2000). Although vegetative cell walls of *Chlamydomonas* are strong (Carpita, 1985), they can be ruptured by high pressure techniques, such as the French press or Yeda press (Chapter 8), whereas zygospore walls are so resistant to breakage and also to enzymatic digestion as to hinder research on processes occurring within the maturing zygote. Catt (1979) reported that zygospore walls contain essentially the same sugar residues as vegetative walls, but in different proportions. Glucose in particular accounted for 50% of the total sugars, in contrast to its relatively low content in vegetative cell walls. It appears primarily in the form of (1→3)-beta-D-glucan, which is present in all zygospore wall layers but absent from intracellular regions (Grief et al., 1987).

Zygospore walls of *Chlamydomonas monoica* contain a substance resembling sporopollenin (VanWinkle-Swift and Rickoll, 1997), previously known as a component of the walls of plant pollen grains and also reported in algae from the lineage leading to land plants. Blokker et al. (1999) characterized this material as an aliphatic polymer of long-chain alcohols and carboxylic acids connected by ester and ether bonds. There are no reports to date of a similar polymer in *C. reinhardtii*. Formation of the zygospore wall will be considered in more detail in Chapter 5, and methods for its disruption in Chapter 8.

F. Evolution of cell wall structure

Mattox and Stewart (1977) proposed that the noncellulosic walls of *Chlamydomonas* and related Chlorophyceaen algae evolved from scales of green flagellates (Prasinophyta) similar to the modern genera *Heteromastix* and *Tetraselmis*, which resembles *Heteromastix* in many respects but has a wall-like theca surrounding the cell body. *Tetraselmis* also has a collapsing telophase spindle and a phycoplast microtubule structure (see below, and Chapter 3) similar to those seen in *Chlamydomonas*. Mattox and Stewart (1977) concluded from their studies of these and other primitive green flagellates that the *Chlamydomonas* type of cell wall and the phycoplast developed coordinately in a single evolutionary line. This entire line of evolution is a separate branch and is not in the main progression leading to the Charophyceae and on to higher plants. While the overall plan of this evolutionary sequence is probably correct (Lewis and McCourt, 2004), later analysis showed that prasinophyte scales are composed primarily of acidic polysaccharides containing unusual 2-keto sugar acids (Becker and Melkonian, 1992; Becker et al., 1994), rather than hydroxyproline-rich glycoproteins similar to those found in walls of the Volvocales. Hence the derivation of the glycoprotein wall is likely to be more complex than was previously hypothesized. For further discussion of the walls of scaly green flagellates, see Becker et al. (1994, 1998), Steinkötter et al. (1994), and Simon et al. (2006).

Within the Volvocales, natural wall-deficient forms such as *Dunaliella* are presumed to have evolved by secondary loss of the wall from a *Chlamydomonas*-like ancestor (Buchheim et al., 1996). Among the unicellular algae, the class II wall structure (see above) is probably the ancestral form (Goodenough and Heuser, 1988a, b). *Volvox* has a complex extracellular matrix consisting primarily of hydroxyproline-rich glycoproteins, some of which are close homologues of *Chlamydomonas* wall proteins (Sumper and Hallmann, 1998; Hallmann, 2003). The wall of the large unicellular flagellate *Gloeomonas* resembles that of *C. reinhardtii* in its tripartite organization but the outer layer is more complex, and contains a lattice of polysaccharide fibers (Domozych et al., 1992). This may represent an intermediate stage in development of the extracellular matrix seen in the colonial Volvocales.

G. Wall-deficient mutants

Mutants with altered cell walls were isolated and analyzed genetically by Davies and colleagues (Davies and Plaskitt, 1971; Hyams and Davies, 1972; Davies and Lyall, 1973), but detailed biochemical characterization of these mutants was not done, and the genes affected by these mutations have still not been identified. Seventy-nine mutants were classified into three morphological groups (Table 2.3).

Mutants in all three groups display a characteristic "soupy" morphology when grown on agar, and can be lysed in liquid suspension with non-ionic detergents such as TritonX-100 or Nonidet. In class C mutants, vesicles containing electron-dense materials appear to be secreted through the plasma membrane. Similar vesicles are seen in wild-type cells, both within the plasma membrane and between the membrane and the cell wall, but when in the latter position these no longer contain electron-dense material. A recombination matrix of 32 of the 79 mutants suggested that they fell into 19 allelic groups (Hyams and Davies, 1972). Three mutants showed aberrant segregation patterns, producing some wild-type progeny when self-crossed. Wall-deficient mutants are more fragile than wild-type cells (see Chapter 8) and therefore often give poor viability in crosses. For this reason the various *CW* loci were never mapped.

Davies (1972) reported that two class A mutants, *cw2* and *cw19*, have only the central triplet of the wall structure, the inner (W1) and outer (W7) layers being absent. The widely used *cw15* mutant is placed in class C (Davies and Plaskitt, 1971), although it does retain some wall components. Monk et al. (1983) found that this mutant fails to assemble the central triplet (W2–W6 layers) but that it has fibers resembling the outer layer of the wild-type wall (Figure 2.5). Some wall-associated proteins were not seen in surface iodination experiments with this mutant (Monk et al., 1983) or in LiCl extracts of total wall proteins (Voigt, 1985a). Zhang and Robinson

Table 2.3	Representative mutants with cell wall deficiencies
Mutants	**Description**
cw1, cw2, cw4, cw6, cw9, cw14, cw17, cw19, cw51, cw177	Class A: walls produced in more or less normal quantities, but not attached to the plasma membrane; walls are shed into the medium and may have abnormal structure.
cw8, cw18, cw20	Class B: walls appear normal in EM, but cells show a typical "amoeboid" shape and colonies have flat appearance characteristic of wall-deficient mutants.
cw3, cw10, cw15, cw92	Class C: cell walls are absent or produced in greatly reduced quantity compared to wild-type cells.

(1986a) compared *cw2* and *cw15* with wild-type cells in an ultrastructural study centered on the endoplasmic reticulum and Golgi. In wild-type cells the endoplasmic reticulum adjacent to the *cis* pole of the Golgi dictyosomes swells markedly during the period of cell wall deposition. In *cw2*, some swelling is also observed, but it is much less extensive than in wild-type cells, and in *cw15* none is seen. Both *cw2* and *cw15* appear to release the same set of glycoprotein products into the culture medium (Zhang and Robinson, 1990; Voigt et al., 1991). Voigt et al. (1997) reported that 18 mutants representing all three classes formed the same set of intracellular wall precursors, although quantities were reduced in some. Most released elevated amounts of wall proteins into the culture medium compared to wild-type cells. The *cw92* mutant, a class C mutant included by Davies among those showing aberrant inheritance in crosses, has very little residual wall material and for this reason has been favored for preparation of DNA to be used for library construction and sequencing. Additional wall-deficient mutants were isolated by Loppes and Deltour (1975, 1978). Some

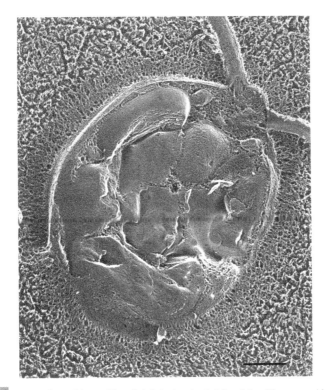

FIGURE 2.5 *Cell surface of the cw15 wall-deficient mutant of C. reinhardtii, prepared by quick-freeze, deep-etch technique. A layer of cell wall fibers extends from the etched cell membrane and surrounds the flagellar bases. The flagellar collar and dense central layer of wild-type cells are absent. The disposition and morphology of the cw15 fibers most resemble the outer layer of the wild-type wall. From Monk et al. (1983); original figure by J.E. Heuser.*

of these were recovered in attempts to isolate mutants deficient in alkaline phosphatase; loss of cell wall structure led to leakage of this enzyme from the periplasmic space, producing a phenocopy of a phosphatase-deficient mutant. One such mutant, cw_{ts1}, formed a cell wall at 25°C but not at 35°C. A wall-deficient mutant of *C. smithii* was isolated by Matagne and Beckers (1987). Fuentes and VanWinkle-Swift (2003) characterized a wall-deficient mutant of *C. monoica*. Wall-deficient mutants have also been described in *Haematococcus* (Wang et al., 2005).

H. Lytic enzymes

Chlamydomonas reinhardtii has two distinct types of enzyme capable of digesting the cell wall. A vegetative cell lysin (V-lysin) releases newly formed cells after mitosis (Schlösser, 1966, 1981; Spessert and Waffenschmidt, 1990), while a gamete lysin (G-lysin) digests the walls of fusing gametes prior to zygote formation (Claes, 1971; see also Chapter 5). Schlösser (1966) found that the vegetative enzyme was released only by young cells, shortly after cytokinesis. It is specific for glycosylated proteins of the inner framework of the mother (sporangial) cell walls and is inactive on synchronously grown cells isolated at other phases of the cell cycle (Waffenschmidt et al., 1988; Spessert and Waffenschmidt, 1990; Vogeler et al., 1990). The enzyme is a 34 kD serine protease that has a pH optimum between 7.5 and 8.5 and requires divalent cations for activity (Jaenicke and Waffenschmidt, 1981; Jaenicke et al., 1987; Matsuda et al., 1995). Schlösser (1976, 1984) reported that 56 strains representing 45 species of *Chlamydomonas* could be placed into 15 groups based on mutual sensitivity to vegetative cell lytic enzymes (Table 2.4).

In contrast to this vegetative cell "hatching enzyme," the gamete lytic enzyme is effective on cells at all stages of the life cycle with the exception of the zygospore and can be used to prepare protoplasts from non-dividing vegetative cells (Claes, 1971; Schlösser et al., 1976; Buchanan and Snell, 1988). This enzyme is a zinc metalloproteinase, one of several such enzymes identified in the *Chlamydomonas* genome (Tamaki et al. 1981; Matsuda et al. 1985a; Kinoshita et al., 1992; Kubo et al., 2001). Related enzymes have been identified in *Volvox* (Hallmann et al., 2001).

Gamete lysin is present in vegetative cells in an insoluble, inactive form, probably outside the plasma membrane (Claes, 1977; Matsuda et al., 1978; Buchanan et al., 1989). Matsuda et al. suggested that it may be stored in the W2 layer of the wall and that the stored enzyme may be equivalent to periplasmic concanavalin A binding sites identified by Millikin and Weiss (1984a). During gametogenesis, the enzyme is changed to a soluble, active form by the action of a serine protease (Snell et al., 1989). It appears to act specifically on a few wall polypeptides, notably the W2 layer or "framework" protein (Imam and Snell, 1988). Crude preparations of the lysin

Table 2.4	Isolates sharing vegetative lytic enzyme susceptibility with *C. reinhardtii*[a]
Isolate	**SAG number**
C. reinhardtii mt[+]	11-32a
C. reinhardtii mt[-]	11-32aM[b]
C. reinhardtii mt[+]	11-32b
C. reinhardtii mt[-]	11-32c
C. reinhardtii mt[+]	73.72
C. reinhardtii mt?	77.81
C. smithii mt[+]	54.72
C. globosa mt[-]	81.72
C. incerta mt[-]	7.73
Unidentified species	11-31

[a]*Autolysin Group 1 from Schlösser (1984).*
[b]*Hatching-deficient mutant.*

from mating cultures are useful for temporary removal of cell walls prior to transformation with exogenous DNA (see Chapter 8).

I. Palmelloid colonies

Species of *Chlamydomonas* vary in the extent to which they form "palmelloid" colonies of adherent, nonmotile cells in nature and in culture, with some species appearing to exist primarily in this state. Lurling and Beekman (2006) have suggested that palmelloid formation has an adaptive advantage in protecting *Chlamydomonas* cells from predation by rotifers and other organisms. The palmelloid condition is often accompanied by production of gelatinous extracellular material, consisting mainly of acidic polysaccharide (Crayton, 1982), which has had commercial applications as a soil conditioner (see Chapter 1). The mucilage is also rich in hydroxy-proline-containing glycoprotein similar to the cell wall constituents (Lewin, 1956a; Miller et al., 1974). Lewin (1956a) identified the following as highly mucilaginous species: *C. acidophila, C. applanata, C. callosa, C. debaryana, C. inflexa, C. mexicana, C. parvula, C. peterfii, C. sphagnophila, C. sphagnophila* var. *dysosmos, C. ulvaensis.* Lewin (1975; see also Schulz-Baldes and Lewin, 1975), also described a palmelloid mucilage-producing species, *C. melanospora*, that produces zygospores encrusted with a brittle, dark reddish-brown material that appears to be a manganese oxide.

In laboratory cultures of *C. reinhardtii* and *C. eugametos*, formation of palmelloids or groups of nonflagellated cells is not uncommon, and it is usually a nuisance when it occurs in liquid culture (see Chapter 8 for suggested remedies). In these cases, microscopic examination reveals clusters of four or eight cells encased in common walls, and aggregates of four-cell groups adherent to one another (Figure 2.6). The individual cells appear to be normal in structure, and in some cases they even have very short but structurally normal flagella (Nakamura et al., 1978). One concludes that the palmelloid condition, at least in these cases, is a failure of hatching or release from the mother cell wall and does not result from intrinsic cellular defects, nor from aggregation of previously free-swimming cells. This is also the mechanism postulated for the evolution of colonial algae such as *Gonium*, which forms a mat of eight cells, and more complex multicellular genera whose cell numbers are always powers of two (Kirk, 1998).

Olsen et al. (1983) observed a tendency for *C. reinhardtii* cells to form palmelloid colonies in chemostat cultures. These became attached to the walls of the culture vessel and subsequently showed a much slower growth rate than free-swimming vegetative cells. Palmelloid colony formation was deliberately induced in *C. eugametos* by treatment with 0.05 chloroplatinic acid, a platinum compound that blocks cell division in *E. coli* and inhibits growth of mammalian tumors (Nakamura et al., 1974, 1976), and in *C. reinhardtii* by deficiency of Ca^{2+}, by addition of chelating agents (EDTA or citrate), by high (>20 mM) phosphate concentrations and by the presence of non-metabolizable organic acids (Iwasa and Murakami, 1968, 1969). Olsen et al. (1983) found that even when phosphorus was limiting in the medium, palmelloid cells had a high ratio of polyphosphate (storage phosphorus) to total phosphorus and had a low level of alkaline phosphatase activity. Palmelloid-forming mutants have been isolated in *C. reinhardtii* (Schlösser, 1966; Warr et al., 1966), *C. eugametos* (Gowans, 1960; Nakamura et al. 1978), and *C. moewusii* (Lewin, 1952b). Mutants with flagellar defects also often form palmelloid colonies. A mutant specifically deficient in *O*-glycosylation forms palmelloid colonies, presumably

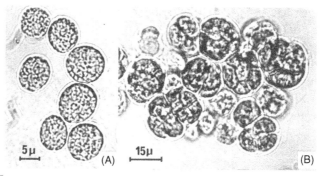

FIGURE 2.6   *Light micrographs of C. eugametos wild-type cells (A) and of a palmelloid-forming mutant strain (B). The palmelloid group shown was disrupted from a much larger aggregate by squashing. From Nakamura et al. (1978).*

because its non-glycosylated wall proteins are not affected by the vegetative lytic enzyme (Vallon and Wollman, 1995). Palmelloids induced by chloroplatinic acid have massive layers of abnormal cell wall material that appear to imprison the cells (Nakamura et al., 1975) and may therefore not be strictly comparable to those induced by variations in culture media or in mutants, whose cell walls appear normal. Abnormal multilayered walls are also seen in *C. reinhardtii* cells treated with the intercalating dyes acriflavin or ethidium bromide, which seem to act primarily on mitochondrial DNA (Alexander et al., 1974). A mutant in which sporangial lysis can be delayed until the beginning of the light period in synchronous cultures was isolated by Mergenhagen (1980) and further characterized by Voigt et al. (1989, 1990). When grown under suboptimal conditions this mutant also showed multilayered walls.

J. The flagellar collar

Chlamydomonas reinhardtii has specialized cell wall regions through which the flagella protrude. These flagellar collars (Figure 2.7) consist of cylinders approximately $0.5 \times 0.5\,\mu m$ in size lined with very regular $11\,nm \times 500\,nm$ fibers in parallel arrays (Ringo, 1967a; Roberts et al., 1972; Goodenough and St. Clair, 1975; Snell, 1983). The collars can be isolated by differential centrifugation from the flagellar fraction following deflagellation by pH shock and contain a major protein of at least 225–250 kD and several minor protein species (Snell, 1983). Flagellar collars are not seen in all *Chlamydomonas*

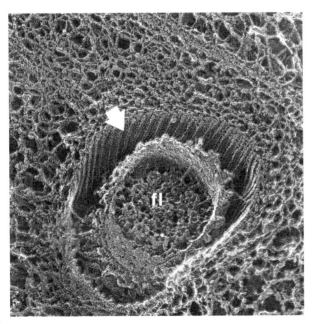

FIGURE 2.7 *Flagellar collar of C. reinhardtii. Courtesy of Ursula Goodenough.*

species, and their presence or absence is correlated with the type of crystalline array seen in cell wall preparations (Roberts, 1974; Roberts et al., 1981; Cann and Pennick, 1982). As discussed above, the protein structure of the collar of species in cell wall class I resembles the wall structure of species in class II, for example *C. eugametos* (Goodenough and Heuser, 1988b).

IV. BASAL BODIES, FLAGELLAR ROOTS, AND CELLULAR FILAMENTS

Two flagella emerge from basal bodies at the anterior end of the *Chlamydomonas* cell. The flagellar axoneme has the 9 + 2 structure characteristic of cilia and flagella of eukaryotic cells, and will be discussed at greater length in Chapter 4 and in Volume 3. The basal bodies and flagellar root system will be described here, since they are involved in microtubule organization of the entire cell, including formation of the mitotic spindle.

The "neuromotor apparatus" of *Chlamydomonas* cells, comprising the basal bodies and the fibers that connect them to the nucleus, was described by Kater in 1929, but then largely ignored for more than 50 years (for historical review, see Salisbury, 1988). In cross-section the basal bodies show the "cartwheel" arrangement of microtubules typical of centrioles and basal bodies in all organisms (Geimer and Melkonian, 2004; see also Figure 4.1). The transitional region between the flagella and basal bodies shows a stellate morphology similar to that of other algae and also sperm cells of land plants, but not seen in protozoa or animal cells (Melkonian, 1982). The basal bodies are connected to each other by the distal striated fiber (Figures 2.8 and 2.9; Goodenough and Weiss, 1978; Wright et al., 1983; O'Toole et al., 2003), which contains the 20 kD contractile, calcium-responsive protein centrin (Salisbury et al., 1988; Salisbury, 1989). Centrin is also found in the nucleus–basal body connector or rhizoplast (Wright et al., 1985, 1989), the stellate portion of the flagellar transitional region, and within the basal body lumen (Ruiz-Binder et al., 2002; Geimer and Melkonian, 2005). Because basal bodies are segregated semi-conservatively at cytokinesis (Chapter 3), each cell always contains one "old" and one "new" basal body (Adams et al., 1985; Gaffal, 1988; Holmes and Dutcher, 1989). In uniflagellate mutants, two basal bodies are present, but the one bearing a flagellum is always the one *trans* with respect to the eyespot (Huang et al., 1982; also see below). Huang et al. postulated that the *trans* basal body is always the older of the two, and this hypothesis was confirmed by Holmes and Dutcher (1989) by tracking the basal body positions with reference to the eyespot through mitosis and during mating. They showed in addition that *C. reinhardtii* cells have an asymmetry with respect to the plane in which the flagella beat, such that the eyespot is positioned clockwise to the closest flagellar root, and opposite the side of the cleavage plane formed in mitosis. The distinction between the *cis* and *trans* flagella is critical to

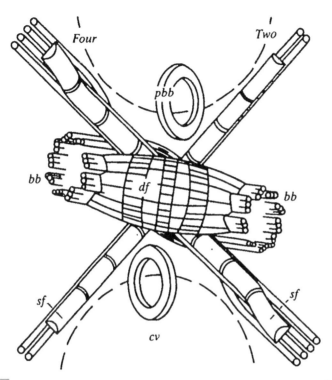

FIGURE 2.8 *Schematic diagram of the flagellar apparatus of Chlamydomonas as seen from the anterior end of the cell and demonstrating the 180° rotational symmetry of the flagellar apparatus. (bb), mature basal bodies; (pbb), probasal bodies that assemble the daughter basal bodies prior to mitosis. Their connections to the flagellar apparatus are omitted. (df), distal striated fiber that connects the basal bodies. Two smaller proximal striated fibers are mostly obscured in this diagram. (cv), contractile vacuoles whose location is represented by the broken circles. Two types of microtubule rootlets emanate from the flagellar apparatus, composed of two and four microtubules, respectively. (sf), striated fibers associated with both rootlet types. All four microtubule rootlets terminate in the region beneath the distal fiber, although the striated component of the two-membered rootlets is continuous. From Holmes and Dutcher (1989).*

the mechanism of phototactic orientation, which is controlled by differential beating in response to changes in calcium concentration (Chapter 4). Representative mutants affecting the basal bodies and flagellar root system are listed in Table 2.5. Details of basal body formation, the role of basal bodies as centrioles, and proteomic analyses are discussed by Dutcher and by Marshall in Volume 3, Chapters 2 and 14 respectively.

The flagellar root system consists of four sets of microtubules descending from the region of the basal bodies (Figure. 2.10; see also Ringo, 1967a). In *Chlamydomonas* species the rootlet microtubules are apportioned in a 4-2-4-2 pattern (Melkonian, 1977; Goodenough and Weiss, 1978; Stewart and Mattox, 1978; Moestrup, 1978; Katz and McLean 1979). In other

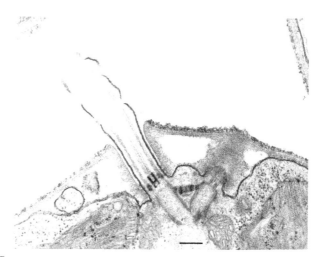

FIGURE 2.9 *Flagellar insertion region of wild-type Chlamydomonas cell, showing distal striated fiber between the two basal bodies. Bar = 0.2 μm. Courtesy of R.L. Wright and J.W. Jarvik.*

Table 2.5	Representative mutations affecting basal bodies and flagellar roots		
Gene	**Mutant**	**Description**	**References**
BLD1 or *IFT52*	*bld1*	Defective in intraflagellar transport protein IFT52 (homologue of *C. elegans osm-6*)	Pasquale and Goodenough (1987); Brazelton et al. (2001); Deane et al. (2001)
BLD10	*bld10*	Deficient in basal body protein of cartwheel structure; lacks basal bodies and flagella	Matsuura et al. (2004)
UNI1	*uni1*	Most cells uniflagellate, form only the *trans* flagellum; normal centrin and stable nucleus–basal body connector	Huang et al. (1982)
UNI3 or *TUD*	*uni3*	Deficient in delta tubulin	Dutcher and Trabuco (1998); O'Toole et al. (2003); Fromherz et al. (2004)
VFL1	*vfl1*	Variable number of flagella; protein localizes to basal bodies	Tam and Lefebvre (1993)
VFL2	*vfl2*	Centrin-deficient; variable number of flagella	Wright et al. (1985); Koblenz et al. (2003)
VFL3	*vfl3*	Variable number of flagella; defects in striated fibers associated with the basal body	Wright et al. (1983)

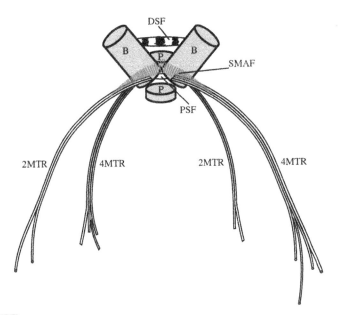

FIGURE 2.10 *Diagram showing the two- and four-membered microtubule roots and their association with the basal body apparatus. Modified from Preble et al. (2001); courtesy of Susan Dutcher.*

algae the "cruciate" pattern of four sets is constant, but the number of microtubules per set can vary. Each set of four microtubules is arranged in a "3-over-1" configuration. The sets of two roots lie at right angles to the plane of the basal apparatus and make a 50° angle with the 3-over-1 roots. Goodenough and Weiss (1978) and Katz and McLean (1979) found daughter basal bodies lying within each of the 50° angles subtended by these roots. Rootlet microtubules contain acetylated tubulin (LeDizet and Piperno, 1986; Holmes and Dutcher 1989). Additional cortical microtubules run from the flagellar apparatus to the base of the cell. Studies by Weiss (1984) show details of the fibers connecting the basal bodies to the flagellar roots (Figure 2.11). The major component of these fibers is the 30 kD protein SF-assemblin (Lechtreck and Silflow, 1997; Lechtreck, 1998; Lechtreck and Melkonian, 1998; Lechtreck et al., 2002).

Basal bodies and flagellar roots take diverse forms among the green algae and have been the basis for taxonomic studies (Manton, 1965; Pickett-Heaps, 1975; Moestrup, 1978; Melkonian, 1984; Mattox and Stewart 1984). In mitosis in *Chlamydomonas*, the spindle collapses during telophase and is replaced by a phycoplast, a system of microtubules lying in the plane of cytokinesis and separating the two daughter nuclei (see Chapter 3). The phycoplast and the cruciate flagellar root system are constant in one entire major branch of green algae, the Chlorophyceae, and distinguish these from the Charophyceae, which eventually gave rise to higher plants. Both groups are presumed to have arisen from primitive scaly green flagellates represented today by such genera as *Pyramimonas* and *Tetraselmis*. Mattox and Stewart (1977; see also Stewart and Mattox, 1978) proposed that the phycoplast evolved coordinately with the evolution of cell walls from scales. The prasinophytes are now put in a separate radiation (Lewis and McCourt, 2004).

Several studies hint at the presence in *Chlamydomonas* cells of proteins related to animal intermediate filaments, but characterization of these proteins is incomplete. A monoclonal antibody prepared to human intermediate filaments was found to label a 66 kD protein and two or three smaller proteins in *C. reinhardtii* (Miller et al., 1985). In a continuation of this study, Parke et al. (1987) reported that monoclonal antibodies to four proteins from a high salt and detergent-insoluble *Chlamydomonas* fraction recognized intermediate filaments in mammalian tissue culture cells and also reacted with proteins in onion root tips. Fan et al. (1994) described a network of filaments in the pyrenoid of an unidentified *Chlamydomonas* species that reacted to antibodies to animal keratin. Hendrychová et al. (2002) reported finding proteins in *C. eugametos* that resemble plectin, the large protein that links intermediate filaments in mammalian cells. However, a search of the annotation of the version 3.0 genomic sequence for the major intermediate filament protein classes retrieved no hits.

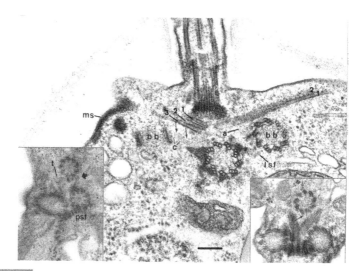

FIGURE 2.11   *Lateral striated fiber (lsf) in an unactivated minus gamete. The fiber appears on the right between the daughter (bb) and parent basal bodies and presumably associates with the two-member root at site (a). On the left, cross-striations (c) extend from the daughter basal body toward the compound root and are connected by filaments (arrows). The two-member root microtubules on the right are labeled 1 and 2. The 3-over-1 root microtubules, labeled 1-3, and the mating structure (ms) appear to the left. Bar = 0.2 μm. The left inset shows a lateral striated fiber (arrow) between daughter and parent basal bodies, short tubules (t), and proximately striated fiber (psf). The right inset shows the connection of a fiber cross-striation (arrow) to the 3-over-1 roots. From Weiss (1984).*

V. THE NUCLEUS

Median sections of *Chlamydomonas* cells in G1 phase show a prominent nucleus about 2–4μm in diameter, with a central nucleolus about
0.5 × 0.9μm (see Figure 2.1). Sager and Palade (1957) described regions of
local differentiation in the nucleolus, with tightly packed 100–150Å particles on the periphery and finer particles and light inclusions in the center.
Tremblay and Lafontaine (1992) distinguished RNA-containing bodies associated with the nucleolus, possibly representing RNA satellites or nucleolar
organizer regions, and other dense granular bodies that might be equivalent
to snRNP particles. Other more elongate bodies were also seen, occasionally aligned on the sides of a chromatin mass (Tremblay et al., 1992).

The double nuclear membrane is continuous with the endoplasmic
reticulum, and one to four Golgi bodies are usually seen situated nearby
(Figure 2.12). A rim of heterochromatin surrounds the nucleus, just
beneath the membrane, and nuclear pore complexes are concentrated primarily on the posterior side of the nucleus (Colón-Ramos et al., 2003)
(Figure 2.13).

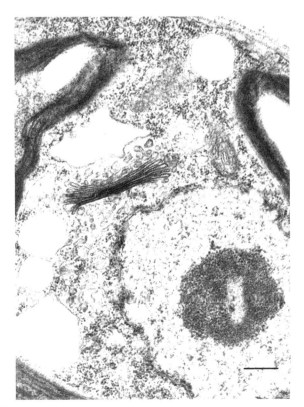

FIGURE 2.12 *Section of wild-type cell showing nucleus and Golgi apparatus. Bar = 0.5μm.
Courtesy of D.G. Robinson.*

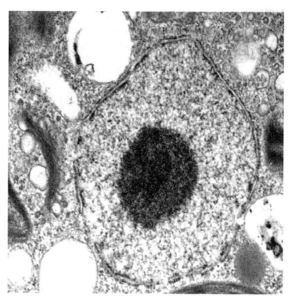

FIGURE 2.13 *Medial section of a nucleus of C. reinhardtii, showing nuclear pores and distribution of heterochromatin in patches uniformly dispersed around the nuclear periphery.*
From Colón-Ramos et al. (2003), with thanks to Mariano García-Blanco. See this paper for the demonstration that heterochromatin redistributes to the anterior side of the nucleus immediately following flagellar excision.

Chromosome structure and changes in cell nuclei during mitosis and meiosis will be discussed in Chapter 3.

VI. GOLGI AND ENDOPLASMIC RETICULUM

The Golgi apparatus appears prominently in sections of *C. reinhardtii* cells, usually at the anterior end of the cell and close to the nucleus. Zhang and Robinson (1986a) documented transfer of vesicles from the endoplasmic reticulum to the Golgi. Swelling of the portion of the endoplasmic reticulum facing the *cis* poles of the Golgi dictyosomes was correlated with periods of maximal synthesis of cell wall constituents in synchronized cells and was diminished or absent in wall-deficient mutants. The interaction of the Golgi apparatus with the plasma and flagellar membranes is discussed further in Volume 3, Chapter 11.

Whereas sections of *C. reinhardtii* typically show only a single Golgi structure, cells of the Chlamydomonad alga *G. kupfferi* usually have 16 or more, surrounding the nucleus and surrounded in turn by a layer of peripheral vacuoles. This alga has therefore been a particularly good system for investigation of endomembrane systems (Domozych, 1989; Domozych and Nimmons, 1992).

Voigt et al. (2001) used polyclonal antibodies to a recombinant 14-3-3-beta-galactosidase fusion protein to identify four 14-3-3 proteins having distinct patterns of association with the endoplasmic reticulum, dictyosome, and plasma membrane fractions. As discussed above, one 14-3-3 protein appears to be specifically involved in cross-linking of hydroxyproline-rich glycoproteins in formation of the cell wall (Voigt and Frank, 2003), and another is associated with the endoplasmic reticulum (Voigt et al., 2004). These proteins are likely to be useful markers for further studies of the cellular membrane systems in *Chlamydomonas*.

In a phylogenetic study of endomembrane proteins, Dacks and Doolittle (2002) sequenced a syntaxin gene from *C. reinhardtii*. Diaz-Troya et al. (2008) reported target of rapamycin and LST8 proteins associated with the endoplasmic reticulum and peri-basal body region. Using *C. noctigama*, Hummel et al. (2007) have identified COP I and COP II proteins in the Golgi periphery and Golgi–endoplasmic reticulum interface, respectively, and have investigated the response of these structures to Brefeldin A.

VII. CONTRACTILE VACUOLES

Most freshwater *Chlamydomonas* species have two contractile vacuoles, a few species have only a single one, and some have four or more (Ettl, 1976a; Luykx, 2000). They are not seen in marine species, or in freshwater species maintained in hypertonic medium. The number of contractile vacuoles was used as a primary taxonomic criterion to divide the subgenera by Gerloff (1940), but this practice was not followed by Ettl (1976a). In species with two vacuoles, for example *C. moewusii* (Guillard, 1960) or *C. reinhardtii* (Luykx et al., 1997a), the vacuoles pulsate alternately, usually at intervals of roughly 10–15 seconds depending on conditions.

Luykx et al. (1997a) made a thorough study of both the structural and the dynamic aspects of contractile vacuoles in *C. reinhardtii*. Using video microscopy, they established three stages in the approximately 15-second vacuole cycle in hypotonic medium. In the first 3 seconds (early diastole), small vesicles ranging from 70 to 120 nm in diameter appear, which then fuse with one another in the 6-second middle stage to form the vacuole (Figure 2.14). Additional small vesicles form and fuse with the vacuole in the remaining 6 seconds of the cycle (late diastole), and the vacuole contacts the plasma membrane. The culmination of the process (systole) is a rapid (0.2 second) discharge of the vacuole contents into the medium. In hypertonic medium, small vesicles similar to those of early diastole are seen, but they do not progress to fusion and systole (Denning and Fulton, 1989a; Hellebust et al., 1989).

Weiss et al. (1977a) described groups of particles visible in freeze-fracture preparations of the plasma membrane overlying the contractile vacuole region. Aggregation of these particles into circular arrays in both the plasma membrane and the underlying contractile vacuole membrane

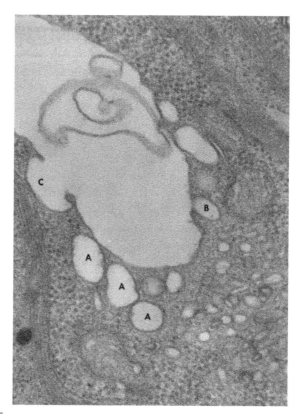

FIGURE 2.14 *Progressive stages of vacuole fusion in the contractile vacuole: a bed of small vesicles is seen in the lower right; larger vacuoles (A) are closer to the contractile vacuole, and some (B) appear to be fusing into the larger vacuole. Note the presence of a thin diaphragm in vacuole B. Section C probably represents a later stage, at which the membrane has begun to smooth out following fusion. From Gruber and Rosario (1979).*

appeared to take place at the time of vacuole discharge. The two arrays came into very close proximity and the membrane took on a puckered appearance, but no openings were observed in the vacuole membrane. Weiss et al. inferred that discharge of water from the vacuole probably occurs in hydrophilic channels created by the membrane contacts. Arrays of particles were also observed by Bray and colleagues (Bray et al., 1983; Bray and Nakamura, 1986) in the plasma membrane of *C. eugametos*. Luykx et al. (1997a) described a distinct contact zone approximately 0.2–0.4 μm wide between the vacuole and the plasma membranes, and in a few sections, a 15 nm pore in this region.

Gruber and Rosario (1979) proposed that vesicles arise from the Golgi, grow in size, and eventually fuse either with the contractile vacuole or with the plasma membrane. The same vesicles were also postulated to be involved in formation of new membrane during cytokinesis. No contractile elements were seen in association with the vacuole, prompting these authors to suggest that discharge of the vacuole should be viewed as a collapse at the cell

Table 2.6	Phenotypes of contractile vacuole (CV) mutants
Mutant	**Description**
osm1	CVs difficult to see; when seen, they are very small, form slowly and contract infrequently
osm3	Multiple CVs, grow slowly, contract infrequently
osm4	CV contraction frequency similar to wild type, but contraction events are slow
osm7	Abnormally large CVs, contract infrequently

Based on mutant descriptions in Luykx et al. (1997b).

surface rather than a repeated contraction, to be followed by formation of a new vacuole by fusion of cytoplasmic vesicles.

Weiss (1983a) described coated vesicles associated with the *minus* mating structure that appear to arise from the contractile vacuole region, and Denning and Fulton (1989b) characterized clathrin-coated vesicles purified by sucrose gradient fractionation. Denning and Fulton (1989a) suggested that these vesicles may be involved in recycling the vacuole membrane. Domozych and Nimmons (1992) used *G. kupfferi* to study uptake of cationic ferritin into the endomembrane system, and concluded that the contractile vacuole may also function in endocytosis.

A mutant of *C. moewusii* lacking contractile vacuoles was described by Guillard (1960). This strain survives in media with high osmotic pressure (0.1 M concentrations of any of a variety of sugars and salts sufficed) but lyses in 0.1 M glycerol, urea, ethanol, or ethylene glycol. Denning and Fulton (1989a) showed that cells of this mutant form small anterior vesicles that fuse to form larger vacuoles, but these do not interact with the plasma membrane and do not complete the systolic discharge. Treatment of wild-type cells with the calcium chelator EGTA to block membrane fusion produced a similar accumulation of vacuoles that did not complete systole.

Luykx et al. (1997b) selected mutants of *C. reinhardtii* that require hyperosmotic medium for survival. On transfer to medium of low osmotic strength, these mutants show abnormal contractile vacuole activity, or in some cases no activity at all. Four distinct phenotypes were found, and were shown to result from mutations at four different genetic loci (Table 2.6).

VIII. MICROBODIES AND OTHER MEMBRANE-BOUND STRUCTURES

Small, membrane-bound bodies such as lysosomes, peroxisomes, and glyoxysomes are found in most eukaryotic cells. Although such structures appear in sections of *Chlamydomonas* cells, they have received relatively little experimental attention in this organism. In an abstract for the

American Society for Cell Biology meeting in 1977, O'Kane et al. reported lysosome-like bodies in *C. reinhardtii* containing acid hydrolase, arylamidase and esterase, cathepsin A and D, and acid and neutral phosphatase activities, but no full-length paper on this topic ever appeared. Badour et al. (1973), working with *C. segnis*, noted the presence of granular organelles bounded by a single membrane, which they speculated could be peroxisomes or glyoxysomes, but did not pursue cytochemical characterization of these *Chlamydomonas* species.

Diaminobenzidine (DAB) staining of plant and animal cells reveals microbodies or peroxisomes containing catalase, which carries out oxidation of various substrates with H_2O_2 as reductant. Giraud and Czaninski (1971) reported that microbodies of mixotrophically grown *C. reinhardtii* did not react with this reagent, and concluded that catalase was absent from these cells. This was not unreasonable in view of contemporary reports that algae lacked glycolate oxidase, the principal source of H_2O_2 in higher plant cells. DAB staining of microbodies was reported in *Chlamydomonas dysosmos*, however (Silverberg and Sawa, 1974; Silverberg, 1975). Subsequently catalase was identified in *C. reinhardtii* but was found to be localized in mitochondria rather than in peroxisomes (Kato et al., 1997), as was glycolate dehydrogenase, the enzyme that generates H_2O_2 in many green algae (Stabenau, 1974; Beezley et al., 1976; see also Stabenau et al., 1993). Glycolate and glyoxylate metabolism will be discussed further in Chapter 6.

Cytoplasmic bodies filled with dense granular material were described by Sager and Palade (1957) as "vacuoles" distinct from the contractile vacuoles. These structures are closely associated with the Golgi (Wolfe et al., 1997; Park et al., 1999b), and were found to consist mainly of polyphosphate (Coleman, 1978; Siderius et al., 1996; Komine et al., 2000), which is released into the culture medium by exocytosis. These bodies may also contain proteins, including HSP70, apoproteins of the photosynthetic light-harvesting complexes, and a 70 kD protein that may be a cell wall component (White et al., 1996; Wolfe et al., 1997; Komine et al., 2000). Park et al. (1999b) examined the relationship of the chloroplast to the cytoplasmic endomembrane system, and speculated that these vacuolar bodies might serve as a means of degrading excess chloroplast proteins. Hoffman and Grossman (1999) proposed an alternative explanation for the presence of chloroplast proteins in these bodies, that the vacuoles serve as sites to which chloroplast proteins may be rerouted under conditions where import into the chloroplast is limited.

IX. MITOCHONDRIA

Mitochondria occupy only 1–3% of the cell volume of phototrophically grown *C. reinhardtii*, in contrast to the approximately 40% occupied by the chloroplast (Boynton et al., 1972; Schötz et al., 1972). Typical mitochondrial

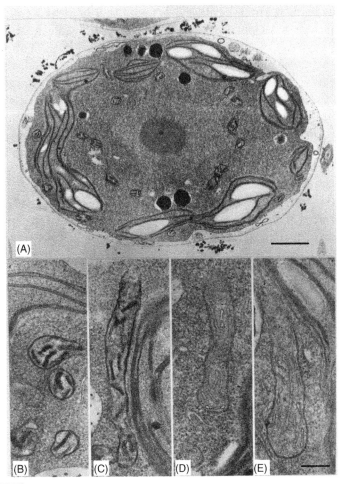

FIGURE 2.15   *Diaminobenzidine (DAB) staining of mitochondria from phototrophically grown wild-type cells. (A) Median section through a cell treated with DAB to indicate the activity of cytochrome oxidase. Bar = 1.0 μm. (B, C) mitochondria stained with DAB; (D, E) mitochondria pretreated with cyanide and then stained with DAB to demonstrate the inhibition of staining by cyanide, an inhibitor of cytochrome oxidase. (B–E) bar = 0.2 μm. Courtesy of Andrew Wiseman.*

profiles show oval or elongate organelles 0.2–0.3 μm in cross-section, bounded by a double membrane and having distinct interior membranes (cristae) projecting into a relatively dark matrix (Sager and Palade, 1957; Figure 2.15). These structures are disrupted or deranged in mutants unable to use acetate for growth in the dark (Wiseman et al., 1977).

Mitochondria in some sections appear to have an elongate, branching morphology and form an interconnecting network (see Grobe and Arnold, 1975, 1977; Figure 2.16; also Hiramatsu et al., 2006). Arnold et al. (1972) reported that the latter state was in fact more prevalent in cells grown in continuous light. Boynton et al. (1972) found that branching mitochondria were characteristic of phototrophically grown cells, while mixotrophically grown cells typically contained many small mitochondria. The total mitochondrial area was approximately the same under both growth conditions. Osafune et al. (1972a, b, 1975, 1976; also Ehara et al., 1995) suggested that changes in mitochondrial morphology follow a consistent pattern over the cell cycle in synchronously grown phototrophic cells, with small mitochondria fusing to form larger ones at 6–8 hours in the light phase of a 12:12 cycle, concomitant with a decrease in cellular O_2 consumption. Shortly before the giant mitochondria appeared, smaller mitochondria were seen to gather in groups in close proximity to the chloroplast, and chloroplast membranes seemed to protrude into the individual mitochondria. By 10 hours, small mitochondria were again predominant and O_2 consumption had increased. Blank et al. (1980) attempted to resolve the discrepancies of these earlier reports with a study of mitochondrial size in synchronous cells maintained in a chemostat on a 14:10 hour light:dark cycle and in gametes induced from these cells. They concluded that the time sequence reported by Osafune et al. was indeed an accurate description of a typical cell but stressed that mitochondrial morphology in *Chlamydomonas* was continually changing, with fusion and division of mitochondrial units occurring through the cell cycle. Studies by Gaffal (1987) confirmed and extended these conclusions. A basket configuration of interconnected mitochondria alternating in the life cycle with discrete mitochondrial units has also been reported in the colorless Chlamydomonad flagellate *Polytoma* (Gaffal and Schneider, 1978).

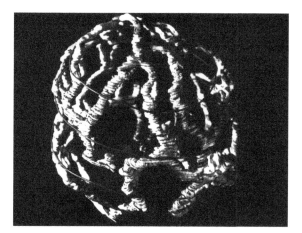

FIGURE 2.16 *Model of a single mitochondrion in one cell, constructed out of about 1000 separate mitochondrial profiles in 110 consecutive sections. Each 20th sectional plane is marked by a surrounding wire. The mitochondrial branches predominantly lie between the external chloroplast membrane and the plasma membrane, and are strongly reticulated, as the model shows. From Grobe and Arnold (1975).*

Effects of inhibitors of DNA and protein synthesis on mitochondrial ultrastructure were documented by Arnold and colleagues (Behn and Arnold, 1974; Blank and Arnold, 1980, 1981; Gercke and Arnold, 1981a, b) and by Boynton, Gillham, and collaborators (Boynton et al., 1972; Alexander et al., 1974; Conde et al. 1975). Mitochondrial DNA and its inheritance will be discussed in Chapter 7 and in Volume 2, Chapter 12.

X. THE CHLOROPLAST

From serial sections, Schötz et al. (1972) concluded that the shape of the single chloroplast of *C. reinhardtii* could vary from a relatively simple trough- or cup-shaped configuration to a complex basket morphology, and this conclusion has been borne out by immunofluorescence studies using antibodies to chloroplast proteins (Figure 2.17; Uniacke and Zerges, 2007). Cross-sections typically show a U-shaped structure surrounding the nucleus, with a broad basal area containing a single prominent pyrenoid (Figure 2.18A). The thylakoid membranes appear in electron micrographs as flat vesicles (discs), which can be either single or arranged in stacks of 2–10 discs but are not differentiated into multidisc grana as in higher plants. Instead, the stacks of discs merge and bifurcate in an anastomosing pattern along the length of the thylakoids (Goodenough and Levine, 1969; Goodenough and Staehelin, 1971; Kretzer, 1973; see also Figure 2.18B). Mutant strains may have characteristic abnormalities in disc arrangement, such as wide stacks of discs, long stacks of two discs each, or long, single discs. Stacking, as well as lipid composition, of thylakoid membranes can be influenced by growth conditions in some mutants and by treatment with some photosynthetic inhibitors. Chloroplast regions free of thylakoids,

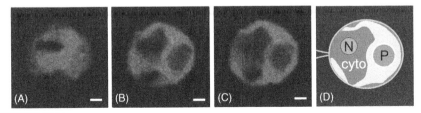

FIGURE 2.17 *Chloroplast morphology as revealed by fluorescence confocal microscopy. Immunofluorescence from the chloroplast ribosomal protein L12 is seen in three serial optical sections of 0.2 µm. (A) A grazing optical section shows the forward most chloroplast lobe. A tangential optical section acquired midway between the center and periphery (B) and a central section (C) show progressively more of the nucleocytosolic region (left-hand/apical) and the pyrenoid (right-hand/basal). (D) An illustration based on the optical section in C, and results in Uniacke and Zerges (2007), shows the chloroplast with its pyrenoid (P) and surrounding starch plates (white). Also shown are the approximate locations of the nucleus (N), the cell wall, and the non-chloroplast cytosolic compartments (cyto). Bars = 1 µm. Courtesy of James Uniacke and Bill Zerges.*

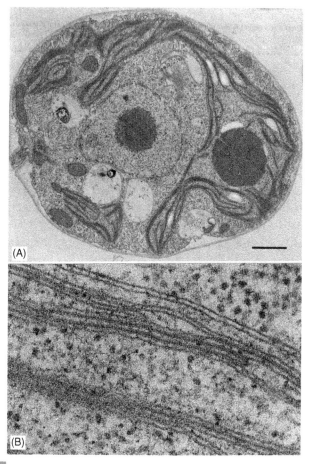

FIGURE 2.18 *(A) Section through a typical wild-type cell of C. reinhardtii grown mixotrophically (in the light, but with acetate as supplementary carbon source). The cup-shaped chloroplast, with a well-developed lamellar system, surrounds the periphery of the cell. Bar = 1 µm. (B) A portion of the same cell at higher magnification, showing the chloroplast envelope, stacked and unstacked thylakoid membranes, and cytoplasmic and chloroplast ribosomes. From Bourque et al. (1971); courtesy of The Company of Biologists Limited.*

which are collectively known as the stroma, contain soluble enzymes as well as chloroplast ribosomes.

Freeze-fracture studies (Ojakian and Satir, 1974; Wollman et al., 1980; Melkonian et al., 1981; Olive et al., 1981) revealed additional information about distribution of the particles believed to be associated with the photosystems and light-harvesting pigment–protein complexes. Exoplasmic (E) and protoplasmic (P) fracture faces are readily distinguished by the characteristic size and spacing of membrane-associated particles (Figure 2.19).

Easily visible by light microscopy, the pyrenoid was an important character in traditional taxonomy of *Chlamydomonas* species (see Chapter 1).

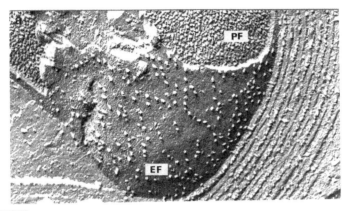

FIGURE 2.19 *Freeze-fractured chloroplast membranes in the absence of MgCl$_2$ from broken cell preparation of wild-type C. reinhardtii grown under 4000 lux. The P faces are covered with closely packed particles (6500/µm^2) about 80Å wide, while the E faces have lower densities of particles (1000/µm^2) about 120Å wide. From Olive et al. (1981).*

Most *Chlamydomonas* species have a single, basally located pyrenoid, but some have two or more, and species of the genus *Chloromonas* as traditionally defined lack an organized pyrenoid structure although based on molecular criteria they are closely related to *Chlamydomonas* species (Morita et al., 1998; Nozaki et al., 2002; see also Chapter 1). Among algae in general, pyrenoids are correlated with presence of a carbon-concentrating mechanism (Badger et al., 1998). The granular pyrenoid region (Figures 2.1 and 2.18A) appears to consist primarily of ribulose-1,5-bisphosphate carboxylase/oxygenase (Rubisco; Vladimirova et al., 1982; Lacoste-Royal and Gibbs, 1987; Kuchitsu et al., 1988; Markelova et al., 1990; Süss et al., 1995; Morita et al., 1997) but has also been reported to contain Rubisco activase (McKay et al., 1991a; He et al., 2003). Nitrate and nitrite reductases (López-Ruiz et al., 1985, 1991) and ferredoxin-NADP$^+$ reductase (Süss et al., 1995) have also been reported in pyrenoids, but would appear to be associated primarily with the periphery of the pyrenoid (Süss et al., 1995), and are absent from the matrix, which consists primarily of Rubisco (McKay and Gibbs, 1991a). Phosphoribulokinase appears in stromal inclusions within the pyrenoid, but not in the pyrenoid matrix (Kuchitsu et al., 1991; McKay and Gibbs, 1991b). The ferredoxin-like iron–sulfur protein FRXB has also been reported to have a specific association with vesicles adjacent to the pyrenoid (Zhang and Wu, 1993).

The fraction of the total Rubisco complement of the cell that is present in the pyrenoid, as opposed to the chloroplast stroma, varies with growth conditions and in particular with expression of the carbon-concentrating system (Badger et al., 1998; Borkhsenious et al., 1998; Morita et al., 1999; Fukuzawa et al., 2001). Mutant strains lacking Rubisco (Rawat et al., 1996) or the CCM1 protein that regulates the carbon-concentrating system (Fukuzawa et al., 2001) also lack pyrenoids, as do mutants deficient in chloroplast ribosomes (Goodenough and Levine, 1970; Boynton et al., 1972; Harris ct al., 1974; scc also Fig. 7.8). Starch deposits typically surround the pyrenoid, but their presence is highly dependent on growth conditions (Ramazanov et al., 1994; see also Chapter 6).

XI. THE EYESPOT

The bright orange eyespot, or stigma, of Volvocacean algae was recognized by Mast (1916) as the organelle by which cells orient their swimming with respect to light. In 1928 Mast postulated that the eyespot was part of the "neuromotor system" but it was many years before electron microscopy revealed its full structure and association with the flagellar roots. Gruber and Rosario (1974) found microtubules near the eyespot of *C. reinhardtii* but were uncertain whether these implied a functional relationship in connecting photoreception to motility. Melkonian and Robenek (1980) confirmed that this association is specific and that the eyespot lies between the distal extremities of the flagellar roots, near the cell equator. The eyespot is displaced from the plane of the flagella by 45° (Rüffer and Nultsch, 1985). The two flagella can therefore be designated as *cis* and *trans* with respect to the eyespot position.

This distinction becomes important in understanding the mechanism by which a light stimulus causes differential flagellar beat frequency to produce a change in the direction of swimming (Chapter 4). It is interesting that an early consequence of colchicine treatment is dissociation of the eyespot from its usual location toward the anterior portion of the chloroplast, such that it is seen further toward the cell posterior (Walne, 1967).

In *C. reinhardtii*, the eyespot consists of two or more layers of regularly arranged electron-dense granules approximately 80–130 nm in diameter. Each layer is subtended by a thylakoid membrane (Morel-Laurens and Feinleib, 1983), and the entire structure is closely appressed to the plasma membrane. The eyespot of *C. eugametos* has a single layer of somewhat smaller granules (Figure 2.20; Lembi and Lang, 1965; Walne and Arnott, 1967; Nakamura et al., 1973). The arrangement of carotenoid pigments within the eyespot has also been examined by polarizing microscopy and Raman scattering analysis (Yang and Tsuboi, 1999; Kubo et al., 2000; Tsuboi, 2002).

Freeze-fracture studies (Nakamura et al., 1973; Bray et al., 1974; Melkonian and Robenek, 1980) showed that both the plasma membrane and the outer chloroplast membrane overlying the eyespot are specialized, containing a different distribution of membrane particles than are found elsewhere in the cell (Figure 2.21). Subsequent studies confirmed that this specialized region is the photoreceptor for phototaxis and the photophobic stop response (Boscov and Feinleib, 1979; Melkonian and Robenek, 1979; see Chapter 4). Foster and Smyth (1980) postulated that the eyespot acts as a quarter-wave plate that reflects and intensifies light of a specific spectral range. The carotenoid layers reflect incident light back to the actual photoreceptor in the plasma membrane, and shield it from light passing through the cell body. The absorption and reflection properties of the eyespot have been further characterized by confocal microscopy and microspectrophotomery of single cells by Schaller and Uhl (1997). The discovery that the photoreceptor is a rhodopsin, and the nature of the signal transduction pathway that modulates flagellar beating to produce phototactic and photophobic

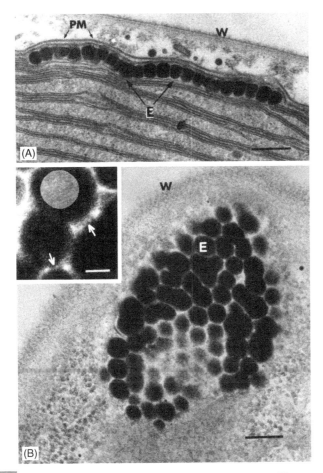

FIGURE 2.20 *Eyespot morphology in C. eugametos. (A) Stigma or eyespot (E) composed of a monolayer of approximately 23 granules lying below the plasma membrane (PM, arrows) and the chloroplast envelope and subtended by a chloroplast lamella. Note modification of plasma membrane in the stigma region, and lack of contiguity of PM with the chloroplast envelope (bar = 0.2 μm). (B) Tangential section of wall (W) and eyespot (E) showing paracrystalline plate of about 100 granules. In some places interconnections are seen (bar = 0.2 μm). Inset shows such interconnections (arrows), and substructure of a granule as a fibrous reticulum in an irregular or unordered configuration. Circular area produced by printing in order to demonstrate more clearly the structure (bar = 0.05 μm). From Walne and Arnott (1967).*

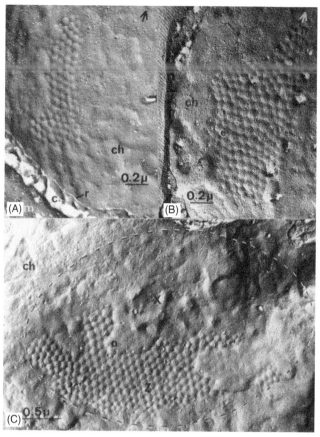

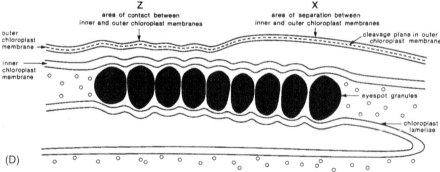

FIGURE 2.21 *Freeze-fracture replicas of the chloroplast envelope (ch) in the region of the eyespot of C. eugametos. Direction of shadowing is indicated by the arrow in the upper right-hand corner of each panel. (A) Convex cleavage of the chloroplast envelope in the eyespot region. The median surface of the plasma membrane (m) and cytoplasm (c) can be seen at the bottom of this panel. A ridge (r) representing the outer surface of the chloroplast envelope has been exposed as a result of etching. (B) Concave cleavage of the chloroplast envelope in the eyespot region showing the hexagonal shape of the granules which have been removed by the fracture plane. Some depressions are larger than others. The striated region at the left side of the panel is the outer aspect of the cell wall. (C) Replica of an eyespot containing about 220 bulges (Z) and obscured areas where the eyespot granules have not affected the membrane contour (X). The dotted outline (2 × 3 μm) delineates the suspected area of the entire eyespot as determined by light microscopy. Parallel rows of granules appear to be continuous on either side of one obscured area. (D) The physical association of the eyespot granules with the chloroplast membrane system. The subtending lamella is shown as also being modified by the pattern of granules, but no cleavage pattern was found to verify this assumption. The regions X and Z correspond to the respective regions in the micrograph above. From Nakamura et al. (1973).*

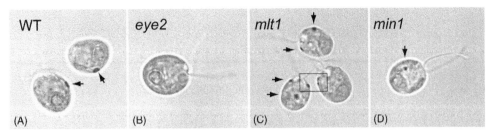

FIGURE 2.22 *Micrographs of eyespot mutants, as labeled: (A) wild-type control showing single, normal eyespot; (B) eye2, no discernable eyespot in any cell; (C) mlt1, two or more eyespots in each cell; (D) min1, eyespot is smaller than normal. Modified from Dieckmann (2003), with thanks to Telsa Mittelmeier.*

responses will be summarized in Chapter 4 in the current volume, and are discussed in much more detail by Hegemann in Volume 3, Chapter 13.

A mutant (*ey-1*, now *eye1*) lacking an eyespot was among the first mutants to be isolated in *C. reinhardtii* (Hartshorne 1953, 1955), and several additional mutants at the same locus were eventually isolated (Morel-Laurens and Feinleib, 1983; Morel-Laurens and Bird, 1984; Kreimer et al., 1992). These cells lack eyespots during logarithmic growth but form them as they reach stationary phase (Figure 2.22; Table 2.7). The *eye1* mutants retain the ability to respond phototactically, but their orientation is less precise than that of wild-type cells. Mutants at the *EYE2* and *EYE3* loci also fail to form eyespots (Lamb et al., 1999). Roberts et al. (2001) used insertional mutagenesis to identify the *EYE2* gene product as a chloroplast-targeted protein of the thioredoxin superfamily that may be required for eyespot assembly. *EYE3* encodes a putative ABC1 kinase (Merchant et al., 2007). In *min1* mutants, the eyespots are small and incorrectly organized, and the chloroplast and plasma membranes at the eyespot region lose their close association (Dieckmann, 2003). The *MIN1* gene encodes a novel protein with an N-terminal C2 domain that does not appear to be essential for its function in eyespot assembly. Mutants that form multiple eyespots include *ptx4* (Pazour et al., 1995) and its allele *mlt1* (Lamb et al., 1999; Dieckmann, 2003; Figure 2.22), and *mes-10* (Nakamura et al., 2001), which differs from *mlt1* in having a negative phototactic response. The *fn68* and *lts1-30* (CC-2359) mutants are alleles at the *LTS1* locus corresponding to the *PSY1* gene encoding phytoene synthase (McCarthy et al., 2004). Mutations at this locus prevent synthesis both of the carotenoid pigments of the eyespot and of the retinal chromophore. As discussed above, incorporation of retinal analogues can restore light perception in these strains. Lawson and Satir (1994) found that although no orange eyespot is visible in these mutants with the light microscope, the granular structure is still present.

A proteomic analysis of eyespots purified by sucrose gradient centrifugation and floatation has led to identification of 202 proteins associated

Gene	Mutant	Description	References
Table 2.7 Representative mutants affecting eyespot structure and function			
EYE1	*eye1*	Several alleles; lack eyespots during logarithmic growth, phototactic orientation impaired	Hartshorne (1953); Morel-Laurens and Bird (1984); Lamb et al. (1999)
EYE2	*eye2*	Eyespots not formed; defect in thioredoxin-like protein	Lamb et al. (1999); Roberts et al. (2001)
EYE3	*eye3*	Eyespots not formed; defect in putative ABC1 kinase	Lamb et al. (1999)
LTS1 or *PSY1*	*lts1*, *fn68*	Defective in phytoene synthase; deficient in carotenoids, including the rhodopsin photoreceptor; many alleles	McCarthy et al. (2004)
	mes-10	Forms multiple eyespots, negative phototaxis	Nakamura et al. (2001)
MIN1	*min1*	Small eyespots, incorrectly organized; gene product is novel protein	Dieckmann (2003)
MLT1 or *PTX4*	*mlt1*, *ptx4*	Forms multiple eyespots	Pazour et al. (1995); Lamb et al. (1999); Dieckmann (2003, 2004)

with the eyespot or the chloroplast and plasma membranes where it resides (Schmidt et al., 2006, 2007). Among the proteins thought to be intrinsic to the eyespot itself are several with PAP-fibrillin domains typical of conserved proteins involved in carotenoid sequestration in plants. Schmidt et al. postulated that these hydrophobic proteins may stabilize the very regular arrangement of the carotenoid globules of the eyespot. Proteins of the signal transduction pathway of phototaxis were also identified. The analysis by Renninger et al. (2006) of eyespot globule proteins in the green alga *Spermatozopsis similis* will also be of interest to persons researching *Chlamydomonas*.

CHAPTER 3

Cell Division

CHAPTER CONTENTS

I. INTRODUCTION

Although division was observed in *Chlamydomonas* cells in the 19th century, the cell geometry, with the chloroplast surrounding the nucleus, and small chromosome size precluded detailed descriptions of mitosis and meiosis before the age of electron microscopy. This chapter reviews the classic structural studies of these processes and subsequent work on cell cycle controls based on studies of synchronously growing cells.

II. MITOSIS

Light microscope studies of mitosis in *Chlamydomonas* species include the early work of Dangeard (1888, 1899), Belar (1926), and Kater (1929); the chromosome studies of Schaechter and DeLamater (1955), Wetherell and Krauss (1956), and Buffaloe (1958); the study of synchronized cells by Bernstein (1964); fluorochrome labeling of DNA by Coleman (1982a); and the investigation of microtubules through the cell cycle, using antibodies to tubulin, by Doonan and Grief (1987). The paper by Johnson and Porter (1968) remains the definitive electron microscopic description of mitosis in *C. reinhardtii*, and the following discussion is taken largely from that classic work. Differences between *C. reinhardtii* and *C. moewusii* as reported by Triemer and Brown (1974) are noted where appropriate.

Sedova (1998a, b, c, 2001a, b, 2002) published studies of mitosis in other *Chlamydomonas* species representative of Ettl's major Hauptgruppen. Useful comparisons may also be made with work on mitosis in primitive green algae (Stewart and Mattox, 1975; Mattox and Stewart, 1977), the colorless alga *Polytoma papillatum* (Wolf, 1995), and *Dunaliella bioculata* (Grunow and Lechtreck (2001).

Nuclear changes in *C. reinhardtii* begin in late G1 phase with loss of a recognizable nucleolus and apparent dispersal of nucleolar granules through the cytoplasm. The entire nucleus appears to change its position relative to the chloroplast and other organelles and comes to lie very near the plasma membrane, surrounded by endoplasmic reticulum. Harper et al. (1990) found that this change was concurrent with the beginning of reactivity with a monoclonal antibody raised against certain mitotic phosphoproteins of HeLa cells. Staining of the nuclear envelope with this antibody increased through metaphase and anaphase, and disappeared rapidly in telophase.

Basal body replication follows the nuclear migration in prophase, and the basal bodies migrate to the vicinity of the nuclear poles to assume their roles as centrioles (see also Coss, 1974; Ehler et al., 1995). As discussed in Chapter 2, Holmes and Dutcher (1989) established the very precise positioning of the basal bodies and their subsequent partition at cytokinesis (see also Volume 3, Chapter 14). By metaphase the nucleus has assumed a spindle shape, with spindle microtubules oriented toward the poles (Figure 3.1). Actin is localized along the rootlet microtubules, and forms a ring around the spindle (Ehler et al., 1995). Centrin becomes concentrated at the spindle poles, as does a kinesin-like calmodulin-binding protein (KCBP) that eventually is redistributed to the phycoplast (Dymek et al., 2006). The mitotic apparatus in *Chlamydomonas* is "closed," as in some other algae and many protozoa and fungi; that is, the nuclear envelope persists through mitosis. Johnson and Porter reported that in *C. reinhardtii* openings (fenestrae) of 300–500 nm appeared at the nuclear poles in a region of cytoplasm with low electron density, containing few ribosomes or other structures. Spindle microtubules seem to terminate in this region. Although sections of *C. moewusii* showed a similar ribosome-free region, Triemer and Brown found no evidence for fenestrae.

Rudimentary basal bodies can be seen in early G1 phase cells prior to the onset of mitosis, and full development of new basal bodies appears to take place late in G1. By prophase, four mature basal bodies are evident, grouped into two pairs each consisting of an old and a new basal body (Huang et al., 1982; Gaffal, 1988; see also Volume 3, Chapters 2 and 14). Typically flagellar resorption is one of the first visible indicators of incipient mitosis, but this does not necessarily always occur (Johnson and Porter, 1968). Even when the flagella persist into mitosis, dividing cells are nonmotile, and flagellar connections with the basal bodies are lost, detachment occurring at the point where the transition from triplet to doublet microtubules occurs (see Figure 2.8).

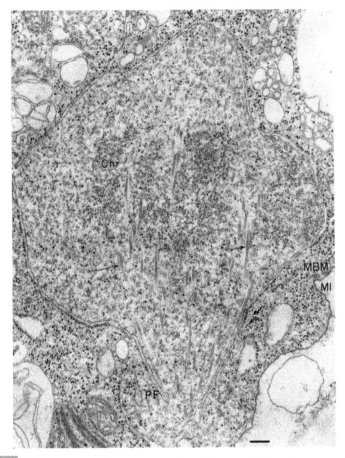

FIGURE 3.1   *Late metaphase-early anaphase nucleus of C. reinhardtii. Spindle microtubules (arrows) extend toward chromosomes (Chr) and toward a polar fenestra (PF). A membrane invagination (MI) and associated metaphase band microtubules (MBM) are seen. From Johnson and Porter (1968). Reproduced from The Journal of Cell Biology, 1968, 38, 403–425 by copyright permission of The Rockefeller University Press.*

Chromosomes align at the metaphase plate but are small and are poorly resolved (see below). At anaphase the nucleus elongates, and the chromosomes move to the nuclear poles as the spindle microtubules disappear (Figure 3.2). Triemer and Brown (1974) described the kinetochores at this stage as three-layered structures, an outer electron-dense region on which the microtubules converge, a central transparent layer, and another darkly staining layer adjacent to the chromosomes. As nuclear division proceeds, a band of endoplasmic reticulum is seen between the newly formed daughter nuclei; Johnson and Porter suggested that this is derived from expansion and folding of the nuclear envelope. New nuclear membrane, with ribosomes attached, appears in telophase as the daughter nuclei are formed,

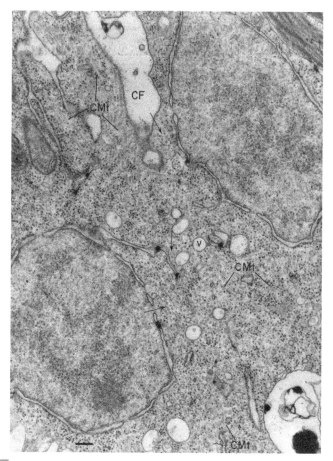

FIGURE 3.2 *Early cleavage in C. reinhardtii. Two daughter nuclei are separated by an array of internuclear microtubules (arrows), sectioned transversely. The nascent cleavage furrow (CF) is lined by cleavage microtubules (CMt) that extend deep into the cytoplasm. Small vesicles (V) lie along the plane of cleavage. From Johnson and Porter (1968). Reproduced from The Journal of Cell Biology, 1968, 38, 403–425 by copyright permission of The Rockefeller University Press.*

and ribosomes then begin to appear around the entire nuclear envelope of the daughter cells except in the vicinity of the basal bodies (Triemer and Brown, 1974). Nucleoli re-form in the daughter cells at this time, and the endoplasmic reticulum surrounding the nucleus becomes fragmented.

Several studies have focused on the relationship of the basal apparatus to the nucleus and the changing configuration of cytoplasmic microtubules during the mitotic process (Doonan and Grief, 1987; Gaffal, 1988; Gaffal et al., 1992; Salisbury, 1995; Ehler and Dutcher, 1998). A "metaphase band" of four microtubules was described by Johnson and Porter (1968) as lying on one side of a cell membrane invagination over the nuclear midline, roughly perpendicular to the spindle axis, but not completely circling the cell. Doonan and Grief (1987) showed that this band consisted of two distinct sets of microtubules, and based on analysis of serial sections of four cells, Gaffal and el-Gammal (1990) concluded that it is actually a transient association of the two four-membered flagellar roots of the parental cell. "Spindle" microtubules within the nucleus terminate in a specialized region near the polar fenestrae of the nuclear membrane but are not associated with any recognizable structures in this region. These microtubules disappear at telophase, and new "internuclear" microtubules appear between the daughter nuclei at right angles to the spindle axis. The fourth set, "cleavage microtubules," appear along the cleavage furrow at cytokinesis, perpendicular to both the spindle and internuclear microtubules and apparently passing between the latter microtubules. Together the cleavage and internuclear microtubules constitute the phycoplast, a formation characteristic of the Chlorophyceae and an important consideration in traditional algal taxonomy. Cross sections through the cleavage furrow suggest that the cleavage microtubules are not strictly in parallel array but converge toward the basal bodies at the anterior end of the cleavage furrow.

The plane of cleavage in most *Chlamydomonas* cells at the first mitotic division appears to be longitudinal with respect to the position of nucleus and chloroplast, but in some species (including *C. reinhardtii*) the entire protoplast may rotate within the cell wall, so division appears to be transverse with respect to the apical papilla and flagellar insertions on the original cell wall (Ettl, 1976a; Figure 3.3; see also Holmes and Dutcher, 1989). In most species under typical laboratory conditions, two rounds of mitosis occur sequentially within a single mother cell wall, followed by release of four daughter cells. In these cases, orientation of the second division with respect to the first may be a recognizable characteristic of the species (Ettl, 1976a, 1979, 1988). However, Johnson and Porter found that the second cleavage plane appeared to be longitudinal in some cells of *C. reinhardtii* and equatorial in others. They postulated that there might be a consistent difference in cleavage pattern among cells that were to end the division process with four daughters and those that would go on into a third round of division to form eight progeny. If equatorial division is to occur, the basal bodies need to migrate to the new anterior cleavage furrow, and Johnson and Porter were in fact able to document the presence of basal bodies in intermediate locations in cells fixed between divisions.

Johnson and Porter described cleavage in *C. reinhardtii* as deduced from electron micrographs. The cleavage furrow first appears at one side of the anterior end of the cell in the region of the metaphase plate and extends from there both laterally around the cell and medially toward the interior (Figures 3.4 and 3.5). Actin is concentrated along the cleavage furrow and phycoplast microtubules (Ehler et al., 1995), and myosin staining also becomes more intense along the cleavage furrow rootlets during cytokinesis (Ehler and Dutcher, 1998). The kinesin-like calmodulin-binding protein

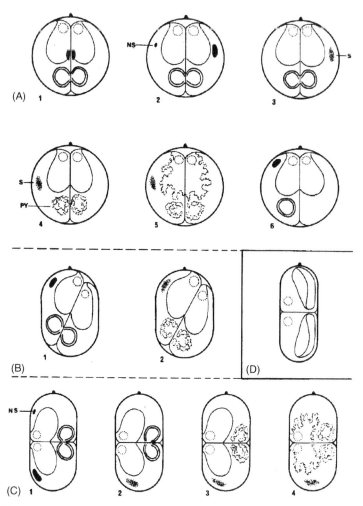

FIGURE 3.3 *Various types of protoplast division seen in Chlamydomonas species. Numbers indicate individual modifications of the corresponding types. (A) True longitudinal division: 1, eyespot and pyrenoid divided (e.g., C. olifanii); 2, chloroplast and pyrenoid divided, but original eyespot remains in one daughter cell and a new one is formed in the other (C. angulosa); 3, original eyespot disintegrates and new ones reappear in daughter cells (C. proboscigera); 4, pyrenoid and eyespot both disintegrate and re-form later in daughter cells (C. concylindrus); 5, eyespot, chloroplast, and pyrenoid all lose discrete structure and re-form in daughters (C. rotula, C. gerloffii); 6, rare form, in which one pyrenoid and eyespot persist in one daughter and new ones form in the other (C. praecox). (B) Oblique divisions, with (1, C. ovalis) or without (2, species uncertain) persistence of original eyespot and pyrenoid; (C) Simulated transverse division, in which the protoplast rotates 90°. Organelle division proceeds as in (A), but perpendicular to the vertical axis of the original mother cell: 1, C. obversa, 2, C. lewinii, 3, C. svitaviensis, 4, C. geitleri, C. bilatus. (D) True transverse division, in which a lateral chloroplast in a narrow cell hinders protoplast rotation (C. gloeophila). S, eyespot; NS, new eyespot; PY, pyrenoid. From Ettl (1976a).*

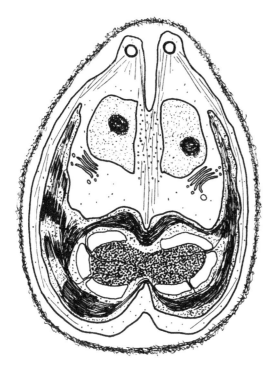

FIGURE 3.4 *Diagrammatic representation of cytokinesis in C. reinhardtii. The cleavage furrow, circumferential around the cell, grows fastest through two sets of microtubules, which together constitute the phycoplast. One set (the cleavage microtubules as well as peripheral microtubules) is oriented toward the basal body complexes at the edge of the cell. The other set (the internuclear microtubules) is shown as small circles, oriented perpendicular to the cleavage microtubules. From Pickett-Heaps (1975).*

also relocates from the centrosomes to the phycoplast microtubules at this time (Dymek et al., 2006). Although the cleavage plane initially extends from the anterior to posterior cell poles and passes through the flagella, protoplast rotation during cytokinesis causes the final division plane to lie about 45° from the flagellar insertion. Treatment of *Chlamydomonas* cells with oryzalin or vinblastine, which depolymerize microtubules, leads to abnormal distribution of actin and misplacement of the cleavage furrow, and subsequent failure to complete cytokinesis (Ehler and Dutcher, 1998). Mutants have been isolated that show similar abnormalities (Table 3.1, Figure 3.6). Mutants with aberrant division were also described in *C. moewusii* by Lewin (1952b, 1953a) and in *C. eugametos* by Gowans (1960).

Goodenough (1970) found that chloroplast division in *C. reinhardtii* did not begin until cytokinesis was under way, but there is some evidence that the timing of chloroplast division relative to nuclear division can vary

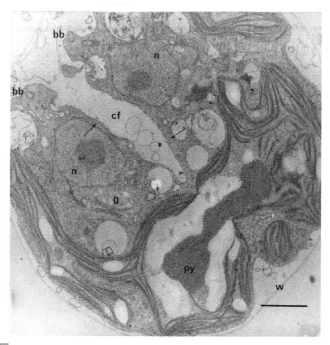

FIGURE 3.5 *C. reinhardtii cell in mid-cytokinesis. A cleavage furrow (cf) separates the two daughter nuclei (n) and has grown inward all around the cell, but fastest between the basal bodies (bb) and daughter nuclei. It is still lined by cleavage microtubules (arrows). The chloroplast is also furrowing the plane of cell cleavage, partitioning the pyrenoid (py) into daughter cells. From Goodenough (1970); relabeled by Pickett-Heaps, 1975).*

among algal species (Ettl, 1976b). In *C. reinhardtii*, the pyrenoid elongates dramatically as the cleavage furrow progresses through the cell (Figure 3.6), and then divides. The chloroplast constricts in the plane of the cleavage furrow and is separated as cytokinesis is completed. Gaffal et al. (1995) used ultrathin serial sections to reconstruct the changes in shape undergone by the chloroplast prior to, during and after cytokinesis. Chloroplast DNA is contained within 8–10 discrete nucleoids in the *C. reinhardtii* chloroplast, each probably containing several copies of the chloroplast genome, which increase in size, become dumbbell shaped, and then divide prior to chloroplast division (Kuroiwa et al., 1981; see also Chapter 7). It is assumed that they are partitioned more or less equally among the daughter cells as the chloroplasts divides. Replication of chloroplast DNA does not appear to be tightly coupled to nuclear DNA synthesis (see Chapter 6).

Like other algae and land plants, *C. reinhardtii* retains the evolutionarily conserved FtsZ proteins essential for cell division in bacteria, a heritage from the endosymbiotic cyanobacterial ancestor of the plastid (Wang et al., 2003; see also Stokes and Osteryoung, 2003, and Miyagishima et al.,

Gene	Mutant	Description	Reference
Table 3.1		Representative mutants with abnormalities in cell division	
BLD2 or *TUE*	*bld2*	Mutation in epsilon tubulin; defects in flagellar assembly and in positioning of the mitotic spindle and cleavage furrow	Goodenough and St. Clair (1975); Preble et al. (2001); Dutcher et al. (2002); Ehler et al. (1995)
VFL2	*vfl2*	Centrin-deficient, variable number of flagella; also shows loss of basal bodies from spindle poles	Wright et al. (1985); Taillon et al. (1992); Zamora and Marshall (2005)
	cmu1-1	Defect in cytoplasmic microtubules leads to abnormal position of spindle	Horst et al. (1999)
FLA10	*fla10*	Mutation in kinesin-2, which is involved in intraflagellar transport, but localizes to basal bodies during mitosis, may affect spindle pole function	Lux and Dutcher (1991); Vashishtha et al. (1996a)
	apm1	Mutant resistant to the microtubule inhibitor amiprophos methyl; may affect microtubule nucleation	Lux and Dutcher (1991)
	cyt mutants	Mutants with defects in cytokinesis	Warr (1968) Ehler and Dutcher (1998)
	oca1 and *oca2*	Incomplete cleavage furrows	Hirono and Yoda (1997)

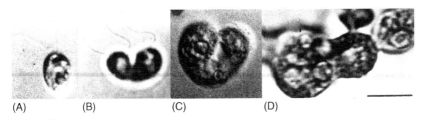

(A)   (B)   (C)   (D)

FIGURE 3.6   *Phase contrast images of cellular morphologies present in cyt- strains. (A) Wild-type morphology. (B) One incomplete cleavage furrow. (C) Doughnut-shaped cell. Through-focusing of this cell showed that the "hole" was continuous. (D) Multiple incomplete furrows. This cell appeared highly "branched" and some portions of it extend out of the plane of focus. Bar = 10 μm. From Ehler and Dutcher (1998).*

2003, for general information on the evolution of these proteins). Genes encoding homologues of the bacterial division site determinants MinD and MinE have also been identified in the *C. reinhardtii* nuclear genome. These genes are chloroplast-encoded in some algae, including *Chlorella vulgaris* (Wakasugi et al., 1997).

III. MEIOSIS

Observations of meiosis in *Chlamydomonas* are complicated by the thick zygospore wall. Lewin (1957b) found that *C. moewusii* zygotes kept in the dark formed a thinner wall than those matured in the light, and Schaechter and DeLamater (1956) also took advantage of this phenomenon for their light microscopic study. They described changes in nuclear staining consistent with chromosome condensation, but the chromosomes appeared smaller than those seen in preparations from mitotic cells, and they were unable to resolve individual chromosomes sufficiently well to make an accurate count (see below). A metaphase plate was clearly discerned at the end of prophase I. This divided equatorially to produce two daughter plates which dispersed to a filamentous state prior to prophase II. Levine and Folsome (1959) sampled *C. reinhardtii* zygotes at intervals from mating through 6 days' maturation. No special provisions were made to minimize zygospore wall formation, but pictures were obtained of meiotic nuclei stained either with Azure A or with Feulgen. Again, chromosome counts were somewhat uncertain.

Triemer and Brown (1976) provided the first detailed ultrastructural study of nuclear events in *C. reinhardtii* meiosis. Trypsin treatment of zygospores prior to fixation greatly improved the permeability to glutaraldehyde used as fixative. In micrographs from this work, the pre-meiotic zygote shows a starch-filled plastid, few mitochondria, and very little endoplasmic reticulum. Leptotene is marked by condensation of chromosomes and appearance of axial cores, which by early zygotene are associated with the nuclear envelope. Tubular structures (25 nm) also appear in the perinuclear space in this region. As zygotene progresses, association between the chromosomes and the nucleolus is also seen. Synaptenemal complexes are visible in pachytene and were studied in detail by Storms and Hastings (1977). These were atypical, consisting of two lateral components 30–40 nm in diameter and separated by about 110 nm (Figures 3.7 and 3.8). No central component was detected. Chromatin appeared to be condensed around each lateral component in a U shape. By diplotene these complexes appear to have degenerated, and the chromosomes are then seen as dense bodies with apparent chiasmata visible in some sections. The nucleolus is no longer obvious at this stage or at diakinesis.

Metaphase, anaphase, and telophase of meiosis I and II resemble the corresponding stages of mitosis with respect to spindle and phycoplast microtubules, polar fenestrae in the nuclear envelope, and behavior of basal bodies. A brief interphase separates the two divisions. During this period daughter nuclei and nucleoli appear completely formed, and basal bodies seem to replicate. The second meiotic division ends with the four tetrad products still encased in the zygospore wall; a mitotic division may follow in some zygotes to produce eight cells prior to hatching.

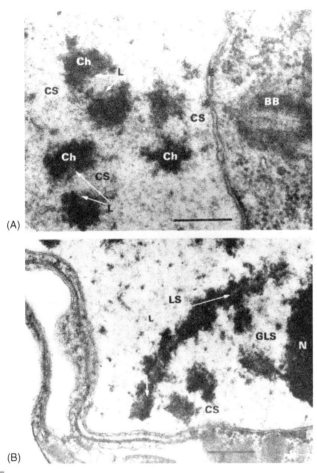

FIGURE 3.7   *Synaptonemal complexes of C. reinhardtii. (A) Cross-sectional (CS) view, in which chromatin (Ch) forms a U shape around the lateral component (L). The nuclear envelope (E) separates the basal body (BB) from the nucleus. (B) Complexes in grazing long section (GLS) and cross section (CS). Also present is a longitudinal section (LS) through one homologue of a bivalent pair showing the lateral component (L) (bar = 0.4 μm). From Storms and Hastings (1977).*

Fusion of *C. reinhardtii* gametes is followed within a few hours by flagellar regression. Cavalier-Smith (1974) reported that basal bodies, striated fibers, and flagellar roots all disappeared in the young zygote during the period of nuclear and chloroplast fusion (Chapter 5) and were absent throughout the succeeding 5-day zygospore maturation period. When mature zygospores are transferred to germination medium, these structures were re-formed. New basal bodies were occasionally visible in Cavalier-Smith's sections about 6 hours into the germination process, at the beginning of meiotic prophase, but were not seen in most sections until 9–10 hours. Cavalier-Smith reported rings of nine singlet microtubules at this stage, similar to those seen in vegetative cells by Johnson and Porter (1968).

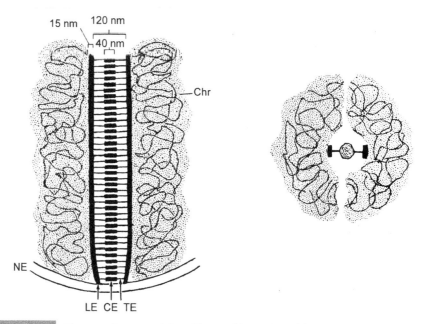

FIGURE 3.8 *Diagrammatic representation of the possible structure of the synaptonemal complex in transverse and longitudinal sections. Two lateral elements (LE, 18–24 nm) associate with the bivalent chromosomes (Chr) which are separated by a central space of 120–130 nm. A central element (CE, 30–40 nm) is present along specialized regions of the synapsed bivalents. Transverse elements (TE, ~7 nm) extend from the lateral elements to the central element in regions where the complex is formed. Note the association of the complex with the nuclear envelope (NE). From Triemer and Brown (1976).*

IV. CHROMOSOME CYTOLOGY

Unfortunately chromosomes are difficult to visualize in *C. reinhardtii* by light microscopy (Figure 3.9), and debate raged for years over the true number present. Attempts to resolve them by electrophoresis have also been unsuccessful. As the number of genetic linkage groups increased with the mapping of additional markers, so did the estimated number of chromosomes. For example, Buffaloe (1958) and Levine and Folsome (1959) believed the haploid number of chromosomes to be 8, consistent with the 7 known linkage groups at the time, and attributed previous reports of 16 or more chromosomes (Schaechter and DeLamater, 1955; Wetherell and Krauss, 1956) to temporary polyploidy or uncondensed prophase chromosomes. By 1962, however, the genetic data indicated at least 11 linkage groups, and by 1965 there appeared to be 16 (Ebersold et al., 1962; Hastings et al., 1965). McVittie and Davies (1971) corroborated the genetic map with cytological studies of meiotic zygotes indicating 16 bivalents at diakinesis. In 1976, however, Maguire published light and electron microscopic

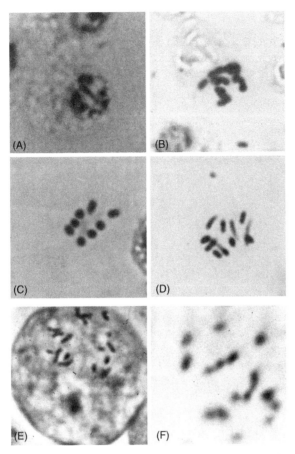

FIGURE 3.9   *Chromosomes of C. reinhardtii. (A–D), from Maguire (1976), showing eight chromosomes per nucleus. Haploid vegetative cells were harvested at mitotic metaphase, about 5 hours into the dark cycle. (A and B) Intact cells, (C and D) nuclear contents separated from cell during preparation (bar = 5 μm). (E, F), from Loppes et al. (1972), early metaphase cells at low and high magnification, showing 16 chromosomes.*

studies supporting a haploid number of 8, and discounted the evidence for 11 or more linkage groups as based on insufficient data. The issue was finally settled to most researchers' satisfaction by Storms and Hastings (1977), who reconstructed models of pachytene nuclei from serial sections of germinating zygotes. Sixteen individual bivalents were identified, and four additional bivalent arms, which could represent two to four individual bivalents, were attached to a mass of chromatin. This is consistent with the current genetic map of 17 linkage groups (Dutcher et al., 1991; Kathir et al., 2003), which in turn is supported by the assembly of genomic sequence data (Merchant et al., 2007).

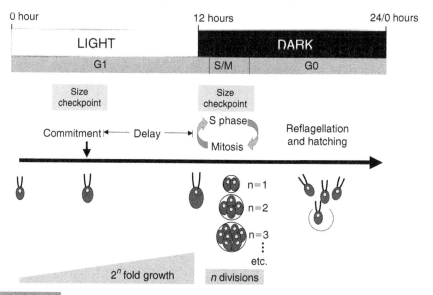

FIGURE 3.10 *Diagram of the multiple fission cell cycle in C. reinhardtii cells grown on a 12:12 hour light–dark cycle. Cells in G1 commit to a future division when a critical size is reached, but continue to grow until the beginning of the dark phase. At the end of G1, a rapid series of alternating S and M cycles generates 2^n daughter cells of uniform size, enclosed within a single mother cell wall. The value of n is variable and is related to mother cell size. Daughter cells are released synchronously during the latter portion of the dark phase, and re-enter G1 when light is restored. Courtesy of Jim Umen.*

V. THE CELL CYCLE

A. Synchronous cultures

Synchronous cell division of *Chlamydomonas* is usually achieved in the laboratory by alternation of light and dark periods (Bernstein 1960; see also Chapter 8).

Log-phase cultures maintained on a 12:12 hour light–dark cycle at 22–25°C typically undergo 2 or 3 doublings in a 24-hour period (Bernstein, 1964; Jones, 1970). All the divisions in a cycle take place in rapid succession during the dark period (Figure 3.10). Each round of DNA synthesis is followed immediately by mitosis, so no cell ever has more than the 2c quantity of DNA (Coleman, 1982a), and the G2 phase is essentially undetectable (Jones, 1970). Typically all the daughter cells are retained within the mother cell wall until the entire division period is complete (Figure 3.11), and are then released simultaneously by the action of the vegetative lytic enzyme (see above), usually shortly before the end of the dark period. In Figure 3.10, this period is designated as G0, since growth does not resume until light is restored.

The number of divisions in one 24-hour cycle is highly dependent on growth conditions. In bright light at 25°C, log-phase cells typically divide

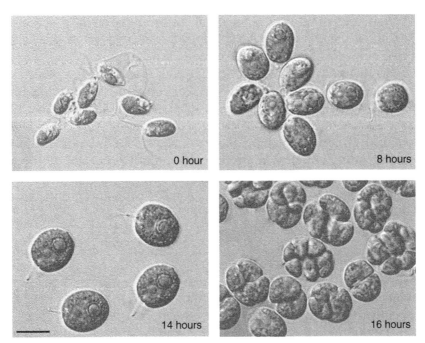

FIGURE 3.11 *Cells grown in synchronous culture on a 12:12 hour light–dark cycle, showing increase in size from the beginning of the light period (0 hour) through division into 8 daughter cells between 2 hours and 4 hours into the dark cycle (14 and 16 hours). Bar = 10 μm. From Bisova et al. (2005).*

three times to produce eight daughters, but at lower temperatures or lower light intensity, two divisions are more usual (see Chapter 8). The control of division number will be discussed in section V.B.

Photosynthetic components are actively synthesized midway through the light phase in cultures grown on a light–dark cycle. A burst of synthesis of tubulin occurs at about the time of cell division. However, these abrupt accumulations of cell components are not necessarily an essential part of the cell cycle. Rollins et al. (1983) found that in light–dark synchronized cells transferred to continuous illumination in a turbidostat, synthesis of tubulin, chloroplast membrane proteins and several hundred soluble proteins all became continuous throughout the cell cycle.

John (1984) suggested that most of the cell cycle periodicities observed in macromolecule synthesis in synchronously grown cells are environmentally induced and are not intrinsically related to control of the cell cycle. Certain fluctuations, for example tubulin synthesis, may be directly related to cell division but can still be modulated by environmental factors (Rollins et al., 1983), while others, such as adaptive changes in the photosynthetic apparatus and in intermediary and biosynthetic metabolism, are more likely to be indirectly related through the general optimization of cell performance in the current environment. From the point of view of cell cycle control, then, most metabolic fluctuations may in fact be side effects, but this in no way diminishes the value of synchronized cells in studying synthesis of specific cell components for their own sake, and many such studies have been made over the years (for review of the early literature, see Harris, 1989). Additional papers of this type published since that time include studies of expression of carbonic anhydrase (Toguri et al., 1989), chlorophyll *a/b*-binding proteins (Jasper et al., 1991), the nitrate-reducing and ammonium assimilation systems (Martinez-Rivas et al., 1991), and ferredoxin-nitrite reductase (Pajuelo et al., 1995). Mittag et al. (2005) have summarized studies of transcription and mRNA accumulation from the perspective of circadian rhythms.

B. Cell cycle controls

Howell and Naliboff (1973) isolated temperature-sensitive *C. reinhardtii* mutants with blocks at specific points in the cell cycle. They assumed that at the restrictive temperature (33°C), a gene product normally required for completion of the cell cycle would be non-functional in mutant cells. In a population shifted to the restrictive temperature, only cells that had completed this function would be able to complete division. Thus the increase in cell number that occurs after transfer of an asynchronous population to restrictive conditions should allow one to calculate the point at which the cell cycle is blocked. Howell and Naliboff derived an equation for this purpose that has been extensively used in studies of the yeast cell cycle but, paradoxically, is difficult to apply accurately to *Chlamydomonas* because of the uncertainty in whether the division number would have been 2, 4, 8, or even 16 at the restrictive temperature if all the cells had succeeded in completing their current cell cycle. Howell and Naliboff assumed from the data available to them that the division number is invariably four, but it is now clear that this is incorrect and that higher temperature allows faster growth under the same conditions of light intensity and composition of the culture medium (Donnan and John, 1983; Donnan et al., 1985).

Despite these difficulties of interpretation, Howell and Naliboff clearly established that block points can be distributed throughout the cell cycle, and additional temperature-sensitive cell cycle mutants were obtained in other laboratories (Table 3.2). For a more extensive review of the properties of these mutants, as well as mutants with structural defects that affect mitosis, see Harper (1999).

Commitment to divide in a given cell cycle is determined first by reaching a critical minimal size during the G1 period (Spudich and Sager, 1980; Oldenhof et al., 2007; Figure 3.10). Having passed this point, a cell will divide regardless of whether further growth occurs or not, but DNA synthesis and mitosis are delayed for 5–8 hours while G1 continues. A second

		Table 3.2 Representative mutants with abnormalities in cell cycle control	
Gene	**Mutant**	**Description**	**Reference**
		Collection of temperature-sensitive mutants with blocks in the cell cycle	Howell and Naliboff (1973)
	cdb	Another group of temperature-sensitive cell cycle mutants	Harper et al. (1995); Harper (1999)
	uvs11	UV-sensitive mutant, possibly bypasses a checkpoint for DNA repair	Vlček et al. (1987); Slaninová et al. (2002)
	met1	Metaphase arrest mutant	Harper et al. (2004)
FA2	fa2	Gene encodes NIMA family kinase required for flagellar excision; mutant has altered cell cycle progression	Mahjoub et al. (2002)
MAT3	mat3	Very small cell size, resulting from alteration in retinoblastoma (RB) tumor suppressor-related protein	Armbrust et al. (1995); Umen and Goodenough (2001a); Fang et al. (2006)
DP1	dp1	Recessive mutation; suppresses mat3 phenotype, producing large cells	Fang et al. (2006)
E2F1	e2f1	Dominant mutation; suppresses mat3 phenotype and restores normal cell size	Fang et al. (2006)

commitment point, more directly analogous to the START point in yeast or the restriction point in mammalian cells, controls progression into S phase and then mitosis. A cell cycle in which multiple fission events occur is not fundamentally different from others, but rather is a case in which the commitment to divide occurs more than once per cell cycle. This is a consequence of the attainment of more than a twofold increase in cell mass in the G1 period (Craigie and Cavalier-Smith, 1982; Donnan and John, 1983). Each commitment is the major rate-limiting control point for the subsequent DNA doubling, mitosis and cytokinesis, which follow in the same time interval regardless of illumination or growth (John, 1984; Donnan et al., 1985). Donnan and John (1983) found that the duration of the timer adheres reasonably well to the relationship:

Time to commitment (hr) = 5 + (0.75 × time required for mass doubling)

Since the time period after first commitment, during which any additional commitments occur and the committed divisions are executed, was close to 6 hours in light-limited cells cultured between 21°C and 30°C, they calculated the total cycle time:

Mean generation time (hr) = 11 + (0.75 × time required for mass doubling).

Donnan and John (1983) found that further commitments were undertaken after the first if the mass apportioned to committed daughters remained more than 27 pg protein per cell. Thus the minimum daughter size is about 13 pg protein, and the mean 20 pg. Larger daughter sizes can of course result if mother cells continue to grow after the final commitment. Umen and Goodenough (2001a) calculated that the minimum volume required for commitment to divide is $178\,\mu m^3$. Using direct observation of single cells cultured in microchambers, Matsumura et al. (2003) reported a normalized ratio of $2.2\times$ the starting cross-sectional area as the critical size to trigger commitment to divide. They also found that the minimal and maximal cell sizes in their system were not altered by differences in light intensity.

The delay between the first and the second commitment points and the time of division can vary with culture conditions and with the strain being used. Under optimal growth conditions in light–dark synchronized cultures, DNA synthesis typically begins early in the dark phase as shown in Figure 3.10 (Jim Umen, personal communication). In some of the early literature, however, the S/M phase is shown occurring later in the dark phase.

Studies with inhibitors confirmed that initiation of mitosis is dependent on completion of DNA synthesis (Harper and John, 1986; John, 1987). Progress toward cytokinesis is independently regulated, at least in part, since in cells in which nuclear division is blocked there is still initiation of a phycoplast and development of a cleavage furrow in the normal plane, where it is obstructed by the undivided nucleus. A second cleavage plane is sometimes initiated which again encounters the single undivided nucleus and produces a four-lobed cytoplasmic configuration (Harper and John, 1986).

Although there has long been a consensus that the size attained by the mother cell in a given cycle determines the number of divisions that will occur, debate has persisted on the nature and role of timer mechanisms in controlling the cycle. A classic functional test for the operation of a biological timer is that regulated events hold their timing in spite of changes in temperature. Donnan and John (1983) reported that the cycle phases prior to commitment to division and following commitment both have timer properties. The duration of the precommitment period was constant despite changes of temperature in the range 20–30°C, and the postcommitment period regained its duration after a lag of one cell cycle following a temperature change in this same range. These stabilizations of duration occurred while metabolic processes were halved or doubled in rate by the temperature change.

The effect of such temperature compensation is to time division to occur during the night, as might also be achieved by an endogenous circadian oscillator. The essential difference from an oscillator is that the precommitment time period can begin whenever growth begins in daughter cells and is then followed by the postcommitment time period (John, 1984). Because the first timed period in the *Chlamydomonas* cell cycle begins following the stimulus of becoming autonomous and initiating growth, and the second timed period follows the stimulus of attaining commitment to divide, Donnan and John (1983) suggested that these timers are in the "hourglass" category of biological timer, measuring time from a discrete point of stimulus (analogous to inversion of the hourglass).

Circadian rhythms are well known in *C. reinhardtii* (see below) but there has been disagreement on the extent of their involvement in the cell cycle. The ease with which the cycle can be shifted away from the previous 24-hour periodicity either by blocking photosynthesis or by a prolonged period in darkness both argue against a circadian control (Spudich and Sager, 1980; McAteer et al., 1985). In these cases synchrony is assumed to arise as a consequence of interruption of growth because divisions to which cells have become committed proceed to completion but no new divisions are started. Zachleder and van den Ende (1992) came to similar conclusions in studies with *C. eugametos*.

Goto and Johnson (1995) presented arguments in favor of a circadian rather than an hourglass timer. They found that the liberation of daughter cells persisted on a circadian rhythm when cells were grown in constant light, and that a mutant with a long (26.9 hour) rhythm for phototactic accumulation showed the same rhythm for daughter cell release. They repeated the dark-shift experiments of Spudich and Sager (1980) and McAteer et al. (1985), and then monitored the cell division rhythm over several more cycles following return to constant light, comparing this to the endogenous rhythm of phototaxis. They concluded that the shift from light to dark could be an entraining stimulus that reset the circadian clock, but that the rhythm subsequently persisted for many cycles, and that it did so regardless of growth rate. Their results were not inconsistent with operation of commitment timers as proposed by Donnan and John (1983) but they argued that the ultimate timer that controls cell cycle periodicity must be circadian. It should be noted that lysis of the mother cell wall, which Goto and Johnson used as their marker for the cell cycle, has long been regarded as an event with a circadian rhythm (Straley and Bruce, 1979; Mergenhagen, 1980), whereas the "hourglass" timers as defined by Donnan and John are the ones that regulate the intervals between commitment to divide and actual division.

Work continues toward identification of specific proteins involved in cell cycle regulation, with a focus on finding homologues of known yeast and mammalian control proteins. The *MAT3* locus in *C. reinhardtii* was first identified as a site of mutations that altered the usual uniparental pattern of inheritance of chloroplast genes (Gillham et al., 1987b). Its linkage to the mating type locus reinforced the assumption that its primary function was in the process controlling inheritance from the *plus* parent (see

Chapter 7). The increase in biparental inheritance when *mat3 plus* cells were crossed to *minus* was subsequently determined to be a consequence of unusually small cell size (Armbrust et al., 1995). By complementation of a stable deletion allele at this locus, Umen and Goodenough (2001a) established that it encodes a protein homologous to the retinoblastoma (RB) protein of mammalian cells. First identified as tumor suppressors, retinoblastoma proteins function in the control of progression from G1 to S phase. Cells of the *Chlamydomonas mat3* mutant commit to divide at a smaller size than wild-type cells, and also undergo more rounds of mitosis during the division period of the dark phase of the cycle. They differ from RB mutants of mammalian cells in that the timing of division is not altered. G1 phase in *mat3* cells is of normal length, and they do not initiate S phase prematurely. In an analysis of suppressors of *mat3* mutations, Fang et al. (2006) found recessive mutations in *DP1* and dominant mutations in *E2F1*. The products of these genes form a heterodimeric transcription factor that interacts with RB-related proteins in other organisms.

Umen and colleagues have undertaken a genome-wide survey of the cell cycle genes in *C. reinhardtii* (Bisova et al., 2005). Besides additional proteins in the RB-E2F pathway, *Chlamydomonas* cells have cyclins, cyclin-dependent kinases (see also Oldenhof et al., 2004a), an orthologue of the wee1 kinase, and homologues of CDC25. The product of the *FA2* gene, a NIMA family kinase, is required for flagellar excision and also appears to have a role in flagellar resorption prior to mitosis (Mahjoub et al., 2002). Quarmby and Parker (2005) have reviewed the associations between ciliary signaling, resorption and assembly, and cell cycle progression, both in *Chlamydomonas* and in mammalian cells (see also Volume 3, Chapter 3).

Another gene with a likely role in cell cycle control in *Chlamydomonas* is represented by the *UVS11* locus. Mutants at this locus are UV-sensitive (Vlček et al., 1987), but unlike most mutants with that phenotype, seem to function normally in excision of pyrimidine dimers and in recombination repair (Slaninová et al., 2002). The observation that these mutants usually go on to divide at least once after UV irradiation suggests that they may bypass a normal checkpoint for DNA damage and divide before repair is completed (Nagyová et al., 2003; Slaninová et al., 2003).

C. Endogenous circadian oscillators

Although Spudich and Sager (1980) and Donnan et al. (1985) argued against a circadian oscillator as the primary control for cell division, endogenous rhythms do appear to be operative in other cellular processes, and *Chlamydomonas* is one of relatively few organisms in which mutants with altered rhythms can conveniently be studied (see Mittag, 2001; Mittag et al., 2005, Breton and Kay, 2006; Iliev et al., 2006b, for review; also Johnson, 2001, for a broader view that includes cyanobacteria and other photosynthetic organisms). Bruce (1970–1974) identified a circadian rhythm of phototactic behavior and isolated mutants with altered rhythm periods. He observed a 24-hour cycle of increasing and decreasing phototactic activity that could be quantified by measuring the decrease in light falling on a photocell as *Chlamydomonas* cells swim into a light beam. This cycle could be initiated by transferring cultures grown in continuous light to testing conditions (continuous darkness except for the narrow test beam, which was turned on for 24 minutes every 2 hours). The rhythm could be entrained to a light–dark cycle, and its frequency appeared to be compensated for temperature variations. Additional mutants with altered periods were described by Mergenhagen (1980, 1984).

The nature of the endogenous circadian oscillator or oscillators and the extent to which external factors influence the biological clock were the focus of several early studies (Goodenough and Bruce, 1980; Goodenough et al., 1981; Hoffmans-Hohn et al., 1984), some of which were suggestive of a multiple oscillators with different frequencies, rather than a single master clock. To test the question whether fluctuations in environmental conditions such as gravity, cosmic radiation or magnetic fields provide an external timer, or Zeitgeber, for circadian rhythms, Mergenhagen and Mergenhagen (1987, 1989) compared phototactic accumulation rhythms of *Chlamydomonas* cells on the ground and in space. Under zero gravity conditions, both wild-type cells and short-period mutants showed rhythms very similar to those of control cultures over a 70-day test. The amplitude of the response was considerably greater at zero gravity, probably because swimming under these conditions was less energy-consuming.

In the course of their studies on phototactic rhythms, Bruce and his collaborators observed what appeared to be cyclic variations in stickiness of cells to glass containers. Straley and Bruce (1979) investigated this observation more thoroughly and found that the rhythms of phototaxis, stickiness, cell division, and hatching were correlated both for wild-type cells (24 hours) and for a long-period mutant (*per4*, 26–27 hours) transferred to heterotrophic (dark) conditions. The rhythms of phototactic behavior and stickiness were also seen in nondividing cells maintained on minimal medium in the dark.

Other phenomena in *C. reinhardtii* with apparent circadian control include chemotaxis to ammonium and its uptake (Byrne et al., 1992), starch accumulation (Ral et al., 2006), and sensitivity to UV irradiation (Nikaido and Johnson, 2000). All these rhythms continue when cells are transferred to constant light or constant darkness, but in their normal diurnal phases they are consistent with the daily life of a soil alga in its natural environment: phototaxis is maximal in daylight hours, starch content is greatest in the middle of the dark phase, nitrogen assimilation peaks toward the end of the dark phase, when cells might seek a nutrient-rich environment, and UV sensitivity is maximal at the transition from the light to the dark phase, when DNA synthesis is beginning.

Action spectra for circadian responses in *C. reinhardtii* show peaks in both the blue and red light regions (Johnson et al., 1991; Kondo et al., 1991). Blue light photoreceptors in *Chlamydomonas* include cryptochromes (Small et al., 1995) and phototropins (Huang et al., 2002). These will be discussed further in Chapter 5 in relation to gametogenesis and in Chapter 6 with regard to other processes. The red light effect is probably related to photosynthesis, as no evidence for phytochrome has been found in *C. reinhardtii*. Red/far-red reversibility is not seen (Kondo et al., 1991), nor have any phytochrome-related proteins been identified in the genome (Mittag et al., 2005). However, *Chlamydomonas* does show a seasonal photoperiodic response with respect to zygospore germination. Suzuki and Johnson (2001, 2002) noted that zygospores maintained on a short day regime (not more than 8 or 9 hours of light in a 24-hour period) showed poor germination rates, in comparison to those on long days. The effect was specific for the length of the light period, not the fluence, and is presumably an adaptive response to prevent germination occurring in winter. Based on studies with other organisms, the perception of day length is probably also under circadian control.

Studies of transcription of specific genes (summarized by Mittag et al., 2005; see also Jacobshagen et al., 2001, and references cited therein) and microarray analysis (Kucho et al., 2005) have revealed that many genes in *C. reinhardtii* are expressed differentially in light or dark phases, and that this expression is under circadian control. Chloroplast gene expression appears to be regulated by a nuclear-encoded sigma factor that is also expressed in a circadian manner (Carter et al., 2004; Misquitta and Herrin, 2005; see also Volume 2, Chapter 25). Mittag and collaborators identified an RNA-binding protein which they called CHLAMY 1 that seems to be specifically involved in a clock-related translational control of many proteins, especially ones involved in nitrogen and carbon metabolic pathways (Mittag, 1996, 2003; Iliev et al., 2006a). This protein is an analogue of the clock-controlled translational regulator (CCTR) protein of the dinoflagellate *Lingulodinium polyedra*, which has been one of the most important model systems for investigation of circadian rhythms. CHLAMY 1 binds to mRNAs containing a UG repeat of at least seven units in the 3′ UTR (Waltenberger et al., 2001). The C3 subunit of CHLAMY 1 is a homologue of a mammalian protein involved in myotonic dystrophy (Zhao et al., 2004).

Mittag et al. (2005) searched the *C. reinhardtii* genome sequence for homologues of known clock-related proteins in other organisms, and found many matches, especially to specific kinases and phosphatases. Proteins similar to enzymes in melatonin biosynthetic pathway were also identified. A proteomic search for clock-related proteins in *Chlamydomonas* (Wagner et al., 2004, 2005) has focused attention on a protein disulfide isomerase and a tetratricopeptide repeat protein that may also function in circadian control.

Matsuo et al. (2006; see also Kiaulehn et al., 2007) have developed a reporter system for circadian rhythms in *Chlamydomonas*. The firefly luciferase gene was optimized to the *C. reinhardtii* codon bias (see Chapter 8) and was coupled to the circadian-controlled promoter for the chloroplast *psbD* gene and transformed into the chloroplast genome. Cells with this construct can be monitored directly in microwell dishes for changes in luminescence over periods of several days.

Motility and Behavior

I. INTRODUCTION

Chlamydomonas is an outstanding system for investigation of flagellar biogenesis and function, with direct application to research on ciliary disorders in humans. No other model organism offers such a favorable combination of simple cellular structure, ease of manipulation, and the potential for genetic and biochemical analyses. The present chapter provides an introduction to flagellar structure and biogenesis, motility, and cell behavior including phototaxis, with emphasis on the historical background for our present knowledge. For a much more comprehensive review of these topics and current directions in research, see Volume 3.

II. FLAGELLAR STRUCTURE AND ORGANIZATION

A. The axoneme

Early studies of flagella at the light and electron microscopic level were limited by the resolution then achievable but revealed some details of flagellar structure (Kater, 1929; Lewin and Meinhart, 1953; Gibbs et al., 1958). Ringo's classic papers published in 1967 defined the organization of the cellular microtubules and flagellar root system, the basal bodies, and axoneme (Figure 4.1). As discussed in Chapter 2, the flagella arise from a pair of basal bodies at the anterior end of the cell. Between the basal body proper and the flagellar axoneme is the transition zone, whose stellate morphology is similar to that of other algae and also sperm cells of land plants but is not seen in protozoa or animal cells (see Melkonian, 1982; also Volume 3, Chapter 2).

Above the transition zone, the axoneme consists of a central pair and nine outer doublet microtubules, which are longitudinally continuous (Figure 4.2). The central pair end above the transition zone, and only the nine doublet microtubules connect directly to the basal bodies. Each of these nine microtubules consists of a fused pair of tubules, designated A and B, with 13 and 11 protofilaments respectively. Radial spokes connect to all nine A tubules, and inner and outer dynein arms arise at regular intervals from the A tubules of eight of the outer doublets; the remaining outer doublet lacks outer dynein arms and has other specializations that distinguish it (Huang et al., 1979).

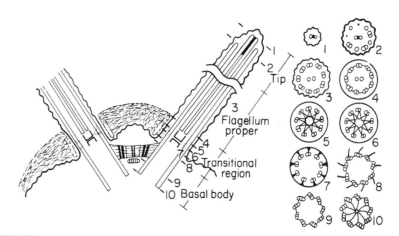

FIGURE 4.1   *Schematic representation of basal bodies and flagella, showing cross sections of microtubules at ten levels from flagellar tip to basal body. Note that the two single central microtubules terminate in the transitional region and are not templated by the basal body. In the transitional region, two of the fused triplet microtubules of the basal body become continuous with the fused doublets of the flagellum, and here is situated the stellate structure characteristic of certain plant flagella. From Pickett-Heaps (1975), redrawn from Ringo (1967a).*

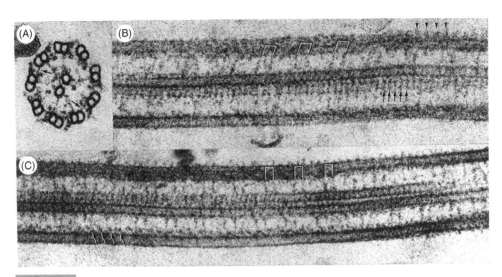

FIGURE 4.2 *Electron micrographs of isolated axonemes of wild-type C. reinhardtii. (A) Cross section in which the inner and outer dynein arms, the radial spokes and the nexin links (arrows) are clearly visible. The central sheath is present but is difficult to distinguish from the spoke heads. (B) Longitudinal section. One of the projections which make up the central sheath can be seen extending from the central tubule at intervals of 165Å (arrows). The outer dynein arms (arrowheads) are evident along the outside edge of the left doublet microtubule, and the radial spokes (brackets) appear as paired projections extending from the outer doublets toward the central tubules. This axoneme bent to the right immediately below the region shown, and the "tilted spoke" configuration seen here may represent an initial stage in bend formation. (C) Longitudinal section. The radial spokes are grouped into pairs (brackets) which repeat at intervals of 1000Å. The inner dynein arms (arrows) are visible along the inner edge of the right doublet microtubule. From Witman et al. (1978). Reproduced from* The Journal of Cell Biology, *1978, 76, 729–747 by copyright permission of The Rockefeller University Press.*

Both tubules of the central pair consist of 13 protofilaments, but they are distinguishable from one another in several aspects of morphology and chemical stability (Dutcher et al., 1984; Mitchell and Sale, 1999; see also Volume 3, Chapter 8). The central pair complex is seen only in sections distal to the transition zone, and extends beyond the outer doublet microtubules to end in a membrane-associated cap (Dentler and Rosenbaum, 1977; see Volume 3, Chapter 10). Identification of genes encoding proteins of the central pair and analysis of its function has been facilitated by mutants in which this complex fails to assemble (Table 4.1). As in other protists, but not in metazoan cells, the central pair complex of *Chlamydomonas* cells is twisted and can rotate during propagation of the flagellar bend (Omoto et al., 1999; Mitchell, 2003; Mitchell and Nakatsugawa, 2004). Projections from the central pair interact with the radial spoke heads to generate signals that ultimately control dynein arm activity. Several signaling proteins and kinesins are associated with the central pair.

Table 4.1		Representative mutations affecting flagellar structure and assembly	
Gene	**Mutant**	**Description**	**References**
CPC1	*cpc1*	Central pair protein with an adenylate kinase domain; mutants have altered flagellar beat frequency	Zhang and Mitchell (2004)
PF6	*pf6*	Protein associated with the C1 central pair tubule; mutant lacks the C1a projection; flagella twitch but don't beat	Rupp et al. (2001); Wargo et al. (2005)
	pf10	Gene not identified; mutant has ineffective flagellar beat pattern	Dutcher et al. (1988)
PF15	*pf15*	Katanin p80 subunit, required for assembly of central pair microtubules; mutants lack central pair and are paralyzed	Dymek et al. (2004)
PF16	*pf16*	Armadillo repeat protein of central pair; mutation causes instability of the C1 microtubule	Dutcher et al. (1984); Smith and Lefebvre (1996)
PF20	*pf20*	WD-repeat protein of flagellum central pair, associated with intermicrotubule bridge; mutants have paralyzed, rigid flagella	Smith and Lefebvre (1997a)
BOP5 or *IC138*	*bop5*	WD-repeat protein IC138 of inner dynein arm I1; suppresses abnormal beat of *pf10* mutant, restoring normal phenotype	Dutcher et al. (1988, 1991); Hendrickson et al. (2004)
IDA2 or *DHC1-beta*	*ida2*	Flagellar inner arm dynein 1 heavy chain beta; mutants swim more slowly than wild type	Kamiya et al. (1991); Perrone et al. (2000)
IDA4	*ida4*	Flagellar inner arm dynein light chain p28; mutants swim more slowly than wild type	Kamiya et al. (1991); LeDizet and Piperno (1995)
IDA5	*ida5*	First described by a mutation affecting the flagellar inner dynein arm; subsequently identified as the structural gene for actin; mutants swim more slowly than wild type and mating efficiency is very low	Kato-Minoura et al. (1997)
IDA7	*ida7*	Flagellar inner dynein arm I1 intermediate chain IC140; mutants swim more slowly than wild type and phototaxis is impaired	Perrone et al. (1998)
DHC1 or *PF9*	*pf9*	Deficient in inner dynein arm heavy chain alpha; mutants swim slowly but smoothly	Porter et al. (1994); Myster et al. (1997, 1999)
FLA14 or *LC8*	*fla14*	Outer dynein arm light chain 8; mutants have short, immotile flagella	Pazour et al. (1998)

(Continued)

Gene	Mutant	Description	References
Table 4.1 *Continued*			
ODA11	*oda11*	Outer dynein arm heavy chain alpha; mutants swim slightly more slowly than wild type	Sakakibara et al. (1991)
ODA2	*oda2*, *pf28*, *sup$_{pf}$2*	Outer dynein arm heavy chain gamma; mutants swim slowly and are defective in backwards swimming	Mitchell and Rosenbaum (1985); Rupp et al. (1996)
ODA4	*oda4*, *sup$_{pf}$1*	Outer dynein arm heavy chain beta; mutants swim more slowly than wild type; *sup$_{pf}$1* suppresses some radial spoke and central pair mutations	Porter et al. (1994)
ODA12	*oda12*	Outer dynein arm light chain 2, Tctex2; mutants show slow, jerky swimming	Pazour et al. (1999); DiBella et al. (2005)

The molecular architecture of the dyneins has received detailed analysis and is discussed in length in Volume 3, Chapter 6. Three-dimensional models for the inner and outer dynein arms, the dynein regulatory complex, and the linkers that connect these structures have been made based on cryoelectron tomography (Nicastro et al., 2006; Oda et al., 2007; also see Chapter 9 in Volume 3). The outer doublets are connected by flexible filaments known as nexin, or inter-doublet links, which appear in a repeating pattern of 96 nm corresponding to four dynein arm repeats (Witman et al., 1978). From studies of mutant strains, Piperno et al. (1992) postulated the existence of a dynein regulatory complex that functions to coordinate the activity of the radial spokes and the inner dynein arms. Gardner et al. (1994) confirmed the location of this complex at the base of the second radial spoke, closely associated with the inner dynein arms.

The docking complex connects each outer dynein arm to its binding site on the doublet microtubule, and comprises three protein subunits, each of which has been identified by a mutation that impairs motility (Table 4.1; Koutoulis et al., 1997; Takada et al., 2002; Casey et al., 2003a, b).

A two-part bridge structure extends from the A tubule of the doublet designated 1, to the adjacent B tubule, and beaklike structures project into the lumen of the B tubule of doublets 1, 5, and 6 (Witman et al., 1972; Hoops and Witman, 1983; Figure 4.3). By analysis of serial sections in which flagellar orientation relative to the basal bodies could be established, Hoops and Witman showed that the number 1 doublets of the two flagella face one another and postulated that this asymmetry is important in the forward swimming stroke.

Radial spokes arise from all nine A tubules, in pairs with an alternate periodicity of 29 nm and 62 nm (Figures 4.4 and 4.5, Yang et al., 2006; and see Volume 3, Chapter 7). The 30 nm stalk of each radial spoke is anchored

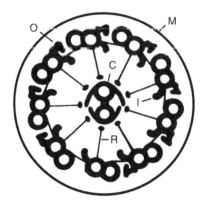

FIGURE 4.3   *Diagram of cross section of flagellum, similar to Figure 4.2A, showing the flagellar membrane (M), inner (I) and outer (O) dynein arms, radial spokes (R), and the central pair (C). Courtesy of George Witman.*

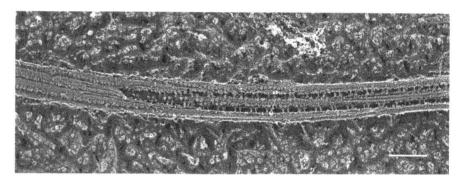

FIGURE 4.4   *Platinum-carbon replica of quick-freeze, deep-etch preparation of a Chlamydomonas axoneme, showing radial spoke structure. Where the fracture enters the axoneme, pairs of spoke heads originating from the uppermost doublet are visible; as the fracture goes deeper, radial spokes are seen attached to lateral doublets and extending in toward the central microtubules. Bar = 0.2 μm. Preparation by Harold Hoops, figure courtesy of George Witman.*

adjacent to the inner dynein arms, and terminates in an enlarged spoke head that interacts with projections from the central pair microtubules. The radial spoke has proved particularly amenable to genetic and subsequently proteomic analysis, and comprises 23 proteins; structural gene mutations have been identified in 8 (Table 4.1; Yang et al., 2006a; Volume 3, Chapter 7).

In the region of the flagellar tip, radial spokes are lost, the nine doublet microtubules become single, and the single tubules terminate individually at different levels, with the A tubules ending in paired filaments 4.2 nm in diameter and 95 nm long. The central pair persists, terminating in a specialized structure, the central microtubule cap, which is attached to the flagellar membrane (Figure 4.6; Ringo, 1967a; Dentler and Rosenbaum, 1977; Dentler, 1980, and Volume 3, Chapter 10). The role of the flagellar

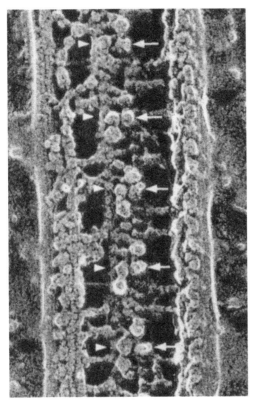

FIGURE 4.5   *Helical organization of the radial spoke system of Chlamydomonas. Adjacent sets of two spokes form a right-handed helix as they ascend around the central pair, with an incremental rise of 32 nm. S_1 spokes of the leftward set are indicated by arrowheads and S_2 spokes of the rightward set are indicated by arrows. From Goodenough and Heuser (1985a). Reproduced from The Journal of Cell Biology, 1985, 100, 2008–2018 by copyright permission of The Rockefeller University Press.*

tip in mating is discussed in Chapter 5 of the present volume, and in Volume 3, Chapter 12.

B. The flagellar membrane

The membrane covering the axoneme is continuous with the plasma membrane but differs from it in several important features. Unlike the plasma membrane, the flagellar membrane is covered with a fuzzy sheath, the glycocalyx (Ringo, 1967a), which consists of surface carbohydrate linked to the principal membrane glycoprotein, FMG-1B (see below). The glycocalyx binds the lectin concanavalin A (Millikin and Weiss, 1984a, b) and antibodies to the carbohydrate epitopes on the FMG-1B protein (Bloodgood et al., 1986). Similar flagellar membrane glycoproteins have been found in *C. eugametos* and *C. moewusii* (Musgrave et al., 1979b; McLean et al., 1981). Mastigonemes, small hair-like structures approximately 0.9-μm long, form

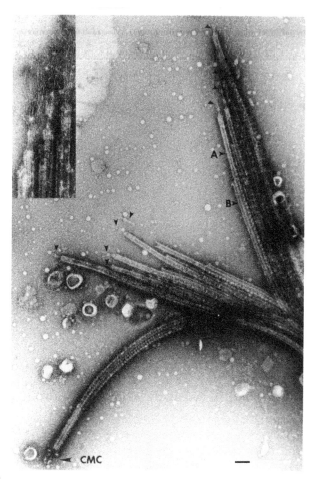

FIGURE 4.6 *Detergent-extracted and negatively stained distal tip of a Chlamydomonas flagellum. The central microtubule cap (CMC) at the tip of the central microtubules and distal filaments (arrowheads) protrude from the distal tips of the A tubules (A) of each outer doublet microtubule. Paired distal filaments can be seen on the third doublet from the bottom of the micrograph. (B) B tubule. Bar = 0.1 μm. Inset: frayed distal filaments at the tips of the A tubules. From Dentler (1980).*

two rows on opposite sides of the distal portion of the flagellar surface (Ringo, 1967a; Witman et al., 1972a; Nakamura et al., 1996; Figure 4.7). A single mastigoneme is a chain of identical subunits, each consisting of a single 220-kD glycoprotein (see Volume 3, Chapter 11).

Mastigonemes and the glycocalyx are found on both vegetative and gametic cells and do not seem to be directly involved in sexual recognition (Bergman et al., 1975), nor do they seem to have a role in gliding motility (Bloodgood, 1977; Nakamura et al., 1996; see below). The primary function of mastigonemes may be to increase the flagellar surface area (see Volume 3,

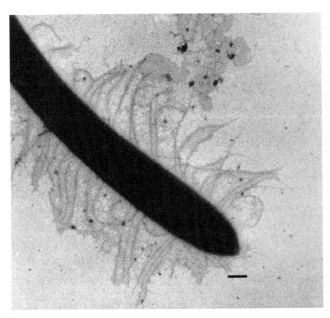

FIGURE 4.7 *Negatively stained flagellum from an unmated minus gamete showing a particularly abundant array of mastigonemes. Bar = 0.2 μm. From Bergman et al. (1975). Reproduced from The Journal of Cell Biology, 1975, 67, 606–622 by copyright permission of The Rockefeller University Press.*

Chapter 11). Sexual agglutinins, an essential part of the flagellar membrane in gametes, are discussed in Chapter 5, and in Volume 3, Chapter 12.

Electron microscopy, and in particular freeze-fracture freeze-etch techniques, reveal several distinctive arrays of particles within the flagellar membrane (see Volume 3, Chapter 11). Rows of particles running longitudinally for the length of the flagellar membrane appear to be oriented over the axonemal doublet microtubules (Bloodgood, 1987; Dentler, 1990) and may be connected to the motor complexes involved in gliding motility and other flagellar surface movements (Bloodgood, 1977). Three circular rows of intramembrane particles form the ciliary/flagellar necklace at the level of the basal body transition zone (Weiss et al., 1977; see Volume 3, Chapter 11). The necklace structure is seen in cilia and flagella of other organisms, whereas another intramembrane particle array, the flagellar bracelet, has been described only in the Volvocales (Weiss et al., 1977; Melkonian, 1982).

The role of the membrane in gliding motility and particle transport will be discussed in section IV.B.

C. Separation of axonemal and membrane polypeptides

Isolation of flagella and their fractionation into membrane, mastigoneme, matrix, and microtubule components was achieved by Witman et al. (1972a), who identified major proteins of these fractions using one-dimensional gel electrophoresis. Two-dimensional electrophoresis of total solubilized flagellar components revealed some 250–300 distinct polypeptides (Piperno et al., 1977; Piperno and Luck, 1979b), and proved to be a very useful technique for identifying missing or altered proteins or complexes in mutants with impaired motility. Huang (1986) summarized the known deficiencies in axonemal polypeptides exhibited by the best characterized mutants of the time. Studies over the succeeding 20 years, culminating in sequence analysis and characterization of the flagellar proteome, have of course extended this work; the flagellum is now believed to contain over 500 proteins, about half of which are highly conserved in vertebrates (Pazour et al., 2005). Table 4.1 lists representative mutants with defects in specific flagellar structures. More detailed descriptions of these and related mutants can be found in Volume 3.

In many of the mutants examined in these electrophoretic studies, an entire complex was found to be lost, but the specific polypeptide defective in the mutant could often be identified by a dikaryon rescue experiment: Lewin (1954a) observed that the quadriflagellate cells formed immediately after mating in *C. reinhardtii* often had four active flagella even when one of the parent strains was paralyzed. This suggested that polypeptides contributed by both parents can be assembled into all four flagella after mating. For some mutants, motility was restored immediately after cell fusion, while in other cases a lag period was required, and for some mutants no restoration was observed. Luck and colleagues mated [^{35}S]-labeled cells of several paralyzed mutant types with unlabeled wild-type cells, allowed restoration of flagellar function in the presence of anisomycin to inhibit new protein synthesis, and then screened polypeptides from isolated flagella for incorporation of labeled proteins (Luck et al., 1977; Huang et al., 1981; Dutcher et al., 1984). Normal polypeptides from the mutant cell were incorporated into flagella and appeared as labeled spots on a two-dimensional gel, whereas a single polypeptide presumed to be deficient in the mutant appeared only in the unlabeled form derived from the wild-type parent. A non-radioactive version of this dikaryon rescue experiment makes a good classroom exercise: cells of two non-motile mutants, *pf1* and *pf14*, are mated, and students watch for restoration of motility in the quadriflagellate zygotes as the wild-type genes complement the mutant alleles.

Another approach used to identify specific proteins affected in motility mutants was to isolate revertant cells showing improved flagellar function. Some of these revertants produced polypeptides that were functional but showed altered electrophoretic mobility. Demonstration that a series of revertant strains all showed effects on the same polypeptide was strong presumptive evidence, in the days before DNA sequencing was routine, that this polypeptide was the gene product affected by the original mutation (Luck et al., 1977; Huang et al., 1981).

The alpha and beta tubulins are highly homologous to those from plant and animal microtubules (Olmsted et al., 1971) and account for about 70%

of the axonemal proteins labeled with radioactive sulfate in continuous culture (Piperno et al., 1977). Gamma tubulin is specifically associated with the microtubule organizing center (Silflow et al., 1999). Delta tubulin, which was first discovered in *Chlamydomonas* (Dutcher and Trabuco, 1998), is required for basal body assembly. Mutants with defective epsilon tubulin are "bald" (lack flagella) and have disorganized cytoskeletons (Goodenough and St. Clair, 1975; Ehler et al., 1995; Dutcher, 2003). Deletion of the gene encoding epsilon tubulin is lethal in haploid cells, but the lethality can be suppressed by an extragenic mutation (Preble et al., 2001).

The flagellar membrane can be solubilized by detergent treatment and its components fractionated (Monk et al., 1983; Witman, 1986). The major protein of the *Chlamydomonas* flagellar membrane is FMG-1B, whose mass was originally estimated as 350 kD (Goodenough et al., 1985; Bloodgood et al., 1986). The predicted amino acid sequence gives a 410 kD protein (Volume 3, Chapter 11). A 17-amino acid cytoplasmic domain is predicted to be a binding site for a phosphoprotein involved in signal transduction (Bloodgood and Salomonsky, 1994, 1998). FMG-1B is heavily glycosylated (Bloodgood et al., 1986; Bloodgood, 1990).

The flagellar membrane presumably also contains the voltage-gated calcium channels involved in phototaxis and photoshock responses (see Volume 3, Chapter 13), but these have not yet been identified biochemically in membrane fractions (Volume 3, Chapter 11). The phototaxis-related AGG2 protein also seems to be membrane-associated (Iomini et al., 2006).

Gealt et al. (1981) found that the fatty acid composition of flagellar membranes differed from that of the cell body, whereas the sterol components seemed to be similar, but very little work has been done on other lipid components of the flagellar membrane. Membrane lipids in general are discussed in Chapter 6, and in Volume 2, Chapter 2.

III. ASSEMBLY AND MAINTENANCE OF THE FLAGELLA

A. Detachment and resorption

Flagella are readily removed from *Chlamydomonas* cells by mechanical shearing, a shift to low pH (Randall et al., 1967; Rosenbaum et al., 1969), treatment with the anesthetic dibucaine (Witman et al., 1978), ethanol (Lewin and Lee, 1985), or other stimuli (see Volume 3, Chapter 3). Flagellar excision occurs precisely at the distal end of the transition zone (Lewin and Lee, 1985; Lohret et al., 1999). The outer flagellar membrane detaches from the axoneme, and a cuff-like widening of the axoneme, surrounded by an electron-dense core, appears just above the transition zone (Figure 4.8). Above this region, the microtubule doublets separate from the base, and below the cuff the flagellar membrane constricts, eventually pinching off the axoneme and cuff and closing the remaining stump.

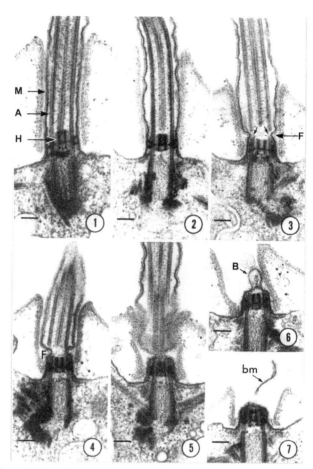

FIGURE 4.8   *Sequential stages of flagellar autotomy. (1) Control: untreated cell. Median longitudinal section of base of flagellum showing the H-shaped cylinder (H) in the transition region. Note that the flagellar membrane (M) is closely apposed to the axoneme (A). (2) Section showing undulation of flagellar membrane above the transition region, presumably due to rupture of membrane–microtubule bridges. (3) Section revealing fracture of microtubule doublets (arrow), and formation of an annular furrow (F) above the transition region. (4) Further constriction of the furrow (F). The region between the proximal end of the central microtubules and the transition cylinder becomes more electron transparent. (5) Flagellar shaft ready to separate. (6) Blebbing (B) of flagellar membranes is occasionally seen after the shaft is severed. Note narrowing of distal end of transition cylinder. (7) Section showing sealing of membrane over flagellar stump and release of bleb membrane (bm). Bar = 200 nm in each panel. From Lewin and Lee (1985).*

The immediate response to mechanical or chemical deflagellation stimuli is activation of a calcium-mediated signaling pathway which culminates in flagellar excision. Mechanical shear in calcium-free medium is lethal (Cheshire et al., 1994), presumably because the flagella break randomly and the plasma membrane cannot reseal. Chemical deflagellation agents appear to activate the same signaling pathway by their effect on membranes (see Volume 3, Chapter 3). Acid shock, the method most frequently used to remove flagella in the laboratory, induces an influx of calcium (Quarmby, 1996). A mutant defective in calcium influx (adf1; Finst et al., 1998) does not deflagellate in response to pH shock but can do so when the plasma membrane is permeabilized by detergent treatment in the presence of calcium.

Deflagellation of wild-type cells is accompanied by contraction of centrin fibers internal to the flagellar transition zone (Sanders and Salisbury, 1989, 1994), but studies with the centrin mutant vfl2 leave some doubt as to whether centrin contraction is the primary force for axonemal severing (see Volume 3, Chapter 3). The microtubule-severing protein katanin also appears to have a role in deflagellation (Lohret et al., 1998, 1999).

Lewin and Burrascano (1983) isolated a mutant strain, fa1 (impaired flagella autotomy), that did not respond to chemical or mechanical deflagellation stimuli, but otherwise had no obvious structural differences from wild-type cells. The FA1 gene product is a 171 kD protein with coiled-coil and Ca/CAM-binding domains (Finst et al., 2000). A second class of mutants with similar deflagellation-defective phenotypes (fa2) have alterations in a Nek kinase (Mahjoub et al., 2002; Quarmby and Mahjoub, 2005). After deflagellation of wild-type cells, the FA1 protein remains with the basal body/transition zone fraction (Finst et al., 2000), whereas the FA2 protein is associated with the severed flagella (Mahjoub et al., 2004).

Chlamydomonas cells normally disassemble and resorb their flagella on entering mitosis (Cavalier-Smith, 1974). Resorption of flagella can be also induced by treatment with caffeine or isobutyl methylxanthine (Hartfiel and Amrhein, 1976; Lefebvre et al., 1980), by some microtubule-binding herbicides (Weeks et al., 1977; Quader and Filner, 1980), or by cation chelation or removal of Ca^{2+} from the medium while raising the concentration of monovalent cations 10-fold over the normal level (Lefebvre et al., 1978; Quader et al., 1978). Resorption induced by these treatments is reversible on removal of the inducing agent, and the resorbed protein can be reused. Curiously, although fa mutants do not excise their flagella in response to pH shock or other stimuli, they resorb and then reassemble them when treated in this way (Parker and Quarmby, 2003). This observation and several others suggest a fundamental link between the deflagellation and resorption processes (see Volume 3, Chapter 3).

B. Flagellar assembly

After deflagellation by mechanical shear or pH shock, or on restoration to normal medium after chemically induced flagellar resorption, vegetative cells immediately begin flagellar regeneration, and full-length (10–12 μm) flagella are restored in 1–2 hours (Randall et al. 1967; Rosenbaum et al., 1969). In deflagellated gametes there appears to be a short (15 minute) lag time before flagellar outgrowth is seen (Weeks and Collis, 1976; Lefebvre et al., 1978). Newly formed (quadriflagellate) zygotes are also capable of regeneration after deflagellation, but this ability is lost within the first 90 minutes after zygote formation (Weeks and Collis, 1979).

A kinetic analysis of regeneration in vegetative cells was published by Randall et al. (1967), and mathematical models for elongation were proposed by Levy (1974). Rosenbaum et al. (1969) showed that the process is accompanied by new synthesis of flagellar proteins: in the presence of cycloheximide, some regeneration occurs, but the flagella reach only about half the normal length. This suggested that the intracellular tubulin pool may not be sufficient for synthesis of two full-length flagella; immuno-chemical assays of tubulin by Piperno and Luck (1977) were consistent with this view. Lithium ion also affects regeneration (Periz et al., 2007).

Microtubule-binding drugs such as colchicine, vinblastine, and oryzalin also inhibit flagellar regeneration (Quader and Filner, 1980). Another inhib-itor of flagellar regeneration is the volatile anesthetic halothane (Telser, 1977). Unlike the local anesthetic dibucaine, halothane does not promote deflagellation. Inhibitors of cyclic nucleotide phosphodiesterase (amino-phylline, caffeine, dibutyryl cAMP) also block flagellar regeneration (Rubin and Filner, 1973).

Classic autoradiographic studies of flagella regenerating in the pres-ence of radioactive precursors showed that incorporation of new pro-tein occurs mainly at the flagellar tip (Rosenbaum et al., 1969; Witman, 1975). A polarity of assembly was also observed with *Chlamydomonas* flagellar microtubules *in vitro* (Allen and Borisy, 1974; Rosenbaum et al., 1975). The flagellar cap structure formed early in the assembly process and remained attached to the central microtubules throughout regeneration (Dentler and Rosenbaum, 1977).

Lefebvre et al. (1978) reported that individual flagellar proteins were synthesized with different kinetics; for example, some proteins were labeled with ^{35}S only in the first 30 minutes after deflagellation, while others con-tinued to be synthesized for several hours. At the time of this work only the tubulins, a dynein fraction, and a flagellar membrane protein could be identified specifically. Further exploration of this question by Remillard and Witman (1982) showed that proteins of a given complex, for example the radial spoke, dynein arms, central pair microtubules, etc., were often syn-thesized with similar kinetics. Relatively little new synthesis was observed for the central tubule protein CT1 and an actin-like component (Piperno and Luck, 1979a), which were apparently assembled largely from pre-existing pools. In general, however, some 35–40% of assembled proteins appeared to be newly synthesized. Tubulin synthesis was typical in this regard, reaching a maximum between 40 and 90 minutes after deflagella-tion and declining to low levels by 180 minutes. Weeks and Collis (1976) calculated that tubulin probably accounts for 10–15% of total cellular protein synthesis during flagellar regeneration. Although microtubules of intact flagella are relatively stable compared to cytoplasmic microtubules, tubulin assembly and disassembly at the distal tip occur continuously (Marshall and Rosenbaum, 2001; Song and Dentler, 2001).

Similar sets of proteins were induced in regenerating cells whether flagella were detached or induced to resorb (Lefebvre et al., 1980). Comparison of vegetative and gametic cells after mechanical deflagellation also showed few differences in the pattern of flagellar proteins synthesized during regeneration (Silflow et al., 1981), apart from synthesis of a few gamete-specific proteins.

In synchronously growing cells, tubulin synthesis is induced 1.5–2 hours before cytokinesis and continues at relatively high levels (5–10% of total cellular protein synthesis) through the entire period of cell division (Piperno and Luck 1977; Weeks and Collis, 1979; Ares and Howell, 1982). Nicholl et al. (1988) showed that some of the other mRNAs characteristic of flagellar regeneration are also accumulated at this time. Once flagellated daughter cells are liberated from the mother wall, little or no synthesis of tubulin can be detected. However, synthesis can be induced by deflagellation at any time during the non-dividing phase, up to a point just before cytokinesis.

C. Intraflagellar transport

Assemblies of proteins are moved to and from the flagellar tip in the space between the doublet microtubules and the flagellar membrane, a process first discovered in *Chlamydomonas* (Kozminski et al., 1993) but subsequently found to be universal in cilia and flagella, and now known as intraflagellar transport or IFT. Anterograde movement is driven by a heterotrimeric complex containing a kinesin homologue (Cole et al., 1998; Piperno et al., 1998), and retrograde movement by a cytoplasmic dynein (Pazour et al., 1998). The particles moved by IFT contain at least 17 proteins organized in two complexes, complex A and complex B; these particles carry a cargo of axonemal proteins from the cell body, where protein synthesis occurs, to the tip of the growing flagellum, where they are assembled onto the axoneme. Several mutants originally identified by temperature-sensitive defects in flagellar assembly or stability have been found to be deficient in components of either the kinesin or the dynein motors, and another mutant with a similar phenotype is deficient in the microtubule plus end-tracking protein EB1, which in wild-type cells is localized to the flagellar tip and basal bodies (Table 4.1; see also Sloboda and Howard, 2007). Steady-state turnover of microtubules requires IFT for assembly at the flagellar tip. When cells of the *fla10* mutant are shifted to the restrictive temperature, IFT stops (Kozminski et al., 1995), but since disassembly of microtubules continues under these conditions, the flagella gradually shorten (Huang et al., 1977; Marshall and Rosenbaum, 2001). IFT is discussed at greater length in Volume 3, Chapter 4.

D. Size control

Brief mechanical shearing in a homogenizer produces a cell population in which as many as 15–20% of the cells have lost only one of their flagella. Rosenbaum et al. (1969) observed that in such cells the remaining flagellum shortened as the missing one was restored. In some cases, the shortened flagellum remained at an intermediate length until the new flagellum reached the same length, and both flagella then elongated to a final length somewhat shorter than in control cells. In other cases, the remaining flagellum regressed virtually entirely and then began to elongate again. Meanwhile, the amputated flagellum grew to an intermediate length and remained there until the regressing flagellum reached zero length and re-elongated. Both flagella then grew coordinately from the intermediate length to a slightly subnormal final length. These results indicate that flagellar length is under tight control (see Volume 3, Chapter 5; Wemmer and Marshall, 2007). When new protein synthesis was inhibited with cycloheximide, the total flagellar length achieved was greater than that reached by similarly treated cells from which both flagella had been removed, suggesting that proteins from the regressing flagellum as well as the cytoplasmic pool are utilized for regeneration of both flagella (Coyne and Rosenbaum, 1970).

An inverse correlation between tonicity of the medium and flagellar length was reported by Solter and Gibor (1978b), but no such relation was found in experiments by Jarvik et al. (1984). Mutations with altered flagellar length are frequently encountered (Table 4.1; see also Volume 3, Chapter 5). Both long- and short-flagella mutants have been obtained, but in general they share the property that both flagella on a given cell are equal in length.

Mutants with long flagella define four nuclear gene loci, three of which encode proteins that are part of structures identified in the cell body and named the Length Regulatory Complex (Tam et al., 2007). The *LF2* gene product is a protein kinase, which appears to interact with the products of the *LF1* and *LF3* genes. Although *lf2* mutants were originally identified by their long flagella, null mutants at the *LF2* locus have short flagella of unequal length, suggesting that this kinase functions both in flagellar assembly and in length control. The *LF4* gene encodes a MAP kinase whose function in wild-type cells is predicted to be flagellar shortening.

The size of the pool of flagellar precursors must not determine ultimate flagella length, since flagella can regenerate to appreciable length in the absence of protein synthesis (Rosenbaum et al., 1969). This is also true of short-flagella mutants, which must therefore not be limiting for precursors. Kuchka and Jarvik (1987) confirmed that all short-flagella mutants they tested contained a pool of unassembled flagellar protein. Also, Kuchka and Jarvik (1982) reported that mutants having variable numbers of flagella had different pool sizes, correlated with the number of flagella formed, but all made flagella of the same length.

In matings of the short-flagella mutant *shf1* to wild-type cells, quadri-flagellate cells are formed which initially have two normal and two short flagella, but the short flagella rapidly grow out to nearly full length while the normal flagella remain constant (Jarvik et al., 1984). This implies that mobilization of the flagellar protein pool does not produce a generalized lengthening of flagella (in which case the normal flagella would also have elongated), but that there is a more specific length control mechanism. Jarvik et al. (1984) speculated that this control might reside in the basal bodies.

Several of the short-flagella mutants resorb their flagella when transferred from minimal medium to medium containing acetate (Jarvik et al., 1984; Kuchka and Jarvik, 1987). Jarvik et al. (1984) discussed several possibilities for this effect, including unusual intracellular pH sensitivity or a relation to the acetylation of alpha tubulin described by L'Hernault and Rosenbaum (1983).

As mentioned earlier, IFT is required for assembly at the flagellar tip, but not for disassembly. Marshall et al. (2005) have proposed a "balance point" model for flagellar length control in which the rate of disassembly is independent of length, but the assembly rate varies because a fixed quantity of IFT complexes take increasingly longer times to reach the tip as the flagellum elongates. A steady-state length is reached at a point where the assembly and disassembly rates are equal. The model correctly predicts the observation that in mutants with variable numbers of flagella, flagellar length decreases with increasing flagellar number.

IV. MOTILITY

A. The mechanism of flagellar motion

When viewed by light microscopy and high-speed flash photography, the forward swimming stroke appears as a biphasic breast-stroke motion (Ringo 1967a), with the two flagella normally moving coordinately (Figure 4.9). This is sometimes referred to as a "ciliary" type of motion because of its resemblance to the beat patterns seen in ciliated organisms. Sleigh (1981) compared beat patterns among many types of flagellated organisms, and Johnson et al. (1984) reviewed the early research on this topic, comparing motility in *Chlamydomonas* with *Tetrahymena* and sea urchin sperm.

In the power stroke of forward swimming in *Chlamydomonas* the flagella remain relatively straight over their length, bending mostly at the base. On the return stroke, a bending wave is propagated from the base toward the tip. These motions were investigated by high-speed cinemicrography by Rüffer and Nultsch (1985, 1987). By tracking the orientation of the eyespot, they determined that cells rotate counterclockwise during forward swimming with a frequency of 1.2–2 Hz due to a lateral component of three-dimensional beating during the power stroke. About once in 20 beats,

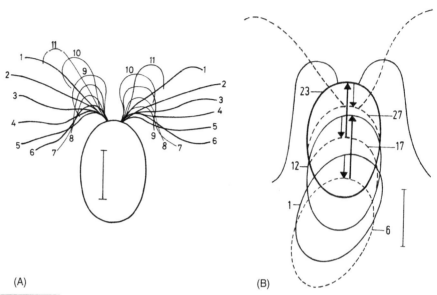

(A)   (B)

FIGURE 4.9 *Diagrams of swimming strokes. (A) Mutant 622 E (described as a strain that swims more steadily and smoothly than wild type at the very high light intensities required for high-speed cinematography). Tracing of one beat cycle of both flagella from a phase-contrast, high-speed film (500 ft/sec), with flagellar positions corresponding to frame numbers. Cell body schematic. (B) Position of another cell during three beats. The solid line marks the cell at the end of the forward motion, the dashed line at the end of the backward motion. For the last beat the position of the flagella at the beginning of the effective (dashed line) and the recovery stroke (solid line) is given. Arrows indicate direction and force of movement; numbers are frame numbers. Bar = 5 μm. From Rüffer and Nultsch (1985).*

the outward-directed flagellum beats alone, producing an asynchrony that results in a helical swimming path overall. The average forward swimming speed is about 100–200 μm/sec (Rüffer and Nultsch, 1985; see also Racey and Hallett, 1983b; Ojakian and Katz, 1973, for additional techniques for estimating swimming speed).

Racey and Hallett (1983a, b) also used cinematography and quasi-elastic light scattering techniques to develop mathematical models describing cell motility. Brokaw and colleagues (Brokaw et al., 1982; Brokaw and Luck, 1983) examined flagellar bending patterns by computer-assisted analysis of high-speed (300 Hz) photomicrographs of swimming cells of the uniflagellate mutant *uni1* (Figure 4.10). Although the flagellar stroke in this mutant is normal, the cells rotate in place with little forward motion and can be photographed more easily than wild-type cells. Brokaw and Luck (1985) extended this analysis to radial spoke head- and central pair-deficient mutants in combination with the *uni1* mutation, and Brokaw and Kamiya (1987) analyzed dynein mutants.

Chlamydomonas cells respond to sudden exposure to bright light or to mechanical stimulation by swimming backward, which is accomplished

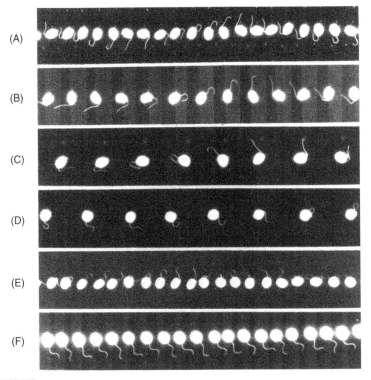

FIGURE 4.10   *Photographs used by Brokaw et al. (1982) for analysis of the movement of Chlamydomonas flagella. All prints show time sequences from left to right. (A) Wild-type bending pattern: flash rate 69.0 Hz. (B) A cell containing the suppressor mutation sup_pf3: flash rate 49.3 Hz. (C) A cell containing the suppressor mutation sup_pf1: flash rate 26.7 Hz. (D) A cell containing the paralyzed flagella mutation pf17 and sup_pf3: flash rate 23.9 Hz. (E) A cell containing the paralyzed flagella mutation pf17 and sup_pf1: flash rate 70 Hz. (F) Bending pattern during the flagellar reversal reaction: wild-type cell: flash rate 56.2 Hz. From Brokaw et al. (1982). Reproduced from The Journal of Cell Biology, 1982, 92, 722–732 by copyright permission of The Rockefeller University Press. [The sup_pf3 mutation is now known to affect the dynein regulatory complex (Piperno et al., 1994). sup_pf1 is a mutation in a regulatory domain in the beta dynein heavy chain (Porter et al., 1994). The pf17 mutation alters radial spoke 9 (Yang et al., 2006a).]*

by undulatory movements of the type seen in sperm flagella (Ringo 1967a; Figure 4.10). As discussed below, this response appears to be mediated by the same rhodopsin photoreceptor as phototaxis, and like phototaxis, is sensitive to changes in calcium levels (Bessen et al. 1980; Segal and Luck, 1985). At low external calcium concentration, a backward swimming response cannot be elicited, and agents that block calcium influx or membrane excitability also inhibit backward swimming. Bean et al. (1982) reported that cells suspended in 1 mM Sr^{2+} in low-calcium medium swim backward only and do not show normal phototactic or photophobic responses. Forward swimming was restored by excess Ca^{2+}, but not by Mg^{2+} or other cations. Certain mutants deficient in outer dynein arms are impaired in this backward swimming response.

Motility can be studied in isolated flagella under appropriate conditions. Allen and Borisy (1974) showed that axonemes prepared by demembranating isolated flagella with the nonionic detergent NP-40 could be reactivated by addition of ATP to undergo bending motions similar to those seen *in vivo*. Lindemann and Mitchell (2007) have used a rapid fixation technique to analyze axonemal distortions during the beat cycle, and to calculate the transverse stress across the axoneme during bending.

The combination of proteomic analysis with a much more detailed view of flagellar structure is rapidly increasing our understanding of how the flagellar components work together to regulate motion, turning, and reversal. Current models for these interactions are summarized in Volume 3, Chapter 9.

B. Gliding movement and flagellar surface motility

Flagellated *Chlamydomonas* cells are capable of creeping or gliding along the surface of solid media (Lewin, 1952a; Bloodgood, 1981a, b; see also Volume 3, Chapter 11). This ability is obviously advantageous for an organism living in soil. Gliding motion results from movement of the flagellar membrane and is mechanistically equivalent to the transport of bound particles along the flagellar surface (Bloodgood, 1977; Bloodgood et al., 1979; Hoffman and Goodenough, 1980). Flagellar surface movement is also involved in the mating process, when adhesions of paired gametes first move up and down along the flagellar length and then migrate to the flagellar tips (see Chapter 5). Movement is bidirectional, at an average speed of 1.6 µm/sec, with the cells moving 30–40 µm in one direction, pausing, then resuming gliding either in the same or in the opposite direction (Bloodgood, 1981a). When cells are gliding, the flagella are adherent to the substratum and are held 180° apart. The motive force always appears to come from the leading flagellum and is calcium-dependent. Bowser and Bloodgood (1984) summarized evidence indicating that this movement does not occur by a "surf-riding" mechanism such as has been postulated for some other organisms. The movement occurs, and indeed is best observed, in paralyzed cells such as the mutant *pf18*, which has a rigid axoneme. Non-gliding mutants have been described both in *C. reinhardtii* and in *C. moewusii* (Lewin, 1982).

V. CELL BEHAVIOR

A. Phototaxis and the photophobic response

In their review on light antennas in algal phototaxis, Foster and Smyth (1980) noted that the phototactic response of algae was reported as early as 1817 and received considerable experimental attention during the 19th and early 20th centuries as a model of behavioral response. A definitive paper was published by Buder in 1919, demonstrating that swimming algal cells, including *Chlamydomonas*, orient their swimming with reference to a light beam. Buder observed that there was a switch from positive taxis (toward the light) at low intensity to negative taxis (away from the light) at high intensity, resulting in aggregation at a particular intensity, and recognized the role of a differential response of the two flagella in orienting the cells. Early work implicated the eyespot as the most likely site of light perception for oriented movement (see Chapter 2, also Melkonian and Robenck, 1984, for review). Foster and Smyth (1980) discussed the nature of the photoreceptor and antenna structures in several experimental systems, including *Chlamydomonas*, and Nultsch (1983) and Feinleib (1984) specifically reviewed the early *Chlamydomonas* literature. Kreimer (1994) provided a lengthy and more general review of phototaxis in flagellate algae (for coverage of the more recent literature, see Hegemann et al., 2001; Kateriya et al., 2004; Sineshchekov and Spudich, 2005, and Volume 3, Chapter 13).

One can generalize that at moderate light intensities (roughly $<10^3$ ergs/cm^2 sec), *Chlamydomonas* cells accumulate in a light beam, but at high light intensity, they avoid the light. At intermediate light intensities the response of the population as a whole may appear to be neutral, or cells may accumulate at the perimeter of the illuminated spot. However, when individual cells are observed, there appears to be an abrupt transition from positive to negative response, with no neutral range (Feinleib and Curry, 1971; Nultsch, 1977). Two responses to light stimulation are seen: a non-oriented photoshock or photophobic stop response to sudden changes in light intensity, and an oriented swimming (phototaxis) toward or away from the stimulus source (Figure 4.10). Either or both responses together may lead to photoaccumulation in an illuminated spot. The extent of the response observed and the threshold light intensity (inversion intensity) required to produce a shift from positive to negative phototaxis are subject to considerable environmental influence and show variation with culture medium, temperature, stage of life cycle, etc., as well as with the strain being investigated (Nultsch, 1977, 1979). Mayer (1968) reported that preadaptation to darkness or to a given light intensity also affected whether a positive or negative response ensued on a shift in light intensity.

Feinleib and Curry (1971a) reported that cells from freshly inoculated cultures were negatively phototactic over most of the range of light intensity tested, while cells from older cultures were positively phototactic over much of the same range. Stavis and Hirschberg (1973) found that overall phototactic behavior (either positive or negative) was maximal during exponential growth and declined markedly in stationary-phase cultures, to a greater extent than did motility. Prolonged growth in darkness resulted in diminished motility but did not impair the phototactic response of those cells that were motile. Experiments by Bruce (1970, 1973) demonstrated a circadian rhythm of phototactic activity (Chapter 3). A shift in phototactic behavior on mating was reported by Adams (1975); unmated gametes were observed to accumulate in the illuminated side of a petri dish under 6000-lux fluorescent light, but newly mated pairs became strongly negatively phototactic.

Qualitative estimates of photoaccumulation can readily be made with minimal equipment, and a great deal of information was obtained in the early years from direct microscopic observation (see Foster and Smyth, 1980). Motility and phototaxis of individual cells can be measured much more precisely using video and electrical recording systems (Pfau et al., 1983, Uhl and Hegemann, 1990; Moss et al., 1995; see also Volume 3, Chapter 13).

The action spectrum for the photoshock and phototactic responses suggested that the receptor pigment might be a rhodopsin-like molecule (Foster and Smyth, 1980). Foster et al. (1984) confirmed this hypothesis by incorporating analogues of the animal rhodopsin chromophore 11-*cis*-retinal into the carotenoid-deficient, non-phototactic mutant *fn68*, and showing that phototaxis was restored with an action spectrum corresponding to that of the particular retinal analog supplied. The assumption that this mutant produces normal amounts of the opsin apoprotein, which then combines with the exogenously added retinal, was confirmed by Foster et al. (1988). Crescitelli et al. (1992) further defined the action spectrum of the *Chlamydomonas* eyespot using microspectrophotometry, and compared this with their previous study of *Euglena gracilis*. By the mid-1990s several papers had been published showing that the native *Chlamydomonas* chromophore was an all-*trans* retinal (Derguini et al., 1991; Hegemann et al., 1991; Lawson et al., 1991), and characterizing an abundant eyespot-associated protein resembling invertebrate opsins (Beckmann and Hegemann, 1991; Deininger et al., 1995). This protein was given the name chlamyopsin. A related protein was discovered in *Volvox* (Ebnet et al., 1999). The *Chlamydomonas* protein originally analyzed is actually a mixture of two alternatively spliced products of the same gene, now designated *COP1/2* (Fuhrmann et al., 2003). However, both these *Chlamydomonas* and *Volvox* opsins were shown to be highly charged and lacked the seven transmembrane helices expected for a photoreceptor, leading to the speculation that the true photoreceptor was yet to be found. Fuhrmann et al. (2001) confirmed this prediction by showing that when expression of the chlamyopsin gene was silenced with an antisense RNA construct, phototaxis and the photoshock response were not impaired. Subsequent analysis of cDNA sequences revealed that *Chlamydomonas* cells also contain two proteins related to microbial rhodopsins, and these appear to be the true photoreceptors. These are large (76 kD) proteins with the expected transmembrane structure that act as light-gated ion channels, and have been referred to as channelrhodopsins (Nagel et al. 2002, 2003), *Chlamydomonas* sensory rhodopsins (Sineshchekov et al., 2002) or archaeal-type rhodopsins (Suzuki et al., 2003). Their function is discussed in more detail in Volume 3, Chapter 13, and in the reviews by Kateriya et al. (2004) and by Sineshchekov and Spudich (2005).

Using published estimates of the threshold intensity, or lowest light intensity at which a phototactic response occurs, Foster and Smyth (1980) estimated a minimum number of photoreceptor molecules per cell on the order of 2×10^5. A similar figure is obtained if one assumes that the photoreceptor lies in a single membrane with the same area as the eyespot and that the concentration of rhodopsin per unit area is similar to that in animal eyes.

Foster and Smyth (1980) proposed a view of the eyespot and the photoreceptor area as a directional antenna analogous to a conical scanned radar system used for tracking airplanes in World War II. In their model, as the cell swims forward, it rotates, so that the antenna scans the incident light around the cell's path. The antenna does not point directly at the target but rather scans a circle about the target, so that the rotation axis is aligned with the target. If movement of the *Chlamydomonas* cell is directed parallel to the light direction, the antenna receives essentially constant illumination, but if the swimming path deviates from the direction of incident light, the antenna perceives a periodic fluctuation in light intensity. By altering the flagellar beat patterns, the cell can correct its swimming path to restore constant light intensity.

The basis for sensory transduction, or transmission of the signal received through the eyespot–photoreceptor complex to the motor apparatus, has gradually been elucidated over years of investigation. A hypothetical view of this process, proposed by Nultsch in 1983, is supported by subsequent research (Figure 4.11).

The essential role of calcium in the sensory transduction responses of both phototaxis and photoshock was apparent even in early experiments (Stavis and Hirschberg, 1973; Stavis, 1974; Schmidt and Eckert, 1976; Nultsch, 1979). Nultsch et al. (1986) used calcium channel blockers to differentiate between calcium-mediated responses affecting motility and those specific to the phototactic and photoshock responses. Two different inward ion currents have been characterized, a rapidly appearing "photoreceptor current" (<0.5 ms) associated with the eyespot and a subsequent "flagellar current" appearing at the flagellar membrane 30–40 ms after a flash of bright light (Harz and Hegemann, 1991; Harz et al., 1992). The photoreceptor current can be separated into two kinetically overlapping components that are followed by H^+ currents (Gradmann et al., 2002; Ehlenbeck et al., 2002; see also Volume 3, Chapter 13). A measurable flagellar current is seen only after the photoreceptor current reaches a threshold level, and it is therefore sometimes referred to as an "all-or-none" current.

Kamiya and Witman (1984) showed that the flagella *cis* and *trans* to the eyespot respond differentially to calcium concentration over the range

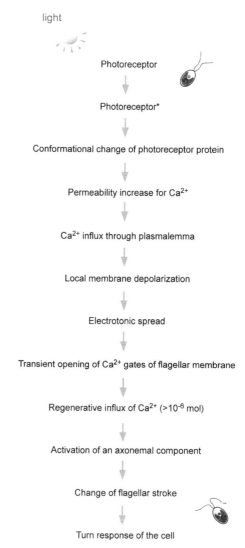

light

Photoreceptor

Photoreceptor*

Conformational change of photoreceptor protein

Permeability increase for Ca^{2+}

Ca^{2+} influx through plasmalemma

Local membrane depolarization

Electrotonic spread

Transient opening of Ca^{2+} gates of flagellar membrane

Regenerative influx of Ca^{2+} ($>10^{-6}$ mol)

Activation of an axonemal component

Change of flagellar stroke

Turn response of the cell

FIGURE 4.11 *Scheme of the phototactic reaction chain of Chlamydomonas, adapted from Nultsch (1983).*

of 10^{-9} to 10^{-6} M, the *trans* flagellum beating more slowly with lower levels of internal Ca^{2+} ($<10^{-8}$ M) and the *cis* flagellum beating more slowly with higher levels (10^{-7} or 10^{-6} M). The net effect of these changes is to produce a turning motion in one direction if the calcium concentration changes. From 10^{-6} to 10^{-4} M the backward swimming response ensues. The intraflagellar free Ca^{2+} in the absence of photostimulation appears to be in the range of 10^{-8} M, a concentration at which both flagella are active (Kamiya and Witman, 1984). Thus, in a positively phototactic cell, a step-up stimulus (eyespot moving to face the light) would result in a moderate increase in intraflagellar Ca^{2+} (to $>10^{-8}$), causing the *cis* flagellum momentarily to beat less effectively than the *trans* flagellum, so that the cell turns toward the side with the eyespot (Witman, 1993). Conversely, a step-down stimulus (eyespot moving away from the light) may result in a decrease in intraflagellar Ca^{2+} (to $<10^{-8}$ M) causing the *trans* flagellum to momentarily beat more effectively, with the effect that the cell again turns toward the light. The observation that the (*trans*) flagellum of a uniflagellate mutant slows in response to increased light under positive phototaxis conditions (Smyth and Berg, 1982) is consistent with this model. By recording swimming paths of individual cells in response to sudden changes in the direction of a light source, Isogai et al. (2000) showed that under conditions of positive phototaxis, the *cis* flagellum was dominant over the *trans* flagellum in steady swimming, and decreased its dominance in response to perception of a change in the light signal. Studies in which cells were held on micropipettes while flagellar beating was monitored have also confirmed and extended understanding of the differential responses of the *cis* and *trans* flagella to light stimulation (Rüffer and Nultsch, 1990–1998; Holland et al., 1997; Josef et al., 2005a, b, 2006). The specific roles of the two channelrhodopsins in effecting these responses are still being elucidated and are discussed at greater length in Volume 3, Chapter 13. How the cell switches from positive to negative phototaxis is still not understood.

The photoshock or photophobic stop response occurs when cells perceive a sudden flash of light, or a sudden increase in light intensity. The external calcium concentrations must be above 3×10^{-4} M for this response to occur, and at higher calcium concentrations the duration of the response is longer (Hegemann and Bruck, 1989; Holland et al., 1997). The behavioral response consists of a sudden stop in which the flagella are extended forward, followed by a brief period of backward swimming, in which the flagella show a symmetrical, undulating waveform. The cells then stop briefly and resume their normal flagellar beat patterns.

As summarized in Table 2.7, the phototactic response can be blocked by mutations affecting eyespot assembly and structure (Lamb et al., 1999) and carotenoid synthesis (Foster et al., 1984; McCarthy et al., 2004). Mutants in which the circadian rhythm of phototaxis is altered have also been characterized (Bruce, 1972; Nultsch and Throm, 1975; see also Chapter 3).

Mutants are also helping greatly in understanding the signal transduction pathways underlying these responses (Pazour et al., 1995; Okita et al., 2005). The *ptx1* mutant swims normally and has a normal photoshock response, but was reported to be non-phototactic (Horst and Witman, 1993). Analysis of flagellar beat patterns showed that both flagella react to a light stimulus, but that they respond identically, as if both were the *trans* flagellum (Rüffer and Nultsch, 1997). Okita et al. (2005) compared *ptx1* to the *ida1* mutant, which lacks the inner arm dynein I1 and was previously reported to have impaired phototaxis (King and Dutcher, 1997), and to a newly isolated mutant *lsp1* that is defective in the calcium-dependent shift in flagellar dominance and in which both flagella have beat frequencies similar to the *cis* flagellum of wild type. Okita et al. found that *lsp1* and *ptx1* cells were able to orient their swimming parallel to the light direction but showed no preference for movement toward or away from the light source, whereas *ida1* cells showed a weak positive phototaxis. These results are consistent with previous suggestions that inner arm I1 is essential for the change in flagellar waveform that normally results from a light stimulus. Okita et al. concluded that the phototactic response may involve multiple signal transduction pathways, some of which might be activated only within certain ranges of light intensity.

The other *ptx* mutants isolated by Pazour et al. (1995) have been less thoroughly characterized: *ptx5*, *ptx6*, and *ptx7* have normal photoreceptor and flagellar currents, but the flagella do not respond to changes in intraflagellar calcium concentration, whereas *ptx2* and *ptx8* lack the flagellar current and are presumed to be deficient in this calcium channel. The *ptx3* mutant has a normal photoshock response but lacks phototaxis, and shows a lower photoreceptor current than wild-type cells.

In contrast, Matsuda et al. (1998) isolated four mutants (*ppr1–ppr4*) that have normal phototaxis but do not show the photophobic response. The photoreceptor current appears to be normal in these mutants, but the flagellar current is absent, implying the existence of a voltage-dependent calcium channel in the flagellar membrane that is specific for the photophobic reaction.

The *agg1* mutant (formerly *np*, Smyth and Ebersold, 1985) shows negative phototaxis even at low light intensity. When suspended in a test tube containing 2–10 ml of culture medium and left on a shelf below fluorescent lights, *agg1* cells form a tight pellet at the bottom of the tube while *agg1*+ cells are dispersed throughout the medium (Figure 4.12). This mutation

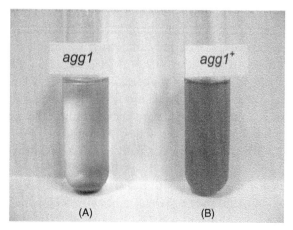

FIGURE 4.12   *Phenotypes of two alleles at the AGG1 locus. Cell suspensions were left undisturbed under a 40-watt cool-white fluorescent bulb for several hours. (A) agg1 cells aggregate at the bottom of the tube. (B) agg1+ cells are dispersed. Photograph by Matt Laudon.*

was originally found as a spontaneous variant, and in fact is carried by the Ebersold-Levine wild-type mt^- strain (CC-124, UTEX 2243). Since many other mutant strains have been either isolated in this background or crossed to it at some point in their history, the *agg1* phenotype is widespread through laboratory cultures. Awareness of this phenotype may prevent misinterpretation of tube assays for motility and mating that might otherwise be confusing.

Mutants at two additional loci have phenotypes similar to that of *agg1* (Iomini et al., 2006). The *AGG2* gene encodes a protein that localizes to the proximal flagellar membrane near the basal bodies. The *AGG3* gene product is a flavodoxin and is more widely distributed in the flagellar matrix but is concentrated near the AGG2 protein. Iomini et al. postulated that these genes function in a pathway that blocks transmission of a light signal in the *cis* flagellum, and reinforces the signal in the *trans* flagellum. Mutation at any of the *AGG* loci would allow the *cis* flagellum to become dominant.

B. Chemotaxis

Hirschberg and Rodgers (1978) reported that *Chlamydomonas* cells were attracted to $CoCl_2$ and $MnSO_4$ in capillary tube assays but showed a negative response to L-arginine. Sjoblad and Frederikse (1981) found that of many ions and organic chemicals tested, only NH_4^+ showed chemoattractant activity. They were unable to determine if the difference in their results and those of Hirschberg and Rodgers was due to assay conditions or to strain differences. The response to ammonium shows a circadian rhythm and is lost in gametes (Byrne et al., 1992; Ermilova et al., 2003, 2007). Conversely, chemotaxis toward tryptone was seen in gametes but not in vegetative cells (Govorunova and Sineshchekov, 2003, 2005). The relationship of this response to the signal transduction pathways of photo taxis is under investigation (Govorunova et al., 2007).

As discussed in Chapter 5, chemotactic attraction has also been considered in the context of mating (see Tsubo, 1957, 1961; see also Bean, 1977 for review), but no compelling evidence has been found for pheromones or other sexual attractants in *C. reinhardtii*. The reports of Moewus on this subject are generally discredited (Hagen-Seyfferth, 1959; Gowans, 1976b). *C. allensworthii*, which is only distantly related to *C. reinhardtii*, does have a sexual pheromone, however (Coleman et al., 2001).

Ermilova et al. (1993) reported that *C. reinhardtii* cells showed chemotaxis toward sucrose and maltose, neither of which can support growth of this species in the absence of photosynthesis (Chapter 6). Conversely, acetate, which does support dark growth, was not a chemoattractant, nor was glucose. Addition of sucrose or maltose to the culture medium did not affect phototaxis. Mutants were isolated that were unable to respond to each of four single compounds (maltose, sucrose, xylose, or mannitol) but reacted normally to the other three (Ermilova et al., 1996, 1999). Another mutant could not respond to either maltose or sucrose. Additional mutants have been obtained by insertional mutagenesis, including two (ctx4 and ctx5) that are insensitive to all the tested compounds and therefore indicative of a block in a signaling component downstream of the chemoreceptor point (Ermilova et al., 2000). The adf1 mutant, which does not deflagellate in response to acid, has been postulated to be defective in a proton-activated Ca^{2+} channel in the plasma membrane (Evans and Keller 1997a, b; Finst et al., 1998). Ermilova et al. (2000) reported that like ctx4 and ctx5, adf1 is blocked in chemotaxis but has normal phototactic and photophobic responses, and proposed a model for an integrated set of signal transduction pathways, taking into consideration the various mutant phenotypes obtained.

C. Gravitaxis and gyrotaxis

Like many other free-swimming unicellular organisms, *Chlamydomonas* cells left for 1–2 hours in a capillary tube in the dark tend to accumulate at the top of the tube (negative gravitaxis, sometimes called geotaxis; see Bean, 1977, 1984). The rate of movement (4–8 μm/sec) is considerably less than overall motility (measured as a population average of 50–70 μm/sec in these experiments) and reorientation of swimming appears to involve long, slow turns over hundreds of micrometers rather than abrupt turning angles. The process is energy-dependent, but photosynthesis is not required; neither is operation of the normal phototactic system, since the non-phototactic mutant ptx1 still shows negative gravitaxis, and calcium channel blockers do not appear to affect the reaction (Kam et al., 1999). Kam et al. concluded that gravitaxis in *Chlamydomonas* has a physical basis and does not rely on a signal transduction mechanism. Roberts (2006) reached the same conclusion based on calculations of cell shape and buoyancy. However, Sineshchekov et al. (2000) reported a blue light effect on gravitaxis, and Yoshimura et al. (2003b) were able to isolate two mutants specifically impaired in gravitaxis but indistinguishable from wild-type cells in terms of body size, shape, specific gravity, and sedimentation velocity, thus arguing in favor of a chemical basis for the phenomenon. These mutants did have somewhat abnormal responses to mechanical stimulation and increased light sensitivity in the photoshock reaction. Yoshimura et al. noted that the phototaxis-deficient mutant ptx3, which has an abnormally low photoreceptor current, also shows diminished gravitaxis. From these observations they postulated that the gravitaxis mutants have a hyperpolarized membrane potential that sensitizes the photoreceptor and desensitizes the flagellar mechanoreceptor.

Gyrotaxis, the directed orientation of cells in a flowing fluid, was investigated in *Chlamydomonas* by Kessler (1985), who proposed its use in concentrating and separating motile cells. Later studies with *Chlamydomonas* and other unicellular flagellates have modeled swimming trajectories in terms of fluid flow, shear, gravity, cell shape, and other parameters (Pedley et al., 1988; Pedley and Kessler, 1990; Kessler, 1991; Kessler et al., 1992; Czirók et al., 2000) and this analysis has recently been extended to the multicellular Volvocales (Short et al., 2006; Solari et al., 2006a, b).

The Sexual Cycle

I. INTRODUCTION

This chapter reviews the literature on sexual reproduction in *Chlamydomonas*, with particular attention to *C. reinhardtii* and *C. moewusii/C. eugametos*, the species on which the most extensive work has been done. After a discussion of the diversity of reproductive styles within the genus and some notes on consistency in designating mating types of laboratory strains, the sequence of events in reproduction in *C. reinhardtii* and *C. moewusii* is described in detail, beginning with gametogenesis and concluding with maturation and germination of

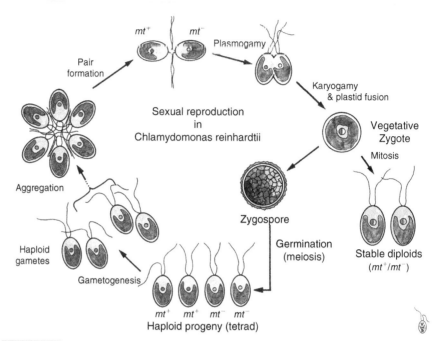

FIGURE 5.1 *Sexual reproduction cycle in C. reinhardtii. Courtesy of Karen VanWinkle-Swift.*

zygospores. The chapter concludes with the genetic control of sexuality in *C. reinhardtii*, including analysis of the mating type locus and mutants in which the mating process is perturbed. Flagellar adhesion and signal transduction in mating are covered more thoroughly in Volume 3, Chapter 12.

II. TYPES OF SEXUAL REPRODUCTION WITHIN THE GENUS *CHLAMYDOMONAS*

A. Isogamy to oogamy

The most widely studied laboratory species, *C. reinhardtii* (Figure 5.1), and *C. moewusii* (including the strains traditionally known as *C. eugametos*; see Chapter 1) are heterothallic and isogamous. That is, mating type (*plus* or *minus*) is genetically determined in a cell line, behaving as a single Mendelian locus in crosses, and *plus* and *minus* gametes are similar in size and superficial appearance, although they may differ at the ultrastructural level. Within the genus as traditionally defined, however, there are also homothallic species, in which both mating types are formed in a single population, and a diversity of mating styles (Figure 5.2). These include anisogamy (morphologically similar gametes that differ markedly in size) and various degrees of heterogamy (morphologically distinct gametes, generally with a pronounced difference in size), culminating in true oogamy (small motile "sperm" cells and large nonmotile, wall-less "eggs"). Table 5.1

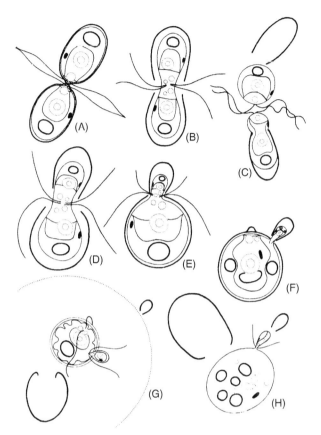

FIGURE 5.2 *The various types of gamete copulation in sexual reproduction of Chlamydomonas species. All figures are somewhat schematic. Gamete walls are indicated with a thick line. (A) Atactogamy (although not clearly indicated in this figure, refers to formation of structurally indistinguishable gametes but with a wide range of sizes, within a single mating population; gametes form pairs regardless of their size). (B) True isogamy (gametes similar in size and indistinguishable in structure at the light microscope level). (C) Isogamy with prior shedding of the wall by one of the gametes. (D) Anisogamy (pairs always consist of one large and one small gamete, but gametes are similar in structure). (E) Heterogamy (pairs of large and small flagellated gametes that are structurally distinguishable prior to copulation; microgametes have reduced chloroplasts). (F) Oogoniogamy (both gametes retain their walls until after fusion, but the larger gamete loses its flagella and forms an oogonium). (G) Oogamy (suboogamy) (flagellated but immotile "egg" is released from the cell wall and fuses with a flagellated "sperm"). (H) True oogamy (small motile "sperm" cell and large, unflagellated "egg"). From Ettl (1976a), with thanks to Thomas Proschold for help with the figure legend.*

lists the species for which a sexual cycle has been described. Reproductive behavior has been observed in only about a quarter of the species named by Ettl (1976a) and was not used by him as a taxonomic criterion. Even so, it is clear that the various forms of reproduction are not evenly distributed among the groups that Ettl defined: isogamy predominates in species in the Euchlamydomonas and Chlamydella groups, whereas most heterogamous

Table 5.1   Sexual species of *Chlamydomonas*

Species	Equivalent strain	Type	References
Species usually described as isogamous			
C. acidophila	Isolate from Ohio, not in SAG or UTEX collection	Heterothallic	Rhodes (1981)
C. applanata (formerly *C. aggregata*)	SAG 2.72; UTEX 969	Homothallic	Deason and Bold (1960)
C. applanata (formerly *C. dysosmos*)	SAG 11-36a; UTEX 2399	Homothallic	Ettl and Schlösser (1992)
C. applanata	SAG 6.72; UTEX 230	Homothallic	Pringsheim (1930); Ettl and Schlösser (1992)
C. chlamydogama	SAG 11-48a,b; UTEX 102, 103	Heterothallic	Bold (1949a)
C. culleus (formerly *C. frankii*)	SAG 18.72, 19.72; UTEX 1057, 1058	Heterothallic	Smith (1950); Hoshaw (1965)
C. culleus (formerly *C. elliptica* var. *britannica*)	SAG 64.72, 65.72; UTEX 1059, 1060	Heterothallic	Hoshaw (1965)
C. debaryana	UTEX 344	Heterothallic	
C. debaryana	SAG 6.79, 7.79	Heterothallic	
C. eugametos	UTEX 9, 10, and others	Heterothallic	See text
C. geitleri (also known as *C. noctigama*)	SAG 6.73; UTEX 2289	Homothallic	Nečas and Pavingerova (1980); Zársky et al. (1985)
C. gloeophila	SAG 12-4, 12-5; UTEX 607, 608	Heterothallic	
C. hindakii	SAG 22.72; UTEX 1338	Homothallic	Burrascano and VanWinkle-Swift (1984)
C. hydra	SAG 11-6a,b,c; UTEX 4, 5, 6	Heterothallic	
C. indica (also known as a *C. moewusii* isolate)	SAG 11-11; UTEX 223	Heterothallic	Mitra (1950)
C. melanospora	SAG 22.83, 23.83; UTEX 2021, 2022	Heterothallic	
C. mexicana (also known as *C. oblonga*)	SAG 11-60a,b; UTEX 729, 730	Heterothallic	Lewin (1957a)
C. microhalophila	Not in SAG or UTEX collection	Homothallic	Bischoff (1959)
C. minutissima	UTEX 1055, 1056, 1063, 1064	Heterothallic	Smith (1946, 1950); Hoshaw (1965)
C. moewusii	UTEX 96, 97, and others	Heterothallic except for var. *monoica*	See text
C. moewusii var. *monoica*	SAG 6.98, UTEX 2020	Homothallic	Deason and Ratnasabapathy (1976)

(Continued)

Table 5.1 *Continued*

Species	Equivalent strain	Type	References
C. monoica (now *C. noctigama*)	SAG 33.72; UTEX 220	Homothallic	Strehlow (1929); VanWinkle-Swift and Bauer (1982)
C. noctigama	SAG 35.72; UTEX 114	Homothallic	Burrascano and VanWinkle-Swift (1984)
C. oblonga	SAG 37.72, UTEX 839	Homothallic	
C. parallestriata	SAG 2.73, 14.88	Homothallic	
C. philotes	SAG 11-53 as *Tetracystis*; UTEX 2024	Heterothallic	Lewin (1957a)
C. pinicola	*C. noctigama*: SAG 40.72, UTEX 1339	Homothallic	Burrascano and VanWinkle-Swift (1984)
C. radiata (formerly *C. archibaldii*)	SAG 1.75; UTEX 1795	Homothallic	Uhlik and Bold (1970)
C. reinhardtii	See Table 1.1	Heterothallic	See text
C. segnis (formerly *C. gymnogama*; now *Lobochlamys segnis*)	SAG 2.75, UTEX 1638	Homothallic	Deason (1967); Pröschold et al. (2001)
C. simplex	SAG 13.79; UTEX 2456, 2457	Heterothallic	
C. sphagnophila var. *dysosmos*	CCAP 11/31 as *C. sphagnophila*; formerly UTEX 206	Homothallic	Lewin (1954c)
C. spreta	SAG 50.91, 11/93; CCAP 11/124, 11/125	Heterothallic	
C. surtseyiensis	SAG 3.75; UTEX 1796	Homothallic	Uhlik and Bold (1970)
Species usually described as anisogamous or variably iso/anisogamous			
C. allensworthii	SAG 27.98, 28.98, 29.98; UTEX 2717, 2718, 2719		Starr et al. (1995); Coleman et al. (2001)
C. nivalis	Isolate from Tatra Mountains, not in SAG or UTEX collection		Kawecka and Drake (1978)
C. planoconvexa	SAG 11-42		
C. segnis	SAG 1.79; UTEX 1919		Badour et al. (1973)
C. segnis (formerly *C. intermedia*)	SAG 11-13; UTEX 222		Smith (1946); Hoshaw (1965)
Species usually described as heterogamous or oogamous			
C. capensis	*Oogamochlamys gigantea*: SAG 22.98, UTEX 1753, and others		Pröschold et al. (2001)
C. gigantea	SAG 44.91		
C. monadina	SAG 31.72, UTEX 493, and others		
C. pseudogigantea	*Oogamochlamys gigantea*: SAG 21.72, UTEX 848, and others		VanDover (1974); Pröschold et al. (2001)

(Continued)

Table 5.1	*Continued*		
Species	**Equivalent strain**	**Type**	**References**
C. zimbabwiensis	*Oogamochlamys zimbabwiensis*: SAG 45.91; UTEX 2213, 2214, and others		Heimke and Starr (1979); Pröschold et al. (2001)
Reproduction reported but not fully described			
C. peterfii	*Heterochlamydomonas lobata*: SAG 38.72, UTEX 728	Homothallic	Deason (1967)

Information from Ettl (1976a), the SAG and UTEX catalogs (http://www.epsag.uni-goettingen.de/html/sag. html, http://www.utex.org), and other references as cited. Ettl described sexual behavior in many other species for which no equivalent strain is available from the major algal collections; see Harris (1989) for a complete list of these.

or oogamous species belong to Ettl's group Pleiochloris (Skuja, 1949; Ettl, 1976a; Heimke and Starr, 1979). Pröschold et al. (2001) reclassified members of the Pleiochloris group into the new genus *Oogamochlamys* (see Chapter 1, Figure 1.2).

Chlamydomonas is a natural model system for investigation of both the genetic control of mating type and the evolution of oogamous reproduction, as seen in multicellular algae. This was recognized early on (Kalmus, 1932) and figured importantly in subsequent essays and theoretical treatments (Lewin, 1954c; Hartmann, 1955; Scudo, 1967; Bell, 1978; Wiese, 1981; Kirk, 2006; Haag, 2007). Two opposing trends are seen to affect gamete size: while there may be a numerical advantage in producing as many small gametes as possible, there is also good reason to produce large gametes that will contribute cytoplasmic mass to the zygote. Together these tendencies are energetically favored and may encourage development of anisogamy from an initially isogamous condition. Wiese et al. (1979) showed that anisogamy could be mimicked in three normally isogamous species (*C. reinhardtii*, *C. moewusii*, and *C. chlamydogama*) by using different gametogenesis conditions for the two mating types. Cells taken from old plates were large but capable of mating as soon as they regenerated flagella, whereas gametes induced in synchronous liquid culture were small products of two successive mitotic divisions. Mixtures of unequal-sized gametes produced by the two methods mated readily. Thus gametogenesis in normally isogamous species can be made to resemble the process in heterogamous species, in which microgametes arise by division and macrogametes by differentiation of a vegetative cell. Wiese et al. (1979; also Wiese, 1981) suggested that evolution of occasional anisogamy into a necessary condition could occur by mutations that block one of these two pathways of gametogenesis in a given line of cells. Matings between gametes of unequal size would be favored over those between equal ones, since two small gametes might lack the necessary mass to ensure zygote survival, and two large, relatively immotile gametes would be less likely to contact one another. True oogamy would develop by loss of flagella from the macrogamete. Essentially the same mechanism can work toward evolution of anisogamy and oogamy in homothallic species. Whereas Parker's model (1971) assumed that all gametes in a population were identical except for size, and could therefore mate freely with one another, Wiese believed it likely that homothallic species produce gametes of two distinct mating types just as heterothallic ones do, and that the mating type produced is determined in a given cell by a switching mechanism. The size of the parent cell might be a determinant of which pathway of gametogenesis (direct differentiation or division) would be followed by a given cell.

B. Homothallic vs. heterothallic reproduction

In homothallic species, two possibilities must be considered: either any gamete is capable of mating with every other gamete, or gametes are differentiated into two physiologically distinct mating types analogous to those of heterothallic species, the only difference from the latter being that both mating types can arise from a single progenitor cell. The idea of two mating types even in homothallic species was put forward by Hartmann and collaborators as early as 1932, and found support in experiments by Hämmerling and others (reviewed by Hartmann, 1955). Their "residual gamete" assay was based on the premise that precisely equal numbers of the two mating types will not be present in a single culture. When sexual pairing has gone to completion, some unmated gametes will remain, all of one mating type, which will be able to mate with testers from another culture. This was shown to be true for several different types of homothallic algae and was assumed by Hartmann as a general principle. Work with *C. monoica* supports this hypothesis (VanWinkle-Swift and Aubert, 1983; VanWinkle-Swift and Hahn, 1986; see also Charlesworth, 1983).

Wiese (1981) suggested that the fundamental processes of gamete recognition, contact, and fusion may be common to all *Chlamydomonas* species, with the difference between homothallic and heterothallic species lying solely in the control of sex expression. Each cell would carry the genetic information for both mating types, and some control mechanism would exist to activate only one set of genes in a given cell. Heterothallic species could be described as those in which one set of mating type genes is stably activated, and homothallic species those in which the activated set is switched at appreciable frequency. This led to the speculation that the difference between homothallism and heterothallism might not be absolute, but rather represents variation in the frequency with which a given set of mating-type-specific genes is functional. One might then expect to find intermediates between the absolutely heterothallic or homothallic types within the genus *Chlamydomonas*. A possible example of this situation is the species *C. zimbabwiensis*, described by Heimke and Starr (1979), and later redescribed as the new genus *Oogamochlamys* by Pröschold et al. (2001). Three clonal isolates from the same soil sample were all homothallic but they differed in the ratio of macrogametes and microgametes produced. One isolate gave approximately equal numbers of the two types, but the other two clones produced nearly 100% products of a single mating type.

Goodenough (1985) suggested that heterothallism predates homothallism, on the grounds that only heterothallism offers the potential of increasing genetic diversity through meiosis. Homothallism would have arisen in species that found means to circumvent the constraints of heterothallic pairing, and was perhaps selected for in cases of nonmotile cells (e.g., yeast) with restricted opportunities to find a partner of opposite mating type. Both forms of sexual union offer the advantages of creation of a diploid state and the potential for forming spores that can survive adverse environmental conditions.

The discovery of the *MID* gene in *C. reinhardtii* (Galloway and Goodenough, 1985) was a breakthrough in understanding the control of sexual expression in this species. MID, now known to be a transcription factor active in gametogenesis (Lin and Goodenough, 2007), is encoded in the mating type (*MT*) locus in *minus* strains, but not in *plus* (Ferris et al., 2002, and see below). Vegetative diploid cells, which are formed by a small fraction of matings (see below, and Chapter 8) mate as *minus* (Ebersold, 1967), hence the name *MID* for *Minus* Dominance. Loss of *MID* in a haploid strain causes a cell to mate as *plus*, even though the mating type locus retains the sequence configuration specific to *minus* cells. While such conversions are not true homothally, they suggest a mechanism by which it might evolve in nature. Also consistent with Wiese's speculations about the evolution of the sexual cycle among the Volvocales, orthologs of the *MID* gene have been found in male strains of the colonial algae *Pleodorina starrii* (Nozaki et al., 2006) and *Gonium pectorale* (Hamaji et al., 2008).

Bell (2005) reported apparent development of homothallism in a culture of *C. reinhardtii*, begun with a mixture of both *plus* and *minus* mating types and maintained for 5 years under conditions in which zygospore formation was enhanced. Cultures derived by clonal growth of individual cells from these experiments were found to be able to mate with testers of either mating type, and to produce zygospores with no added testers, indicating that a population of mixed mating types had arisen from a single cell. Bell acknowledged Ursula Goodenough for the suggestion that this seemingly homothallic behavior could result from transposition of an unstable copy of the *minus*-specific *MID* gene to another chromosome, unlinked to mating type, as occurred in strain CC-421 (Ferris and Goodenough, 1997; see below).

In subsequent crosses, inheritance of this copy of *MID* in a cell with the *plus* allele at the mating type locus would convert its phenotype to *minus*, and excision of the unstable element would restore the *plus* phenotype. Whether this was in fact what had happened in Bell's cultures was not determined, however. Another possible explanation is suggested by the *iso1* mutation (Campbell et al., 1995), which permits genetically *minus* gametes to express the *plus* agglutinin. In a population of *iso1* cells, some agglutinate as *plus* and others as *minus*. However, the "*plus*" gametes do not form a completely normal *plus* mating structure, and no fusion occurs, so no zygospores are formed, whereas Bell did observe zygospore formation.

C. Designation of mating types in heterothallic species

In oogamous species the identification of "male" and "female" cells is simple, based on analogy to multicellular plants and animals. In isogamous species, however, it is a somewhat arbitrary distinction. Mating types *plus* and *minus* of the laboratory strain of *C. reinhardtii* were assigned, presumably at random, by Smith, who isolated heterothallic pairs of this and several other species; subsequent isolates have been typed with reference to this strain. (Note, however, that the mating types of the UTEX stocks 89 and 90 were inadvertently reversed relative to the Sager and Ebersold-Levine 137c strains, and that this was not annotated in their catalog until after 1984; see Chapter 1 and Pröschold et al., 2005).

Subsequent studies (see below) showed that the partner designated *plus* in *C. reinhardtii* is the one which produces a fertilization tubule upon activation by flagellar agglutination (Friedmann et al., 1968; Goodenough and Weiss, 1975; Weiss et al., 1977b) and which transmits its chloroplast genome to the meiotic progeny in greater than 90% of zygotes under normal laboratory conditions (see Chapter 7 for full discussion). By analogy to chloroplast transmission through the female line in higher plants, *plus* cells of *C. reinhardtii* have conventionally been designated the maternal parent. Friedmann et al. (1968), however, made the counter-argument that on anatomical grounds, *plus* cells of *C. reinhardtii* should be called male.

The situation is similarly confused in the *C. moewusii* group of strains. As in *C. reinhardtii*, *plus* and *minus* mating types were arbitrarily assigned in the *C. moewusii* strain (UTEX 96, 97), which was isolated by Provasoli in 1948 (Lewin, 1949). Mated cells of this group form swimming pairs, persisting for several hours, in which only one set of flagella is active (Figure 5.3). The flagella inactivated are always those of the same parent, and the convention was established by Moewus for *C. eugametos* (Gowans, 1963) that the swimming "active" gamete be called male and the paralyzed "passive" gamete be designated female. The *C. moewusii plus* strain is the actively motile partner (Lewin, 1952a), and crosses have confirmed that

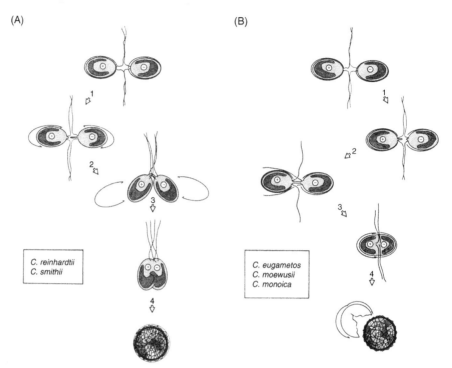

FIGURE 5.3 *Two styles of isogamous mating. (A) C. reinhardtii, C. smithii: (1) Gametic flagellar agglutination between cells of opposite mating type is followed by flagellar tipping, activation of anterior mating structures (differentiation of plasma papillae), and activation of lytic enzyme to promote shedding of gametic cell walls. (2) Activated mating structures fuse to form a cytoplasmic bridge, flagella de-adhere, and the cells immediately "jackknife" to begin the final stages of cell fusion. (3) Gametes fuse "shoulder-to-shoulder" to produce the quadriflagellate zygote. (4) Deposition of the multilayered zygospore wall completes the process. (B) C. eugametos, C. moewusii, C. monoica: (1) Gametic flagellar agglutination promotes activation of anterior mating structures (differentiation of plasma papillae) and limited, anterior site-specific dissolution of the cell wall to allow extension of the fertilization tubule. (2) Activated mating structures fuse to form a cytoplasmic bridge, gametic cell walls are retained, and mating-type specific flagellar paralysis allows directed movement of the tandem gamete pair ("vis-à-vis pair"), which may persist for several hours. (3) The gametic cell walls loosen, and the cell bodies fuse following a shortening and widening of the cytoplasmic bridge. (4) The multilayered zygospore wall is assembled and a primary zygote wall is released carrying the loosely attached gametic cell walls. Courtesy of Karen VanWinkle-Swift.*

this is the same mating type as "male" *C. eugametos*. In this group chloroplast genes are inherited as in *C. reinhardtii*, predominantly from the *plus* parent, which is however by the above convention, the "paternal" parent (McBride and McBride, 1975; Mets 1980; Lemieux et al., 1980). For the remainder of this chapter *plus* and *minus* will be used in discussion of both species groups, as delineated in Table 5.2.

Table 5.2	Designation of mating types in laboratory strains of *Chlamydomonas*		
Species	*plus*	*minus*	**Notes**
C. reinhardtii	CC-125; UTEX 90; Sager 21 gr	CC-124; UTEX 89; Sager 6145	See Table 1.1 for additional strains
	Forms fertilization tube		
	Transmits chloroplast DNA	Transmits mitochondrial DNA	Boynton et al. (1987)
	Tunicamycin sensitive	Tunicamycin resistant	Wiese and Mayer (1982)
C. moewusii and *C. eugametos*	UTEX 9; UTEX 97	UTEX 10; UTEX 96	
	Flagella active in *vis-à-vis* pairs	Flagella inactive in pairs	
	Transmits chloroplast DNA		Lemieux et al. (1980)
	Transmits mitochondrial DNA		Lee et al. (1990)
	Tunicamycin sensitive	Tunicamycin resistant	See Wiese et al. (1983) for additional inhibitors with differential effects

D. Sexual incompatibility within the *C. moewusii* group

As discussed in Chapter 1, the UTEX 9 and 10 cultures originally identified as *C. eugametos* are interfertile with *C. moewusii* UTEX 96 and 97, although poor zygote viability is encountered in some crosses (Gowans, 1963; Cain 1979; Lemieux et al., 1980). Morphologically these two pairs of strains are indistinguishable from one another, but they differ in some physiological respects, including the ability to be induced to gametogenesis by nitrogen deprivation (see below). Also, chloroplast DNAs from these strains show quite different restriction digest patterns (Lemieux et al., 1980; Mets, 1980). Some other isolates that resemble *C. moewusii* structurally seem to be incompatible sexually with either *C. moewusii* (UTEX 96 and 97) or *C. eugametos* (UTEX 9 and 10), which were identified by Wiese and Shoemaker (1970) as "syngen I" of this group. The Lewin isolates originally called *C. moewusii* syngen II (UTEX 792 and 793) are now designated *C. moewusii yapensis*, from their isolation site on the island of Yap in the South Pacific (Wiese et al., 1983). Other pairs of isolates, including the varieties *rotunda* and *tenuichloris*, and *C. moewusii* UTEX 2018 and 2019, appear to mate only with themselves (Tsubo, 1961; Wiese and Wiese, 1977; Wiese et al., 1983). A homothallic variety, *C. moewusii* var. *monoica*, has also been described (Deason and Ratnasabapathy, 1976).

III. STAGES IN THE REPRODUCTIVE PROCESS

A. Gametogenesis

Gametogenesis in algae often appears to be triggered by adverse environmental conditions, and in the laboratory is usually induced by transfer of cells to distilled water or nitrogen-free medium (see Coleman, 1962; also Coleman and Pröschold, 2005, for general discussion, and Beck and Haring, 1996, for review of the literature pertinent to *Chlamydomonas*). In *C. reinhardtii*, *C. moewusii* var. *rotunda*, and *C. chlamydogama*, nitrogen deprivation seems to be the most important inducing factor (Sager and Granick, 1954; Bernstein and Jahn 1955; Tsubo, 1956; Trainor, 1958, 1959). Bernstein and Jahn (1955) reported that this was also true for *C. eugametos*, but Trainor (1975) found that the laboratory strain of *C. eugametos* was capable of forming gametes in sterilized river water collected below a sewage treatment plant, and he suggested that nitrogen deprivation was probably not the gametogenetic trigger in nature. The nitrogen level in the sewage effluent (0.3 mg/l) was still much lower than that typically used in culture media, however. Tomson et al. (1985) reported that *C. eugametos* cells acquire gametic competence for a few hours at the end of the exponential phase of growth even when adequate nitrogen is present. Trainor (1959) found that *C. moewusii* (UTEX 96 and 97) is capable of gametogenesis even in the presence of 300 mg/l ammonium nitrate, which inhibits gamete formation entirely in several other species studied. Gametogenesis can be induced in *C. moewusii* by transfer of agar cultures to darkness for 24 hours followed by flooding the culture with distilled water or dilute medium (Lewin, 1953a).

Both chloroplast and cytoplasmic ribosomes are degraded on a large scale during gametogenesis in synchronous cultures (Siersma and Chiang, 1971; Martin et al., 1976). Nuclear DNA per cell appeared to remain constant (Kates et al., 1968), and the gametogenic cell division that followed the period of differentiation was accompanied by new DNA synthesis utilizing the degraded rRNA products as nucleotide precursors (Siersma and Chiang, 1971). Some new ribosomal components are also formed (Martin et al., 1976). Ultrastructural studies of gametes formed from synchronously grown cultures confirmed the loss of ribosomes and show alterations in chloroplast morphology, starch accumulation in the chloroplast, and changes in the nuclear envelope and endoplasmic reticulum (Cavalier-Smith, 1975; Martin and Goodenough, 1975). The chloroplast changes were accompanied by diminution of photosynthetic activity. These changes were presumed to be the direct result of nitrogen deprivation and not intrinsic to preparation for mating. They were less dramatic in cells shifted to darkness after 6 hours of gametogenesis.

Sager and Granick (1954) noted that light facilitated gametogenesis of *C. reinhardtii* but attributed it to a secondary effect of photosynthetic acceleration of nitrogen depletion from the medium, and reported that dark-grown cells on acetate medium could also become sexually active when deprived of nitrogen. The observation that streptomycin, an inhibitor of chloroplast protein synthesis, also blocks gametogenesis (Hipkiss, 1967) was consistent with this hypothesis. Subsequent experiments (Treier et al., 1989; Beck and Acker, 1992) indicated that light was essential in the conversion of pregametes induced by nitrogen starvation into gametes competent for mating. Weissig and Beck (1991) showed that the action spectrum for gamete formation was consistent with a blue light receptor, which was later identified as a phototropin (Huang et al., 2002; Huang and Beck, 2003). The signal transduction pathway that mediates this step can be activated in the dark by treatment with either staurosporine or papaverine (Pan et al., 1996, 1997), and by mutations that block negative control elements (Buerkle et al., 1993; Glöckner and Beck, 1995, 1997; see also Table 5.4). Some isolates of the standard laboratory strain also are capable of gametogenesis in the absence of light, although all recent field isolates tested were found to have light-dependent gametogenesis (Saito et al., 1998).

Gene expression during gametogenesis has been examined by Treier and Beck (1991), von Gromoff and Beck (1993), Merchán et al. (2001), Abe et al. (2004, 2005), and others. Merchán et al. (2001) identified 10 genes induced by nitrogen starvation (*NCG* = *n*itrogen *c*ontrolled *g*ene), some of which are probably involved with scavenging alternative nitrogen sources rather than with gametogenesis *per se*. Using synchronized cells in which nitrogen deprivation was begun early in G1 phase, Abe et al. (2004) classified several previously identified genes and 18 newly identified ones (*NSG* = *n*itrogen-*s*tarved *g*ametogenesis) by their time of expression after removal of nitrogen from the medium. Early genes, which were expressed transiently within the first 2hours, included nitrate reductase, the nitrate transporter *NRT2;1*, the *NCG2* gene described by Merchán et al. (2001), and the newly identified *NSG17* gene, whose product is unknown. The *GAS3* and *GAS96* (*ga*mete-*s*pecific) genes are also expressed early in gametogenesis (von Gromoff and Beck, 1993). Abe et al. (2004) classified a large group of *NSG* and other genes as "middle" expression, in the third and fourth hours after beginning nitrogen starvation. Among these were *FUS1*, which encodes a glycoprotein required for sex recognition in *plus* cells (Ferris et al., 1996), and two other gamete-specific genes, *MTA1* and *MTA2* (Ferris et al., 2001, 2002). *GAS28*, which encodes a hydroxyproline-rich glycoprotein of the zygospore wall (Rodríguez et al., 1999) is expressed late in gametogenesis (von Gromoff and Beck, 1993) together with two more of the *NSG* genes (Abe et al., 2004).

There has also been debate as to whether gametogenesis requires a round of cell division. In markedly anisogamous or oogamous species, microgametes arise from multiple rounds of mitosis in a single progenitor cell, while macrogametes differentiate directly from vegetative cells without division. Wiese et al. (1979) postulated that there are likewise two parallel pathways of gametogenesis in isogamous species such as *C. reinhardtii* or *C. moewusii*, one requiring cell division, the other not. Despite several early studies on induction of gametogenesis in synchronously growing cells (Kates and Jones, 1964a; Chiang et al., 1970; Schmeisser et al., 1973) this question was not answered definitively until methods for inducing synchrony were improved and the cell cycle was better understood. Abe et al. (2004) have shown that *plus* cells transferred to nitrogen-free medium in early G1 phase become gametes within 4 hours, without undergoing division. As discussed below, electron micrographs show the appearance of mating-specific structures during mitosis under nitrogen-deprived conditions, but detailed studies of gene expression have not yet been done on this division-dependent pathway, if in fact it is an alternatively regulated one as Wiese et al. suggested. Mature gametes can remain viable for several weeks without dividing, but resume mitotic growth on transfer to nitrogen-containing medium (Goodenough et al., 2007).

Gametogenesis in *minus* cells is controlled by two genes not found in *plus*, *MID* and *MTD1* (Lin and Goodenough, 2007). The MID protein, a transcription factor in the RWP-RK family, activates expression of the *minus* agglutinin (*SAD1* gene; Ferris et al., 2005) and the *GSM1* gene, encoding a *minus*-specific homeodomain protein that functions in the zygote (Goodenough et al., 2007). MID also represses expression of the *plus* agglutinin, encoded by the *SAG1* gene (Ferris et al., 2005), and the *GSP1* gene, encoding a *plus*-specific homeodomain protein (Kurvari et al., 1998; Wilson et al., 1999; Zhao et al., 2001). Neither *SAG1* nor *GSP1* is part of the mating type locus and therefore both these genes are present in both *plus* and *minus* cells. *MID* is expressed early in gametogenesis, within the first 30 minutes after nitrogen withdrawal, and then decreases by 60 minutes to the basal level of expression seen in vegetative cells. *MID* expression rises again, to a considerably higher level, around 6 hours, after the cells have acquired mating competence.

MTD1, found in the mating type locus in *minus* cells, is expressed beginning at 4 hours, and is sustained thereafter, whereas *SAD1* and *GSM1* reach peak expression around 6 hours, and then decline. Lin and Goodenough (2007) proposed that early *MID* expression activates *MTD1*, which then facilitates the second level of *MID* expression. In their model, the early expression of *MID* represses *plus*-specific genes carried by both mating types, but the later, higher level expression is required to induce the other *minus*-specific genes. Mutants lacking *MID* mate as *plus* rather than *minus*, although they do show some *MTD1* expression. To account for the ability of a transposed copy of *MID* to cause cells to mate as *minus*, although they lack the *MTD1* gene, Lin and Goodenough postulated that *plus* gametes have their own regulatory system equivalent to MTD1, although no similar gene is found in the *plus* mating type region.

Formation of mating-specific structures occurs in both mating types. In *plus* cells, a specialized region is differentiated at the cell anterior which will give rise to the fertilization tubule (Figure 5.4). In cross-section an electron-dense ring appears to be associated with the plasma membrane (Martin and Goodenough, 1975; Triemer and Brown, 1975b). Cavalier-Smith (1975) likened this structure to the intracellular portion of half of a desmosome and named it the gamosome. The ring formation, which he called the gamosomal plaque, is equivalent to the "choanoid body" described in mating cells by Friedmann et al. (1968) and to the "doublet zone" of Goodenough and Weiss (1975). (Goodenough is the only one of these authors to continue studies on the *Chlamydomonas* sexual cycle to the present day, and her terminology has prevailed.) The corresponding region of *minus* cells is also differentiated from that of vegetative cells. The overlying membrane zone is broader and more diffuse than in *plus* gametes, and no doublet zone is seen (Figure 5.4). In freeze-fracture studies (Weiss et al., 1977b), the membrane overlying the *plus* mating structure is seen to have sparse, asymmetrically distributed particles on the P face, whereas the *minus* membrane has symmetrically arranged particles on the E face, which seem to be involved in the fusion of gametes (see below). The mating-specific structures are first seen during the gametogenic mitosis and are associated with the cleavage furrow membrane separating two daughter gametes (Figure 5.5). The time of their appearance in synchronously grown cells undergoing gametogenesis without mitosis has not yet been established. The relationship of the mating structures to the flagellar root system was discussed by Goodenough and Weiss (1978).

(A) (B)

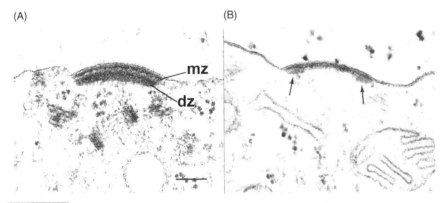

FIGURE 5.4 *Mating structures of plus and minus gametes of C. reinhardtii before activation: (A) Mating structure of plus gamete showing the narrow membrane zone (mz) underlying the plasma membrane and the broad doublet zone (dz) beneath. Fine hairlike projections extend outward from the plasma membrane overlying the mating structure. A young basal body and several microtubules are present in the underlying cytoplasm. (B) Mating structure of minus gamete showing the membrane zone beneath the plasma membrane, which thickens somewhat at both edges of the structure (arrows). Bar = 0.1 μm. From Weiss et al. (1977b). Reproduced from The Journal of Cell Biology, 1977, 72, 144–160 by copyright permission of The Rockefeller University Press.*

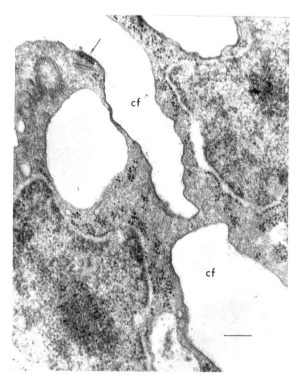

FIGURE 5.5 *Formation of C. reinhardtii plus mating structure in gametogenic division. Cells grown in liquid medium in light were fixed 15 hours after being deprived of nitrogen. Two daughter cells have just completed mitosis and remain connected by a strand of cytoplasm between the cleavage furrows (cf). Cleavage microtubules are evident along the cleavage furrow. A mating structure (arrow) has formed in association with the cleavage furrow membrane. Bar = 0.2 μm. From Martin and Goodenough (1975). Reproduced from The Journal of Cell Biology, 1975, 67, 587–695 by copyright permission of The Rockefeller University Press.*

B. Agglutination

When *plus* and *minus* gametes of *C. reinhardtii* are mixed, there is immediate adhesion of cells by their flagella. Initial contacts and the resulting weak adhesion can be anywhere along the flagellar length, and an individual gamete can adhere to more than one cell of the opposite mating type, with the result that clumps of gametes are formed (Figure 5.6). The agglutination reaction is specific to gametes; the active molecules (agglutinins) are absent from flagella of vegetative cells, and are assembled onto gamete flagella from a reservoir in the plasma membrane of the cell body (Hunnicutt et al., 1990). Agglutinins are hydroxyproline-rich glycoproteins related to those of the cell wall (Chapter 2), sufficiently closely to suggest a common evolutionary origin (Adair, 1985). The agglutinins of both *plus* and *minus* cells of *C. reinhardtii* have a terminal head region that is distinguishable in the two mating types, a shaft which in *minus* joins the head in a "shepherd's crook"

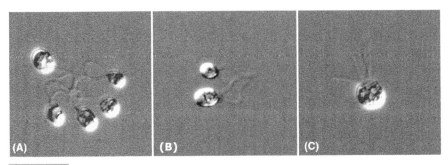

FIGURE 5.6 *(A–C) Clump of agglutinating gametes, gamete pair, and quadriflagellate zygote of C. reinhardtii. Courtesy of Qian Wang and William Snell.*

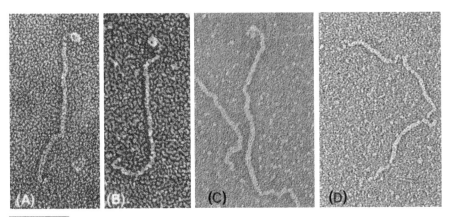

FIGURE 5.7 *Agglutinins of Chlamydomonas cells. (A) C. reinhardtii plus; (B) C. reinhardtii minus; (C) C. eugametos plus; (D) C. eugametos minus. (A, B) Quick-freeze deep-etch preparations of purified agglutinins adsorbed to mica; courtesy of Ursula Goodenough and John Heuser. (C, D) Negative stained preparations, courtesy of Alan Musgrave. Also see Volume 3, Chapter 12, Figure 3 for additional, more detailed micrographs.*

configuration, and a hook region which attaches to the flagellar surface (Figure 5.7; Goodenough et al., 1985). The genes encoding these proteins have been sequenced, and the predicted amino acid sequences correlated both with the observed structures, and with the postulated interactions between the *plus* and *minus* agglutinins in mating gametes (Ferris et al., 2005; see also Volume 3, Chapter 12). Goodenough et al. (1985) described three additional gamete-specific flagellar surface proteins that appear to share structural and immunological similarities with the agglutinin molecules but do not themselves have adhesive activity. All three are fibrillar molecules with distinctive shapes (short canes, loops, and crescents) and all are found in both *plus* and *minus* gamete preparations. They are also present in extracts from nonagglutinating mutant strains that lack the larger agglutinin molecules. Agglutinins have also been characterized structurally, but not yet sequenced, from *C. eugametos* and *C. moewusii* (Klis et al., 1985, 1989; Crabbendam et al., 1986; Samson et al., 1987a, b; Versluis et al., 1993).

Flagellar adhesion is not an essential component of mating in all *Chlamydomonas* species, but some recognition mechanism must nevertheless be operative. In oogamous forms, the macrogamete is formed without flagella, and the microgamete fuses either at the apical end or at any site on the macrogamete cell periphery (Ettl, 1976a). There is one early report (Mitra, 1950) of two isogamous *Chlamydomonas* species, *C. indica* and *C. iyengarii*, and a *Carteria* species in which gametes fused at the posterior cell ends without prior flagellar contact. Lewin (1954c) thought it more likely that the pairs observed by Mitra were the result of incomplete cell division, not posterior fusion, and Ettl (1976a) also questioned this report.

Apart from the largely discredited work of Moewus, there are relatively few reports of sexual pheromones distinct from agglutinins in *Chlamydomonas* species. Pascher (1932) described chemotactic attraction in *Chlamydomonas paupera*, later reclassified by Ettl (1970) as a *Chloromonas* species. Tsubo (1957, 1961) reported evidence for a chemotactic attractant in *C. moewusii* var. *rotunda*, apparently a volatile substance of low molecular weight. The attractant was mating type-specific in the *rotunda* cells but elicited a chemotactic response in both mating types of other strains of *C. moewusii* and *C. eugametos*. Since agglutination did not occur with any of the latter strains, Tsubo concluded that the attraction and agglutination were separate processes. Kochert (1978) briefly reviewed this work and suggested further experimentation, but no one appears to have taken up this challenge. Tomson et al. (1986) found no evidence for pheromonal attraction between *C. eugametos plus* and *minus* cells; they suggested instead that repeated collisions between prospective partner cells may facilitate adhesion, possibly by stimulating relocation of agglutinin molecules into patches or to the flagellar tips.

More persuasive evidence has been found for a pheromone in *Chlamydomonas allensworthii*, a heterogamous species described by Starr et al. (1995). The pheromone, which was given the name lurlene (Jaenicke and Starr, 1996), was identified as a pentosylated hydroplastoquinone, and shown to be secreted by female gametes (Starr et al., 1995; Coleman et al., 2001). Male gametes were attracted to lurlene adsorbed on DEAE-acrylamide beads, confirming its activity and the chemotactic nature of the response. An acid form (lurlenic acid) and the corresponding alcohol (lurlenol) both showed pheromonal activity. Originally described from a soil sample collected in California, *C. allensworthii* was subsequently shown to have a worldwide distribution, with some isolates having greater response to lurlenic acid and others to lurlenol (Coleman et al., 2001).

C. Pair formation and signal transmission

An increase in cyclic AMP (cAMP) is seen within 20 seconds of the beginning of agglutination in *C. eugametos* (Pijst et al., 1984b), preceding all other morphological and physiological changes leading to fusion. This phenomenon was also demonstrated in *C. reinhardtii* by Pasquale and Goodenough (1987), who showed that exogenous dibutyryl cAMP specifically induces all three agglutination-triggered responses: flagellar tip activation, loss of cell walls, and activation of mating structure. Similar effects were seen with inhibitors of cyclic nucleotide phosphodiesterase (e.g., isobutyl methylxanthine), which acted synergistically with dibutyryl cAMP. Cell wall loss and mating structure activation could also be demonstrated after treatment with these agents in mutant cells lacking flagella, indicating that flagellar agglutination can be bypassed.

The *fla10* mutant has a temperature-sensitive defect in flagellar assembly due to loss of the kinesin-2 motor protein that drives intraflagellar transport (Chapter 4; also Chapter 4 in Volume 3) and is also unable to mate when shifted to the restrictive temperature (Piperno et al., 1996). Pan and Snell (2002) demonstrated from studies with this mutant that kinesin-2 is required for the sensory transduction by flagella during mating (see Volume 3, Chapter 12).

Agglutination induces a change in the flagellar tips in both *C. reinhardtii* and *C. eugametos* (see van den Ende, 1985 for review). The tip enlarges, and fibrous material accumulates in a specific region between the nine single A microtubules and the terminal membrane (Figure 5.8). Tip activation can

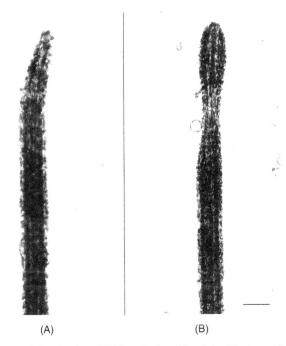

(A) (B)

FIGURE 5.8 *Unactivated (A) and activated (B) flagellar tips of C. reinhardtii minus cells, prepared as described by Mesland et al. (1980). Bar = 0.1 μm. Courtesy of Ursula Goodenough.*

be elicited in gametes of a single mating type with isolated agglutinins or antiflagellar antiserum and is also seen with antibody treatment of nonagglutinating mutants. Crabbendam et al. (1984) showed that in *C. reinhardtii* normal tip morphology is restored in the quadriflagellate stage after cell fusion, while in *C. eugametos* the "activated" tip structure is retained by the non-motile flagella of the *minus* cell. The *minus* flagella also shorten appreciably at the time of pairing in *C. eugametos* (Mesland, 1976).

Studies with the *C. reinhardtii* mutant *pf18*, which has rigid, immobile flagella, showed that membrane transport occurs along the length of the flagellum (Bloodgood, 1977; Bloodgood et al., 1979; Hoffman and Goodenough, 1980). The same phenomenon was recognized in a paralyzed mutant of *C. moewusii* by Lewin (1952a), who correctly equated it with gliding motility seen on agar (see Chapter 4 this volume, also Volume 3, Chapter 11). The initial sites of gamete contact are subject to movement, possibly by this mechanism, and in *pf18* cells appear to migrate back and forth before moving to the tips of the flagella and becoming fixed there (tip locking; Goodenough et al., 1980). The relationship between bead translocation, tipping, and intraflagellar transport is still unresolved (see Chapters 11 and 12 in Volume 3).

The same tipping process occurs in wild-type cells but is much more rapid and therefore difficult to observe. Adherent beads or antiflagellar antibodies are similarly moved to the flagellar tips (Goodenough and Jurivich, 1978; Hoffman and Goodenough, 1980). As tip locking occurs, pairs of cells separate from the group and begin the subsequent steps leading to cell fusion. It is at this stage that the signal is transmitted that results in shedding of the gamete cell walls and activation of mating structures (Volume 3, Chapter 12).

D. Cell wall lysis

In the normal course of mating, cell wall lysis is one of the cAMP-mediated events triggered by flagellar tip activation in *C. reinhardtii*. The gamete lytic enzyme is a metalloproteinase (Buchanan and Snell, 1988; Kinoshita et al., 1992), and is entirely different from the enzyme that liberates vegetative cells after mitosis (see Chapter 3). It can act on both vegetative and gametic cells, but not on zygospore walls (Claes, 1971; Schlösser, 1976). In wild-type cells, the lytic enzyme is stored as an inactive precursor protein in the periplasmic space beneath the cell wall (Buchanan et al., 1989).

This temporal sequence is not obligatory, however. Wall lysis can also occur during gametogenesis in *C. reinhardtii* (Matsuda, 1980), and cell wall-deficient mutants mate normally. In some species (e.g., *C. gymnogama*; Deason, 1967) gametes lack walls prior to contact, while in others (*C. chlamydogama*; Bold, 1949a), walls are not completely shed until cell fusion is under way.

Cell wall lysis in *C. moewusii* and *C. eugametos* occurs initially only at the tips of the fusing papillae, and the swimming pairs of gametes retain their walls until very shortly before cytoplasmic fusion (Brown et al., 1968). The lytic enzyme is not excreted into the medium as in *C. reinhardtii*, and the culture medium contains pieces of what appear to be undegraded cell walls (Musgrave et al., 1983).

Concanavalin A binding sites were localized in *C. reinhardtii* gametes by Millikin and Weiss (1984a) using fluorescein isothiocyanate–concanavalin A staining and ferritin–concanavalin A labeling for electron microscopy. A crescent-shaped region binding concanavalin A was seen in the periplasm at each end of wild-type gametes of both mating types. Although also visible in vegetative cells, these regions were particularly prominent in gametes prior to fusion and cell wall lysis. They were not seen in cell wall-deficient mutants or in young zygotes. Since the gamete lysin is strongly bound to concanavalin A affinity columns, Millikin and Weiss suggested that the periplasmic concanavalin A binding regions represented concentrations of lysin, or a precursor thereof, that would be released in response to the calcium-mediated signal (Snell et al., 1982, 1983).

Wall lysis, perhaps in conjunction with osmotic stress, activates expression of three genes, *GAS28*, *GAS30*, and *GAS31*, all of which encode hydroxyproline-rich glycoproteins with Ser(Pro)x domains (Hoffmann and Beck, 2005). Accumulation of *GAS28* transcripts late in gametogenesis was previously described (von Gromoff and Beck, 1993; Abe et al., 2004). Using mutants blocked at various stages in the mating process, Hoffmann and Beck (2005) showed that *GAS28*, *GAS30*, and *GAS31* did not accumulate unless agglutination occurred, but did occur in mutants blocked in fusion of the mating structures, and in vegetative cells treated with gamete lysin.

E. Activation of mating structures

The fertilization tubule connecting mating gametes of *C. reinhardtii* was first described by Friedmann et al. (1968). In 1975 three laboratories independently published studies showing formation of this structure from the apical region of *plus* cells (Cavalier-Smith, 1975, Goodenough and Weiss, 1975; Triemer and Brown, 1975b). The first state in this process is formation of a bud, with separation of the electron-dense membrane zone from the underlying doublet zone (Figures 5.9 and 5.10). In the second stage of activation, microfilaments appear to radiate into the growing tubule. Detmers et al. (1985) demonstrated that these consist of actin filaments (Figure 5.11) and that their extension, as expected, is inhibited by cytochalasin D. They suggested that the actin filaments nucleate at specific sties in the doublet zone (Figure 5.12) and found no evidence for a concentrated cytoplasmic pool of actin from which the filaments of the

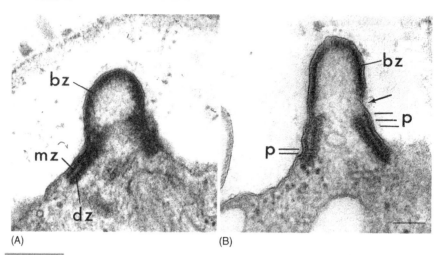

FIGURE 5.9 *Activation of mating structures of C. reinhardtii plus gametes. (A) Early stage in activation. The bud interior is clear and contains some reticulate material. A zone of dense aggregated material adheres to the inner surface of the bud membrane (bz); an amorphous material extends from the membrane's outer surface. An intact cell wall overlies the bud. mz, Membrane zone; dz, doublet zone (compare to Figure 5.3). (B) Late bud stage in activation. The bud has lengthened, and the bud interior is somewhat denser and more fibrous than before. Amorphous material continues to associate with the external surface of the bud membrane and bud-zone material (bz) with its internal surface. Arrow points to a discontinuity between the membrane zone and the bud zone; the membrane zone has also opened up medially. Some periodically distributed material (p) extends from the doublet to the membrane zones. Bar = 0.1 μm. From Goodenough and Weiss (1975). Reproduced from The Journal of Cell Biology, 1975, 67, 623–637 by copyright permission of The Rockefeller University Press.*

tubule were assembled (Figure 5.13). As fusion progresses, the doublet zone appears to detach from the plasma membrane and become free in the cytoplasm, and the microfilaments appear to extend to the nucleus of the *minus* cell. Friedmann et al. (1968) described tubular projections extending from the surface of the fertilization tubule, but these were not seen by any of the later observers and were attributed by Triemer and Brown (1975b) to possible artifacts.

Mating structures of *minus* cells also change their appearance upon activation (Figure 5.13; Weiss et al., 1977b; Goodenough et al., 1982). The initial small bud enlarges, and the material of the membrane zone transforms into a domed shape with extracellular "fringe" material along its surface (Figure 5.14). In freeze-fracture preparations, densely clustered particles are seen in a central dome on the E face, surrounded by a zone free of particles. Distinctive coated vesicles appear to arise from the contractile vacuole and fill the region below the mating structure (Weiss, 1983a). Treatment of *plus* gametes with cytochalasin D allows formation of actinless pseudotubules which can still make contact but not fuse with the mating

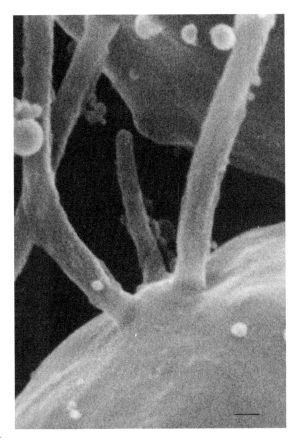

FIGURE 5.10 *Activated wild-type C. reinhardtii plus mating structure as seen by scanning electron microscopy. Bar = 0.1 μm. From Forest et al. (1978). Reproduced from The Journal of Cell Biology, 1978, 79, 74–84 by copyright permission of The Rockefeller University Press.*

structure of *minus* cells (Goodenough et al., 1982; Detmers et al., 1983). In these matings, contact appears to be specific to the two regions of fringe (Figure 5.15). Goodenough et al. (1982) postulated that fusion of mating structures can be separated into two stages: a recognition phase involving the fringe regions followed by membrane fusion. The *plus* fringe has been identified as a glycoprotein encoded by the *FUS1* gene (Ferris et al., 1996). The gene encoded the *minus* fringe protein has not yet been identified. Since nonagglutinating mutants can mate normally if artificially activated (Pasquale and Goodenough, 1987), while other mutants are known that agglutinate normally but fail to fuse (see below), two separate gamete recognition systems must be operative in *C. reinhardtii* mating.

The *FUS1* gene, present in the mating type locus only in *plus* cells, encodes a protein required for fusion. The *plus* mating structure in *fus1* mutants appears normal except that it lacks the fringe coating.

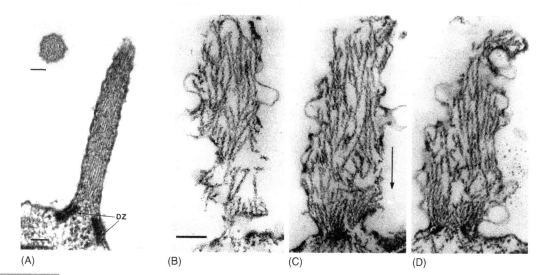

FIGURE 5.11 *(A) Longitudinal thin section through a fertilization tubule of C. reinhardtii showing the arrangement of the microfilaments forming the core of the process. Although the filaments were tightly packed, no regular cross striations were observed. The section catches the middle of the doublet zone (DZ) revealing the medial discontinuity at the apex of this cone-shaped structure (bar = 0.1 µm). (B, C) Cross-section through a fertilization tubule showing that the microfilaments are packed in a random arrangement (bar = 0.1 µm). (D) Three serial sections through a fertilization tubule following S-1 (myosin subfragment-1) decoration. S-1 arrowheads are clearly visible on all filaments within the fertilization tubule, and all arrowheads point away from the tip of the process. Arrow indicates the polarity of the filaments (bar = 0.2 µm). From Detmers et al. (1983). Reproduced from The Journal of Cell Biology, 1983, 97, 522–532 by copyright permission of The Rockefeller University Press.*

Immunofluorescence studies showed that the FUS1 protein is concentrated in an apical patch in unactivated *plus* gametes, and is redistributed over the entire surface of the mating structure after activation (Misamore et al., 2003).

F. Cell fusion

Once adhesion and fusion of the fertilization tubule and the *minus* mating structure occur in *C. reinhardtii*, the tubule rapidly shortens, bringing the apical ends of the two cells into apposition (Friedmann et al., 1968). The fusing cells then bend, jackknife fashion, and fusion occurs laterally from anterior to posterior (Figure 5.16) to form a motile cell with four flagella (planozygote or quadriflagellate cell, abbreviated QFC; Figure 5.17). This stage is found in many isogamous green algae of other genera as well (Friedmann et al., 1968), although Ettl (1976a) stated that most *Chlamydomonas* species seem to form nonmotile zygotes directly without a motile stage. Flagellar movement in *C. reinhardtii* QFCs initially is uncoordinated; acquisition of directional motility by the QFC is possibly associated with connection of the flagellar roots of the two component cells (Friedmann et al., 1968).

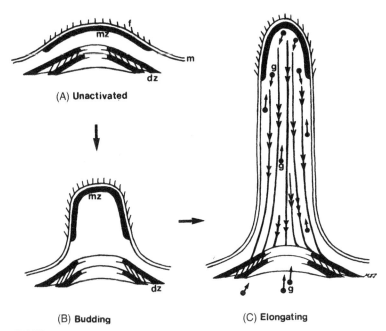

(A) **Unactivated**

(B) **Budding**   (C) **Elongating**

FIGURE 5.12 *Model for elongation of the fertilization tubule in C. reinhardtii. (A) In the unactivated plus mating structure an electron-dense membrane zone (mz) overlies a cone-shaped doublet zone (dz). The exterior surface of the mating structure is provided with "fringe" (f) which appears to mediate recognition of and binding to the minus mating structure prior to fusion. (B) Following receipt of mating signals, the membrane zone separates from the central portion of the doublet zone, and a bud is formed. (C) Nucleation of actin polymerization then occurs at the doublet zone, assuring the uniform polarity of the filaments, which grow by monomer addition to the barbed end (g, monomers adding in this manner)). When elongation is complete the barbed ends of the filaments are embedded in the membrane zone, which may help promote stability of the filaments. From Detmers et al. (1983). Reproduced from The Journal of Cell Biology, 1983, 97, 522–532 by copyright permission of The Rockefeller University Press.*

During the early QFC stage the two gametic chloroplasts and nuclei remain separate. Nuclear fusion precedes chloroplast fusion in *C. reinhardtii* (Blank et al., 1978; Cavalier-Smith, 1970, 1976). Grobe and Arnold (1977) reported that the large, branched mitochondria seen in gametes were replaced in young zygotes by as many as 50 small individual mitochondria per cell; mitochondrial fusion was not observed. The QFC remains motile for about 2 hours. Ultimately, the flagella regress gradually over a 30-minute period, becoming shorter and finally disappearing (Randall et al., 1967). The basal bodies persist during flagellar regression but disintegrate shortly thereafter, and the flagellar roots and connecting fibers are also lost, to be reformed at the time of germination (Cavalier-Smith, 1974).

In *C. eugametos*, agglutinating flagella of the *plus* and *minus* gametes pair along their entire length prior to cell fusion rather than transferring their

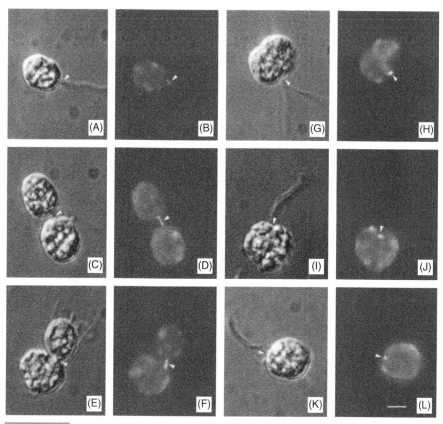

FIGURE 5.13 *Mating gametes of C. reinhardtii stained with antibody to actin. Gametes mating for 2 minutes (A–F) or 5 minutes (G–L) were selected to represent a rough time course of the mating reaction. Criteria for placing the figures in this order were described by Friedmann et al. (1968). (A, B) Plus gamete with fertilization tubule (arrowhead) activated by the presence of minus gametes (not shown); (C–F) post-fusion gametes with a bridge between them (arrowheads); (G, H) early zygote (quadriflagellate cell) with a remnant of the fertilization tubule (arrowhead); (I, J) zygote with defunct fertilization tubule (arrowhead). (K, L) zygote in which the fertilization tubule has partially disassembled (arrowhead). A, C, E, G, I, and K, Nomarski optics; B, D, F, H, J, and L, corresponding fluorescent images. Bar = 4 μm. From Detmers et al. (1985).*

primary adhesion to the flagellar tip, although a specialized tip structure is formed. The paired flagella then orient around the cell body of one gamete, thereby appearing to bring the papillar regions of the mating cells into direct contact (Figure 5.18). By labeling gametes of one mating type, Musgrave et al. (1985) showed that the flagella always wrap around the *minus* gamete, and that clumps of gametes are formed in which the *minus* cell bodies are always in the center ,with the paired flagella making contact with those of adjacent cells (Figures 5.19 and 5.20). Fusion begins at the papillar ends, to form a cytoplasmic bridge which remains as the sole connection between the mated pairs for several hours (Lewin and Meinhart, 1953;

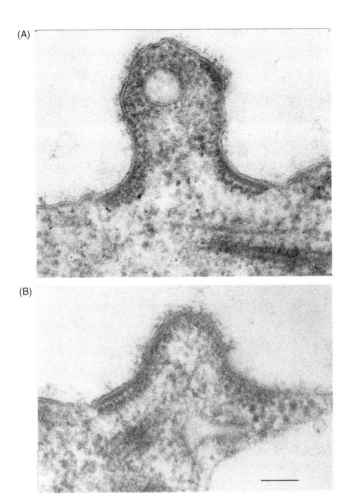

FIGURE 5.14 *(A) Bud stage of wild-type C. reinhardtii minus mating structure activation elicited by 2 minutes of flagellar agglutination with the nonfusing mutant imp1. The bipartite sector of the membrane zone lies at the base of the bud. The bud interior contains a membrane vesicle, particulate material, and patches of submembranous dense material. The surface fringe is discontinuous, being absent, for example, from a sector of membrane at the apical end of the bud. (B) Fully activated wild-type minus mating structure from a 2-minute agglutination with the nonfusing mutant imp11. Some particulate material remains in the interior but the structure has converted from a bud to a dome-shaped structure and has a smaller surface area. The fringe is concentrated over the continuous layer of submembranous dense material, the central zone; the fringe-free bipartite sectors of the original membrane zone lie on either side. Bar = 0.1 μm. From Goodenough et al. (1982). Reproduced from The Journal of Cell Biology, 1982, 92, 378–386 by copyright permission of The Rockefeller University Press.*

Gibbs et al., 1958). About 8 minutes after papillar fusion, the tips of the *plus* gametes deactivate, the flagella separate, and the *vis-à-vis* pair resumes motility, with the *plus* flagella providing the motive force. The *minus* flagella remain immotile, with the tips still activated, and eventually shorten by about one-third. The flagella of the two gametes always align with the base

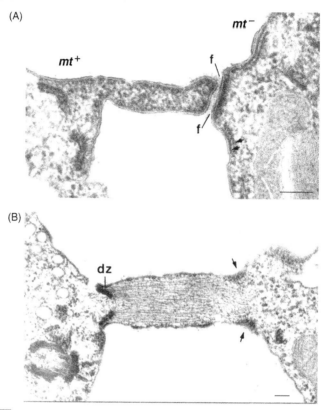

FIGURE 5.15 *(A) Pre-fusion interaction between tip of C. reinhardtii plus fertilization tubule (left) and activated minus mating structure (right), whose bipartite membrane zone sectors are marked by double arrows. The fertilization tubule contains no actin because cytochalasin D was present in the mating mixture. The surface fringes (f) of the fertilization tubule and the minus central zone are enmeshed along a broad flattened region of contact. (B) Cytoplasmic bridge between plus (left) and minus (right) gametes of C. reinhardtii. The doublet zone (dz) marks the plus end of the bridge. Fusion has occurred in the region marked by the two arrows. The dense material in the fusion zone is most likely a hybrid of plus membrane zone and minus central zone. Bars = 0.1 µm. From Goodenough et al. (1982). Reproduced from The Journal of Cell Biology, 1982, 92, 378–386 by copyright permission of The Rockefeller University Press.*

of the right flagellum of one gamete (as viewed from the posterior) above the left flagellum of its partner. This orientation apparently results from a 90° counterclockwise rotation of the cells relative to one another as the flagella de-adhere and separate (Mesland, 1976).

The *"vis-à-vis"* configuration is also seen in *C. moewusii, C. monoica,* and *C. philotes* (Strehlow, 1929; Lewin, 1957a; see Figure 5.3). Triemer and Brown (1975a) suggested that paralysis of the *minus* flagella in *C. moewusii*

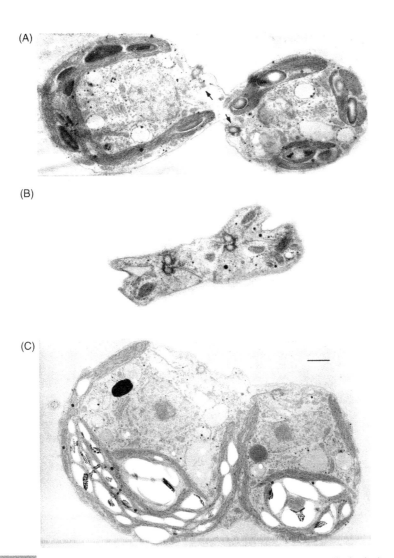

FIGURE 5.16 *Fusion of C. reinhardtii gametes. (A) Early zygote in stage just before the beginning of jackknifing. The fertilization tubule is no longer evident. This pair is in the trans position; note basal bodies. (B, C) Cross-sectional and longitudinal views respectively of zygotes in the early stage of flagellar coordination. The basal bodies of the flagellar pairs deriving from the two gametes are still relatively far apart. Within the common zygotic plasma membrane the region of protoplasmic confluence is extensive, but both gamete protoplasts retain their original identities and, as seen in lower figure, the cup-shaped plastid of each gamete, with starch granules and pyrenoid, still surrounds the nucleus of that gamete. The plus gamete is to the right in each figure. Bar = 1 μm. From Friedmann et al. (1968).*

may be related to altered orientation of the basal bodies, which become parallel and appear to lose their connection with one another. After 6–8 hours, motility ceases and cell walls are shed. Fibrous material (probably actin) in the connecting bridge disappears, the bridge widens, and the two gamete

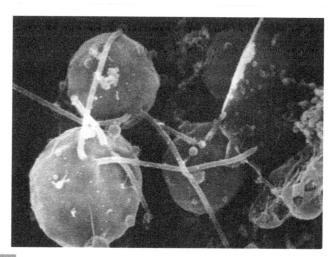

FIGURE 5.17 *Quadriflagellate cells of C. reinhardtii, resulting from fusion of gam5 plus cells with wild-type minus cells. The long flagella contributed by the wild-type cell are approximately 7.4 μm long, while those from gam5 are approximately 3.2 μm. From Forest (1982).*

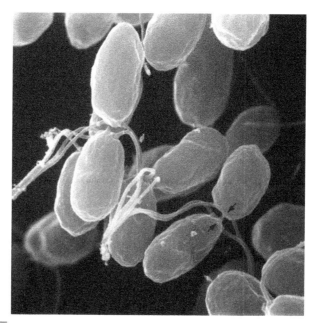

FIGURE 5.18 *Scanning electron micrograph of C. eugametos gametes just before cell fusion. Papillae are indicated by arrows. From Homan et al. (1987).*

cytoplasms begin to fuse (Triemer and Brown, 1975a). Chloroplast fusion in *C. moewusii* may precede nuclear fusion (Brown et al., 1968). Endoplasmic reticulum connecting the bridge to the outer nuclear membranes appears to shorten until the nuclei are juxtaposed and fuse.

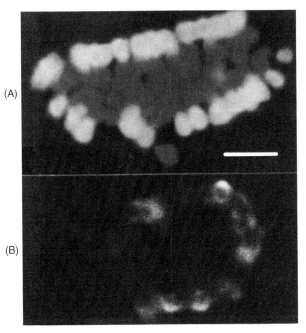

FIGURE 5.19 *Fluorescent images of C. eugametos plus and minus gametes during sexual agglutination in clumps. In both cases the plus gametes were labeled with fluorescein, but whereas in (A) all the cells are visible due to the autofluorescence of the chloroplasts, in (B) the autofluorescence was filtered out to reveal the positions of the plus cells only. The images are blurred due to the movement of the living cells and relatively long exposure times. Bar = 20μm. From Musgrave et al. (1985).*

G. Zygospore formation, maturation, and germination

Zygote-specific polypeptides are synthesized beginning in the QFC stage (Minami and Goodenough, 1978; Weeks and Collis, 1979). Ferris and Goodenough (1987) identified five zygote-specific mRNA transcripts that appear within 5–10 minutes of gamete fusion and a sixth transcript that appears after 90 minutes. New protein synthesis was required only for synthesis of the late transcript. In crosses between *C. smithii* and *C. reinhardtii*, a cDNA clone for one of the early transcripts hybridized to bands showing a restriction fragment length polymorphism with very tight linkage to the mating type locus. Some of the genes expressed early in zygote formation appear to have roles in the uniparental inheritance of chloroplast DNA (Ferris et al., 2002; Chapter 7).

Initiation of zygospore development is under control of the *plus*-specific GSP1 and *minus*-specific GSM1 homeodomain proteins, which form a heterodimer. When GSP1 is expressed ectopically in unmated *minus* gametes, zygote-specific genes are expressed, including *ZSP2*, which encodes a cell surface adhesion molecule (Zhao et al., 2001). Similarly, *plus* cells

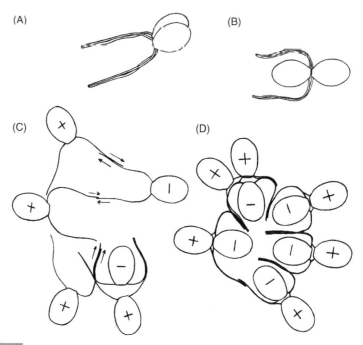

FIGURE 5.20 *Diagram illustrating how tipping and flagellar orientation in C. eugametos mating result in the alignment of papillae and the asymmetric orientation of the mating types in clumps. Tip-to-base contacts between flagella (C) are labile and likely to be broken by vigorous motion, whereas tip-to-tip and base-to-base contacts are more stable, allowing further adhesions over the rest of the flagellar surface and aligning the papillae. Alignment of plus and minus flagella does not permit contact between the papillae if both gametes hold their flagella forward (A). However, when the minus flagella flex back around their own cell body (B), papillar contact is facilitated. This characteristic orientation during the later stages of agglutination has the consequence that minus gametes become confined to the interior of agglutinating clumps and plus gametes on the outside (D). Since all the flagella of the pairing gametes are now oriented around the minus cell bodies, any secondary contacts made with other gametes automatically fix the minus gametes within the center of the clump. Agglutination eventually leads to cell fusion. From Musgrave (1987).*

transformed with *GSM1* driven by a constitutive promoter also express zygote-specific genes (Goodenough et al., 2007). Cells carrying constitutively expressed *GSP1* and *GSM1* express zygote-specific genes without nitrogen starvation or expression of other gamete-specific genes, indicating that these two proteins are sufficient to activate zygospore differentiation. Furthermore, vegetative diploid strains carrying both the *GSP1* and *GSM1* constitutively-expressed transgenes are able to form zygospores that then undergo normal meiosis after the usual maturation period (Goodenough et al., 2007; Lee et al., 2008).

The zygospore wall of *C. reinhardtii* forms by accumulation of fibrous material, with thick fibers being seen next to the cell surface, overlain by a continuous "dense layer," and with thin filaments connecting adjacent

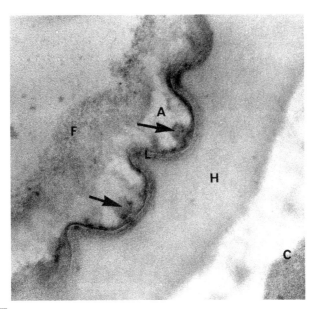

FIGURE 5.21 *Transverse section of a mature zygospore wall of C. reinhardtii. Features identified include a homogeneous layer (H) separated from the cell surface (C) by a pale area, three dense and two pale layers of the central lamina (L), dense knobs (arrows), an alveolate layer (A) and a fibrous layer (F). Triply stained gray section. Bar = 0.1 μm. From Cavalier-Smith (1976).*

zygospores (Minami and Goodenough, 1978). At low magnification a mass of zygospores appears as a reticulate layer (pellicle) on the surface of liquid medium. Observation of this very characteristic structure confirms mating, and is used as the basis of mating type tests (see Figure 8.3). Zygospores of *C. reinhardtii* increase greatly in volume during the first 24 hours after mating.

During the next 4–6 days of zygospore maturation, chloroplasts appear to disintegrate, chlorophyll is lost, and orange lipid-storage granules are accumulated. *C. reinhardtii* zygotes contain only a single pyrenoid (Cavalier-Smith, 1970), whereas in *C. moewusii* two are seen (Brown et al., 1968). Cavalier-Smith (1970) suggested that one of the two gametic pyrenoids of *C. reinhardtii* is probably degraded in the zygote but did not rule out the possibility that they could fuse.

In the light microscope, the zygospore walls of *C. reinhardtii* seem smooth (as first described by Goroschankin, 1891), whereas those of *C. moewusii* appear rough (Lewin, 1949). At the ultrastructural level, however, they are similar in organization, the most striking difference being in the spacing of the ridges (Figure 5.21), which accounts for their different appearance at low magnification (Cavalier-Smith, 1976). The zygospore wall of the homothallic species *C. monoica* has a more complex structure (Figure 5.22). This species is particularly suitable for genetic analysis of wall formation, since recessive mutations affecting wall formation can be

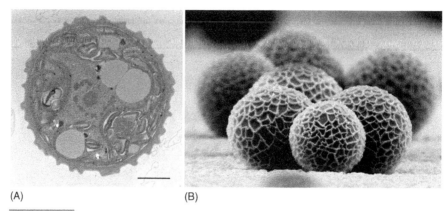

(A) (B)

FIGURE 5.22 *Transmission (A) and scanning (B) electron micrographs of mature zygospores of C. monoica. Scale bar = 2 μm. The left figure is similar to one published by VanWinkle-Swift and Rickoll (1997); the scanning electron micrograph is by Patricia Daniel and Karen VanWinkle-Swift, unpublished.*

identified by failure of a culture to produce normal zygospores (VanWinkle-Swift and Rickoll, 1997; VanWinkle-Swift et al., 1998).

IV. VEGETATIVE DIPLOIDS AND CYTODUCTION

Under normal laboratory conditions, as many as 1–5% of mated gamete pairs of *C. reinhardtii* do not form meiotic zygotes, but instead divide mitotically as vegetative diploids (Ebersold, 1967; see also Chapter 8). These cells begin to divide shortly after mating, without the maturation period needed for meiotic zygotes, and are recognizable under a dissecting microscope as discrete colonies, 3 or 4 days after plating a mating mixture. They can be selected deliberately using complementing auxotrophic markers and plating on minimal medium on which neither parental type can grow. While diploid strains have been very useful for geneticists, their utility to the alga is questionable, since they are effectively a terminal state, unable to undergo subsequent zygospore formation in response to environmental deprivation, and unable to mate with one another (since presence of the *MID* gene causes all naturally formed diploid cells to mate as *minus*).

A third alternative fate for a mated pair of gametes is cytoduction, the division of the newly formed zygote before nuclear fusion occurs (Matagne et al., 1991). In such cases, haploid cells can be recovered that carry the nuclear genome of one partner in the cross and the chloroplast or mitochondrial genome of the other. Such cells can be selected using appropriate genetic markers at a somewhat lower frequency than vegetative diploids. Bennoun et al. (1992) used this technique to transfer mutations in mitochondrial DNA into recipient strains with other genetic markers.

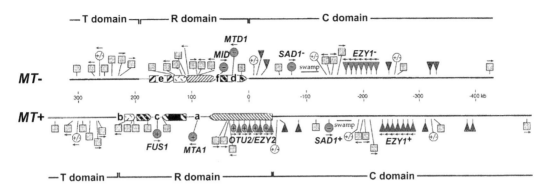

FIGURE 5.23 *Diagram of the mating type loci of minus and plus cells of C. reinhardtii, showing the telomere-proximal (T), rearranged (R), and centromere-proximal (C) domains. Letters (a–f) indicate regions within the R domain that are unique to either minus or plus cells. Genes indicated by circles are related to the mating process and are described further in Table 5.3; squares indicate the positions of genes expressed in vegetative cells, and triangles indicate genes expressed exclusively in zygotes. Short arrows show the direction of transcription where known. Modified from a figure in Goodenough et al. (2007) which in turn was based on the one by Ferris et al. (2002).*

V. GENETIC CONTROL OF SEXUALITY

A. The mating type locus

Mating type was recognized in early genetic studies with *C. reinhardtii* as a trait showing simple Mendelian inheritance, and was mapped by Ebersold et al. (1962) to linkage group VI. The observation that several mutations whose phenotypes had no obvious relationship to the sexual cycle were tightly linked to mating type led to the speculation that this might be a region of suppressed recombination (Gillham, 1969; Smyth et al., 1975; Ferris and Goodenough, 1994). A chromosome walk through 1.1 megabases confirmed that the *plus* and *minus* mating type loci differed by translocations, inversions, duplications, and deletions within a region of about 200 kb (Ferris and Goodenough, 1994), and a transcriptional map of the entire locus in both mating types was constructed (Ferris et al., 2002; Figure 5.23). Recombinational suppression extends beyond the rearranged (R) domain in both directions. Each mating type contains regions not found in the other. In addition to genes with essential roles in the mating process and in uniparental inheritance of organelle genes (Table 5.3), the mating type loci contain many genes expressed in vegetative cells (see Ferris et al., 2002, for a comprehensive list).

B. Genes not linked to mating type

As discussed above, genes encoded in the *plus* and *minus* mating type loci control the expression of genes present elsewhere in the genome. Some of

Table 5.3	Mapped genes involved in gametogenesis and mating		
Gene	**Linkage group**	**Protein function**	**References**
MID	VI; MT minus	Transcription factor, controls expression of minus genes	Ferris and Goodenough (1997); Lin and Goodenough (2007)
MTD1	VI; MT minus	Membrane protein, minus specific, exact function uncertain	Lin and Goodenough (2007)
SAD1−	VI; MT minus	minus agglutinin	Ferris et al. (2005)
EZY1−	VI; MT minus	Expressed in both mating types very soon after gamete fusion; chloroplast transit sequence	Armbrust et al. (1993); Ferris et al. (2002)
FUS1	VI; MT plus	Fringe glycoprotein of plus mating structure	Ferris et al. (1996); Misamore et al. (2003)
MTA1	VI; MT plus	plus specific, expressed in gametes and very young zygotes	Ferris et al. (2002)
OTU2	VI; MT plus	Related to otubain cysteine protease, expressed in plus gametes	Goodenough et al. (2007), citing unpublished work
EZY2	VI; MT plus	plus specific, expressed very soon after gamete fusion; chloroplast transit sequence	Ferris et al. (2002)
SAD1+	VI; MT plus	minus agglutinin; gene present in MT plus locus, but not expressed	Ferris et al. (2002)
EZY1+	VI; MT plus	Expressed in both mating types very soon after gamete fusion; chloroplast transit sequence	Armbrust et al. (1993); Ferris et al. (2002)
GCS1	Probably XVIII	minus-specific expression; transmembrane protein required for cell fusion	Mori et al. (2006)
GSM1	VII	minus gamete-specific homeodomain protein	Lin and Goodenough (2007); Goodenough et al. (2007)
GSP1	II	plus gamete-specific homeodomain protein	Kurvari et al. (1998); Wilson et al. (1999); Zhao et al. (2001)
SAG1	VIII	plus agglutinin	Ferris et al. (2005)

The mating type (MT) locus is on linkage group VI. For more detailed discussion and descriptions, see Goodenough et al. (2007), Ferris et al. (2002), and Volume 3, Chapter 12. This list excludes the NCG and NSG genes identified by expression during gametogenesis.

these sex-specific, but not sex-linked, genes were identified first by mutations (Table 5.4) and in some cases later characterized at the molecular level. Among these is the *SAG1* gene encoding the *plus* agglutinin, which maps to linkage group VII. *GCS1* (Generative Cell Specific) is a conserved gene encoding a transmembrane protein essential for fertilization in angiosperms, but also identified in *Chlamydomonas*, where it is expressed more strongly in *minus* cells than in *plus* (Mori et al., 2006). *GSP1* and *GSM1* are also outside of the mating-type locus, but are expressed only by *plus* and *minus* cells respectively (Table 5.3).

VI. PERTURBATIONS OF THE MATING PROCESS

Understanding of the interactive steps involved in mating has been greatly enhanced by chemical and enzymatic treatments that block specific events and by characterization of mutants that are unable to complete the sexual process. Deflagellation (by pH shock) prior to gamete mixing prevents agglutination, as would be expected. The regenerating flagella are motile and can agglutinate when about 25% of their final length has been reached, but complete mating (as measured by carbohydrate accumulation in the medium as a result of cell wall lysis) does not occur until 50% of flagellar length is attained (Solter and Gibor, 1977; Ray et al., 1978). Deflagellation after the first 2 minutes of mating in *C. reinhardtii* (by which time tip activation has occurred) does not block the remainder of the mating process. Formation of flagellar agglutinins requires protein synthesis and is sensitive both to proteolytic enzymes and to glycosylation inhibitors. Flagella can be artificially agglutinated with antiflagellar antibodies or concanavalin A and by isolated flagella or glutaraldehyde-fixed cells of the opposite mating type. Cells agglutinated by any of these means can proceed with cell wall lysis (Claes, 1977; Kaska and Gibor, 1982). Cells agglutinated by antibodies or isolated flagella can also activate mating structures, whereas those treated with concanavalin A or fixed cells cannot (Goodenough and Jurivich, 1978; Musgrave et al., 1979b; Mesland et al., 1980).

Treatment of gametes with colchicine or vinblastine does not affect flagellar motility or agglutination but blocks movement of adhesion sites to the flagellar tips, the morphological changes in the flagellar tips, and the subsequent cell wall lysis and mating structure activation (Mesland et al., 1980; Hoffman and Goodenough, 1980). Brief chymotrypsin treatment of gametes allows normal agglutination, tip locking, and cell wall lysis, although it inhibits flagellar membrane binding of polystyrene beads (Hoffman and Goodenough, 1980) and blocks mating structure activation (Mesland et al., 1980). Prolonged chymotrypsin exposure appears to destroy agglutinins (Wiese and Hayward, 1972; Mesland et al., 1980). Cytochalasins B and D also block activation, adhesion, and fusion of the mating structures (Mesland et al., 1980; Detmers et al., 1983). Treatment with sulfhydryl reagents interferes chiefly with adhesion and fusion of the mating structures (Forest, 1985).

Triple fusions are occasionally encountered in microscopic examination of *Chlamydomonas* matings, and a strain in which more than 5% of the mating events appeared to be triple fusions was described by Sager (1955). No systematic study of this aberration appears to have been made.

Table 5.4   Representative mutations affecting gametogenesis and mating

Gene	Mutant	Description	References
FUS1	imp1, fus	*plus* specific, unable to fuse because of defect in *plus* fringe protein	Goodenough et al. (1976); Matsuda et al. (1978, 1981); Saito and Matsuda (1984a); Ferris et al. (1996); Wang and Snell (2003)
SAG1	imp2,5,6,7, and 9	*plus* agglutinin	Goodenough et al. (1976, 1978); Adair et al. (1983)
SAG2	imp8, gag1	O-glycosylation of *plus* agglutinin and of cell wall proteins	Goodenough et al. (1978); Vallon and Wollman (1995)
	gam1	Expressed in *minus*; cells form a mating structure that appears normal but fails to activate	Forest and Togasaki (1975); Forest et al. (1978); Forest and Ojakian (1989)
	gam5	Unable to mate at restrictive temperature; flagellar axonemal structure is abnormal	Forest (1982, 1983b)
	gam10	Expressed in *minus*; cells can agglutinate and mating structures adhere but fusion does not occur	Forest (1983a); Forest (1987); Forest and Ojakian (1989)
SAD1	imp10, imp12, agl1	*minus* agglutinin	Hwang et al. (1981); Matsuda et al. (1988); Ferris et al. (2005)
MID	mid1, imp11	Mutation in *minus* mating type locus; mutant cells mate as *plus*	Goodenough et al. (1982); Galloway and Goodenough (1985); Ferris and Goodenough (1997)
	iso1	Sex-limited mutation unlinked to mating type; some but not all *minus* cells in a culture agglutinate as *plus* but form an abnormal *plus* mating structure and cannot fuse	Campbell et al. (1995)
	dif1, dif2	Mutants with temperature-conditional block in gametogenesis	Saito and Matsuda (1991)
LRG5	lrg2	Defective in blue light response step of gametogenesis, delayed in pregamete to gamete conversion	Buerkle et al. (1993); Glöckner and Beck (1995, 1997); Ermilova et al. (2003)
	lrg1	Mutant cells do not require light exposure for gametogenesis, and gametes do not respond chemotactically to ammonium	Buerkle et al. (1993); Glöckner and Beck (1995); Ermilova et al. (2003)
	lrg4	Mutant cells do not require light exposure for gametogenesis; also zygospores homozygous for *lrg4* do not require light for germination	Glöckner and Beck (1995)

Our understanding of the mating process in *Chlamydomonas* could not have advanced without the insights and often intriguing mysteries offered by mutants that fail to complete gametogenesis, agglutination, fusion, or zygospore formation. Goodenough et al. (1976) began their genetic analysis of mating with the isolation of a group of mutants in a *plus* background, which they named *imp* (for impotent) because they were unable to form zygotes (Table 5.4). A by-product of this screen was a group of mutants unable to mate because they lacked flagella (*bald*, now *bld* mutants), which have turned out to be very valuable for research on basal body formation and flagellar biogenesis (see Chapter 2, Table 2.5). Some of the original *imp* mutants were defective in agglutination and were sex-limited in expression; that is, although the mutated gene was not part of the mating type locus, the mutant phenotype was expressed only in *plus* cells. One group of five allelic *imp* mutants defined the *SAG1* locus, which was later identified as the structural gene for the *plus* agglutinin (Goodenough et al., 1978; Adair et al., 1983; Ferris et al., 2005). A second sex-limited locus, *SAG2*, represented by the original *imp8* mutant, proved to encode a protein required for O-glycosylation of the *plus* agglutinin (Goodenough et al., 1978; Vallon and Wollman, 1995). The *gag1* mutant, an allele at this locus, was identified by Vallon and Wollman by its altered cell wall proteins. *SAD1*, the gene encoding the *minus* agglutinin, was originally identified by the *imp10* and *imp12* mutations (Hwang et al., 1981).

To facilitate subsequent genetic analysis, Forest and Togasaki (1975) adopted the strategy of seeking temperature-conditional gametogenesis mutants. The first such mutant isolated, *gam1*, was sex-limited in expression; *minus* cells formed mating structures that appeared normal but were unable to activate. Additional *gam* mutants were described in subsequent publications by Forest and colleagues (Table 5.4).

Mutations that obviate the light requirement for gametogenesis have been isolated by Beck and colleagues (Buerkle et al., 1993; Glöckner and Beck, 1995). In several of these mutants, gametes also show loss of a chemotactic response to ammonium (Ermilova et al., 2003).

As discussed above, many additional genes have been identified by their expression during gametogenesis and mating. Targeted gene inactivation, insertional mutagenesis, and other tools will undoubtedly produce informative functional disruptions and mutants in these loci as well.

The Life of an Acetate Flagellate

I. INTRODUCTION

Pringsheim (1937) proposed the term "acetate flagellate" to describe the colorless *Polytoma*, which grows well on acetate as sole carbon source but cannot use glucose. Later authors (Hutner and Provasoli, 1951; Lloyd and Cantor, 1979) extended the designation to include an assortment of both green and colorless cells, some of which can also use pyruvate or lactate. The acetate flagellates generally have plasma membranes with low permeability to most organic substrates, with only small, lipid-soluble molecules showing good penetration. As a group, the acetate flagellates are able to tolerate low O_2 tension and high levels of CO_2. Pringsheim (1946b) used pieces of cheese covered with soil and water to enrich for colorless acetate flagellates, which grew well in the relatively high levels of fatty acids and alcohols produced by bacteria in this milieu. In nature, acetate flagellates are found in similarly metabolite-rich environments, including sewage, and species of *Chlamydomonas* are among these. As discussed in Chapter 1, the isolates of *C. reinhardtii* currently in laboratory use all came from soil, in most cases garden or agricultural soil that can be assumed to have been rich in organic compounds, some of them potentially toxic. Limited permeability to larger molecules, versatility in assimilation of elements such as nitrogen, phosphorus, and sulfur, and ability to tolerate large variations in hydration, are all clearly adaptive advantages in such an environment.

This chapter presents an overview of metabolic processes in *Chlamydomonas* from the perspective of its adaptation to this natural habitat. The chapter begins with a discussion of the nuclear genome, and the surprising diversity of genes that have been identified. Transcription and translation are reviewed briefly, to lay the foundation for understanding gene expression under varying environmental conditions. The second part of the chapter covers metabolism of carbon, nitrogen, sulfur, and other elements, and synthesis of starch and lipids. These topics are reviewed in much more detail in Volume 2, as are photosynthesis, respiration, and hydrogen production, which conclude this chapter.

II. NUCLEIC ACIDS AND PROTEIN SYNTHESIS

A. DNA species in the *Chlamydomonas* cell

Total DNA was extracted from *Chlamydomonas* cells in 1960 by Sueoka, in a paper that is cited much more frequently today as the reference for HS medium, and by Schwink, whose paper is noteworthy for its anticipation of the desirability of developing a transformation system for *Chlamydomonas*. Sueoka (1960) reported that the density and the denaturation temperature of total *C. reinhardtii* DNA were high, indicative of high GC content. This inference was confirmed for nuclear DNA by Sager and Ishida (1963), who calculated the GC content as 62.1%, a value that is in excellent agreement with the 64% determined from analysis of the genomic sequence (Merchant et al., 2007). Other *Chlamydomonas* species from which genes have been sequenced also have high-density DNA (http://www.kazusa.or.jp/codon/; query "Chlamydomonas"). However, Tetík and Zadrazil (1982) reported the GC content for *C. geitleri* nuclear DNA to be only about 44%. Presumably this information is of phylogenetic significance, but so few species have been analyzed that drawing conclusions is difficult. The high GC content of *C. reinhardtii* affects expression of foreign genes inserted by transformation (Chapter 8, section IX.D.4) and also complicates sequencing and PCR reactions.

The classic paper of Ris and Plaut (1962) provided the most convincing evidence at that time for the existence of chloroplast DNA in any organism by showing that chloroplasts of *C. moewusii* contained DNase-sensitive, Feulgen-positive regions. Acridine staining produced a yellow–green fluorescence, typical of DNA, rather than the orange color expected for RNA. Soon afterward Chun et al. (1963) demonstrated the presence in *Chlamydomonas* cells of a satellite band of lower density than main band DNA and showed that a similar satellite was enriched in chloroplasts from higher plants. Sager and Ishida (1963) reported that a chloroplast fraction from *C. reinhardtii* was as much as sevenfold enriched in the satellite band, which they determined had a GC content of 39.3%. As fractionation techniques improved, minor species were distinguished, including mitochondrial DNA (Ryan et al., 1978), a fraction that hybridized with cytoplasmic rRNA (Bastia et al., 1971; Howell, 1972; Sinclair, 1972), and several others whose identity remains uncertain (Chiang and Sueoka, 1967b; Dron et al., 1979; Chiang et al., 1981).

Chiang and Sueoka (1967a) estimated the total DNA content per haploid *C. reinhardtii* cell to be 7.2×10^{10} D, or 1.23×10^{-7} µg/cell. Lemieux et al. (1980) obtained estimates of about 1.8×10^{-7} µg/cell for *C. reinhardtii* and $1.9–2.5 \times 10^{-7}$ for *C. moewusii* and *C. eugametos*. Conflicting estimates were obtained for the fraction of unique vs. repetitive sequences in nuclear DNA (Wells and Sager, 1971; Howell and Walker, 1976). These issues have been largely resolved with the completion of the draft genome sequence. Kapraun (2007) can be seen for comparisons of nuclear DNA content of various algae, but the reader should be aware that outdated chromosome numbers are used for *C. reinhardtii*, and are additionally misleading because a diploid genome is assumed.

Quantitative preparation of chloroplast DNA is not always achieved, but most of the early studies are consistent in estimating that it accounts for about 14% of the total cellular DNA complement in haploid vegetative cells of *C. reinhardtii*, that is, about 1.72×10^{-8} µg/cell, and about half this amount in gametes (see Gillham, 1978; see also Whiteway and Lee, 1977, for data on diploid cells). Some variation in the percentage of chloroplast DNA content can be observed, depending on growth conditions and redox status (Lau et al., 2000). Mitochondrial DNA and the nuclear rDNA satellite probably represent less than 1–2% of the total. The rDNA satellite was estimated to contain on the order of 250–400 copies of the rRNA cistrons (Howell, 1972; Marco and Rochaix, 1980). Based on these figures, the total genome size was estimated to be a little less than 100 megabases. This is in reasonable agreement with the 121 megabases calculated from genome sequencing (Merchant et al., 2007).

B. The nuclear genome

1. Overview

From analysis of the draft genome sequence (Merchant et al., 2007) and EST data (Jain et al., 2007), the *Chlamydomonas* genome is predicted to contain about 15,000 protein-coding genes, which are more or less evenly distributed on the assembled scaffolds. A few gene-poor regions are marked by high AT content. Most genes have introns, which tend to be somewhat longer than those of other protists that have been studied, and may have arisen by transposon insertions. Merchant et al. (2007) calculated an average of 8.3 exons per gene.

Merchant et al. identified 1226 gene families, of which only 26 had 10 or more members. Of the 798 families with only two genes, more than a third consisted of two genes in tandem, a strong indication that they arose by duplication events. Sixty-one classes of simple repeat sequences were found.

2. Histones and chromatin structure

Histone proteins were extracted from isolated nuclei by Rizzo (1985) and characterized further by Waterborg et al. (1995; also Waterborg, 1998) and by Morris et al. (1999). Fabry et al. (1995) reported about 15 clusters of the nucleosomal *H2A*, *H2B*, *H3*, and *H4* genes, which were divergently transcribed in pairs from a short intercistronic region containing conserved promoter elements. The 3′ untranslated regions did not contain polyadenylation signals, and in this respect more closely resembled animal rather than plant histone genes. However, other associated sequences, including enhancer elements, were more similar to those of plants. Lindauer et al. (1993) also found both plant- and animal-like features in *Volvox* histone genes, which are very similar to those of *Chlamydomonas*. Laurens Mets has annotated about 40 histone clusters, and also some variant histones, in the version 3 genome sequence. Proteins of the histone deacetylase complex, histone methyltransferases, proteins of the SWI/SNF chromatin remodeling complex, and other proteins associated with chromatin structure have also been identified in the genome sequence by Carolyn Napoli and others.

3. Genes for rRNA

Marco and Rochaix (1980) identified the nuclear ribosomal DNA genes by hybridizing rRNA to restriction digests of nuclear DNA, and by electron microscopy of RNA/DNA heteroduplexes. The nuclear rDNA was determined to have a unit structure of 5.3×10^6D, consisting of 18S, 5.8S, and 25S rRNA genes separated by internal transcribed spacer sequences. These units appeared to be in tandem arrays, and were estimated to be repeated about 400 times in the nuclear genome. They form a satellite band of distinctive buoyant density on analysis of whole cell DNA (see above). The rDNA genes have been mapped to clusters on linkage groups I, VIII, and XV (Merchant et al., 2007; linkage group VIII was incorrectly stated as VII in this paper). The 5S rRNA of the cytosolic ribosomes is not transcribed as part of the rDNA unit. However, it also appears in three clusters in the nuclear genome, and is present as two distinct species, 121 and 122 nucleotides in length and differing in 17 bases (Darlix and Rochaix, 1981). Both 5S species resemble 5S rRNAs of higher plants rather than those of animals and fungi.

The internal transcribed spacer sequences (ITS1 and ITS2) have been very useful for phylogenetic studies within the Volvocales (Coleman and Mai, 1997) and in comparing isolates of *C. reinhardtii* (Pröschold et al., 2005).

4. tRNAs and codon bias

The high GC content of the nuclear genome of *Chlamydomonas* is of course reflected in the bias of codons in protein-coding genes. The Kazusa Institute maintains tables of codon usage that are periodically updated. The usage for the nuclear genome of *C. reinhardtii* is at http://www.kazusa.or.jp/codon/cgi-bin/showcodon.cgi?species=3055. Codon bias in plastid genes in various plants and algae, including *Chlamydomonas*, has been surveyed by Morton (1998).

The standard genetic code is used in all three genomes – nuclear, chloroplast, and mitochondrial (Kück and Neuhaus, 1986; Boer and Gray, 1988c) – in contrast to some protists in which mitochondrial UGA encodes tryptophan (Hayashi-Ishimaru et al., 1997; Inagaki et al., 1998; Ehara et al., 2000; Simpson et al., 2000). However UGA in a specific nuclear context was found to encode selenocysteine in several proteins in *Chlamydomonas*, including glutathione peroxidase, methionine sulfoxide reductase, and a homologue of animal selenoprotein W (Fu et al., 2002; Novoselov et al., 2002; H.Y. Kim et al., 2006). A Sec tRNA has been described (Novoselov et al., 2002; Rao et al., 2003). These reports were the first evidence for a plant selenoprotein, although UGA-encoded selenocysteine had previously been described in animals (reviewed by Novoselov and Gladyshev, 2003).

Highly expressed genes in unicellular eukaryotes appear to use a set of "preferred" codons, but the early data from which these conclusions were drawn were based largely on organisms with AT-rich genomes. Naya et al. (2001) used *Chlamydomonas* as a model to test whether this observation also held when the genome was GC-rich. As expected, they found the most frequently used codons to be those with C or G in the third position, but they found a significant difference when highly expressed genes were compared to ones expressed at low levels, with C being much more frequent in the highly expressed genes. For example, of the four codons for valine, GUU and GUA are infrequent, GUG is the most common overall, but GUC is used more than twice as often in highly expressed genes than in others, indicating that translational codon preference is operative in *Chlamydomonas* as in *Dictyostelium*, *Plasmodium*, and the other organisms previously studied.

This conclusion is also supported by work with higher plants. Wang and Roossinck (2006) included *Chlamydomonas* in a survey of codon preferences among plant species and reported that preferred codons generally had C or G in the third position, regardless of the overall GC content of the plant. Optimal usage was similar in monocots and dicots, and also conserved to a somewhat lesser extent between plants and *Chlamydomonas*.

In a comparison of coding sequences for 67 proteins from *C. incerta* with their counterparts in *C. reinhardtii*, Popescu et al. (2006) found that codon usage varied among genes in each of these two closely related species, but was highly correlated between the species for individual proteins.

Merchant et al. (2007) tabulated 259 genes encoding tRNAs in the *C. reinhardtii* genome sequence. Detailed information can be obtained in the supplementary material for this paper.

5. Small noncoding RNAs

In 2007, two laboratories independently reported finding microRNAs (miRNA and siRNA) in *Chlamydomonas* (Molnár et al., 2007; Zhao et al., 2007). These studies provided the first evidence of microRNAs in a unicellular eukaryote; none had been found in previous investigations of *Schizosaccharomyces pombe* and *Tetrahymena thermophila* (reviewed by Siomi and Siomi, 2007). Analysis of the *Chlamydomonas* genome sequence confirms the presence of Dicer and Argonaute proteins, the expected components of a microRNA-based silencing mechanism (Schroda, 2006).

Small nucleolar RNAs (snoRNAs) have also been identified in *Chlamydomonas* (Antal et al., 2000) and have been mapped in the genomic sequence (Merchant et al., 2007).

6. Transposons

One of the original motivations for seeking new natural isolates of *C. reinhardtii* was the hope of finding transposons that might be adapted for use in transformation, but in fact most of the transposons discovered to date have been found by accident in the course of doing other studies (Table 6.1). They have, however, proved to be useful as tags for cloning genes and as markers to assess relatedness of strains. Day et al. (1988) found the first *Chlamydomonas* transposon, *TOC1*, in characterizing the 5-kb insertion in the *FUD44* mutant, which interrupts the gene encoding the photosynthetic protein OEE1 (Mayfield et al., 1987a). Ferris (1989) found *Gulliver* as a 12-kb sequence in the mating type locus that appeared to be present in multiple copies in the genome. Both *TOC1* and *Gulliver* are widespread

Table 6.1	Transposons identified in *Chlamydomonas*		
Transposon	**Description**	**Strain in which identified**	**References**
TOC1	LTR retrotransposon, *DIRS01* type	First found in *FUD44* mutant in Ebersold/Levine background; widespread in natural isolates	Day et al. (1988); Day and Rochaix (1991a–c)
TOC2	Class II transposon, atypical	CC-2290 S1 D2 (Minnesota)	Day (1995)
Gulliver	Class II transposon, *Ac/Ds*-like	First found in CC-620 (Ebersold/Levine background); widespread in nature but absent from S1 D2 and North Carolina strains	Ferris (1989)
Pioneer1	Class II transposon, *Ac/Ds*-like	CC-2343 #224 (Florida); absent from the standard laboratory strain	Graham et al. (1995)
CRRE1	LTR retrotransposon, *gypsy*-like	Unknown	Kumekawa et al. (1999)
REM1	LTR retrotransposon, *gypsy*-like	21 gr	Pérez-Alegre et al. (2005)
Tcr1	Class II transposon	A54	Schnell and Lefebvre (1993); Ferris et al. (1996); Kim et al. (2006)
Tcr2	Class II transposon	A54	Schnell and Lefebvre (1993); Wang et al. (1998)
Tcr3	Class II transposon	A54	Wang et al. (1998)
Bill	Class II transposon, *Ac/Ds*-like	4A+	Kim et al. (2005a, 2006)
Dualen	Novel retro-transposon lacking long terminal repeats	Identified in genomic sequence (CC-503 *cw92*)	Kojima and Fujiwara (2005)
MRC1	Miniature retrotransposon, *TRIM* type	4A+	Kim et al. (2006)

For additional information on strain identifications, see Chapter 1. A54 is a nit+ strain ultimately derived from 21 gr. 4A+ is a derivative of the Ebersold/Levine 137c strain that was selected for rapid growth in the dark. CC-503, the wall-deficient strain used for genomic sequencing, is also in the Ebersold/Levine 137c background.

among *C. reinhardtii* isolates of different origin (Zeyl et al., 1994). However, Ferris (1989) found that *Gulliver* was absent from the Minnesota strain S1 C5 (which is indistinguishable from the more frequently studied S1 D2; see Chapter 1). An isolate from North Carolina also lacks *Gulliver* (Patrick Ferris and Clifford Zeyl, personal communications).

In CC-620, the strain in which it was discovered, *Gulliver* mapped to 12 different locations in the genome (Ferris, 1989). A survey of other strains in the same background (Ebersold/Levine strain, 137c) revealed that 11 of these insertion sites were conserved in most, and one was unique to CC-620. *C. smithii* and several other natural isolates of *C. reinhardtii* from diverse locations also had *Gulliver* insertions but in patterns that differed from those of the laboratory strain. However, four putative isolates from other locations (France, Caroline Islands, Florida, and one of unknown provenance) all had *Gulliver* profiles very similar to the standard laboratory strain (Ferris, 1989). Subsequent analysis of ribosomal ITS sequences supports the conclusion that these were incorrectly labeled cultures of the standard laboratory strain (Pröschold et al., 2005).

The minor variations in *Gulliver* location among the strains known to trace back to Smith's 1945 isolate (Chapter 1) suggested that at least some of the elements were mobile within the time frame of laboratory work. With this rationale, Schnell and Lefebvre (1993) screened spontaneous *nit2* mutants for insertion of the *Gulliver* and *TOC1* transposons, and were able to clone the *NIT2* gene using *Gulliver* as a tag. They also found mutants with previously unknown transposons that were later named *Tcr1* through *Tcr3*.

The 2.8-kb element *Pioneer1* was discovered in an isolate from Florida, and shown to be present in low copy number in strains from several other locations, but not in the standard laboratory strain (Graham et al., 1995). K.S. Kim et al. (2006) used the *AMT4* locus to search for additional transposons. Mutations in this gene account for about 80% of isolates resistant to methylammonium, a compound that slows growth of wild-type cells but does not kill them. Dark green resistant colonies appear over a period of weeks on a background of pale parental cells. Mutants can be induced by UV irradiation, but also appear spontaneously under selective pressure. The *Bill* element was discovered among a population of such mutants.

Two transposons related to the plant *gypsy* family were found by Kumekawa et al. (1999) and Pérez-Alegre et al. (2005) respectively. Pérez-Alegre et al. showed that their *REM1* transposon could be activated to amplify its copy number in genetic crosses, and during integration of foreign DNA. *Dualen*, the first representative of a new family of retrotransposons of phylogenetic interest, was identified by Kojima and Fujiwara (2005) by searching the various plant genomic sequences for possible transposons lacking long terminal repeats.

C. DNA synthesis

1. Replication of nuclear DNA in mitosis and meiosis

Sueoka (1960) first reported evidence from ^{15}N–^{14}N density shift experiments that nuclear DNA of *C. reinhardtii* was synthesized during the vegetative cell cycle in the semiconservative manner already described for bacteria by Meselson and Stahl (1958). The next few years saw publication of several important papers describing the details of synthesis of nuclear and chloroplast DNA components in relative to the vegetative cycle, gametogenesis, and meiosis.

Using growth on a 12:12 light:dark cycle under conditions that produced a high degree of synchrony, Kates and Jones (1967) confirmed that nuclear DNA synthesis occurs in a burst early in the dark phase (14–16 hours), just prior to nuclear division. A low rate of synthesis was observed beginning about the 11th hour of the light phase. Kates et al. (1968) observed that when cells underwent division into four daughter cells within a single cycle, DNA synthesis appeared to precede each nuclear division step. They also reported that gametogenesis entailed a program of DNA synthesis similar to that seen in vegetative cells, that is, synthesis of nuclear DNA immediately prior to cell division.

Analysis of DNA synthesis preceding meiosis is complicated by the difficulty of extracting cellular components from the hard-walled zygospore, and early studies left some questions about precisely when replication occurred, especially in zygotes that produced eight rather than four progeny cells on generation, presumably resulting from meiosis and a subsequent round of mitosis (Chiang and Sueoka, 1967a, b). These conflicts were resolved by Tan and Hastings (1977), and are reviewed by Harris (1989).

Although proteins with homology to DNA polymerase subunits have been identified in the *Chlamydomonas* genome sequence, as of this writing no thorough study has been published. Considerably more work has been done in characterizing enzymes involved in DNA repair (see below).

2. Replication of chloroplast DNA

In early experiments with ^{15}N–^{14}N density shifts, Chiang and Sueoka (1967a) had reported that most chloroplast DNA replication occurred during the light period and that replication was semiconservative. Catto and Le Gal (1972) similarly reported a peak of thymidine incorporation into relatively AT-rich DNA between 7 and 8 hours into the light period. Experiments by Lee and Jones (1973) indicated that in fact the density shift was gradual and that synthesis did not occur in a single burst. Subsequent work by Chiang and his colleagues (Chiang, 1971; Grant et al., 1978) showed that some incorporation of radioactive precursors occurred in both the light and the dark periods, but suggested that the incorporation in the dark period was the result of DNA repair rather than new synthesis. However, Turmel et al. (1980, 1981) found that net accumulation of chloroplast DNA occurred throughout the cell cycle with maximum synthesis taking place in the dark period, coincident with chloroplast division and cytokinesis.

The extent of dispersive labeling observed in their density transfer experiments led Turmel et al. (1981) to postulate that chloroplast DNA molecules engage in repeated heteroduplex formation whereby homologous single-stranded segments are exchanged. This process would also account for chloroplast DNA recombination in vegetative cells (see Chapter 7). Analyses of chloroplast DNA in crosses in which the parental DNAs are distinguishable by restriction site and length polymorphisms are consistent with this model: recombination appears to involve discrete, continuous segments of chloroplast DNA, rather than multiple interchanges between a single pair of genomes (Lemieux et al., 1984a, b; Newman et al., 1992).

Two replicative origins were identified in a single 0.42-kb region of the chloroplast genome by Wang et al. (1984) and were localized by electron microscopy of replicating molecules by Waddell et al. (1984). These were further characterized by Wu and colleagues (Wu et al., 1986a, b; Hsieh et al., 1991; Chang and Wu, 2000; see also Volume 2, Chapter 24). *Chlamydomonas* chloroplast sequences that promote autonomous replication in yeast were also identified (Loppes and Denis, 1983; Vallet et al., 1984; Houba and Loppes, 1985), and are distinct from the origins of replication (Vallet and Rochaix, 1985). Additional recombination-dependent origins were identified by Woelfle et al. (1993).

Two distinct chloroplast DNA polymerase activities were identified in *Chlamydomonas* (Keller and Ho, 1981; Wang et al., 1991), but were not been fully characterized in functional terms (see Volume 2, Chapter 24). Other proteins that have been identified include a DNA primase (Nie and Wu, 1999), and topoisomerases (Thompson and Mosig, 1985; Woelfle et al., 1993).

3. Recombination

As discussed in Chapter 3, *Chlamydomonas* chromosomes are difficult to visualize by light microscopy, and analysis of the events of meiosis is further complicated by the nearly impenetrable zygospore wall. Studies of recombination beginning in the 1950s focused therefore on inferences from segregation data in crosses. An early claim by Moewus that crossing over occurred *only* at the two-strand stage of meiosis was clearly refuted (Ebersold, 1956; Levine and Ebersold, 1958a, b) but questions remained about the extent of chiasma and chromatid interference (see Chapter 8, section VII.C.4).

Lawrence and colleagues published a series of papers between 1965 and 1970 describing the effects of various types of radiation on recombination frequencies (Lawrence, 1965a, b, 1970; Lawrence and Holt, 1970) and also the effects of inhibitors of DNA and protein synthesis (Davies and Lawrence, 1967; Lawrence and Davies, 1967).

Storms and Hastings (1975) found that two specific periods during zygospore germination were critical for recombination of nuclear genes. The first of these periods appears to be before meiotic prophase, probably at the time of DNA replication. The second critical period corresponds to meiotic pachytene, presumably a time of DNA repair in crossing over. Treatment of cells with nalidixic acid, hydroxyurea, or 5-fluorodeoxyuridine during this second period inhibited incorporation of radioactive phosphate or adenine into DNA, and reduced the recombination frequencies observed among the progeny (Chiu and Hastings, 1973). Rosen et al. (1980) reported enhanced recombination when zygotes were treated with caffeine during this period. Caffeine also improved survival of wild-type cells that had been irradiated with UV light, but had no effect on several UV-sensitive mutants now thought to be deficient in recombinational repair (see below).

Evidence for mitotic recombination in heterozygous diploid strains has been reported several times (Matagne and Orbans, 1980; Smyth and Ebersold, 1985; and in two earlier abstracts). Frequencies for the appearance of unselected mitotic recombinants appear to be on the order of 0.1–1% of colonies scored, depending on the selection conditions.

Relatively little has been published on nuclear recombination in *Chlamydomonas* since about 1990, but renewed interest in this topic seems likely as analysis of the genome reveals the probable proteins involved. A gene encoding a homologue of RAD51C, the eukaryotic equivalent to bacterial RecA, has been cloned and characterized (Shalguev et al., 2005).

4. DNA damage and repair

Use of *Chlamydomonas* as a system for study of DNA repair has been thoroughly reviewed by Vlček et al. (2008). This publication includes a complete list of homologues of known DNA repair genes that were identified in the version 3 nuclear genomic sequence as of summer 2007.

Beginning with work in the 1960s (Davies et al., 1969; Bryant and Parker, 1977, 1978; and see Harris, 1989, for review), survival curves were determined for various types of radiation, and mutants were isolated that were deficient in repair, especially of UV-induced damage (Davies, 1967a; Rosen and Ebersold, 1972; Vlček et al., 1981, 1987; Table 6.2). Other early studies included investigation of photosensitizing agents such as furocoumarins as mutagens (Schimmer, 1983; Schimmer and Abel, 1986; see also Schimmer and Kühne, 1990).

UV radiation produces pyrimidine dimers, which are removed primarily by photoreactivation, a light-dependent repair process mediated by enzymes known as photolyases. The *CPH1* gene encodes a plant-like class I photolyase, also known as cryptochrome, which has been studied primarily for its role as a blue light photoreceptor (Reisdorph and Small, 2004; Immeln et al., 2007; see also section II.E.2 and Volume 3, Chapter 13). A second cryptochrome, UVR3, appears to be more like animal proteins and may function in light-dependent DNA repair. Two class II photolyases, PHR1 and PHR2, are also implicated in repair of UV damage (Petersen and Small, 2001). PHR1 was identified initially by mutants deficient in photoreactivation (Cox and Small, 1985; Munce et al., 1993; Table 6.2) and appears to be necessary for full activity of PHR2 (Petersen and Small, 2001). The *PHR2* gene was cloned by Petersen et al. (1999).

Table 6.2	Representative mutants with alterations in DNA repair processes	
Mutant	Description	References
phr1	*PHR1* gene, encodes photolyase; mutant is defective in photoreactivation	Cox and Small (1985); Miadoková et al. (1991); Munce et al. (1993); Petersen and Small (2001)
rex1	*REX1* gene; UV-sensitive, also sensitive to methyl methanesulfonate; blocked in excision of pyrimidine dimers	Cenkci et al. (2003)
uvs9, uvs12, others	Blocked in excision of pyrimidine dimers	Vlček et al. (1995, 1997)
uvs10	Blocked in recombinational repair	Rosen and Ebersold (1972); Vlček et al. (1987, 1995)
uvs11	UV-sensitive, probable defect in cell cycle checkpoint	Vlček et al. (1981, 1987); Slaninová et al. (2002, 2003)
uvs13, uvs14	Probably blocked in mismatch repair	Vlček et al. (1995, 1997)
uvs15	Deficient in error-prone repair	Vlček et al. (1997)
uvsE1	Shows reduced recombination frequency in crosses, implicated in recombinational repair	Rosen and Ebersold (1972); Portney and Rosen (1980); Vlček et al. (1991)
uvsE5	Implicated in recombinational repair but not associated with meiotic recombination	Rosen and Ebersold (1972); Portney and Rosen (1980); Vlček et al. (1991)
uvsE6	Implicated in recombinational repair	Rosen and Ebersold (1972); Portney and Rosen (1980); Vlček et al. (1991)
uvr1	UV-resistant	Vlček et al. (1981)

DNA damage by alkylating mutagens such as nitrosoguanidine (MNNG), methylmethanesulfonate, and methylnitrosourea is reversed in bacteria, yeast, and animals by alkyltransferases that seem to be lacking in plants and in *Chlamydomonas* (Frost and Small, 1987). A DNA methyltransferase homologue has been identified in the *Chlamydomonas* genome (Vlček et al., 2008) but has not yet been demonstrated to have a role in this repair process. The discovery that some UV-sensitive mutants are also sensitive to alkylating agents suggests that these bases may be removed by excision repair instead (Sweet et al., 1981; Miadoková et al., 1994).

Several glycosylases that could be involved in base excision repair have been identified, but this process has not yet been fully characterized in *Chlamydomonas* (Vlček et al., 2008). Nucleotide excision repair, requiring strand separation, excision of an oligonucleotide, and repair by DNA synthesis, has been investigated more extensively. Beginning with isolation of mutants defective in "dark repair" (Davies, 1967a, b), at least eight genetic loci have been identified that probably represent components of this process (Table 6.2, Vlček et al., 2008), but most of these have not yet been correlated with specific proteins. The *REX1* gene was cloned by Cenkci et al. (2003) after isolation of an insertional mutant, *rex1*, that was blocked in excision of cyclobutane pyrimidine dimers and was also sensitive to alkylating agents. The gene is actually dicistronic, encoding an 8.1-kD protein (REX1-S), and a 31.8-kD protein (REX1-B). Both proteins have homologues in other organisms, but they are encoded by separate genes. Other genes likely to be involved in nucleotide excision repair have been identified in the genome (Vlček et al., 2008) but are as yet incompletely characterized. Among a group of mutants hypersensitive to alkylating agents, Sarkar et al. (2005) identified one deficient in cytosolic thioredoxin h1, previously implicated in responses to oxidative stress (see Volume 2, Chapter 11).

Other aspects of DNA repair are also under investigation. The mutants *uvs13* and *uvs14* have been implicated in mismatch repair pathways (Vlček et al., 1997). Both mutants map to linkage group I, as does the gene encoding the MutL homologue *MLH1*, but an equivalence has not yet been demonstrated. Some of the first UV-sensitive mutants isolated also showed deficiencies in recombination, and have therefore been implicated in recombinational repair (Rosen and Ebersold, 1972; Rosen et al., 1980; Portney and Rosen, 1980; Vlček et al., 1987). At least four distinct loci have been identified (Podstavková et al., 1994; Table 6.2). Moharikar et al. (2006, 2007) have characterized UV-C-induced apoptotic cell death in *Chlamydomonas*.

5. Recombination and repair of chloroplast DNA

By transforming *Chlamydomonas* cells with wild-type and mutant versions of the *E. coli recA* gene flanked by *Chlamydomonas* sequences, Cerutti et al. (1995) provided convincing evidence that chloroplast recombination is mediated by a RecA-type system, and postulated that its major function is in repair. A nuclear-encoded prokaryotic-type RecA homologue has since been found, and shown to function in the chloroplast (Nakazato et al., 2003). Recombination of chloroplast DNA is discussed in a genetic context in Chapter 7, and with regard to chloroplast transformation in Chapter 8.

D. The machinery of transcription and translation

1. RNA polymerases

Dusek and Preston (1988) partially purified an RNA polymerase activity and obtained mutants resistant to amatoxin, but this work was not continued. Genes for the expected RNA polymerase I, II, and III proteins have been annotated in the version 3 genome sequence but as of this writing no detailed descriptions have been published. In a search for intein-coding sequences in RNA polymerase II genes, Goodwin et al. (2006) identified one such sequence in *Chlamydomonas*, the first nuclear-encoded intein to be discovered outside of the fungi. An intein had previously been characterized from the chloroplast *clpP* gene of *C. eugametos* (Wang and Liu, 1997, and see Volume 2, Chapter 19). The intein in the *Chlamydomonas RPB2* gene was found by a BLAST search of the EST database with the *Candida tropicalis* ThrRS intein as a query, and the *RPB2* gene was then characterized more fully.

2. Cytosolic ribosomes

A definitive proteomic analysis and cryoelectron microscopic study of the *Chlamydomonas* 80S ribosome was published by Manuell et al. (2005). Both the overall composition and the sequences of individual proteins are highly conserved with comparison to yeast, mammalian, and plant ribosomes, and Manuell et al. adopted the nomenclature system used in mammals, as has also been done for yeast and *Arabidopsis*. This supersedes an earlier convention for *Chlamydomonas* proteins based on electrophoretic mobility (Fleming et al., 1987a, b). A complete set of orthologues to the mammalian ribosomal proteins was identified either by proteomic analysis or, for a few proteins, from EST and genomic sequences. The protein set includes the signaling protein RACK1, which had previously been described from *Chlamydomonas* as the G-protein beta subunit Cblp (Schloss, 1990). The acidic stalk protein P3, found in higher plants, is missing in *Chlamydomonas*.

The high degree of structural similarity between *Chlamydomonas* 80S ribosomes and those of both plants and animals must also be manifest in function. Genes encoding initiation and elongation factors have been annotated in the genome, primarily by Sara Zimmer, but as of this writing no descriptive paper has been published.

3. Chaperones and protein targeting

Wild-type *Chlamydomonas* cells tolerate temperatures as high as 32–34°C without obvious detriment, but many mutant strains are known that are unable to grow at this temperature, and others are viable but have flagellar defects or blocks in mating. Temperatures in the 38–42°C range induce heat shock responses, and can be lethal on prolonged exposure. As in other organisms, high temperature induces synthesis of a characteristic set of heat shock proteins. These were first described by Kloppstech et al. (1985), who characterized a 22-kD protein, now known as HSP22A. Later work showed that this protein does not enter the chloroplast, as was originally thought, but becomes associated with the chloroplast outer envelope during the heat shock response, and later dissociates (Eisenberg-Domovich et al., 1994). Eight *HSP22* genes have been identified in the genome sequence. The HSP22E and HSP22F proteins are chloroplast targeted and can be induced by starvation for sulfur (Zhang et al., 2004) or phosphorus (Moseley et al., 2006), or by oxidative stress (Fischer et al., 2005). A single gene has been found in *Chlamydomonas* for HSP33, another protein likely to be induced by oxidative stress.

Genes for four proteins of the HSP60 family have been identified. CPN60A, CPN60B1, and CPN60B2 are chloroplast targeted, whereas CPN60C is targeted to the mitochondria. Balczun et al. (2006) showed that CPN60A could interact with group II intron RNA, and suggested that it might be a general organelle splicing factor.

Genes for three light-regulated HSP70 family proteins were cloned by von Gromoff et al. (1989). One of these genes, *HSP70A*, was characterized further by Müller et al. (1992) and its light induction and regulation have been investigated (Kropat et al., 1995, 2000; Kropat and Beck, 1998). It appears to be the only cytosolic DnaK-type HSP70 in *Chlamydomonas*. Use of the *HSP70A* promoter for transgene expression is discussed in Chapter 8, section IX.D.3 and in Volume 2, Chapter 19. HSP70B functions in the chloroplast, and can be induced by heat shock or by high light intensity. It appears to function in photoinhibition (see Chapters 19 and 23 in Volume 2). HSP70C, also inducible by heat shock or light, is associated with the mitochondria. The *HSP70E* and *HSP70G* genes encode proteins of the HSP110/SSE family, which interact with cytosolic DnaK-type HSP70s. Genes designated *HSP70D* and *HSP70F* have also been annotated in the genome, but are not supported by EST data. A phylogenetic analysis of HSP70 proteins from several diverse algae, including *Chlamydomonas*, was published by Renner and Waters (2007). Co-chaperone proteins interacting with HSP70 proteins are discussed in Volume 2, Chapter 19.

Three genes encoding HSP90 proteins have been identified. HSP90A is cytosolic and was found in the flagellar proteome (Pazour et al., 2005). HSP90B is probably targeted to the endoplasmic reticulum, and the gene encoding this protein is flanked in the genome by *BIP1* and *BIP2*, which code for ER-targeted HSP70 homologues (Schroda, 2004). HSP90C functions in the chloroplast, where it interacts with HSP70B (Willmund and Schroda, 2005).

The *Chlamydomonas* genome contains two *HSF* genes encoding transcription factors regulating synthesis of heat shock proteins; one of these, *HSF1*, has been cloned (Schulz-Raffelt et al., 2007). Downregulation of *HSF1* by RNAi causes cells to become very sensitive to heat stress.

Ten genes have been identified encoding proteins of the HSP100/Clp family: three ClpX/Y, five ClpB, and one each of the ClpC- and D-types. These are discussed at greater length in Volume 2, Chapter 19, as are FtsH, Lon, and DegP proteases, and other enzymes involved in protein processing. Import of proteins into organelles is discussed in Chapter 7, section V.

E. Gene expression

1. Transcriptional controls

Every field of *Chlamydomonas* research of course includes studies on the regulation of expression of individual genes. This section provides only a brief summary of the major features. As a broad generalization, nuclear gene expression is regulated primarily at the transcriptional level, whereas translational control has a more important role in synthesis of chloroplast proteins (see Chapters 25 and 27 in Volume 2).

The Plant Transcription Factor database (http://www.plntfdb.bio.uni-potsdam.de/v2.0/) includes *Chlamydomonas* as its representative chlorophyte. Transcription factors of the MYB family have been described by Rubio et al. (2001) and by Yoshioka et al. (2004). DOF family factors in algae, including *Chlamydomonas*, and in plants were surveyed by Moreno-Risueno et al. (2007) and by Shigyo et al. (2007), and the WRKY family was examined by Zhang and Wang (2005).

Gene silencing in chromatin was studied by van Dijk et al. (2005), and RNA-mediated silencing was reviewed by Cerutti and Casas-Mollano (2006; see also Casas-Mollano et al., 2007); see also section II.B.4. for microRNAs, and Chapter 8, section IX.D.6.

2. Regulation by blue light

The role of blue light in circadian rhythms and in gametogenesis has been mentioned briefly in Chapters 3 and 5, respectively. A more extensive list of processes with blue light control is given in Table 6.3. Two classes of blue light photoreceptors have been identified, cryptochromes (Small et al., 1995), and phototropins (Huang et al., 2002). The cryptochrome CPH1 resembles plant proteins and has been implicated in circadian control (Reisdorph and Small, 2004; Mittag et al., 2005; see also Banerjee and Batschauer, 2005 for review). Phototropins are proteins with a C-terminal serine–threonine kinase and two blue light sensitive domains, called LOV1 and LOV2 (see Volume 3, Chapter 13). Only a single phototropin, PHOT, has been identified in *Chlamydomonas*. This protein is found in the flagellar proteome (Pazour et al., 2005) but appears to have a more general role in controlling gametogenesis and zygospore formation (Huang and Beck, 2003; Huang et al., 2004; and see also Volume 3, Chapter 12). It may also be involved in expression of photosynthesis-related genes (Im et al., 2006).

3. Stress conditions

The ability of *Chlamydomonas* cells to adapt quickly to changing environmental conditions, and the extensive repertoire of enzymes and metabolic pathways that it has preserved through evolution, make it an outstanding model system for studies of gene expression under stress conditions. Studies of stress responses have been greatly facilitated by the construction of cDNA libraries from wild-type cells grown under different stress conditions (listed in the Appendix, see also Grossman et al., 2007), and microarrays prepared from those libraries (Eberhard et al., 2006).

Table 6.4 summarizes some representative papers. Oxidative stress is discussed in more detail in Volume 2, Chapter 11, and high light stress and

Table 6.3	Selected references on regulation by blue light
Process	**References**
Gametogenesis	Glöckner and Beck (1997); Pan et al. (1996, 1997); Dame et al. (2002); Huang and Beck (2003)
Cell division	Münzner and Voigt (1992); Oldenhof et al. (2004a,b)
Chlorophyll synthesis	Matters and Beale (1995); Im et al. (1996)
Carotenoid synthesis	Bohne and Linden (2002)
Chlorophyll and carotenoid synthesis; LHC apoproteins	Im et al. (2006)
Expression of LHC-like *LHL4* gene	Teramoto et al. (2006)
Carbonic anhydrase expression	Dionisio et al. (1989); Dionisio-Sese et al. (1990)
Activation of nitrate reductase	Azuara and Aparicio (1983)
Nitrite transport	Quinones et al. (1999)
Chemotaxis in gametes	Ermilova et al. (2004, 2007)
Gravitaxis	Sineshchekov et al. (2000)
Prevention of dark lethality	Thompson et al. (1985); Voigt and Münzner (1994)

Table 6.4	Selected references on stress responses
Condition	**References**
Oxidative stress	Hanikenne et al. (2001); Yoshida et al. (2003); Baroli et al. (2003, 2004); Cohen et al. (2005); Förster et al. (2005); Molen et al. (2006), Vega et al. (2006), Leisinger et al. (2001); Lodford and Niyogi (2005); Fischer et al. (2006); Ledford et al. (2007)
High light intensity	Fischer et al. (2005, 2006, 2007); Förster et al. (2005, 2006)
Anaerobiosis	Quinn et al. (2002); Mus et al. (2007)
Osmotic stress	Cruz et al. (2001); Yoshida et al. (2004); Hoffman and Beck (2005)
Phosphate deprivation	Shimogawara et al. (1999); Chang et al. (2005); Moseley et al. (2006); Yehudai-Resheff et al. (2007)
Sulfur deprivation	Davies et al. (1996); Antal et al. (2004); Chang et al. (2005); Pollock et al. (2005)
Copper deficiency	Quinn and Merchant (1998); Moseley et al. (2000)
Iron deficiency	Allen et al. (2007b); Fei and Deng (2007); Naumann et al. (2005, 2007)
Manganese deficiency	Allen et al. (2007a)

photoinhibition are covered in Volume 2, Chapter 23. Oxygen deprivation induces fermentative pathways, including production of molecular hydrogen (Volume 2, Chapter 7). Other chapters in Volume 2 deal with shifts in availability of carbon, sulfur, phosphorus, and metals. The review by Grossman (2000a) provides a good summary of responses to nitrogen, sulfur, and phosphorus deprivation.

III. METABOLIC PROCESSES

A. Carbon metabolism

1. Carbon sources

Not all species of *Chlamydomonas* are able to grow in the dark on acetate (Table 6.5), and undoubtedly one of the principal reasons for the ascendancy of *C. reinhardtii* over *C. eugametos* and *C. moewusii* as a research organism was its ability to do so, making possible the isolation of numerous non-photosynthetic mutants. *C. eugametos* and *C. moewusii* are nevertheless capable of using exogenous acetate in the light (Gowans, 1976a), and mutants that require acetate or other organic substrates for growth in the light have been isolated in these species. A few species, such as *C. nakajaurensis* (Bennett and Hobbie, 1972), *C. acidophila* (Erlbaum, 1968), *C. pseudagloë*, and *C. pseudococcum* (Lucksch, 1932), have been reported to be able to use glucose as their sole carbon source, but *C. reinhardtii* cannot do this. Utilization of exogenous ribose by *C. reinhardtii* was reported by Patni et al. (1977), but studies by Sager and Granick (1953) indicated that this compound would not support growth in the dark. The other compounds tested by Sager and Granick that were nontoxic in the light but did not support growth of *C. reinhardtii* in the dark were as follows: glucose, galactose, sucrose, lactose, maltose, mannose, D-xylose, L-arabinose, ethanol, glycerol, mannitol, formate, glycerophosphate, proprionate, butyrate, formaldehyde, oxalate, tartrate, pyruvate, malate, fumarate, succinate, α-ketoglutarate, citrate, *trans*-aconitate, glutamine, glutamate, asparagine, aspartate, and glycine. All were tested at 0.01 M concentration. Since the metabolic pathways to utilize these compounds presumably exist, and genes encoding many of the expected enzymes have been identified in the genomic sequence, the obstacle to their assimilation is probably lack of active uptake into the cell.

Despite all the years that *Chlamydomonas* has been grown on acetate medium in the laboratory, the precise mechanism by which acetate is assimilated is still uncertain. The consensus is that it must first be incorporated into acetyl coenzyme A (acetyl-CoA), from which it could then enter the glyoxylate cycle, in which acetyl-CoA and glyoxylate are condensed to form malate, which is then converted to oxaloacetate. A second acetyl-CoA enters to form citrate, which is converted to isocitrate, from which isocitrate lyase generates succinate and glyoxylate (see Volume 2, Figure 8.2).

Table 6.5	Utilization of acetate for heterotrophic growth by *Chlamydomonas* species
Species	**References**
Growth in the dark on acetate	
C. angulosa [C. debaryana]	Hellebust et al. (1982)
C. debaryana	Lewin (1954b)
C. dorsoventralis	Lucksch (1932)
C. dysosmos	Lewin (1954b)
C. globosa	Pimenova and Kondrat'eva (1965)
C. humicola	Laliberte and De La Noue (1993)
C. komma [C. debaryana]	Harris (1989)
C. monoica	Lucksch (1932)
C. pallens	Pringsheim (1962)
C. pseudagloë	Lucksch (1932)
C. pseudococcum	Lucksch (1932)
C. pulchra [C. callosa]	Lucksch (1932)
C. pulsatilla (New Brunswick)	Hellebust and Le Gresley (1985)
C. reinhardtii	Sager and Granick (1953)
C. spreta	Droop and McGill (1966); Turner (1979); Lucksch (1932)
Poor or no growth in the dark on acetate	
C. chlamydogama	Hutner and Provasoli (1951)
C. elliptica var. *britannica [C. culleus]*	Harris (1989)
C. eugametos	Wetherell (1958)
C. euryale	Coughlan (1977)
C. frankii [C. culleus]	Harris (1989)
C. humicola [C. applanata]	Lucksch (1932)
C. incerta	Harris (1989)
C. intermedia [C. segnis?]	Harris (1989)
C. iyengarii [C. sphaeroides]	Harris (1989)
C. melanospora	Lewin (1975)
C. moewusii	Lewin (1950); Bernstein (1968)
C. orbicularis	Harris (1989)
C. philotes	Lewin (1957a)
C. pulsatilla (Finland, Scotland)	Droop (1961); Turner (1979)
C. smithii	Harris (1989)
C. typica	Harris (1989)

The first species name is the one originally published. Names in brackets are revised nomenclature based on molecular studies of Pröschold et al. (2001) and others; see Chapter 1.

Isocitrate lyase and malate synthase, which are unique to this cycle, have both been characterized in *Chlamydomonas* (Petridou et al., 1997; Nogales et al., 2004). In plants this cycle takes place in specialized structures known as glyoxysomes, but their existence in *Chlamydomonas* remains unconfirmed (Giraud and Czaninski, 1971, and see Chapter 2).

Acetyl-CoA can be formed directly through acetyl-CoA synthetase, and also in a two-step process by the enzymes acetate kinase and phosphate acetyltransferase. Acetate kinase activity was characterized in *Chlamydomonas* by Faragó and Dénes in 1968, but this paper seems to have been forgotten or neglected, and the general assumption was that this enzyme was restricted to bacteria until it was rediscovered in *Chlamydomonas*, several fungi, and some other protists by BLAST searches of genomic sequences (Ingram-Smith et al., 2006). Phosphate acetyltransferase has also been identified in the *Chlamydomonas* genome (see Volume 2, Chapter 8).

2. Uptake of inorganic carbon

Carbon availability as CO_2, HCO_3^-, and CO_3^{2-} fluctuates in the environment depending on pH and hydration, and is also affected by respiratory output and photosynthetic assimilation by living organisms. As an organism that may alternate between soil and aquatic habitats, *Chlamydomonas* has a complex system for using whatever inorganic carbon is available, and the ability to concentrate it internally. These processes, known collectively as the carbon concentrating mechanism (CCM), have been extensively studied. Moroney and Ynalvez (2007) have proposed a model for the function of this mechanism in *Chlamydomonas*. Collins et al. (2006a, b; 2007) have used *Chlamydomonas* as a model for evolutionary adaptation of algae to changing environmental CO_2 levels. For additional review of inorganic carbon metabolism, see Spalding et al. (2002), and Volume 2, Chapter 8.

A shift to limiting CO_2 conditions induces expression of more than 50 genes, encoding proteins directly involved in the CCM process (which include seven carbon transport proteins and four different carbonic anhydrases), an Rh protein thought to act as a CO_2 channel (Soupene et al., 2002, 2004; Kustu and Inwood, 2006), and other proteins that have no obvious link to CO_2 uptake or concentration, but may be stress related (Im and Grossman, 2001; Miura et al., 2004). Starch accumulates around the pyrenoid, and mitochondria become more peripherally located (Geraghty and Spalding, 1996).

Three α-carbonic anhydrases and six β-carbonic anhydrases have been identified in *Chlamydomonas* (see Mitra et al., 2005, also Volume 2, Chapter 8). CAH1 and CAH2 are α-carbonic anhydrases located in the periplasm between the plasma membrane and the cell wall; CAH3 is an α-carbonic anhydrase in the thylakoid lumen; CAH4 through CAH9 are β-carbonic anhydrases, two mitochondrial, one in the plastid stroma, and the others as yet not definitely localized. *CAH1*, *CAH3*, *CAH4*, and *CAH6* are significantly upregulated by limiting CO_2. Three genes encoding putative γ-carbonic anhydrases have also been annotated in the genome sequence.

The *cia5* mutant was first identified by its slow growth in limiting CO_2 (Moroney et al., 1989), and was found to show none of the changes in gene expression associated with adaptation to low CO_2 (Marek and Spalding, 1991; Spalding et al., 1991). Further work showed that it was a mutation in what appeared to be the master gene controlling induction of the carbon concentrating mechanism, which was given the name *CCM1* (Fukuzawa et al., 2001). The *CCM1* gene product is a hydrophilic protein with N-terminal zinc finger motifs, a central Gln repeat region, and a C-terminal glycine-rich region typical of transcription factors (Xiang et al., 2001; Fukuzawa et al., 2001; Kohinata et al., 2008). Other mutations affecting uptake and assimilation of inorganic carbon are listed in Table 6.6.

Photosynthetically assimilated carbon is incorporated by the C3 carbon reduction cycle, in which Rubisco catalyzes the carboxylation of ribulose-1,5-bisphosphate to produce 3-phosphoglyceric acid. When CO_2 is limiting, photorespiration may occur, in which oxygenation of Rubisco produces phosphoglycolate, which is metabolized to produce 3-phosphoglyceric acid (PCO cycle; Tolbert, 1979). Glycolate may also be excreted, especially when photorespiration is first induced (Tolbert et al., 1983; Moroney et al., 1986a).

In contrast to land plants, however, where photorespiration involves glycolate oxidase in peroxisomes, in *Chlamydomonas* this process is accomplished in the mitochondria using glycolate dehydrogenase (Nakamura et al., 2005; Tural and Moroney, 2005). Peroxide is not formed, and electrons enter the mitochondrial respiratory chain by reduction of the ubiquinone pool.

B. Nitrogen metabolism

1. Nitrogen sources

Most algae use ammonium in preference to nitrate as a nitrogen source; that is, if both ions are present, nitrate will not be utilized until the ammonium is exhausted (Syrett, 1962; Thacker and Syrett, 1972a). A few species of *Chlamydomonas* have been reported to show no preferential ammonium utilization (Cain, 1965). Except for the Ebersold/Levine branch of the standard laboratory strain, which carries two mutations blocking nitrate reductase activity (Chapter 1), most *Chlamydomonas* species are in any case capable of assimilating nitrate, nitrite, and ammonium as sole nitrogen sources, as well as any of a number of other compounds (Table 6.7). *C. reinhardtii* will also grow on urea, uric acid, acetamide, glutamine, ornithine, arginine, hypoxanthine, allatoin, allantoic acid, guanine, and adenine (Sager and Granick, 1953; Cain, 1965; Gresshoff, 1981; Pineda et al., 1984). No species in Cain's survey was able to use cytosine, thymine, or uracil as its sole nitrogen source.

	Table 6.6	Representative mutants with alterations in uptake or assimilation of inorganic carbon	
Gene	**Mutant**	**Description**	**References**
CCM1	cia5, C16	"Master regulator" of carbon concentrating mechanism; mutants do not respond to limiting CO_2	Moroney et al. (1989); Marek and Spalding (1991); Spalding et al. (1991); Xiang et al., (2001); Fukuzawa et al. (2001)
CAH1		Periplasmic carbonic anhydrase; null mutant isolated, but was not impaired in photosynthesis or CCM adaptation	Fukuzawa et al. (1998); Van and Spalding (1999)
CAH3	ca1, cia1, cia2, cia3	Carbonic anhydrase of thylakoid lumen; bicarbonate accumulates in mutants, but cannot be assimilated	Spalding et al. (1983); Moroney et al. (1986b); Suzuki and Spalding (1989); Funke et al. (1997); Karlsson et al. (1995, 1998); Hanson et al. (2003)
LCR1	lcr1	Myb-type transcription factor; mutant is defective in regulation of CAH1	Yoshioka et al. (2004)
LCIB	pmp1, ad1	Chloroplast protein; mutants die at atmospheric levels of CO_2, but can grow in high CO_2 and in very low CO_2	Wang and Spalding (2006)
RCA1	rca1	Rubisco activase; insertional mutant requires high CO_2 for growth	Pollock et al. (2003)
GYD1	gdh1	Glycolate dehydrogenase, predicted mitochondrial, PCO cycle; mutant requires high CO_2 for growth	Nakamura et al. (2005)
PGP1	pgp1	Phosphoglycolate phosphatase; mutant requires high CO_2 for growth	Mamedov et al. (2001); Nakamura et al. (2005)
ycf10 (cemA)		Chloroplast-encoded protein required for inorganic carbon transport; insertional mutants show decreased CO_2 and bicarbonate uptake	Rolland et al. (1997)

Isolates of the marine species *C. pulsatilla* from three different locations were reported to be unable to use nitrate as a sole nitrogen source, and only one of these three could use ammonium (Droop, 1961; Hellebust and Le Gresley, 1985). Each could use some amino acids or other organic nitrogen sources. In nature *C. pulsatilla* is found in rock pools below the nesting sites of sea birds, and its extensive organic needs are provided by the bird droppings.

Table 6.7 Nitrogen sources utilized by *Chlamydomonas* species

Species	Compounds utilized				
	Urea	Uric acid	Acetamide	Succinamide	Adenine
C. actinochloris	−	−	−	+	+
C. calyptrata (C. latifrons)	+	+	+	−	+
C. chlamydogama	−	−	−	−	+
C. carrosa	+	−	+	−	+
C. eugametos	+	−	−	−	−
C. gloeogama	+	−	+	−	+
C. gloeopara	+	+	+	+	+
C. inflexa	+	+	+	−	+
C. kakosmos	+	+	+	−	+
C. mcxicana	+	−	+	+	+
C. microsphaera var. *acuta*	+	+	+	−	+
C. minuta	+	+	+	+	+
C. moewusii	+	−	−	−	+
C. moewusii var. *rotunda*	+	+	−	−	+
C. mutabilis	+	−	−	−	+
C. peterfii	+	−	+	−	−
C. radiata	−	−	−	−	−
C. reinhardtii	+	+	+	−	+
C. sectilis	−	−	−	−	−
C. typhlos	+	+	+	−	+

Species	Amino acids							
	Ala	Asn	Glu	Gln	Gly	Lys	Orn	Ser
C. actinochloris	+	+	−	+	+	−	−	+
C. calyptrata (C. latifrons)	−	−	−	−	−	−	−	−
C. chlamydogama	+[1]	−	−	+	−	−	−	−
C. carrosa	+	+[1]			−[1]	−[1]	+	−
C. eugametos	−	−	−	−	−	−	−	−
C. gloeogama	−	−	−	−	−	−	+	−
C. gloeopara	−	+	+	+	−	+	+	−
C. inflexa	+	−	−	+	+	−	+	+
C. kakosmos	−	−	−	+	−	−	−	−
C. mexicana	−	−	−	+	−	+	+	−

(Continued)

Species	Amino acids							
	Ala	Asn	Glu	Gln	Gly	Lys	Orn	Ser
C. microsphaera var. *acuta*	+	−	−	−	−	−	−	−
C. microsphaerella	−	−	−	−	−	−	−	−
C. minuta	−	−	−	+	−	−	−	−
C. moewusii	−	+	−	−	−	−	+	−
C. moewusii var. *rotunda*	−	+	−	−	+	−	−	−
C. mutabilis	+	−	+	+	+	−	+	+
C. peterfii	−	−	−	+	−	−	−	−
C. radiata	−	−	−	−	−	−	−	−
C. reinhardtii	−	−	−	+[2]	−	−	+	−
C. sectilis	−	−	−	−	−	−	+	−
C. typhlos	−	−	−	−	−	−	−	−

Table 6.7 *Continued*

Information primarily from Cain (1965). Species are as described by Ettl (1976a); some have since been renamed, and not all are available from current culture collections. See http://www.chlamy.org/species.html for a comprehensive list of Chlamydomonas species and their availability.
[1]*Some variation among different isolates of this species.*
[2]*Cain (1965) stated that the UTEX 89 and 90 strains of C. reinhardtii did not grow on glutamine. However, the Ebersold/Levine wild-type strain appears to grow quite well using this compound as its sole nitrogen source.*

2. Uptake of inorganic nitrogen

Uptake and assimilation of exogenous nitrate and nitrite have been extensively studied, and are reviewed in Volume 2, Chapter 3 and by Fernández and Galván (2007). The summary here is based largely on these reviews. Nitrate uptake is stimulated by nitrogen starvation and repressed by ammonium or nitrite, and is an energy-requiring process. Nitrate is taken into the cell by specific transport systems, and reduced to nitrite by nitrate reductase in the cytosolic compartment. Transporters then deliver nitrite to the chloroplast, where it is reduced to ammonium by nitrite reductase. Excess inorganic nitrogen is excreted.

Three families of nitrate and nitrite transporters have been characterized, NRT1 (a single gene), NRT2 (six genes), and NAR1 (six genes). These are reviewed in detail in Volume 2, Chapter 3.

A single gene encodes nitrate reductase in *Chlamydomonas* and is usually referred to in the *Chlamydomonas* literature as *NIT1*, although *NIA1* is the preferred name for the corresponding gene in plants. The enzyme is a homodimer of two 105-kD polypeptides, each of which has the prosthetic groups flavin adenine dinucleotide (FAD), heme b_{557}, and molybdenum cofactor (Moco). Two partial activities can be separated using artificial electron acceptors, a diaphorase that oxidizes NAD(P)H, and the terminal nitrate reductase that reduces nitrate to nitrite. Assay of these separate reactions has been used to classify *nit1* mutants (Table 6.8). Gene expression is regulated largely by redox mechanisms, in particular by the redox state of the plastoquinone pool. In the absence of nitrate, the enzyme becomes overreduced and is inactivated. Nitrite reductase, also encoded by a single gene in *Chlamydomonas*, reduces nitrite to ammonium in the chloroplast, using reduced ferredoxin generated by photosynthesis as electron donor.

External ammonium can enter the cell by either of two transport systems, which are differently regulated. The high affinity system (HATS) has been better characterized in *Chlamydomonas*, and involves enzymes of the AMT/MEP family (González-Ballester et al., 2004; Kim et al., 2005). AMT proteins can also act as gas channels for NH_3 (Kustu and Inwood, 2006).

Ammonium is assimilated through the GS/GOGAT cycle to produce L-glutamate, which enters amino acid metabolism by transamination reactions. *Chlamydomonas* has two glutamine synthetases, the cytosolic GS1 and plastid GS2, whose ratio shifts depending on the relative abundance of nitrate and ammonium in the medium. Two GOGAT enzymes (glutamine oxoglutarate amidotransferase) are also present, Fe-GOGAT and NADH-GOGAT, both apparently plastidic. Three glutamate dehydrogenase isozymes were shown to be differentially regulated (Muñoz-Blanco et al., 1989; Moyano et al., 1995), but only two genes have been identified.

3. Assimilation of organic nitrogen compounds

Urea is taken up by an active transport system and is then distributed between a metabolic pool and a non-metabolic one, thought to be localized in the chloroplast. It is hydrolyzed to ammonium and CO_2 by ATP: urea amidolyase, an enzyme complex consisting of urea carboxylase and allophanate hydrolase (Hodson et al., 1975). A cluster containing the genes for these enzymes, two putative urea transporters, and an arginine deiminase has been identified in the genomic sequence on linkage group VIII (Volume 2, Chapter 3). Urate can also support growth in the absence of other nitrogen sources, and is metabolized by urate oxidase (Pineda et al., 1987; Merchán et al., 2001).

With the exception of arginine, for which an active transport system exists, and glutamine, most amino acids are not readily taken up by *Chlamydomonas* cells. However, nitrogen starvation in the presence of acetate as a carbon source induces a periplasmic enzyme, L-amino acid oxidase, which can deaminate some amino acids to produce ammonium and a 2-oxoacid (Muñoz-Blanco et al., 1990; Piedras et al., 1992; Vallon et al., 1993). Nitrogen from histidine can be assimilated after degradation by histidase and urocanase (Hellio et al., 2004).

Table 6.8 Representative mutants with alterations in nitrogen assimilation

Gene	Mutant	Description	References
NIA1 or *NIT1*	*nit1* mutants, many alleles	Nitrate reductase structural gene; mutants cannot grow on nitrate as sole N source	Nichols and Syrett (1978), Sosa et al. (1978); Fernández and Matagne (1984); Kalakoutskii and Fernández (1995)
	nit1-301	Lacks terminal reductase activity	
	nit1-305	Lacks diaphorase activity	
NIT2	*nit2* mutants, many alleles	Transcription factor, regulates *NIT1* and other genes; mutants can't grow on nitrate	Sosa et al. (1978); Schnell and Lefebvre (1993); Quesada and Fernández (1994); González-Ballester et al. (2004, 2005); Camargo et al. (2007)
	nit3-307	Nitrate reductase-deficient, lacks terminal reductase and Moco, not rescued by molybdate. Possible defect in molybdopterin biosynthesis	Fernández and Aguilar (1987); Aguilar et al. (1992a)
	nit4-104	Lacks Moco but can grow on nitrate in presence of high concentrations of molybdate; probable mutation in molybdate processing	Fernández and Matagne (1984); Pérez-Vicente et al. (1991)
	nit5, nit6	Two different loci; double mutant has similar phenotype to *nit4*; resistant to high concentrations of molybdate and tungstate	Fernández and Matagne (1984); Llamas et al. (2000)
	nit7	Similar phenotype to *nit3-307* but not allelic	Aguilar et al. (1992b)
	nit8	Chlorate-resistant, lacks nitrate reductase activity	Nelson and Lefebvre (1995)
	nit9	Regulatory locus linked to *NIT2*	Rexach et al. (1999)
	nit10, nit11	Deficient in Moco biosynthesis	Navarro et al. (2005)
	ma1, ma2	Methylammonium resistant due to defect in transport of ammonium and methylammonium; able to use nitrate in the presence of methylammonium	Franco et al. (1988a,b)
	amt4	Methylammonium resistant	Kim et al. (2005)
CNX1E, *CNX1EG*		Genes required for incorporation of molybdenum into Moco precursor; insertional mutants	Llamas et al. (2007)

(Continued)

Table 6.8	Continued		
Gene	**Mutant**	**Description**	**References**
NIR	*M1, M2, M3*	Nitrite reductase; mutants can take up nitrate but excrete nitrite into the medium	Navarro et al. (2000)
	nar-C2, nar-D2	Deletions of *NAR3, NAR2, NIT1 g*; can grow on nitrite but not nitrate	Quesada et al. (1993)
	nar-F6	Rearrangement in *NIT1* cluster region; cannot grow on nitrate or nitrite	Quesada et al. (1993)
	nar-G1	Deletion in *NAR3, NAR2, NIT1 g* region; cannot grow on nitrate or nitrite	Quesada et al. (1993)

Purines can be assimilated under some conditions (Pineda and Cárdenas, 1996; Piedras et al., 1998), and putative permeases have been identified in the genomic sequence. Xanthine dehydrogenase has been characterized by Pérez-Vicente et al. (1988, 1992).

4. Studies with mutants

Many of the genes involved in nitrogen assimilation in *Chlamydomonas* were first identified through mutants with altered responses to nitrogen availability (Table 6.8). Mutants incapable of using nitrate are often insensitive to chlorate (Nichols and Syrett, 1978). Previously this was explained by the assumption that chlorate was reduced by nitrate reductase to the more toxic chlorite. However, not all chlorate-resistant mutants lack nitrate reductase, and mutants deficient in nitrate reductase are not necessarily resistant to chlorate. Nichols and Syrett (1978) found that the largest group of chlorate-resistant mutants arising from their experiments were able to grow on nitrate in the absence of acetate but not in its presence, but they were unable to explain this result. Prieto and Fernández (1993) in a study of spontaneous chlorate-resistant mutants determined that the mechanism of toxicity is more likely related to transport.

Mutants have been found that lack the terminal reductase but have diaphorase activity, that lack only the diaphorase, and that lack both activities. Some of the mutants deficient in the terminal reductase also lack xanthine dehydrogenase and are unable to use hypoxanthine as a nitrogen source. This observation led to the realization that the molybdenum cofactor (Moco) is shared by nitrate reductase and xanthine dehydrogenase in *Chlamydomonas*, as is also true in fungi and higher plants (Fernández and Cárdenas, 1981). Like nitrate reductase, xanthine dehydrogenase activity is repressed in wild-type cells grown on ammonium, but Moco is still produced. Mutants lacking xanthine dehydrogenase do not appear to be at any growth disadvantage. In wild-type cells grown on urea or aspartate as sole nitrogen source, very little nitrate reductase is synthesized, but xanthine dehydrogenase is still made at substantial levels. Thus these two enzymes are regulated separately. Fernández and Aguilar (1987) further characterized mutants deficient in Moco.

González-Ballester et al. (2005) created an ordered library of 22,000 insertional mutants in a strain carrying the arylsulfatase reporter gene under the control of the *NIT1* promoter (Llamas et al., 2002; Rexach et al., 2002). These mutants were screened to identify regulatory genes with one of four phenotypes: ammonium insensitive (ARS expressed in the presence of ammonium), nitrate insensitive (ARS not expressed in the presence of nitrate), overexpression on nitrate, and no growth on nitrate. Analysis of these mutants to identify the affected genes is still in progress.

C. Sulfur metabolism

Environmental sulfur enters the *Chlamydomonas* cell predominantly as sulfate ion, which is transported by an energy-dependent process (Yildiz et al., 1994). Seven putative sulfate transporters have been identified in the genome sequence, four H^+/SO_4^{2-} co-transporters similar to those of plants, and three Na^+/SO_4^{2-} co-transporters that are more similar to mammalian proteins. Within the cell, sulfate is transported into the chloroplast (Chen and Melis, 2004), where it is activated to adenosine 5′ phosphosulfate (APS) by the enzyme ATP sulfurylase. APS can be phosphorylated by APS kinase to 3′-phosphoadenosine 5′-phosphosulfate (PAPS), or reduced to sulfite by APS reductase. Sulfite in turn can be reduced by sulfite reductase to sulfide, which combines with *O*-acetylserine to form serine (see Volume 2, Chapter 5 for details of these reactions, and for formation of methionine, glutathione, and other sulfur compounds).

Many soil microorganisms, including *Chlamydomonas*, also secrete extracellular enzymes that cleave sulfate esters and sulfonates from organic molecules. A periplasmic arylsulfatase first purified by Lien et al. (1975) and later cloned by de Hostos et al. (1988, 1989) is expressed in response to sulfur starvation. Ohresser et al. (1997) developed this gene (*ARS1*) as a reporter under control of the *NIT1* promoter, assaying its activity with α-naphthylsulfate as substrate coupled with a chromogenic diazonium salt. A nearly identical gene, *ARS2*, is present close to *ARS1* in the genome sequence, in inverted orientation. Other *ARS* genes have also been identified.

Sulfur starvation severely limits photosynthetic electron transport. If respiration continues, the cells become anaerobic, and pathways of fermentation and light-dependent hydrogen production are induced (Wykoff et al., 1998; Forestier et al., 2003; Mus et al., 2007). Sulfur deprivation also induces synthesis of sulfate transporters, sulfatases, and other enzymes

Table 6.9	Mutants deficient in acclimation to sulfur deprivation		
Gene	**Mutant**	**Description**	**References**
SAC1	sac1	Sulfur transporter; mutant does not acclimate to sulfur deprivation; no arylsulfatase activity in S-depleted medium, defective sulfate uptake	Davies et al. (1994, 1996); Yildiz et al. (1996); Zhang et al. (2004)
	sac2	Arylsulfatase deficient	Davies et al. (1994)
SAC3	sac3	Putative serine-threonine kinase; mutant has constitutive arylsulfatase activity in S-replete medium; does not regulate chloroplast transcriptional activity on S starvation	Davies et al. (1994); Ravina et al. (2002); Irihimovitch and Stern (2006)
	ars	Many insertional mutants	Pollock et al. (2005); see also Volume 2, Chapter 5, Table 5.2

(Ravina et al., 2002; Zhang et al., 2004; Pollock et al., 2005). Mutants with altered response to sulfur starvation have been characterized (Table 6.9).

D. Phosphorus

As in the case of sulfur, environmental phosphorus is recovered from organic compounds by extracellular enzymes and then taken up by *Chlamydomonas* cells as the inorganic anion. In early work, five distinct phosphatase activities were identified: a soluble and a bound acid phosphatase, both of which were constitutively expressed, derepressible soluble and bound alkaline phosphatases, and a depressible neutral phosphatase active over a wide pH range (Matagne and Loppes, 1975; Nagy et al., 1981; Bachir and Loppes, 1997). Disintegration of the mother cell wall after division releases phosphatases into the culture medium (Lien and Knutsen, 1973), and wall-deficient cells also secrete phosphatase activity (Loppes, 1976a). Matagne et al. (1976a, b) found by cytochemical staining that the soluble acid phosphatase was located primarily in vacuoles, while the neutral phosphatase was found both in vacuoles and in the periplasmic space beneath the cell wall. Alkaline phosphatase was not localized by this staining procedure but indirect evidence suggested that it too was periplasmic, since mutants selected for alkaline phosphatase deficiency often proved to be wall-deficient mutants (Loppes and Deltour, 1975, 1978, 1981).

Loppes and Matagne (1973) developed a colony assay for acid phosphatase activity based on hydrolysis of α-naphthylphosphate to release α-naphthol, which was coupled to a diazonium salt to produce a red color. (This is similar to the assay for arylsulfatase, described above). Using this assay, they were able to identify phosphatase-deficient mutants. Additional

Table 6.10   Representative mutants with altered phosphate metabolism

Gene	Mutant	Description	References
PSR1	psr1	MYB-type transcription factor; controls level of PNPase expression during P starvation; mutants have low levels of extracellular phosphatase activity	Shimogawara et al. (1999); Yehudai-Resheff et al. (2007)
	psr2	Constitutive expression of extracellular phosphatase	Shimogawara et al. (1999)
LPB1	lpb1	"Low phosphate bleaching"; mutants bleach much more rapidly than wild type when starved for phosphate; protein is conserved, but function is still uncertain	Chang et al. (2005)
SQD1	sqd1	UDP-sulfoquinovose synthase; mutant cannot make sulfolipids, grows poorly in P-deficient medium	Sato et al. (2003); Riekhof et al. (2003)
	P2	Lacks bound acid phosphatase	Loppes and Matagne (1973); Matagne et al. (1976a)
	Pa	Lacks soluble acid phosphatase	Loppes and Matagne (1973); Matagne et al. (1976a)
	PD2	Lacks derepressible neutral phosphatase	Matagne et al. (1976a, b); Loppes et al. (1977); Dumont et al. (1990); Bachir et al. (1996)

mutants with altered responses to phosphorus starvation have also been characterized (Table 6.10).

Phosphate is transported across the plasma membrane by high affinity transporters. As is the case with sulfate transport, *Chlamydomonas* has both plant-like (H^+/PO_4^{3-}) and animal-like (Na^+/PO_4^{3-}) transporters (Shimogawara et al., 1999). After uptake, inorganic phosphate is stored within the cell in the form of polyphosphate, mostly complexed with calcium, magnesium, and zinc in cytosolic granules (Siderius et al., 1996; Komine et al., 2000; Ruiz et al., 2001) and in the cell wall (Werner et al., 2007). Polyphosphate metabolism has also been studied in *C. acidophila* (Nishikawa et al., 2003, 2006b).

Phosphorus deprivation induces synthesis of phosphatases and transporters, as well as mobilization of stored polyphosphate (Dumont et al., 1990; Quisel et al., 1996; Rubio et al., 2001). This topic is reviewed in Volume 2, Chapter 6.

E. Starch synthesis and degradation

Starch deposits surround the pyrenoid of *Chlamydomonas* cells (see Figure. 2.1), and in dark-grown cells, starch granules are also seen scattered throughout the chloroplast matrix (Ohad et al., 1967a). In *Chlamydomonas* cells grown on a diurnal rhythm, starch is accumulated within the chloroplast during the light phase and degraded in the dark (Klein, 1987; Thyssen et al., 2001). The temporal pattern of accumulation and degradation persists in continuous light or darkness, indicating circadian control (Ral et al., 2006). *Chlamydomonas* accumulates both

Table 6.11		Representative mutants with defects in starch synthesis	
Gene	**Mutant**	**Description**	**References**
STA1	*STA1*	Large subunit of ADP-glucose pyrophosphorylase; mutants accumulate very little starch; mutant enzyme in *sta1-1* is less sensitive than wild-type enzyme to activation by 3-phosphoglyceric acid	Ball et al. (1991); van den Koornhuyse et al. (1996); Zabawinski et al. (2001)
STA2	*sta2*	Granule-bound starch synthase I; mutants are "waxy", lack amylose, accumulate structurally modified amylopectin	Delrue et al. (1992); Maddelein et al. (1994); Wattebled et al. (2002)
STA3	*sta3*	Soluble starch synthase III; mutants have high amylose content, low amylopectin, and show 80% decrease in starch on nitrogen starvation; high amylose content	Maddelein et al. (1994); Buléon et al. (1997); Ral et al. (2006)
PHOB	*sta4*	Plastidial phosphorylase; mutant has 60% decrease in starch under nitrogen starvation; high amylose content	Libessart et al. (1995); Ral et al. (2006)
	sta5	Possibly a mutation in *GPM1*; deficient in phosphoglucomutase activity, very little starch when starved for nitrogen, low amylose	van den Koornhuyse et al. (1996)
STA6	*sta6*	Small (catalytic) subunit of ADP-glucose pyrophosphorylase; mutants lack starch	Ball et al. (1991); Iglesias et al. (1994); van den Koornhuyse et al. (1996); Zabawinski et al. (2001).
ISA1	*sta7*	Subunit of isoamylase debranching enzyme; mutants substitute glycogen for starch	Mouille et al. (1996); Dauvillée et al. (1999, 2001a); Posewitz et al. (2004)
	sta8	Isoamylase; not allelic with *sta7*, possibly = *ISA2* or *ISA3* high amylose content, glycogen synthesis	Dauvillée et al. (2001a, b); Ral et al. (2006)
STA11	*sta11*	*DPE1*, 4-alpha-glucanotransferase (disproportionating enzyme); mutants have 90% decrease in starch, accumulate unbranched malto-oligosaccharide	Colleoni et al. (1999a, b); Wattebled et al. (2003)

amylopectin and amylose, but the proportion varies depending on growth conditions. The enzymes involved in synthesis of both, and the studies with mutants that have enhanced understanding of their regulation, are reviewed in Volume 2, Chapter 1. Some representative mutants are summarized in Table 6.11.

Under anaerobic conditions in the dark, starch fermentation leads to production of formate, acetate, and ethanol. Kreuzberg (1984a) and Gfeller and Gibbs (1984) concurred in finding that in *C. reinhardtii* these three compounds were produced in a 2:1:1 ratio, with small amounts of H_2 and CO_2 also being released. In *C. moewusii* large amounts of glycerol were produced (Klein and Betz, 1978a), whereas in *C. reinhardtii* glycerol and D-lactate were seen only in minor amounts and at low pH (Gfeller and Gibbs, 1984; Kreuzberg, 1984a). The relationship of starch fermentation to hydrogen production is discussed in Volume 2, Chapter 7.

F. Lipid metabolism

The lipid composition of *Chlamydomonas* cells was extensively investigated in the 1970s and 1980s, and those studies are summarized here. The enzymes of lipid biosynthesis are reviewed in Volume 2, Chapter 2. Phototrophically and mixotrophically grown cells contain on the order of 18–24 mg of ether-soluble lipid per gram of dry weight ($\sim 10^9$ cells), and heterotrophically grown cells contain about half this amount (Eichenberger, 1976). Of this total, chlorophyll accounts for about 19% in mixotrophically grown cells, monogalactosyl diglycerides (MGDG) about 26%, and digalactosyl diglycerides (DGDG) another 19%. Phosphatidyl ethanolamine, phosphatidyl glycerol, phosphatidyl choline, and sulfolipid together account for about 14% of polar lipids (Eichenberger, 1976; Janero and Barrnett, 1981b). The remaining ether-soluble lipid, 24–30% of the total, comprises carotenoids, sterols, and several other components. Prominent among the latter is an unusual membrane lipid, diacylglyceryl trimethyl homoserine (DGTS), which is thought to function primarily as an acyl carrier for desaturation of oleic and linoleic acids (Schlapfer and Eichenberger, 1983). In plants and in some other algae this function is served by phosphatidyl choline, a compound that is present in relatively low levels in *Chlamydomonas*.

Cellular lipids of *Chlamydomonas* show marked partitioning between the chloroplast and the cytosolic compartments. Glycolipids account for 70–80% of the thylakoid membrane lipid content, with DGTS and phosphatidyl glycerol in approximately equal proportions making up the remainder (Eichenberger et al., 1977; Janero and Barrnett, 1981a, 1982a, b). Phosphatidyl ethanolamine and phosphatidyl choline are not found in association with thylakoid membranes, whereas more than 50% of total cellular phosphatidyl glycerol is chloroplast associated (Janero and Barrnett, 1981b; Mendiola-Morgenthaler et al., 1985b).

C_{16} and C_{18} fatty acids predominate in *Chlamydomonas* lipids (Erwin, 1973; Gealt et al., 1981; Janero and Barrnett, 1981b). The thylakoid MGDGs contain mostly unsaturated C_{18} acids, whereas the DGDGs and sulfolipids contain a much higher proportion of saturated C_{16} acid (Janero and Barrnett, 1981a).

The predominant sterols in wild-type cells are ergosterol and 7-dehydroporiferasterol (Patterson, 1974; Gealt et al., 1981). Bard et al. (1978) also reported finding smaller amounts of several additional sterols. Three mutants isolated by Bard et al. as resistant to nystatin proved to have altered sterol composition.

G. Vitamins and cofactors

Wild-type *C. reinhardtii* cells do not require added vitamins for growth, in contrast to numerous other algae. Of 306 species surveyed, representing seven phyla, Croft et al. (2006) found that more than half required vitamin B_{12} (cobalamin), 22% required thiamine, and 5% required added biotin. Within the genus *Chlamydomonas*, *C. chlamydogama*, *C. mundana*, *C. pallens*, and some isolates of *C. pulsatilla* require cobalamin; none required thiamine or biotin (Provasoli and Carlucci, 1974, and Droop, 1962, summarized by Croft et al., 2006). Watanabe et al. (1991) showed that *C. reinhardtii* cells could take up and accumulate exogenous cobalamin, but were also able to synthesize it.

Genes for thiamine biosynthesis have been characterized by Croft et al. (2006). Thiazole and the pyrimidine moiety of thiamine are synthesized in separate pathways, and then condensed to form thiamine monophosphate, which is phosphorylated to the active form thiamine pyrophosphate. Croft et al. (2007) showed that this process is regulated by riboswitches in *Chlamydomonas*.

As discussed in Chapter 8, section II.A., *Chlamydomonas* has been grown for more than 50 years on suboptimal concentrations of trace elements, and probably will continue to be grown this way in order to preserve consistency with earlier experiments. Merchant et al. (2006) reexamined trace element nutrition in the context of the genomic sequence to identify pathways of uptake and utilization, and their regulation. The result of these studies is a fascinating picture of how *Chlamydomonas* has preserved mechanisms whose closest homologues are spread among prokaryotes, plants, animals, and fungi. More information on metal nutrition is provided in Volume 2, Chapter 10.

H. Amino acids

The enzymes involved in biosynthesis and metabolism of amino acids are reviewed in Volume 2, Chapter 4. Most of our current knowledge of these pathways is based on analysis of the genome, and there are relatively few early publications on this topic. The one exception is the arginine biosynthetic pathway, which because of the availability of mutants (Table 6.12) was one of the earliest metabolic processes to be studied in *Chlamydomonas* (Ebersold, 1962; Hudock, 1962, 1963). Auxotrophic mutants have been described for six of the eight enzymes in this pathway (Loppes and Heindricks, 1986), which resembles the pathway used in yeast and *Neurospora*.

I. Why so few auxotrophs?

The range of auxotrophic mutations identified in *Chlamydomonas* is severely limited (Table 6.12). The only amino acid auxotrophs are mutants requiring arginine, and only a few auxotrophs for vitamins have been identified (nicotinamide, thiamine, p-aminobenzoic acid). Purine-requiring mutants were reported in *C. eugametos* (Gowans, 1960), but not in *C. reinhardtii*. Several possible explanations have been advanced for the paucity of auxotrophs, including fundamental differences in metabolic pathways of plants and algae compared to those of fungi and bacteria, differences in inducibility or repression of biosynthetic enzymes, permeability barriers, and extensive gene duplication.

Kirk and Kirk (1978a, b) were unable to detect active (carrier-mediated) uptake for any amino acid except arginine in *Chlamydomonas*, although Loppes (1969) had reported that leucine at least was taken up sufficiently well to serve as a sole nitrogen source. The rate of uptake of leucine observed by Kirk and Kirk could be accounted for by passive diffusion but was indeed marginally adequate to support growth (a 1 mM solution was sufficient to provide 10 nmol/min/mg of cellular protein). The non-metabolizable amino acid analogue 2-amino-isobutyric acid appears to be actively transported, at least to some extent (Hasnain and Upadhyaya, 1982). Proline utilization in response to high salt concentration was reported by Reynoso and Gamboa (1982). Glutamine can also be used as the sole nitrogen source. Thus there is substantial evidence supporting at least limited uptake of number of amino acids or their analogues. Kirk and Kirk (1978b) suggested that previous searches for auxotrophs failed because amino acids were not provided at sufficiently high concentration in the nonselective medium to which cells were exposed following mutagenesis, but no additional amino acid auxotrophs have been obtained since publication of their paper.

Nakamura et al. (1981) showed that methionine could competitively relieve inhibition by methionine sulfoximine, suggesting that at least some methionine was being taken up. However, they too were unable to isolate an auxotrophic mutant and found in fact that cells on methionine medium were killed by daylight fluorescent lights. They therefore postulated a photodynamically produced toxicity that might inhibit recovery of mutants on

Table 6.12	Auxotrophic mutants		
Gene	**Mutant**	**Description**	**References**
AGK1(?)	arg1	Acetylglutamate kinase; mutant can grow on arginine, ornithine, or citrulline	Ebersold (1956); Loppes and Heindricks (1986)
ARG9	arg9	Acetylornithine aminotransferase; mutant can grow on arginine, ornithine, or citrulline	Loppes and Heindricks (1986)
	arg10	Acetylornithine glutamate transacetylase; mutant can grow on arginine, ornithine, or citrulline	Loppes and Heindricks (1986)
OTC1	arg4	Ornithine carbamoyltransferase; mutants require arginine, cannot use ornithine; in principle would be expected to grow on citrulline, but in fact cannot	Loppes (1969); Loppes and Heindricks (1986)
AGS1	arg8	Argininosuccinate synthetase; mutants require arginine	Loppes and Heindricks (1986)
ARG7	arg7; several alleles, including arg7-8, also known as arg2	Argininosuccinate lyase; mutants require arginine; intragenic complementation can occur among some	Gillham (1965a); Loppes and Matagne (1972); Purton and Rochaix (1995)
	nic1	Requires nicotinamide or 3-hydroxyanthranilic acid; sensitive to 3-acetyl pyridine	Eversole (1956)
	nic2	Requires nicotinamide, uses quinolinic acid very poorly; sensitive to 3-acetyl pyridine	Eversole (1956)
	nic5	Requires nicotinamide; sensitive to 3-acetyl pyridine	Eversole (1956)
NIC7	nic7	Quinolinate synthetase; requires nicotinamide; sensitive to 3-acetyl pyridine	Eversole (1956)
	nic11	Requires nicotinamide; sensitive to 3-acetyl pyridine	Ebersold et al. (1962)
	nic13	Requires nicotinamide; sensitive to 3-acetyl pyridine	Hastings et al. (1965)
	nic15	Requires nicotinamide, difficult to score on 3-acetyl pyridine	Hastings et al. (1965)
	pab1	Requires p-aminobenzoic acid, cannot use shikimic acid, p-hydroxybenzoic acid, or other aromatic compounds; sensitive to sulfanilamide	Eversole (1956)
	pab2	Requires p-aminobenzoic acid; sensitive to sulfanilamide	Eversole (1956)
	thi1	Requires thiamine, cannot use thiazole or vitamin B_1 pyrimidine; sensitive to oxythiamine	Eversole (1956)
	thi2	Requires thiamine, can use thiazole plus vitamin B_1 pyrimidine; not sensitive to oxythiamine	Eversole (1956)
	thi3	Requires thiamine or thiazole; sensitive to oxythiamine	Eversole (1956)
THI4	thi4	Thiazole biosynthetic enzyme; requires thiamine or thiazole; not sensitive to oxythiamine	Eversole (1956); Croft et al. (2007)
	thi8	Requires thiamine or vitamin pyrimidine; sensitive to oxythiamine	Ebersold et al. (1962)
	thi9	Requires thiamine	Levine and Goodenough (1970)
THI10	thi10	Hydroxyethylthiazole kinase; requires thiamine	Ebersold et al. (1962)

The arg1 mutation maps near the location of the AGK1 gene on linkage group I, but its identity with this locus has not yet been confirmed.

medium high in methionine, and by extension, perhaps other compounds as well. A mutant that was hypersensitive to this methionine-mediated light damage was isolated by Nakamura et al. (1978, 1981; Nakamura and Lepard, 1983). Catalase and superoxide dismutase suppressed the methionine photo-inactivation, leading Takahama et al. (1985) to conclude that the mechanism of damage is formation of peroxide and O_2^-. Egashira et al. (1989) ascribed the mutant phenotype to a reduced activity of catalase. Photodynamic toxicity was also proposed as a mechanism for tryptophan sensitivity of some *C. eugametos* and *C. reinhardtii* strains (Nakamura et al., 1979).

In a later publication, Nakamura et al. (1985) reported their failure to obtain arginine auxotrophs in *C. eugametos*. This species is capable of at least limited arginine uptake, since arginine alleviates inhibition by canavanine, but assimilation of radioactive arginine was considerably lower than in *C. reinhardtii*. Thus at least in this case, the failure to isolate auxotrophs probably does result from inefficient uptake of exogenous amino acid.

J. Physiological ecology

1. Environmental adaptations

Most metabolic studies with *Chlamydomonas* have focused on behavior under laboratory conditions, often far removed from the situation to be found in nature, whereas more ecological investigations have tended to concentrate on population dynamics or response to specific environmental agents. Relatively few studies have been concerned with setting the physiological characteristics of the organism into a natural context, apart from work with species such as *C. acidophila* or *C. raudensis* that are native to extreme environments (Visviki and Palladino, 2001; Morgan-Kiss et al., 2005; Spijkerman et al., 2007; Szyszka et al., 2007). *C. reinhardtii* has been used, however, as a model for experimental evolution in continuous culture by Bell and colleagues in a series of papers entitled "The ecology and genetics of fitness in *Chlamydomonas*" (Bell, 1990a, b; Renaut et al., 2006, and references cited therein). These papers consider effects of environmental changes, and of the sexual cycle, on adaptation.

2. Effects of herbicides and pesticides

The use of *Chlamydomonas* as a model system for understanding the metabolism of higher plants, and particularly its use as a vehicle for genetic engineering, requires that its response to herbicides and other inhibitors be similar to that in plants, and indeed this has proved to be the case. The major classes of herbicides affecting photosynthesis, chlorophyll or carotenoid biosynthesis, microtubule assembly, and amino acid biosynthesis are all effective in *Chlamydomonas*, and resistant mutants have been isolated for many (Table 6.13). For some compounds, however, the effective concentration is substantially higher, as much as several hundred or a thousandfold,

Table 6.13 Representative studies on herbicide response

Herbicide	Target	Resistant mutants	References
Atrazine, DCMU (diuron), metribuzin, bromacil	Photosystem II D1 protein (chloroplast *psbA* gene)	Many	Galloway and Mets (1984); Erickson et al. (1985, 1989); Wildner et al. (1990); Govindjee et al. (1992); Wilski et al. (2006)
Chlorsulfuron, imazaquin, sulfometuron methyl	Acetolactate synthase	*csr, imr* mutants, specific mutation unknown but likely to be in *ALS*; lesion is confirmed for *smr1*	Hartnett et al. (1987); Winder and Spalding (1988); Kovar et al. (2002)
Diphenyl ethers	Porphyrin biosynthesis; protoporphyrinogen oxidase	*rs-3*	Oshio et al. (1993); Randolph-Anderson et al. (1998)
Amiprophos methyl	Microtubules	*apm1, apm2*	James et al. (1988); James and Lefebvre (1992)
Oryzalin	Microtubules	*ory1*	James et al. (1988, 1989)
Colchicine	Microtubules	*colR2, colR4* (β-tubulin); *cor1* (not allelic with these)	James et al. (1989)
Isoxaflutole, pyrazolate	Hydroxyphenylpyruvate dioxygenase		Trebst et al. (2004)
NDUA	Lycopene cyclase		Fedtke et al. (2001)
Norflurazon	Phytoene desaturase		Trebst and Depka (1997); McCarthy et al. (2004)

than is found for plant tissue cultures. Reboud et al. (2007) described an experimental design for using *Chlamydomonas* to screen herbicides. The first edition of *The Chlamydomonas Sourcebook* included tables of herbicides tested and toxic levels of various other inhibitors (Harris, 1989). The reader is referred to those tables, and to the papers of Mottley and Griffiths (1977) and McBride and Gowans (1970) for summaries of early work on this topic. *Chlamydomonas* species are among several types of algae that have been used as indicators of pesticide toxicity in the environment. For example, Kent and Currie (1995) included *C. reinhardtii* in a study of responses of 12 different algae to the insecticide fenitrothion. Selected publications on toxicity of some other pesticides and organic chemicals are listed in Table 6.14.

3. Metal toxicity

So far as is known, cadmium, lead, and mercury have no physiological function in *Chlamydomonas*. Copper, while essential (Volume 2, Chapter 10), is also a significant environmental toxin. Of these elements, cadmium has been the most extensively investigated in terms of uptake, toxicity,

Table 6.14 Studies of effects of pollutants on *Chlamydomonas* species

Pollutant	Species studied	References
Arsenic	*C. reinhardtii*	Planas and Healey (1978); Kaise et al. (1999); Kobayashi et al. (2005)
	Unidentified freshwater species	Christensen and Zielski (1980)
Cadmium	*C. reinhardtii*	Lawrence et al. (1989); Behra (1993); Macfie et al. (1995); Nagel et al. (1996); Prasad et al. (1998); Voigt et al. (1998); Vigneault and Campbell (2005); Tüzün et al. (2005); see text for additional citations
	C. noctigama	Cepák et al. (2002)
	C. bullosa	Visviki and Rachlin (1994)
Chromium	*C. reinhardtii*	Rodríguez et al. (2007)
Cobalt	*C. reinhardtii*	Lustigman et al. (1995); Macfie et al. (1995); Macfie and Wellbourn (2000)
Copper	*C. reinhardtii*	Macfie et al. (1995); Prasad et al. (1998); Macfie and Wellbourn (2000); Boswell et al. (2002); Luis et al. (2006); Jamers et al. (2006)
Lead	*C. reinhardtii*	Ahlf et al. (1980); Irmer et al. (1986); Behra (1993); Tüzün et al. (2005)
Mercury	*C. reinhardtii*	Ben-Bassat et al. (1972); Cain and Allen (1980); Weiss-Magasic et al. (1997); Tüzün et al. (2005)
	C. variabilis	Delcourt and Mestre (1978)
Nickel	*C. reinhardtii*	Macfie et al. (1995); El-Naggar (1998); Macfie and Wellbourn (2000)
Sulfur dioxide	*C. reinhardtii*	Wodzinski and Alexander (1978)
Nitrogen dioxide	*C. reinhardtii*	Wodzinski and Alexander (1980)
Chlorine	marine species	Hirayama and Hirano (1980)
Sulfite	*C. reinhardtii*	Stamm (1980)
Naphthalene	*C. angulosa*	Hellebust et al. (1982, 1985); Hutchinson et al. (1981, 1985)
Crude oil extracts	*C. angulosa*	Hellebust et al. (1982, 1985); Hutchinson et al. (1981, 1985)
	C. pulsatilla	Hsiao (1978)
Polychlorinated biphenyls	*C. reinhardtii*	Gresshoff et al. (1977); Mahanty and Gresshoff (1978); Conner (1981) Jabusch and Swackhamer (2004)
	Unidentified freshwater species	Christensen and Zielski (1980)
Various insecticides	*C. reinhardtii*	Birmingham and Colman (1977); Netrawali et al. (1986); Gandhi et al. (1988); Netrawali and Gandhi (1990); Kent and Caux (1995); Wang et al. (2007)
	C. moewusii	Cain and Cain (1984a); Rioboo et al. (2001)
	C. segnis	Weinberger et al. (1987); Kent and Caux (1995)

and development of tolerance mechanisms (Hu et al., 2001; Rubinelli et al., 2002; Hanikenne et al., 2005; Gillet et al., 2006; François et al., 2007) and cadmium-resistant mutants have been isolated (Nagel and Voigt, 1989; Collard and Matagne, 1990; McHugh and Spanier, 1994). Acidophilic *Chlamydomonas* species, which have been found in areas polluted by mining and other industrial activities, have also been studied in this regard (Nishikawa and Tominaga, 2001; Nishikawa et al., 2003, 2006a; Aguilera and Amils, 2005; Spijkerman et al., 2007). Rajamani et al. (2007) have reviewed heavy metal toxicity, and the potential for using genetically engineered *Chlamydomonas* as a means of bioremediation.

IV. PHOTOSYNTHESIS

A. Overview

A diagrammatic representation of the thylakoid membrane components involved in the light reactions of photosynthesis is shown in Figure 6.1. Two photosystems, each with a light-harvesting antenna complex, act in series through an intermediary complex containing cytochromes b_6 and f. The net result is that the redox state of electrons passing through the chain is elevated from a level that brings about the oxidation of water to a

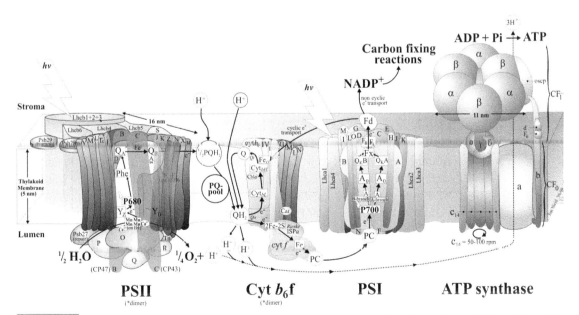

FIGURE 6.1 *A detailed model of the photosynthetic Z scheme including structural information on the organization of the protein complexes involved in electron (e⁻) and proton (H⁺) transport within the thylakoid membrane. This figure appears in color with additional annotation at http://www.queenmaryphotosynthesis.org/nield/psIImages/oxygenicphotosynthmodel.html. Courtesy of Jon Nield.*

level capable of NADP reduction. The two photosystems are physically distinct and can be recognized in freeze-fracture photographs as discrete particles associated with the thylakoid membranes. ATP is generated via the membrane-associated CF_0CF_1 ATP synthase complex. The enzymes that fix carbon from CO_2 into inorganic compounds (the "dark reactions" or Calvin cycle) are not membrane bound, and they appear in the soluble supernatant phase when cells or chloroplasts are fractionated.

Each of these complexes has been thoroughly investigated in *Chlamydomonas*, beginning in the 1960s with isolation of non-photosynthetic mutants and with studies of chloroplast development in the *y1* mutant, which is unable to form chlorophyll in the dark. The literature is huge, and has been reviewed at intervals over the years. For the current state of research, the reader should see Volume 2. Work through the 1990s was covered in the book edited by Rochaix et al. (1998) and by Grossman (2000b). The first edition of *The Chlamydomonas Sourcebook* (Harris, 1989) provided a history of the early studies, including descriptions of many mutants. For the historically minded reader in search of a broader perspective, volume 76 of *Photosynthesis Research* (2003) is also highly recommended. The discussion to follow here therefore provides only a brief introduction to the components of photosynthesis, with a focus on features that are unique to *Chlamydomonas*, or particularly well studied in this alga, and the usual emphasis on mutants.

B. Light-harvesting complexes

Early studies both in *Chlamydomonas* and in land plants focused on electrophoretic separation of polypeptides binding chlorophyll, and on purification of protein complexes released by detergent extraction from thylakoid membranes. Five distinct chlorophyll–protein complexes could be resolved electrophoretically, but their functions were not clearly correlated with the membrane fractions, or with their roles in Photosystems I and II. (Readers of the first edition of *The Chlamydomonas Sourcebook* will be aware just how murky the terminology was in the 1980s.) Fortunately, with proteomic and genomic identification of the individual proteins (Stauber et al., 2003; Elrad and Grossman, 2004) and community efforts to resolve the nomenclature, the newcomer to this field today will find it much easier to understand.

LHCI and LHCII designate the light-harvesting antennae of Photosystems I and II, respectively. Within each of these groups, the proteins are identified as LHCA (LHCI complex) or LHCB (LHCII complex). These include proteins formerly known as Cab, for chlorophyll *a/b* binding. Sequences are similar enough among the various LHC proteins to suggest a common origin.

The LHCI and LHCII complexes collect light energy for photosynthesis and can also regulate the amount of light energy that is transferred to the photosystems. Details of their organization can be found in Volume 2,

Chapter 14. Their role in state transitions, the short-term adaptation to changes in light intensity, is discussed in Volume 2, Chapter 22. Regulation of antenna size, a process likely to become an important research direction in *Chlamydomonas*, is discussed by Tetali et al. (2007).

C. Photosystem II

The core of the Photosystem II reaction center consists of a heterodimer of the proteins D1 and D2, encoded by the chloroplast *psbA* and *psbD* genes respectively, surrounded by two chlorophyll *a*-binding proteins CP43 (*psbC* gene) and CP47 (*psbB*), and several proteins of lower molecular weight, including cytochrome b_{559} (*psbE* and *psbF*). Peripheral to this core are a monomeric antenna comprising the chlorophyll *a/b*-binding CP29 and CP26 proteins and a major trimeric antenna of LHCII proteins. Excitation energy is transferred from CP43 and CP47 to the core reaction center, where charge separation occurs in chlorophyll P680. A transient intermediate P680* is formed, and pheophytin *a* (Pheo$_{D1}$) is reduced, all within a few picoseconds. The charge separation is stabilized by transfer of an electron to a bound plastoquinone, Q_A. This is followed by a two-step reduction of a plastoquinone at the Q_B site, with addition of the second electron being the rate-limiting step. The intermediate semiquinone is tightly bound to the Q_B site but the fully reduced plastoquinol (PQH_2) is weakly bound and leaves the Q_B site to exchange with the plastoquinone pool. P680+ is reduced in a four-step reaction by electrons from oxidation of water to molecular oxygen, mediated by proteins of the manganese-containing oxygen evolving complex which is bound to D1 and CP43 (Figure 6.1). For additional details of these reactions and analysis of the molecular structure of the Photosystem II complex, see Volume 2, Chapter 16; also Kern and Renger (2007).

The 32-kD apoprotein of the secondary quinone electron acceptor B (Q_B) is known as the D1 protein, in plants as well as in *Chlamydomonas*, because of the diffuse bands seen on gels when the Photosystem II reaction center proteins from *Chlamydomonas* were separated electrophoretically (Chua and Gillham, 1977). The D2 protein is similar in size to D1, and shares 27% primary sequence identity with it, but only the D1 side of the reaction center core is active in electron transfer. Each protein has five transmembrane units and together they form a heterodimer that binds the cofactors of electron transfer. D1 has been the subject of intense investigation, especially in *Chlamydomonas* where it is easily altered by site-directed mutagenesis. These studies have been essential to understanding the reaction center structure, especially of the Q_B-binding site, and of the mechanism of herbicide interaction with this site.

Photosystem II is susceptible to damage at high light intensities (photoinhibition), and the D1 protein is the key component in absorbing the damage and in its repair. The rapid turnover of D1 was first demonstrated in

Chlamydomonas (Ohad et al., 1984). Damaged reaction centers are removed from the grana to the stromal lamellae, and D1 is degraded. Newly synthesized D1 protein is incorporated into the reaction center which is then religated into its correct location (see Volume 2, Chapter 23).

D. The cytochrome b_6f complex

The cytochrome b_6f complex is functionally a plastoquinol–plastocyanin reductase, and is the intermediate between Photosytems II and I. It is evolutionarily related to the respiratory bc_1 complex, and comprises four large and four small subunits. The major proteins are cytochrome f, cytochrome b_6, the Rieske iron sulfur protein, and a protein known as subunit IV. These are encoded by the *petA*, *petB*, *PETC*, and *petD* genes, respectively. The small subunits are usually referred to simply as PetG, PetL, PetM, and PetN after their homologues in cyanobacteria. In *Chlamydomonas* the *petG* and *petL* genes are in the chloroplast genome, whereas *PETM* and *PETN* are nuclear. This differs from the situation in land plants, where *petN* is a chloroplast gene. Associated with these eight proteins are seven prosthetic groups: one molecule each of chlorophyll a and β-carotene, whose functions are still unclear; an iron–sulfur [Fe_2S_2] cluster, two b-hemes, and two c-hemes. The crystal structure of the complex was determined by Stroebel et al. (2003). The organization and assembly of the complex are reviewed in detail by Cramer et al. (2005) and in Volume 2, Chapter 17.

The b_6f complex can also participate in cyclic electron transport with plastoquinone and Photosystem I. In plants, two different pathways have been modeled for this process, one involving ferredoxin-quinone oxidoreductase (FQR) and the other requiring NAD(P)H dehydrogenase (NDH). No homologues of the plant *NDH* genes have been found in the *Chlamydomonas* chloroplast or nuclear genomes, however, and it is assumed that only the FQR cyclic pathway is operative. The postulated roles of cyclic electron transport in state transitions and in response to stress are discussed in Volume 2, Chapters 17 and 22.

E. Plastocyanin and cytochrome c_6

The small copper-containing protein plastocyanin transfers electrons from cytochrome f to Photosystem I (see Katoh, 2003, for review). Its synthesis is regulated by copper availability (Li and Merchant, 1995; Li et al., 1996; see also Chapters 10 and 15 in Volume 2). Cytochrome c_6 is a soluble protein with a covalently attached heme. In the early literature it was usually identified as cytochrome c-552, and is equivalent to cyanobacterial cytochrome c-553. Under copper-deficient conditions it substitutes for plastocyanin as electron acceptor from cytochrome f (Merchant and Bogorad, 1987a, b; Howe and Merchant, 1992). Mutations have been identified in five genes involved in attachment of the heme moiety, the chloroplast-encoded *ccsA*, and four nuclear genes, *CCS1–CCS4* (Xie et al., 1998; Table 6.15). For additional information on heme attachment to cytochromes, see Volume 2, Chapter 17.

F. Photosystem I

Photosystem I receives electrons from plastocyanin or cytochrome c_6 on the lumenal side of the thylakoid membrane and uses light energy to transfer them across the membrane to ferredoxin on the stromal side. It can also function in a cyclic electron transport pathway. The core structure is a heterodimer of the PsaA and PsaB proteins, which are encoded by chloroplast genes. As in Photosystem II, light is harvested by antenna complexes, and the primary light reaction is a charge separation beginning stabilized by transfer of an electron to a quinone, but in Photosystem I the terminal electron acceptor is an FeS cluster, which permits reduction of ferredoxin. The peripheral subunits PsaC, PsaD, and PsaE form the docking site for ferredoxin. The primary electron donor, P700, is ultimately reduced by plastocyanin or cytochrome c_6. PsaF and PsaJ are required for docking these proteins. Details of the molecular structure are provided in Volume 2, Chapter 15.

G. The ATP synthase complex

The CF_0CF_1 ATP synthase couples the reducing power generated by photosynthetic electron flow to production of ATP. The CF_0 complex, comprising four subunits (numbered I through IV) is embedded in the thylakoid membrane and translocates protons into the thylakoid lumen. The catalytic complex CF_1, which is extrinsic to the membrane, comprises five subunits (α through ε). CF_1 produces ATP when coupled to CF_0, but can hydrolyze ATP to ADP and phosphate when detached from CF_0. As indicated in Table 6.15, six of the nine genes encoding the CF_0 and CF_1 polypeptides are in the chloroplast genome, and the remaining three are in the nucleus. Mutants have been obtained in all except *ATPG* and *atpH*. Additional nuclear mutations are known that affect synthesis of the chloroplast-encoded subunits (Drapier et al., 1992, 2002; Majeran et al., 2001). The stoichiometry and detailed function of both complexes are reviewed in Volume 2, Chapter 18.

H. Chlororespiration

Bennoun (1982) presented evidence that thylakoid membranes of *Chlamydomonas* after dark adaptation used molecular oxygen to oxidize photosynthetic plastoquinone, which could be reduced by NADH using a thylakoid-bound iron–sulfur oxidoreductase (Godde, 1982; Gfeller and Gibbs, 1985). An electrochemical gradient was shown to form across the

Table 6.15   Major components of photosynthesis, with representative mutants

Gene	Product	Mutant	Comments	References
Photosystem II				
psbA	D1 protein	*FUD7, ac-u-beta*	Deletion of the entire *psbA* gene; many other mutations known, including herbicide resistance	Bennoun et al. (1986); see also Tables 7.1 and 7.2
psbB	CP43 protein	*222E*		Sugimoto and Takahashi (2003)
psbC	CP47 protein	*FUD34*	Mutation in 5′ flank of gene	de Vitry et al. (1989); Rochaix et al. (1989)
psbD	D2 protein	*FUD47*	Duplication producing frameshift	
PSBO	OEE1 protein	*FUD44*	Insertion of *TOC1* transposon	Mayfield et al. (1987a, 1989)
PSBP	OEE2 protein	*FUD39, BF25*		Mayfield et al. (1987b); de Vitry et al. (1989); Rova et al. (1994, 1996)
PSBQ	OEE3 protein			Mayfield et al. (1989)
PSBW				Bishop et al. (1999, 2003)
Cytochrome $b_6 f$ complex				
petA	Cytochrome *f*		Various site-directed mutations	Ponamarev and Cramer (1998); Baymann et al. (1999)
petB	Cytochrome b_6	*F41G, FUD2*	*F41G* is a site-directed point mutation; *FUD2* is a 36-bp duplication; there is also a deletion mutant	Kuras and Wollmann (1994); Ondarroa et al (1996); Finazzi et al. (1997)
PETC	Rieske iron–sulfur protein			de Vitry (1994); de Vitry et al. (1999)
petD	Subunit IV	*FUD6*	Deletion at 5′ end of gene	Sturm et al. (1994)
PCY1, or *PETE*	Plastocyanin	*ac208*		Quinn et al. (1993)
PETF	Ferredoxin			Stein et al. (1993)
petG	Small essential subunit		Deletion mutant	Berthold et al. (1995)
PETH	Ferredoxin NADP reductase			Kitayama et al. (1994); Decottignies et al. (1995)
petL	Small non-essential subunit			Takahashi et al. (1996)
PETM	4-kD subunit, probably non-essential			de Vitry et al. (1996); Ketchner and Malkin (1996)
PETN	Small essential subunit			Zito et al. (2002)

(Continued)

Gene	Product	Mutant	Comments	References
Table 6.15 *Continued*				
PETO	Subunit V, loosely bound transmembrane protein			Hamel et al. (2000)
CCS1	Protein required for heme attachment to c-type cytochromes	ac206, other ccs1 alleles		Inoue et al. (1997); Xie et al. (1998); Dreyfuss et al. (2003)
CCS2	Protein required for heme attachment to c-type cytochromes	ccs2		Xie et al. (1998)
CCS3	Protein required for heme attachment to c-type cytochromes	F18		Xie et al. (1998)
CCS4	Protein required for heme attachment to c-type cytochromes	F2D8		Xie et al. (1998)
CCS5	Protein required for heme attachment to c-type cytochromes	ccs5		Dreyfuss and Merchant (1999)
ccsA	Protein required for heme attachment to c-type cytochromes	ct34.5, ct59		Xie and Merchant (1996); Xie et al. (1998); Hamel et al. (2003)
Photosystem I				
psaA	PsaA reaction center protein		Many nuclear mutants controlling splicing of chloroplast gene	see Tables 7.1 and 7.4
psaB	PsaB reaction center protein	FUD26	4-bp deletion produces frameshift	Girard-Bascou et al. (1987); Xu et al. (1993)
psaC	Iron–sulfur protein		Insertional and site-directed mutants	Takahashi et al. (1991, 1992); Fischer et al. (1999b)
PSAD	Ferredoxin-binding protein			Farah et al. (1995b); Hahn et al. (1996)
PSAE	8.1-kD protein			Franzén et al. (1989b)
PSAF	17-kD protein		Insertional and point mutants	Franzén et al. (1989b); Farah et al. (1995a); Hippler et al. (1997, 2000); Fischer et al. (1999a)
PSAG	10-kD protein			Franzén et al. (1989a)
PSAH	11-kD protein			Franzén et al. (1989a)
PSAI	8.4-kD protein			Franzén et al. (1989a)
psaJ	Small hydrophobic subunit			Fischer et al. (1999a)

Table 6.15 *Continued*

Gene	Product	Mutant	Comments	References
ATP synthase				
atpA	CF$_1$ alpha subunit	FUD16	FUD16 has two mis-sense mutations; deletion mutants also generated	Ketchner et al. (1995); Fiedler et al. (1997); Drapier et al. (2007)
atpB	CF$_1$ beta subunit	FUD50, ac-u-c-2-21	FUD50 and ac-u-c-2-21 delete the entire atpB gene; point mutations have also been isolated	Lemaire et al. (1988); Robertson et al. (1989); Fiedler et al. (1997)
ATPC	CF$_1$ gamma subunit		Insertional mutant	Drapier et al. (2007)
ATPD	CF$_1$ delta subunit		Insertional mutant	Dent et al. (2005)
atpE	CF$_1$ epsilon subunit	FUD17, ac-u-a-1-15	Frameshift mutations	Robertson et al. (1990); Johnson and Melis (2004)
atpF	CF$_0$ subunit I	FUD18		Lemaire and Wollman (1989)
ATPG	CF$_0$ subunit II			
atpH	CF$_0$ subunit III			
atpI	CF$_0$ subunit IV	FUD23		Lemaire and Wollman (1989)
Carbon cycle				
rbcL	Rubisco small subunit			see Table 7.2
RBCS1, RBCS2	Rubisco large subunit		RBCS1 and RBCS2 are linked in the genome; a mutant exists that lacks both	Goldschmidt-Clermont (1986); Goldschmidt-Clermont and Rahire (1986); Khrebtukova and Spreitzer (1996b)
PRK1	Phosphoribulokinase	F60, ac214		Roesler et al. (1992)
Chlorophyll biosynthesis				
CHLH	Magnesium chelatase	chl1, brs1	Accumulate protoporphyrin	Chekounova et al. (2001)
POR	Light-dependent NADPH: protochlorophyllide oxidoreductase	pc1		Ford et al. (1981, 1983); Li and Timko (1996)
CAO	Chlorophyll a oxygenase (chlorophyll b synthase)	cbn1	Deficient in chlorophyll b	Chunayev et al. (1987); Mirnaya et al. (1990); Polle et al. (2000); Lohr et al. (2005)
		y mutants	Yellow in the dark; several loci, genes not yet determined	Ford and Wang (1980a,b)
Carotenoid biosynthesis				
PSY1	Phytoene synthase	lts1 mutants		McCarthy et al. (2004)

Genes with names in lower case (psbA, atpB etc.) are encoded in the chloroplast genome; see Chapter 7, Table 7.1 for additional references. Additional mutants in this genome are listed in Table 7.2, and some nuclear mutations that affect expression of chloroplast genes are given in Table 7.4. Other nuclear genes for which there are no published descriptions have been omitted.

thylakoid membrane as a result of this NADH oxidation and by reverse functioning of chloroplast ATP synthase. Bennoun suggested that this process, called chlororespiration, would recycle ATP and NAD(P)H generated by glycolysis. Mutants blocked in photosynthesis between plastoquinone and Photosystem I still showed the chlororespiration phenomenon, suggesting that some other electron carriers might be involved in this process (Bennoun, 1983). Subsequent studies by Kreuzberg (1984b), Gfeller and Gibbs (1985), and Maione and Gibbs (1986b) confirmed the likelihood of a plastoquinone-mediated chloroplast oxygen consumption. Kreuzberg also identified alcohol dehydrogenase and lactate dehydrogenase activities in the chloroplast fraction which he suggested might be used in reoxidation of reducing equivalents. The process of chlororespiration has been reviewed by Peltier and Cournac (2002; see also Cournac et al., 2002).

I. Chlorophyll and carotenoids

The biosynthetic pathway for chlorophyll was worked out from biochemical and genetic studies with bacteria, plants, and algae, including *Chlamydomonas*. Although presumed to take place within the chloroplast, the process in algae and plants is largely under the control of nuclear genes.

The first committed step in chlorophyll biosynthesis is formation of δ-aminolevulinic acid (ALA), which in plants is derived from glutamate via glutamate-1-semialdehyde. (This is in contrast to the synthesis of heme porphyrins in animal mitochondria and bacteria, which begins with ALA formation by condensation of glycine and succinyl CoA.) Two ALA molecules are joined by ALA dehydratase to form porphobilinogen, four molecules of which condense to form a tetrapyrrole ring, uroporphyrinogen. Decarboxylation and oxidation steps produce protoporphyrin (PROTO), which by incorporation of iron becomes heme, and by incorporation of magnesium becomes MgPROTO, the precursor of protochlorophyllide. Genes encoding all the enzymes of this pathway have been identified in the *Chlamydomonas* genome, and mutants are known in several (Table 6.15).

Protochlorophyllide can be reduced to protochlorophyll in *Chlamydomonas* either by light exposure or by enzymatic reduction in the dark. The latter process does not occur in angiosperms, but is seen in gymnosperm seedlings, some ferns, and other algae (Wang, 1978). Yellow-in-the-dark mutations block the dark reduction step. The first of these, *y1*, was isolated by Sager and Palade (1954) and has been widely used as a system for studying chloroplast development in response to light (see Chapter 7, section V).

Details of chlorophyll and also heme synthesis are reviewed in Volume 2, Chapter 20.

The principal carotenoid pigments of *Chlamydomonas* are β-carotene, violaxanthin, neoxanthin, and lutein, all of which are found in land plants and other algae, and loroxanthin (19-hydroxy-lutein), which has been detected only in some families of green algae (Fawley and Buchheim, 1995; Angeler and Schagerl, 1997). All are found in the chloroplast, in association with the photosynthetic reaction centers. Under conditions of high light stress some of the violaxanthin is converted to zeaxanthin through the xanthophyll cycle (see Chapters 21 and 23 in Volume 2). β-Carotene is found in the eyespot, and the rhodopsin chromophore retinal is also a β-carotene derivative (Chapter 2). Additional carotenoids, in particular ketocarotenoids, are found in zygospores. All the genes encoding enzymes of carotenoid biosynthesis are in the nuclear genome (see Volume 2, Chapter 21 for details).

Mutants blocked early in carotenoid biosynthesis (Table 6.15) are very light-sensitive and fail to accumulate chlorophyll or form normal chloroplasts. When grown in the dark they are typically pale green or nearly white. Phenocopies can be produced by treatment of wild-type cells with pyridazinone herbicides, which block carotenoid synthesis. Mutants at the *LTS1* locus have been shown to affect the gene encoding phytoene synthase (McCarthy et al., 2004). Although difficult to maintain because of their extreme light sensitivity, these mutants have been very useful for investigation of the rhodopsin light receptor for phototaxis (see Chapter 4, and Volume 3, Chapter 13). Other mutants are known that have blocks later in the biosynthetic pathway.

V. RESPIRATION

Difficulties in isolating purified mitochondria from *Chlamydomonas* hampered early attempts at physiological investigation of respiration, although extensive ultrastructural studies of mitochondria were made in the 1970s (see Chapter 2). In many respects, the colorless alga *Polytomella* has been a better system for research on oxidative phosphorylation and other respiration-related processes (Antaramian et al., 1996; Reyes-Prieto et al., 2002; van Lis et al., 2007), although its electron transport pathway differs in some respects from that of *Chlamydomonas* (van Lis et al., 2005). With the development of better methods for preparing mitochondria from *Chlamydomonas* free of chloroplast contamination (Eriksson et al., 1995; Nurani and Franzén, 1996; Funes et al., 2007), and the availability of proteomic data (Cardol et al., 2005) and the genome sequence, this field of research is expected to grow significantly in the coming years.

The mitochondrial electron transport chain comprises five major transmembrane complexes (I–V). The proteins of each complex, as well as those of the alternative oxidase, the ubiquinone biosynthesis pathway and other mitochondrial components, are tabulated in Volume 2, Chapter 13. The mitochondrial genome, which encodes only a few proteins, is described briefly in Chapter 7, and in more detail in Volume 2, Chapter 12.

The mitochondrial electron transport chain of *C. reinhardtii* resembles that of plants in having a classical cyanide-sensitive pathway and also a cyanide-insensitive, salicylhydroxamic acid-sensitive branch that uses an alternative oxidase to transfer electrons from ubiquinol directly to oxygen (Wiseman et al., 1977; Weger et al., 1990a, b). The alternative oxidase (AOX1) is present under normal growth conditions, but diminished at high concentrations of CO_2 (Goyal and Tolbert, 1989). Expression of the *AOX1* gene is activated by a shift from ammonium to nitrate, and by H_2O_2, cold stress, and antimycin A (Molen et al., 2006).

Mutants in mitochondrial functions are unable to grow on acetate in the dark. Wiseman et al. (1977) showed that such dark-dier (*dk*) mutants could be separated into two classes according to their sensitivity to fluoroacetate, an inhibitor of aconitase. Permeability and glyoxylate cycle mutants were not inhibited by this compound, but mutants with defects elsewhere in respiration were sensitive. These mutants have not been thoroughly characterized with respect to their lesions in respiratory pathways. Mutants in the mitochondrial genome have also been isolated (Table 6.16). Deletion of the entire mitochondrial genome by treatment with acriflavine or ethidium bromide produces tiny colonies, called *minutes* by analogy with yeast *petite* mutants, that are lethal after a few generations (Alexander et al., 1974).

Table 6.16 Representative mutants with altered mitochondria

Gene	Mutant	Description	References
	dk mutants	Nuclear mutations, genes not identified	Wiseman et al. (1977)
	minute mutants	Deletions of mitochondrial DNA; lethal after 8-9 generations	Alexander et al. (1974); Gillham et al. (1987a)
cob	*dum* mutants (many)	Deletions and point mutations in *cob* gene	Matagne et al. (1989); Dorthu et al. (1992); Randolph-Anderson et al. (1993); Remacle et al. (2006)
cob	*mud2, bm20, im1*	Myxathiazol resistant	Bennoun et al. (1991, 1992)
cox1	*dum18, dum19*		Colin et al. (1995)
nd1	*dum25*		Remacle et al. (2001a)
nd4	△*nd4, dum22*		Remacle et al. (2001b; 2006)
nd5	*dum5*		Cardol et al. (2002)
nd6	*dum17*		Colin et al. (1995)
	dum24	4.35-kb deletion affecting *cob*, *nd4* and part of *nd5*	Duby and Matagne (1999); Duby et al. (2001)

VI. HYDROGEN PRODUCTION

The ability of green algae to produce molecular hydrogen was first reported by Hans Gaffron and collaborators in 1942 (reviewed by Homann, 2003, and Melis and Happe, 2004), and explored further in *Chlamydomonas* by Frenkel (1949, 1951) and Abeles (1964). Work continued through the 1970s and 1980s, with characterization of hydrogenase activity and its expression under different culture conditions. Under anaerobic conditions in light, and in the absence of CO_2, hydrogenases function to release H_2 and O_2 from water (*biophotolysis*; see Bothe, 1982). They can also catalyze CO_2 reduction in algae illuminated after incubation with H_2 under anaerobic conditions (*photoreduction*) and can reduce various electron acceptors in the dark (see Kessler, 1974, and Maione and Gibbs, 1986a, for review of early work in this area). Evolution of H_2 from anaerobic cells can also occur in the dark through a fermentation pathway coupled with starch degradation. This has been particularly noted in *C. moewusii* (Klein and Betz, 1978b).

Brand et al. (1989) screened cultures in the University of Texas algal collection for ability to produce hydrogen through the biophotolysis reaction, and found that most of their *Chlamydomonas* isolates, as well as some colonial Volvocales, were capable of doing so (Table 6.17).

What might be called the modern age of hydrogen research in *Chlamydomonas* began with presentations at the 2000 AAAS meeting by Tasios Melis of a recently published method for sustaining H_2 production by cyclic sulfur deprivation (Melis et al., 2000) and by Elias Greenbaum and colleagues of prolonged H_2 evolution in cultures under nitrogen. A report on that meeting, entitled "Power from pond scum" (Kaiser, 2000), brought *Chlamydomonas* to the attention of the news media.

Light-dependent H_2 production can occur in two ways. Both pathways ultimately use Photosystem I to reduce ferredoxin, which under anaerobic conditions is coupled to the hydrogenase pathway. In direct biophotolysis the ultimate source of electrons is water oxidation by Photosystem II, whereas in indirect biophotolysis the electrons are derived from nonphotochemical reduction of plastoquinone through NAD(P)H–plastoquinone oxidoreductase (see Volume 2, Chapter 7 for details).

Algal hydrogenases are of the [FeFe] type, rather than the [NiFe] type found in cyanobacteria; these are reviewed in detail in Volume 2, Chapter 7. Hydrogenase activity is induced by transfer to anaerobic conditions, and can be sustained in the light if photosynthetic oxygen production is blocked. This can be achieved by sulfur deprivation (Ghirardi et al., 2000; Melis et al., 2000), which inhibits the rapid turnover of the D1 protein of Photosystem II. Oxygen exposure inactivates the enzyme irreversibly, and the *Chlamydomonas* hydrogenase appears to be unusually sensitive to oxygen.

Table 6.17 Hydrogenase activity in *Chlamydomonas* species

Species	Strain Number	Activity	References
C. actinochloris	UTEX 965	Not detected	Brand et al. (1989)
C. aggregata [now *C. applanata*]	UTEX 969	Not detected	Brand et al. (1989)
C. agloeformis [now *C. debaryana*]	UTEX 231	Not detected	Brand ct al. (1989)
C. akinetos [now *C. applanata*]	UTEX 967	Not detected	Brand et al. (1989)
C. applanata	UTEX 230	Yes	Brand et al. (1989)
C. chlamydogama	UTEX 102	Yes	Brand et al. (1989)
C. debaryana	UTEX 344	Not detected	Brand et al. (1989)
C. debaryana		Yes	Healey (1970a)
C. debaryana var. *cristata*	UTEX 1344	Yes	Brand et al. (1989)
C. dorsoventralis	UTEX 228	Yes	Brand et al. (1989)
C. dysosmos		Yes	Healey (1970a)
C elliptica var. *britannica* [now *C. culleus*]	UTEX 1059, 1060	Not detected	Brand et al. (1989)
C. eugametos	UTEX 9, 10	Yes	Brand et al. (1989)
			Abeles (1964); Ward (1970a)
C. gloeophila var. *irregularis*	UTEX 607	Not detected	Brand et al. (1989)
C. hindakii [now *C. noctigama*]	UTEX 1338	Yes	Brand et al. (1989)
C. hydra	UTEX 4	Yes	Brand et al. (1989)
C. intermedia		Yes	Ward (1970a)
C. melanospora	UTEX 2022	Not detected	Brand et al. (1989)
C. moewusii	UTEX 96, 97	Yes	Brand et al. (1989)
	UTEX 792	Yes	Brand et al. (1989)
	UTEX 2018	High rate	Brand et al. (1989)
	UTEX 2019	Yes	Brand et al. (1989)
			Frenkel (1949, 1951); Frenkel and Rieger (1951); Ward (1970a,b)
C. reinhardtii	UTEX 89, 90	Yes	Brand et al. (1989)
	UTEX 2246	Yes	Brand et al. (1989)
		Yes	Ben-Amotz and Gibbs (1975)
C. smithii [now *C. culleus*]	UTEX 1061	Not detected	Brand et al. (1989)
C. texensis [now *C. zebra*]	UTEX 1904	Yes	Brand et al. (1989)

Hydrogen production is enhanced by addition of acetate to the culture medium (Gibbs et al., 1986; Happe et al., 1994) and can be modulated by changes in CO_2 availability (Cinco et al., 1993; Semin et al., 2003) and by nitrate vs. nitrite concentrations (Aparicio et al., 1985). It is also affected by uncouplers of phosphorylation and inhibitors of ATP synthesis (Lee and Greenbaum, 2003; Kruse et al., 2005). Yields under laboratory conditions can be improved by manipulating conditions of cell growth and by optimizing sulfur concentration (Kosourov et al., 2002, 2005, 2007).

Current research efforts are being directed at modification of the hydrogenase protein to hinder access of oxygen to the active site (Ghirardi et al., 2005; Ghirardi, 2006). Attempts are also being made to construct strains in which photosynthesis or other physiological processes are modified in order to increase electron flow through the hydrogenase pathway. For example, the *stm6* mutant (Schönfeld et al., 2004; Surzycki et al., 2007; see also Volume 2, Chapter 22) is inhibited in cyclic electron transport around Photosystem I and is deficient in active Photosystem II reaction centers, It also accumulates starch and has a higher respiration rate than wild type. This strain shows significantly higher levels of H_2 production than wild-type strains (Kruse et al., 2005; Rupprecht et al., 2006). The *tla1* mutant, which has truncated light-harvesting antennae in both photosystems, is also being developed as a strain for hydrogen production (Tetali et al., 2007).

More information on hydrogen production and its future prospects can be found in Volume 2, Chapter 7, and in Melis (2007), Melis et al. (2007), and Posewitz et al. (2008).

Organelle Heredity

I. INTRODUCTION

Studies of organelle heredity and biogenesis in *Chlamydomonas* are to a very great extent studies on the chloroplast. The mitochondrial genome contains very few genes, and fewer genetic markers, but is nevertheless interesting by virtue of being unusual in size, gene content, and organization. There are several additional topics for discussion that embrace both chloroplasts and mitochondria, and both organelles will therefore be covered in this chapter.

The early literature on inheritance of "non-Mendelian" genes was reviewed in the books by Sager (1972) and Gillham (1978). At the time these books were written, it was clear that these genes must be in chloroplast DNA, but the detailed organization of the molecule was unknown, and there was disagreement on how one should go about constructing a genetic map of the markers that had been identified. By the 1980s, genes were being sequenced (Allet and Rochaix, 1979; Dron et al., 1982a; Woessner et al., 1986; Schneider and Rochaix, 1986; Kück et al., 1987), restriction maps of chloroplast DNA were being constructed (Rochaix, 1978; Grant et al., 1980; Palmer et al., 1985), mutations had been found in chloroplast genes encoding components of the photosynthetic apparatus (Shepherd et al., 1979; Dron et al., 1983), and chloroplast protein synthesis was being characterized (Chua et al., 1973a, c; Mets and Bogorad, 1974; Schmidt et al., 1983, 1984). With the advent of transformation technology in 1988 (see Chapter 8), the modern era began. The present chapter focuses mainly on inheritance of the organelle genomes, but includes descriptions of the chloroplast and mitochondrial DNA molecules, and a summary of transcription and translation of chloroplast genes. The latter topics are covered more fully in Volume 2. Additional reviews of the literature, especially of the 1980s and 1990s, can be found in the book edited by Rochaix et al. (1998).

II. UNIPARENTAL INHERITANCE

A. Inheritance of chloroplast genes in crosses

Chloroplast genetics in *Chlamydomonas* began with Sager's isolation of two streptomycin-resistant mutants that showed distinguishable phenotypes and different patterns of inheritance in crosses (Sager, 1954). The *sr-1* (now *sr1*) mutation conferred resistance to relatively low levels of streptomycin (50–100μg/ml) and was inherited in a 2:2 Mendelian pattern, while the *sr-2* (now *sm2* or *sr-u-sm2*) mutant produced resistance to 1 mg/ml or more streptomycin and was inherited uniparentally from the *plus* parent (Figure 7.1). The latter pattern is often called maternal inheritance by analogy with chloroplast and mitochondrial inheritance in higher plants and mitochondrial inheritance in animals. However, because most *Chlamydomonas* species do not differentiate into distinctive male and female gametes (see Chapter 5), the convention used here will be the one proposed by Mets (1980): UP+ and UP– will refer to uniparental inheritance from the *plus* and *minus* parents, respectively, as defined in Table 5.2, and BP will refer to biparental, but non-Mendelian, inheritance. This convention has the advantage of applying equally to *C. reinhardtii* and to *C. eugametos*/*C. moewusii*, in which traditional "male" and "female" identities have been assigned on different rationales (see Chapter 5). At the time

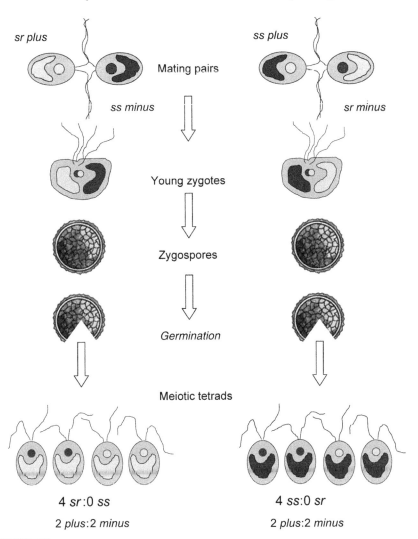

**Nearly all zygospores show
uniparental inheritance of chloroplast genes**

sr plus

ss plus

Mating pairs

ss minus

sr minus

Young zygotes

Zygospores

Germination

Meiotic tetrads

4 sr:0 ss

4 ss:0 sr

2 plus:2 minus

2 plus:2 minus

FIGURE 7.1 *Uniparental inheritance of streptomycin resistance (light-colored chloroplasts) in reciprocal crosses. All four progeny have the resistance phenotype of the plus parent, whereas mating type (indicated as dark vs. light nuclei) segregates in a Mendelian 2:2 fashion regardless of the direction of the cross. Courtesy of Karen VanWinkle-Swift.*

of Sager's early work, however, it was by no means certain that the mutations she was studying were in chloroplast DNA, and they were therefore usually described as non-chromosomal, cytoplasmic, or non-Mendelian, with non-Mendelian eventually emerging as the preferred terminology (see Gillham, 1969, for dicussion of the early literature).

Genetic analysis of uniparentally inherited mutations finally converged with biochemical studies of chloroplast DNA around 1980, when physical markers in chloroplast DNA were shown to display the same UP+ pattern of inheritance long known for antibiotic resistance and other genetic markers (Grant et al., 1980; Mets, 1980; Lemieux et al., 1980).

Analysis of chloroplast gene inheritance in *Chlamydomonas* proceeded along two main lines, purely genetic studies directed at determining the number of segregational and recombinational units of uniparentally transmitted markers, and biochemical analysis of chloroplast DNA throughout the life cycle. Eventually the genetic experiments were superseded by molecular analysis of the chloroplast genome (see below). Rare exceptions to the pattern of uniparental inheritance were observed in the early experiments, and permitted detection of recombinants between parental genotypes (Sager and Ramanis, 1965; Gillham, 1965b). Extensive genetic analysis did not begin until about 1967, however, when Sager and Ramanis reported a major breakthrough, the discovery that UV irradiation of *plus* gametes prior to mating greatly augmented transmission of markers from the *minus* parent.

Both Sager and Gillham and their colleagues exploited this phenomenon to develop methods for quantitative analysis of recombination (reviewed by Gillham, 1969; Sager, 1972; Harris, 1989). In general, Sager favored methods based on analysis of segregation of chloroplast markers in the early postzygotic divisions, while most cells were still heteroplasmic, that is, still carried alleles from both parents, whereas the Boynton-Gillham group primarily worked with recombination frequencies in randomly selected clones from biparental zygotes, chosen many generations after meiosis when segregation was essentially complete. Mets and Geist (1983) proposed a modified zygote clone analysis in which only single progeny meeting specific phenotypic criteria were chosen as representatives of each biparental zygote.

Studies in both the Sager and Boynton-Gillham laboratories resulted in publication of genetic maps for chloroplast genes (Sager and Ramanis, 1976a, b; Singer et al., 1976; Harris et al., 1977), which were however mutually inconsistent. The debate finally ended with the demonstration that nearly all the antibiotic resistance mutations used as genetic markers were in fact point mutations in the genes encoding the chloroplast 16S and 23S ribosomal RNAs (Harris et al., 1989). The one exception was Sager's *sm2* marker, which proved to be a point mutation in the chloroplast gene encoding ribosomal protein S12 (Liu et al., 1989). As in most land plants, in *Chlamydomonas* the rRNA genes are encoded in a region of the chloroplast genome that is present in an inverted repeat. The *rps12* gene lies outside the repeat, but close to the gene encoding 16S rRNA in the nearest repeat region, and the realization that this gene was in the single-copy region of the genome neatly solved the previous problems with interpretation of genetic mapping data regarding the *sm2* marker (Harris et al.,

1989). Thus the markers being studied occupied only a small fraction of the chloroplast genome, which genetically behaved as a linear region (Harris et al. 1977), rather than the circular map proposed by Sager and colleagues (Singer et al., 1976). However, if one looks only at the five antibiotic resistance markers in common between these two alternative maps, the order is the same. Efforts to map acetate-requiring (non-photosynthetic) mutations in the chloroplast genome were unsuccessful, but soon became unnecessary as the genes were sequenced.

Ironically, because so many different laboratories were sequencing individual genes in the *Chlamydomonas* chloroplast genome during the 1980s and 1990s, completion of the entire genome sequence was delayed compared to other algae and many plants and was finally achieved by Maul et al. (2002), who assembled the previously sequenced pieces and filled in the remaining gaps. The chloroplast genome is discussed in more detail in section III.A. and in Volume 2, Chapter 24.

Conflicting models were put forward by Sager (1972; Sager and Ramanis, 1976a, b) and by Gillham, Boynton, and colleagues (reviewed by Gillham, 1974, 1978) for the organization, partitioning and segregation of UP genetic units. Part of the controversy involved the question of how many functional copies of the organelle (chloroplast) genome are present in each cell. Measurements of kinetic complexity suggested a main component of chloroplast DNA of about 2×10^8 D, which could be calculated to be present in 25–50 copies per cell (Wells and Sager, 1971; Bastia et al., 1971). However, Sager argued that her genetic data, obtained using the techniques of allelic segregation in pedigrees and cosegregation frequency analysis, implied the existence of only two chloroplast genomes per *plus* gamete. She suggested that the functional genome was a minor, slow-renaturing component of approximately 8×10^8 D, present only in one or two copies per cell, whereas Boynton, Gillham and coworkers presented data based largely on zygote clone analysis (Gillham et al., 1974; Conde et al. 1975), but also on pedigree studies (Gillham, 1969; Forster et al., 1980), supporting a multicopy model and consistent with a genome number greater than two, but not necessarily as large as 80.

The population of homoplasmic segregants from biparental zygotes and from vegetative diploids is typically biased in favor of the alleles contributed by the *plus* parent (see Birky et al., 1981; Forster et al., 1980; Matsuda et al., 1983; Matagne and Schaus, 1985; Galloway and Holden, 1985, and references cited therein; only Sager reported seeing no such effect). Such a bias can be explained by an unequal input of the two parental genomes in a multicopy system. This is consistent with the observation that treatments that reduce the input from the *plus* parent, such as UV irradiation (Adams 1978) or growth in 5-fluorodeoxyuridine (Wurtz et al., 1977), lead to an increase in the number of BP and UP– zygotes.

Molecular studies eventually resolved the size question by showing that the chloroplast DNA is a circular molecule of about 1.3×10^8 D. Thus, there probably really are about 80 copies of the genome in vegetative cells and perhaps 40 in gametes (Chiang and Sueoka, 1967a; Gillham, 1978). However, the consensus from microscopic studies and theoretical analysis is that these molecules are grouped into a smaller number of functional units.

In cells stained with the DNA-binding fluorochrome DAPI (4',6-diamidino-2-phenylindole), chloroplast DNA fluorescence is concentrated in discrete bodies, or nucleoids, each 0.3–0.5 μm in diameter (Coleman, 1978; Kuroiwa et al., 1981). Nakamura et al. (1986) reported that the configuration of nucleoids changes with cell age and developmental stage; the presence of 8–10 small nucleoids is typical of recently divided cells (Figure 7.2), but condensation into a single large nucleoid has been observed in old cells in stationary phase. Mutants have been isolated in which this condensation does not occur (Nakamura et al., 1994), and also a mutant in which cells have a single chloroplast nucleoid throughout their life cycle, but in which uniparental inheritance of organelle genes appears to be unaffected (Nakamura et al., 2004).

Adams (1978) used data from crosses in which both parents were irradiated with varying doses of UV light to calculate that the choice of chloroplast genome units for transmission was made from a population averaging 27 UP+ and 2 UP– units per cell with each unit perhaps consisting of more

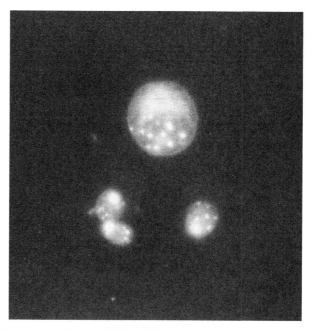

FIGURE 7.2 *Nucleoids of chloroplast DNA in DAPI-stained cells. The three small cells are newly hatched after division and show 7, 7, and 9 nucleoids respectively. The large cell, in which 21 nucleoids can be counted, is nearly ready to divide. Courtesy of Annette Coleman.*

than one genome. VanWinkle-Swift (1980) proposed that nonrandom distribution of nucleoids to daughter cells, a consequence of the arrangement of nucleoids with respect to the plane of division, would further increase rates of segregation and that recombination would occur only when nucleoids fuse. Birky et al. (1981) extended this model by testing the possibility that allelic frequencies within zygote clones were subject to random drift, and proposed that repeated rounds of recombination among randomly paired molecules within a given cell would further change the frequencies of specific alleles, by analogy with phage crosses (Visconti and Delbrück, 1953) and yeast mitochondrial genetics (Dujon et al., 1974).

B. The fate of chloroplast DNA in zygotes

In contrast to the situation in many higher plants, where chloroplasts themselves appear to be transmitted only through the maternal parent (reviewed by Sears, 1980c), gametes of *Chlamydomonas* fuse in their entirety. For uniparental inheritance to occur under these circumstances, the organelle genome of one parent must be destroyed or at least prevented from replicating. The presumption that uniparentally inherited genes were in fact located in chloroplast DNA prompted Chiang (1968) and Sager and Lane (1972) to investigate the fate of this DNA in young zygotes after mating. Chiang's preliminary conclusion that chloroplast DNA from the *minus* parent persists after mating was subsequently revised on the grounds that the labeled precursors used to mark DNA from the two parents appeared in a compartmentalized chloroplast DNA pool during zygotic development and were probably recycled into new molecules (Chiang, 1971). Sager and Lane reported, based on data from reciprocal crosses with ^{14}N and ^{15}N-labeled gametes, that chloroplast DNA from the *plus* parent underwent a density shift within the first 6 hours after gamete fusion, while that from the *minus* parent disappeared.

Ultrastructural studies of the mating reaction demonstrated that the chloroplasts of the *plus* and *minus* gametes fuse between 3 hours and 7 hours after mating (Cavalier-Smith 1970). Microscopic investigation using DAPI staining showed that chloroplast nucleoids from the *minus* parent disappear even before this time, although this does not necessarily indicate that the chloroplast DNA had actually been destroyed. Within the first 40–50 minutes after mating, the chloroplasts of the parental cells appear to be in close apposition, and nucleoids have disappeared from the chloroplast contributed by the *minus* parent (Kuroiwa et al., 1982; Coleman, 1984; Figure 7.3). (Kuroiwa established the identity of the gametic chloroplasts within the quadriflagellate zygote by using a mutant with short flagella for one parent in a cross, and by making crosses between small and large gametes. Coleman prestained gametes of one mating type before mating to unstained gametes). Coleman and Maguire (1983) reported that

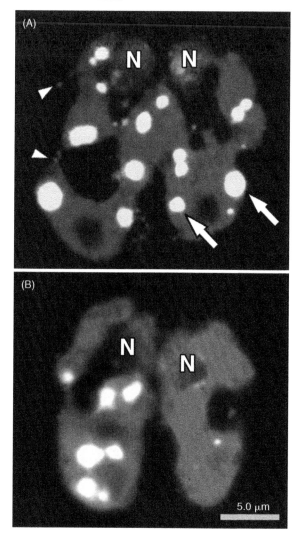

FIGURE 7.3 *Young living zygotes stained with the DNA-specific fluorochrome SYBR Green I, 30 minutes (A) and 60 minutes (B) after mating. "N" is the cell nucleus and white spots (~1–2 μm: arrows) overlapping the gray area (red autofluorescence from chlorophyll) are chloroplast nucleoids. Minute spherical spots (~0.2 μm: arrowheads) outside the chloroplasts are mitochondrial nucleoids. At 30 minutes after mating, almost equal numbers of chloroplast nucleoids were observed in both of the chloroplasts (A). However, after 60 minutes, the chloroplast nucleoids in the minus chloroplast (right) disappeared completely, indicating the degradation of minus chloroplast DNA. Courtesy of Yoshiki Nishimura, modified from Nishimura et al. (1998).*

uniparental destruction of chloroplast nucleoids in *C. moewusii* did not occur until 9–10 hours after mating, at a time when vis-à-vis pairs still persisted and nuclear fusion was just beginning. The decline in visible nucleoids per zygote continued for the next 4–5 hours in zygotes kept in the light but was arrested in zygotes transferred to the dark.

Using optical tweezers and PCR-based analysis on single zygotes formed from transgenic *C. reinhardtii* gametes expressing the bacterial *aadA* gene (Chapter 8), Nishimura et al. (1999) demonstrated that chloroplast DNA of the *minus* parent is actively digested during the period when *minus* nucleoids disappear, rather than simply diffusing throughout the chloroplast.

Destruction of the *minus* chloroplast nucleoids is blocked by nuclease-inhibiting agents, such as aurintricarboxylic acid and ethidium bromide, and by cycloheximide, which inhibits protein synthesis on cytosolic ribosomes, but not by inhibitors of chloroplast protein synthesis (Kuroiwa et al., 1983a). In agreement with the studies of Sager and Ramanis (1967), Kuroiwa et al. (1983b, 1985) also found that UV treatment of the *plus* but not the *minus* parent interfered with nucleoid destruction. Nishimura et al. (2002) identified a calcium-dependent nuclease activity that was localized to chloroplasts from the *minus* parent in newly fused cells, but was first detectable in *plus* gametes during their maturation. This enzyme is active within the first 60–90 minutes after zygote formation, before chloroplast fusion occurs. Nishimura et al. proposed that chloroplast DNA in *plus* chloroplasts is resistant to digestion, and reviewed the arguments for and against a restriction-modification mechanism, a topic to be discussed more fully in the next section.

After chloroplast fusion, the remaining nucleoids coalesce to a final average of two or three per zygotic plastid (Birky et al., 1984; Coleman, 1984). During germination of the zygote, new chloroplast DNA synthesis appears to occur, as judged by quantitative fluorescence measurements indicating that newly released zoospores contain 3.5 times as much chloroplast DNA as gametes (Coleman, 1984).

C. Methylation of chloroplast DNA

Sager and her colleagues proposed that a restriction-modification system is responsible for the selective degradation of chloroplast DNA from the *minus* parent in crosses of *Chlamydomonas* (Sager and Lane, 1972; Sager and Ramanis, 1973, 1974; Sager and Kitchin 1975). They subsequently published several papers suggesting that this system is based on protection of chloroplast DNA of *plus* gametes by methylation, consistent with a shift in buoyant density observed in *plus* gametes and in young zygotes (Sager and Lane, 1972). Burton et al. (1979) reported that chloroplast DNA from *plus* gametes contained a measurable quantity of 5-methyl cytosine, whereas no methylation was detected in chloroplast DNA from *minus* gametes or from vegetative cells of either mating type. These results were confirmed by Sano et al. (1980) using antibodies specific for 5-methyl cytosine. In zygotes assayed 6 hours after mating, the *plus* chloroplast DNA appeared to be heavily methylated, and some methylation was also seen in the *minus* chloroplast DNA, which was largely degraded by this time

(Burton et al., 1979; Royer and Sager, 1979). Feng and Chiang (1984) reported that during gametogenesis methylation of deoxycytidine increased at least 20-fold over the level seen in vegetative cells. Although methylation increased in both *plus* and *minus* cells, it was always at least threefold higher in *plus*. In fully differentiated *plus* gametes, 12.1% of the deoxycytidine residues were methylated, and within 7 hours after zygote fusion this level had risen to nearly 50% (Feng and Chiang, 1984).

Arguing against the hypothesis that methylation protected chloroplast DNA of the *plus* parent from degradation, a very high level of constitutive methylation resulting from the nuclear *me1* mutation did not alter the pattern of chloroplast DNA inheritance (Bolen et al., 1982; see also Dyer, 1982). However additional methylation of chloroplast DNA was observed in *plus* cells of this mutant (Sager and Grabowy, 1983). Feng and Chiang (1984) reported normal uniparental inheritance in crosses where methylation was inhibited by treatment of gametes with L-ethionine or 5-azacytidine (Feng and Chiang, 1984), although Umen and Goodenough (2001b) reported that 5-aza-2-deoxycytidine did alter the inheritance pattern. Taken together, these results suggested that hypermethylation *per se* did not protect *plus* DNA from degradation but that certain specific sites must be methylated if this mechanism is to work. Also, these specific sites would have to be methylated even in the presence of inhibitors of methylation, as in the experiments by Feng and Chiang (1984). Dedifferentiation of gametes by restoration of nitrogen to the culture medium led to gradual loss of methylation at a rate consistent with dilution of methyltransferase activity by cell division (Sano et al., 1984). No rapid loss of methylation was observed, as might have been expected if an enzymatic demethylating activity were present.

A mechanism that absolutely eliminated chloroplast DNA from *minus* gametes would of course result in 100% UP+ inheritance. In Sager's early experiments, this was very nearly the case, with only about 0.1% of zygotes showing BP or UP− inheritance (Sager and Ramanis, 1963, 1967). In experiments in the Boynton-Gillham laboratory, however, up to 5% of zygotes expressed markers from the *minus* parent (Gillham, 1969; Gillham et al., 1974), and there is evidence that even more zygotes may harbor "hidden" copies of the chloroplast genome contributed by the *minus* parent (see below). The discrepancy in exceptional zygote frequencies seen in the two laboratories was attributed to differences in the procedures by which gametes were formed, although strain differences may also have played a role (Sears et al., 1980). To account for the existence of exceptional zygotes in terms of a restriction-modification mechanism, one must assume either that some chloroplast DNA molecules in *minus* gametes are protected, or that the *plus* restriction system is somewhat inefficient. Another observation that should be taken into account is the demonstration by Wurtz et al. (1977) that reduction of the number of copies of chloroplast DNA in *plus* cells by

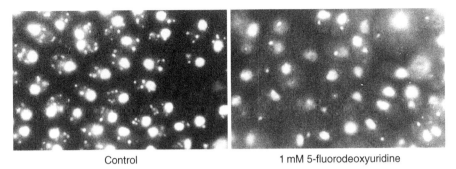

Control	1 mM 5-fluorodeoxyuridine

FIGURE 7.4   *DAPI staining of nucleus and chloroplast nucleoids from wild-type cells. Left, control; right, cells grown for 8 days on agar containing 1.0 mM 5-fluorodeoxyuridine, which reduces copy number of chloroplast DNA. From Matagne and Hermesse (1981).*

treatment with the thymidine analog 5-fluorodeoxyuridine (Figure 7.4) leads to an increase in the number of BP and UP– zygotes. In contrast to UV treatment, which produces mostly BP rather than UP– zygotes at sublethal doses, 5-fluorodeoxyuridine treatment even at the lowest effective doses yields a substantial fraction of UP– zygotes. Some UP+ zygotes are always seen even at higher doses, however, consistent with other lines of evidence suggesting that total elimination of chloroplast DNA is lethal (Liu et al., 1993; Boudreau et al., 1997a).

The presence of DNA methyltransferase activity specific to gametic and zygotic cells was documented by Sano et al. (1981), and a chloroplast-specific enzyme was later characterized (Nishiyama et al., 2002, 2004). Contrary to the prediction that specific sites needed to be methylated to protect chloroplast DNA, this enzyme appeared to be nonselective; that is, methylation of all cytosine residues occurred, regardless of the neighboring sequence of the DNA. When *minus* cells were transformed with the DNA encoding this enzyme, under control of a constitutive promoter, the same nonselective pattern of methylation was observed. When these transgenic cells were crossed to normal *plus* cells, the frequency of UP– and BP zygotes increased substantially. Nishiyama et al. (2004) concluded that their results did support a restriction-modification mechanism for protection of chloroplast DNA.

However, the results of Umen and Goodenough (2001b) and Nishimura et al. (2002) suggest that the explanation is more complex. Umen and Goodenough found that in *plus* cells treated with the methylation inhibitor 5-aza-2-deoxycytidine, the hypomethylated DNA was not degraded in the young zygote, as a simple restriction-modification model would predict, but that it persisted until zygote germination, at which point it failed to replicate at normal levels. They proposed that germination is the critical time at which unmethylated or damaged chloroplast DNA is finally destroyed.

The nuclease characterized by Nishimura et al. (2002) degrades *minus* chloroplast DNA to the nucleotide level, prior to chloroplast fusion, rather than leaving small DNA fragments as would be expected for a restriction endonuclease. Nishimura et al. considered several possibilities for how *plus* chloroplast DNA might be protected from the nuclease, especially in view of the evidence indicating that methylation, either general or site-specific, is probably not sufficient to explain all the results. Since the nuclease requires Ca^{2+} for full activation, they postulated that regulation of calcium levels within the chloroplast is important in the protection process, and called attention to the growing number of papers on zygote-specific gene expression. In particular, the *EZY1* gene encodes a protein that localizes to chloroplast nucleoids immediately after mating (Armbrust et al., 1993) and is therefore a candidate for involvement in the mechanism of protection. Taking into consideration the results of Umen and Goodenough (2001b) showing higher replication rates for methylated *plus* DNA in germinating zygotes, Nishimura et al. suggested that two or more distinct mechanisms might operate cooperatively to bring about uniparental inheritance of chloroplast genes.

D. Inheritance of mitochondrial DNA

Boynton et al. (1987) found that in meiotic zygotes derived from crosses between *C. smithii* (*plus*) and *C. reinhardtii* (*minus*), the mitochondrial genome showed UP– inheritance. That is, while chloroplast genes were inherited from the *plus* parent in more than 95% of zygotes, mitochondrial DNA was inherited from the *minus* parent in more than 99%. The same inheritance pattern was seen in reciprocal crosses, in which a *plus C. reinhardtii* parent was mated to a *minus* strain carrying the chloroplast and mitochondrial genomes of *C. smithii* in a hybrid nuclear background. The pattern of inheritance was not perturbed by UV irradiation of either parent, in contrast to the effect of UV on chloroplast genes discussed in section II.B. Only a single exceptional zygote showing UP+ inheritance of mitochondrial DNA was found among 128 examined. Also using *C. smithi* × *C. reinhardtii* crosses, Beckers et al. (1991) found by Southern blotting that mitochondrial DNA from the *plus* parent disappeared gradually during zygospore maturation, and the UP– pattern of inheritance was also confirmed by crosses of *C. reinhardtii* strains with mitochondrial mutations (Dorthu et al., 1992; Colin et al., 1995).

In crosses of *C. eugametos* × *C. moewusii*, uniparental inheritance of mitochondrial DNA was also observed, but in this case the pattern was UP+, the same as observed for chloroplast genes (Lee et al., 1990, and see Table 5.2).

In apparent conflict with the earlier results, Nishimura et al. (1998) reported biparental transmission in zygotes of mitochondrial nucleoids, which they were able to identify for the first time using the fluorescent dye SYBR Green I. They observed about 20 mitochondrial nucleoids per cell, each containing an estimated 15 genomes. However, Nakamura et al. (2003) used the same staining technique coupled with PCR analysis of individual zygotes in crosses in which the *C. smithii* and *C. reinhardtii* mitochondrial genomes were present in the *C. reinhardtii* nuclear background, and found that mitochondrial nucleoids from the *plus* parent disappeared between 3 and 6 hours after zygote formation. In later experiments, using a restriction site change in the *cox1* gene as a marker, rather than the intron present in *C. smithii*, Aoyama et al. (2006) reported that destruction of mitochondrial DNA from the *plus* parent occurred during meiosis in the germinating zygote.

E. Chloroplast gene transmission in vegetative diploids

In a typical cross up to about 5% of mated pairs do not form a hard-walled zygospore that subsequently undergoes meiosis, but instead divide soon after mating to produce a clone of vegetative diploid cells (Ebersold, 1967, and see Chapter 5). In these cells the mechanism leading to uniparental inheritance of chloroplast genes appears to be inactive, and a large proportion of the colonies arising express chloroplast markers from both gametic parents, although a bias is seen for markers inherited from the *plus* parent (Gillham, 1963). These markers segregate somatically in mitotic divisions, and eventually subclones are recovered that are homoplasmic (pure) for one or the other parental marker or a recombinant type (Gillham, 1963, 1969). These observations were exploited in several classic studies with the ultimate goal of elucidating the basis for uniparental inheritance.

VanWinkle-Swift (1978) showed that the proportion of vegetative diploids expressing chloroplast markers from the *minus* parent dropped significantly if the first mitotic division was delayed by nitrogen starvation or incubation in darkness. Together with experiments by Sears (1980a, b) showing that extended maturation periods for meiotic zygotes also result in a reduction in the proportion of biparental zygotes, these studies suggest that the environment can exert a significant influence on chloroplast gene inheritance. VanWinkle-Swift postulated that polyploidy of the chloroplast genome might be a luxury allowed only in favorable environments and that elimination of chloroplast genome copies would permit energy conservation in times of environmental stress. Sears and VanWinkle-Swift (1994) elaborated further on this idea in their Salvage/Turnover/Repair (STOR) model, which proposed a selective advantage to the organism in re-using nucleotides recovered from DNA degradation. An alternative explanation for the evolutionary selection of uniparental inheritance was advanced by Coleman (1982b), that the elimination of paternal genomes in *Chlamydomonas* and other algae is a manifestation of mechanisms that protect egg cells from infection by foreign organisms, perhaps potential endosymbionts.

Whatever the evolutionary reason for uniparental inheritance, studies with vegetative diploids show unequivocally that its control by mating type can be inactivated under some conditions. Natural mt^+/mt^- diploids arising from sexual crosses mate as *minus* cells (Gillham, 1963a; Ebersold, 1967), as do mt^+/mt^- diploids produced by polyethylene glycol-induced fusion of vegetative or gametic cells (Matagne et al., 1979; Matagne and Hermesse, 1980a). That is, *minus* is dominant to *plus* with respect to sexual recognition and fusion. As discussed at length in Chapter 5, this is controlled by the *MID* (Minus Dominance) gene located in the mating type locus of *minus* cells. (Fusion diploids with the constitution mt^+/mt^+ and mt^-/mt^- mate as *plus* and *minus* respectively; therefore the diploid state itself does not affect genetically determined mating type). Although both natural and fusion-induced diploids show predominantly biparental transmission of chloroplast genes, in sexually induced diploids there is a bias favoring chloroplast alleles from the *plus* parent, whereas in fusion diploids the distribution is essentially random, whatever the mating types of the parent cells (Matagne, 1981; Eves and Chiang, 1984; Tsubo and Matsuda, 1984; Galloway and Holden, 1985). One infers that the process that normally destroys chloroplast genes from the *minus* parent in a sexual cross is still partially active in mitotic diploids arising in such a cross, but that this process is activated only during natural mating. The suggestion that treatment of one parent with isolated flagella of the opposite mating type prior to polyethylene glycol-induced fusion led to biased inheritance of chloroplast genes (Adams, 1982) was subsequently disproved (Galloway and Holden, 1984; Matagne and Schaus, 1985).

When diploid cells are mated with haploid partners, the mating process itself and subsequent zygospore formation proceed normally. Although the ensuing triploid meiosis is marked by a high degree of lethality among the progeny, a sufficient number survive to permit assessment of transmission of chloroplast markers. Gillham (1969) reported that in crosses of haploid *plus* cells to diploid mt^+/mt^-, streptomycin resistance could be transmitted efficiently from either parent. Matagne and Mathieu (1983) confirmed these results in haploid × diploid crosses in which both parents carried chloroplast antibiotic resistance markers. Both natural and fusion-induced diploids were tested. In crosses of haploid *plus* × diploid homozygous *minus*, and in diploid homozygous *plus* × haploid *minus*, inheritance was overwhelmingly UP+. However, in the haploid *plus* × heterozygous diploid crosses, biparental inheritance was seen. In tetraploid crosses of the form homozygous *plus* × homozygous *minus*, inheritance was also UP+, while in crosses of $mt^+/mt^+ \times mt^+/mt^-$, inheritance was predominantly biparental. Thus although *minus* is dominant to *plus* with respect to the mating process itself, some genetic functions of the *plus* mating type locus in the heterozygous diploid are still expressed and prevent destruction of the chloroplast genome from this phenotypically *minus* parent.

F. Persistent heteroplasmicity of chloroplast genes

Immediately after mating of gametes with distinguishable chloroplast markers, a transient heteroplasmic state must always ensue. As described above, chloroplast genomes contributed by the *minus* parent are thought to disappear in most cases within the first 6 hours after mating. In a few meiotic zygotes, and in the majority of vegetative diploid cells, genomes from both parents may continue to be present for several generations. Eventually, however, most of these seem to segregate in mitosis, so after 10–20 divisions nearly all cells appear to be homoplasmic either for one parental genome or for a recombinant genome. That is, no further segregation occurs of cells expressing any other phenotype in subsequent mitotic divisions.

Sager (1972) described occasionally finding cell lines that appeared to be homoplasmic in vegetative growth, but that segregated two distinct phenotypes in subsequent meioses. She called these lines *persistent cytohets*. For example, a streptomycin-resistant clone was isolated after withdrawal of streptomycin from a streptomycin-dependent line of *plus* cells. When these resistant cells, which appeared to be stable and homoplasmic, were crossed to *minus* cells, streptomycin-resistant and -dependent progeny were recovered, in what appeared to be a 50:50 ratio. Other instances of the same phenomenon were observed in multiply antibiotic-resistant progeny from biparental zygotes, in which case apparently homoplasmic cells in a cross to an antibiotic-sensitive strain segregated a large proportion of the two original parental genotypes that had given rise to the multiply resistant strain.

Sager (1972, 1977) interpreted this phenomenon in terms of her two-copy model of chloroplast genome organization and attributed it to a block in recombination, perhaps resulting from a change in orientation of chloroplast DNA molecules. Schimmer and Arnold (1969, 1970a–d), working with streptomycin-dependent segregants from sensitive revertants of a dependent strain, seemed to be observing the same phenomenon. They concluded, however, that they were seeing random segregation of a gene present in many copies. Since Sager and Ramanis (1968, 1970) had already reported that chloroplast genes were present in no more than two copies per cell, Schimmer and Arnold decided that their multicopy genes must therefore be mitochondrial. Sager (1972) dismissed this conclusion and suggested that if Schimmer and Arnold had tested their strains by crossing, they would have realized that their strains were persistent cytohets.

Although the mechanism by which chloroplast genomes might be sequestered in vegetative cells and not expressed is still unclear, the phenomenon itself is much easier to accommodate in a multicopy model than in Sager's two-copy model. Bolen et al. (1980) in an extensive reinvestigation of streptomycin dependence and its reversion presented evidence that the *sd-u-3-18* mutation studied by Schimmer and Arnold was in a chloroplast gene whose product is a constituent of the chloroplast ribosome. They suggested that expression of a minority population of streptomycin-dependent alleles in a cell growing on streptomycin-free medium would probably not be detrimental to cell growth, whereas expression of sensitive alleles during growth on streptomycin could be lethal. Thus "hidden" genomes carrying the dependent allele would not be selected against and could persist for many generations. Possibly, the reduction in chloroplast nucleoid number after fusion creates a "bottleneck" that allows recovery and subsequent amplification of these minority genomes in crosses.

The results of Bolen et al. suggest that the persistent heteroplasmic state may be less rare than was originally supposed, and that it can occur whenever the minority genome is not subjected to severe selective pressure (e.g., growth on antibiotic-containing medium). Adams (1975) reported that apparently UP+ progeny selected soon after meiosis and subjected to a second round of crosses transmitted chloroplast genes of the original *minus* parent, and Chua (1976) demonstrated persistent heteroplasmicity for an electrophoretic variant of a thylakoid membrane polypeptide with no overt phenotype.

Spreitzer et al. (1984, 1985; Spreitzer and Chastain, 1987) found that apparent revertants of a nonsense mutation in the chloroplast *rbcL* gene encoding the Rubisco large subunit resulted from intergenic suppression but appeared to be heteroplasmic, giving rise to both mutant and wild-type cells in crosses or when subcloned from cells grown under nonselective conditions. When "revertant" cells were kept in the light on minimal medium, Rubisco holoenzyme was made, but in less than wild-type amounts, consistent with the idea that each cell contained a mixture of wild-type and mutant genomes. On acetate medium in the dark, cells with the mutant phenotype appeared to have a selective advantage. Spreitzer et al. (1984) postulated that all the chloroplast genomes retained the nonsense mutation in the *rbcL* gene but that a suppressor mutation, possibly an altered tRNA, was present elsewhere in some, but not all, of the chloroplast genomes. The homoplasmic suppressor condition would presumably be lethal on minimal medium.

In practical terms, the possibility of persistent heteroplasmicity prompts a cautionary note: all newly arising chloroplast mutants, segregants from crosses, and transformants should probably be treated as potentially heteroplasmic until proven otherwise (see Gillham, 1969; see also Chapter 8, section IX.E., and Purton, 2007). Maintaining antibiotic-resistant isolates on antibiotic medium from several rounds of transfer appears to be sufficient to produce homoplasmic isolates, but if the "hidden" allele is not selected against, repeated subcloning is essential.

G. Mutations that alter transmission of chloroplast genes

In 1974, Sager and Ramanis described two mutations that affected the pattern of transmission of chloroplast genes. One mutation, *mat-1*, was isolated in a *minus* clone from progeny of a routine cross and appeared to be linked to the mating type locus. Progeny from crosses of this stock to *plus* cells showed a substantial increase in BP and UP– inheritance. Sager et al. (1981) reported that chloroplast DNA of *mat-1* cells became methylated during gametogenesis. Subsequent work suggested that the original *mat-1* stock was in fact a spontaneous diploid isolate. The *minus* mating type, the increased frequency of exceptional zygotes, and the methylation during gametogenesis, as well as poor viability in crosses, noted by Sager and Ramanis, are all consistent with this conclusion, which was eventually verified by microscopy (Sager and Grabowy 1985) and by genetic analysis (Gillham et al., 1987b). A second mutation, *mat-2*, was linked to the mating type locus in a *plus* isolate and led to an unusually high frequency of UP+ transmission; that is, the usual low frequency of BP and UP– progeny was virtually eliminated (Sager and Ramanis, 1974). This stock was not subjected to further study and Sager eventually lost it.

Gillham et al. (1987b) isolated three mutations, also mating type-linked in *plus*, which they designated *mat-3-1* through *mat-3-3*. All had the property of increasing BP and UP– transmission in crosses. Gillham et al. postulated that these mutations interfered with synthesis or function of the gene product that eliminates chloroplast DNA from the *minus* parent. Armbrust et al. (1995) subsequently found however that *mat3* cells were unusually small, and attributed the alteration in inheritance pattern to their reduced amount of chloroplast DNA. Umen and Goodenough (2001a) showed that *MAT3* corresponds to a gene homologous to human retinoblastoma protein, involved in control of cell division (see Chapter 3).

H. Inheritance of chloroplast genes in other *Chlamydomonas* species

Inheritance of antibiotic resistance markers is predominantly uniparental through the *plus* parent in the heterothallic species *C. eugametos* (McBride and McBride, 1975) but is biparental in a substantial fraction of progeny from crosses of the closely related species *C. moewusii* (Lee and Lemieux, 1986). These two species are interfertile, although a high degree of zygotic lethality is observed (Gowans, 1963; Cain, 1979; Lemieux et al., 1980). Among the surviving progeny of a *C. eugametos* × *C. moewusii* cross, Lemieux et al. (1980, 1981, 1984a, b) found a strong bias (>95%) in favor of a streptomycin resistance marker contributed by the *plus* (*C. eugametos*) parent, but determined from restriction digests that recombination of the chloroplast genomes in fact had occurred in most zygotes. In backcrosses of F1 progeny from the interspecific crosses to *C. moewusii*, product survival was improved, and a greater degree of biparentality for the antibiotic resistance markers was observed. This may possibly represent a situation of nuclear-organelle genome incompatibility analogous to plastome incompatibility in higher plant genera such as *Oenothera* (see Sears, 1980c).

In the homothallic species *C. monoica*, VanWinkle Swift and Aubert (1983) showed that chloroplast genes were also inherited uniparentally. Using strains carrying complementary recessive lethal markers, which were thus unable to self-mate, they showed that the inheritance of an erythromycin resistance marker was uniparental in all tetrads examined. In approximately half the tetrads all four products were erythromycin sensitive and in the remaining ones all four products were sensitive.

III. ORGANELLE DNA

A. Chloroplast genome structure and gene content

The 203-kb chloroplast genome of *C. reinhardtii* (Maul et al., 2002; Figure 7.5) is substantially larger than the chloroplast genomes of typical plants (e.g., spinach: 150 kb; tobacco: 155 kb). As in most higher plants, two single-copy regions are separated by inverted repeat sequences containing the ribosomal RNA genes, but the relative positions of most of the genes involved in photosynthetic functions are quite different, and there are some differences in gene content (Table 7.1). The chloroplast genome of *C. eugametos* is even larger (243 kb) and it differs both from the higher plant model and from *C. reinhardtii* in gene placement (Lemieux and Lemieux, 1985). The *C. moewusii* chloroplast genome is essentially colinear with that of *C. eugametos*, with which it is interfertile, but is larger yet as the result of insertions of 21 kb within the inverted repeat region and 5.8 in a single-copy region. In both *C. eugametos* and *C. moewusii* the inverted repeat region contains the *psbA* and *rbcL* genes. In *C. reinhardtii psbA* is in the inverted repeat, but *rbcL* is about 15 kb away in a single-copy region. Whereas in higher plants the single-copy regions are vastly unequal in size, in these *Chlamydomonas* species the single-copy regions are roughly equivalent, and genes that in spinach or tobacco are in the same region may be separated by inverted repeats.

At 203,828 bp, the final sequence reported by Maul et al. (2002) is significantly longer than the earlier predictions of about 196 kb from restriction digests (Rochaix, 1978; Grant et al., 1980). The difference may be attributed in part to inaccurate estimates of sizes of some of the larger restriction fragments, and also to variability in sizes of intergenic regions among the strains used. Table 7.1 summarizes the gene content, which is discussed in more detail in Volume 2, Chapter 24, and Table 7.2 lists some representative mutants in chloroplast-encoded genes. The intergenic regions contain many short dispersed repeat (SDR) sequences, often marked by AatII and KpnI restriction sites (Schneider et al., 1985; Maul et al., 2002). Some of these short repeat sequences can form step-loop structures at the 3′ ends of mRNAs (Jiao et al., 2004), and they may also function in repair of double-strand breaks in DNA (Odom et al., 2008).

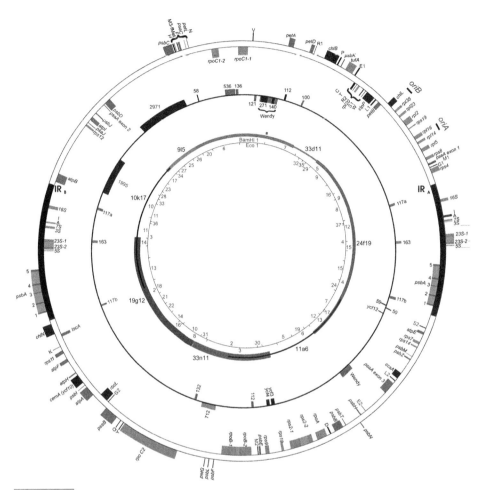

FIGURE 7.5 *Diagram of chloroplast DNA of C. reinhardtii, showing inverted repeats (IR) separating two single-copy regions of approximately equal size. The inner circle shows BamHI and EcoRI restriction fragments mapped according to Rochaix (1978) and numbered according to Grant et al. (1980). Position 0 is at the 12 o'clock position. The second concentric circle indicates seven overlapping BAC clones that span the genome. The third circle shows genes and ORFs of unknown function. The outer circle shows genes of known or presumed function; see Table 7.1. From Maul et al. (2002), and also included (in color) in Volume 2, Chapter 24.*

Contrary to early expectations, however, they do not appear to define hotspots of recombination within the genome (Newman et al., 1992). A possible role in evolutionary rearrangements of chloroplast genomes has also been postulated, but it should be noted that not all *Chlamydomonas* species have chloroplast SDRs, and the repeats defined by AatII and KpnI sites seem to be restricted to *C. reinhardtii* and its closest relatives (Boudreau and Turmel, 1996; see also Volume 2, Chapter 24.

Table 7.1	Genes in the chloroplast genome	
Complex or category	**Genes**	**References**
Photosystem I	*psaA*	Kück et al. (1987)
	psaB	Kück et al. (1987); Bingham et al. (1991)
	psaC	Takahashi et al. (1991)
	psaJ	Fischer et al. (1999a)
Photosystem II	*psbA*	Erickson et al. (1984b)
	psbB	Berry-Lowe et al. (1992)
	psbC	Rochaix et al. (1989); Girard-Bascou et al. (1992)
	psbD	Erickson et al. (1986)
	psbE	Alizadeh et al. (1994)
	psbF	Fong and Surzycki (1992b); Alizadeh et al. (1994)
	psbH	Johnson and Schmidt (1993)
	psbI	Künstner et al. (1995)
	psbJ	Kück et al. (1987); Hauser et al., GenBank
	psbK	Silk et al. (1990)
	psbL	Fong and Surzycki (1992b)
	psbM	Higgs et al. (1998)
	psbN	Johnson and Schmidt (1993)
	psbT	Maul et al. (2002)
	psbZ	Swiatek et al. (2001)
Cytochrome b_6f	*petA*	Bertsch and Malkin (1991); Büschlen et al. (1991); Matsumoto et al. (1991)
	petB	Büschlen et al. (1991); Huang and Liu (1992)
	petD	Büschlen et al. (1991)
	petG	Fong and Surzycki (1992b)
	petL	Takahashi et al. (1996)
ATP synthase	*atpA*	Leu etal. (1992)
	atpB	Woessner et al. (1986)
	atpE	Woessner et al. (1987)
	atpF	Choquet et al. (1992)
	atpH	Fiedler et al. (1995); Rolland et al. (1997)
	atpI	Hauser et al., GenBank
Carbon fixation	*rbcL*	Dron et al. (1982a)
Chlorophyll synthesis	*chlB*	Liu et al. (1993b)
	chlL	Suzuki and Bauer (1992)
	chlN	Maul et al. (2002)
RNA polymerase	*rpoA*	Maul et al. (2002)
	rpoB	Fong and Surzycki (1992a)
	rpoC1a	Maul et al. (2002)
	rpoC1b	Maul et al. (2002)
	rpoC2	Fong and Surzycki (1992a)
Ribosomal proteins	*rpl2*	Huang and Liu, GenBank
	rpl5	Huang and Liu, GenBank
	rpl14	Lou et al. (1989)
	rpl16	Lou et al. (1987)
	rpl20	Yu et al. (1992)
	rpl23	Huang and Liu, GenBank
	rpl36	Maul et al. (2002)
	rps2	Leu (1998)

(Continued)

Table 7.1 *Continued*		
Complex or category	**Genes**	**References**
	rps3	Liu et al. (1993a)
	rps4	Randolph-Anderson et al. (1995a)
	rps7	Randolph-Anderson et al., GenBank
	rps8	Maul et al. (2002)
	rps9	Boudreau et al. (1997)
	rps11	Rochaix, GenBank
	rps12	Liu et al. (1989)
	rps14	Randolph-Anderson et al., GenBank
	rps18	Boudreau et al. (1997); Leu (1998)
	rps19	Maul et al. (2002)
Ribosomal RNAs	*16S*	Dron et al. (1996)
	23S	Rochaix and Darlix (1982); Lemieux et al. (1989)
	7S	Lemieux et al. (1989)
	3S	Lemieux et al. (1989)
	5S	Schneider et al. (1985)
tRNAs		see Maul et al. (2002) for these 29 genes
Other identified genes	*ccsA*	Xie and Merchant (1996)
	cemA	Rolland et al. (1997) as *ycf10*
	clpP	Huang et al. (1994)
	tscA	Goldschmidt-Clermont et al. (1991)
	tufA	Baldauf and Palmer (1990)
	ycf3	Boudreau et al. (1997b)
	ycf4	Boudreau et al. (1997b)
	ycf12	Khrebtukova and Spreitzer (1996a, b)
ORFs	*ORF50*	Khrebtukova and Spreitzer (1996a, b)
	ORF58	Takahashi et al. (1991, 1996)
	ORF59	Khrebtukova and Spreitzer (1996a, b)
	ORF112	Büschlen, GenBank
	ORF140	Maul et al. (2002)
	ORF271	Maul et al. (2002)
	ORF1995	Boudreau et al. (1997a)
	ORF2971	Watson and Purton, GenBank
Transposon	*Wendy*	Fan et al. (1995)

References cited are in most cases for the original sequence or most complete description of a gene. For additional information, see Maul et al. (2002); Volume 2, Chapter 24; and GenBank Accession BK000554.

B. Role of the inverted repeat in chloroplast genome organization

The inverted repeat structure permits drawing chloroplast gene maps in two isomeric orientations (Figure 7.6). Digestion with restriction enzymes that do not cut chloroplast DNA within the inverted repeat demonstrates that the genome exists naturally as a 50:50 mixture of the two isomeric forms (Aldrich et al., 1985; Palmer et al., 1985), implying a mechanism of "flip-flop"

Table 7.2 Representative mutations in chloroplast genes

Gene	Mutant	Description	References
atpA	FUD16	Deficient in ATP synthase	Ketchner et al. (1995)
atpB	ac-u-c-2-21	Deficient in ATP synthase	Woessner et al. (1984)
atpE	FUD17, ac-u-a-1-15	Deficient in ATP synthase	Robertson et al. (1990)
ccsA	ct34.5, ct59	Deficient in cytochrome c assembly, at heme attachment step	Xie et al. (1998)
chlB		aadA disruption mutants; yellow in the dark	Li et al. (1993)
chlL	chlL-JS1000	Yellow in the dark	Suzuki and Bauer (1992)
petA		Site-directed mutations; deficient in cytochrome $b_6 f$ complex	Ponamarev and Cramer (1998)
petB	F41G, FUD2	Deficient in cytochrome $b_6 f$ complex	Ondarroa et al. (1996); Finazzi et al. (1997)
petD	FUD6, others	Deficient in cytochrome $b_6 f$ complex	Lemaire et al. (1986); Kuras and Wollman (1994); Higgs et al. (1998)
psaA		Site-directed mutants deficient in photosystem I; nuclear mutants affecting trans-splicing of this gene have also been studied	Goldschmidt-Clermont et al. (1990); Evans et al. (1999)
psaB	FUD26	Deficient in photosystem I	Girard-Bascou et al. (1987); Xu et al. (1993)
psaC		Insertional mutants; deficient in photosystem I	Takahashi et al. (1991, 1992)
psbA	FUD7, others	Deficient in photosystem II	Bennoun et al. (1986); Minagawa and Crofts (1994); Lardans et al. (1998)
psbA	DCMU4, dr-u-2, MZ2, others	Herbicide-resistant	Pucheu et al. (1984); Erickson et al. (1984a, 1985); Johanningmeier et al. (1987); Heifetz et al. (1997)
psbC	FUD34	Deficient in photosystem II	de Vitry et al. (1989); Rochaix et al. (1989)
psbD	FUD47	Deficient in photosystem II	Erickson et al. (1986)
psbK		Disruption; deficient in photosystem II	Takahashi et al. (1994)
rbcL	10-6C, others	Altered Rubisco activity	Dron et al. (1983); Spreitzer and Ogren (1983b)
rps12	sr-u-sm2	Streptomycin-resistant	Liu et al. (1989)
16S	sr-u, nr-u, spr-u, kr-u mutants	Resistant to streptomycin, neamine, spectinomyin and kanamycin respectively	Harris et al. (1989)
23S	er-u mutants	Resistant to erythromycin	Harris et al. (1989)

Note: These are only a small subset of the many chloroplast mutants that have been isolated.

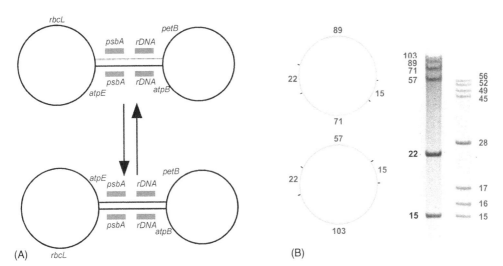

FIGURE 7.6 *Evidence for "flip-flop" recombination within the chloroplast genome. (A) The hypothetical dumbbell configuration in two isomeric forms that would be generated by recombination within the inverted repeat. (B) Experimental data supporting this model: circles indicate the position of inverted repeats (heavy lines) and expected sizes (kb) of fragments from chloroplast DNA cut with the enzyme PvuI. (Size estimates are based on the assumption of a 197 kb total genome length, the accepted prediction at the time this figure was originally made). These bands are evident in the first gel. The gel at the far right shows the same DNA digested with PvuI and BssHII. The top four bands are present at half the concentration of the lower bands, as would be predicted from a 50:50 mixture of the two isomers. Gels are negative prints of ethidium-stained gels, not Southern blots. Experiment by J.D. Palmer.*

recombination within the inverted repeat. Deletion mutations and natural variants affecting the inverted repeat are readily obtained (Grant et al., 1980; Myers et al., 1982; Bennoun et al., 1986). Most prevalent among the induced mutations are simple deletions of 8–10 kb that eliminate the *psbA* gene, and similar deletions coupled with inversion of the rRNA cistrons, such that the 5′ end of the 16S rRNA becomes oriented proximal to what remains of the *psbA* region. Both these types of mutations affect both inverted repeats in a symmetrical fashion. Myers et al. (1982) isolated these mutations in diploid cells and examined chloroplast DNA from the isolates only after they had been restored to the haploid state by crossing to haploid wild-type cells. In later experiments, in which chloroplast DNA was examined soon after mutagenesis, mixed clones were found in which some chloroplast DNA molecules had symmetrical deletions plus inversions and others had only the deletions (Palmer et al., 1985). This suggests that the deletion event occurs first and perhaps destabilizes the inverted repeat to permit inversion in some cells, but not in all, and that the process that spreads alterations from one repeat to the other is highly efficient.

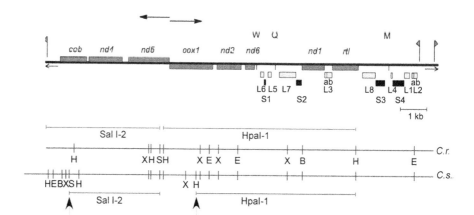

FIGURE 7.7 *Upper part of the figure shows map of the 15.8 kb mitochondrial genome, including positions of genes (see Table 7.3), and terminal inverted repeats. This figure also appears in Volume 2, Chapter 12. Lower part of the figure shows restriction maps of the mitochondrial genomes of the standard laboratory strain of C. reinhardtii (C.r.) and of C. smithii (C.s.), modified from Boynton et al. (1987), who used these differences to demonstrate inheritance of the mitochondrial genome from the minus parent.*

The striking frequency of symmetrical mutations affecting both inverted repeats (Myers et al., 1982) and the lack of sequence divergence between repeats observed in plant genomes (e.g. Shinozaki et al., 1986) suggested that the chloroplast DNA molecule can assume a "dumbbell" configuration in which the inverted repeat regions are paired (Figure 7.6), and that a copy-correction mechanism exists that maintains identity between the repeats. By reciprocal transformations of chloroplast DNA between cells of the standard laboratory strain and the interfertile strain *C. smithii* (Chapter 1), which are easily distinguishable by restriction fragment polymorphisms (Palmer et al., 1985), Newman et al. (1990) confirmed that precise copy correction does occur.

C. The mitochondrial genome

In contrast to the very large mitochondrial DNAs of higher plants, the mitochondrial DNA of *C. reinhardtii* is only 15.8 kb long (Figure 7.7) and contains very few genes (Table 7.3). Although rare open and supercoiled circular molecules were visible with electron microscopy (Ryan et al., 1978), restriction enzyme digestion produced a linear map with unique ends (Grant and Chiang, 1980). The terminal regions were later shown to consist of inverted repeat sequences that were postulated to have a role in replication of the genome (Vahrenholz et al., 1993).

Confirmation that this DNA species is the true mitochondrial genome came from hybridization and sequence analyses, which demonstrated the

Table 7.3	Genes in the mitochondrial genome	
Gene	**Product**	**References**
cob	Apocytochrome *b*	Ma et al. (1990); Michaelis et al. (1990)
cox1	Cytochrome oxidase subunit 1	Vahrenholz et al. (1985); Boer and Gray (1986b)
nd1, nd2, nd4, nd5, nd6	NADH:ubiquinone oxidase subunits	Boer and Gray (1988a); Pratje et al. (1984); Boer and Gray (1986b); Ma et al. (1989b); Pratje et al. (1989); Vahrenholz et al. (1985); Boer and Gray (1986a); Ma et al. (1988); Boer and Gray (1989)
rtl	Reverse transcriptase-like protein	Boer and Gray (1988a)
L1-L8, S1-S4	Segments of ribosomal RNA	Boer and Gray (1988b)
W,Q,M	tRNAs (Trp, Gln, Met)	Boer and Gray (1988c); Ma et al. (1989a)

Note: The nd genes were called nad in the early literature.

presence of genes for ribosomal RNAs, cytochrome oxidase subunit 1, the cytochrome *b* apoprotein, and sequences having homology to the mammalian *URF2* and *URF5* genes, later identified as subunits of NADH dehydrogenase (for review, see Gray and Boer, 1988, and Volume 2, Chapter 12). Only three tRNAs are encoded. The genes encoding the ribosomal RNAs are fragmented and are present in scrambled order (Boer and Gray, 1988b; Figure 7.7). This is also true in at least some other species of *Chlamydomonas*, including *C. eugametos* and *C. moewusii* (Denovan-Wright and Lee, 1994), and in the colorless alga *Polytomella parva* (Fan et al., 2003). Evolution of mitochondrial rRNA genes among green algae is discussed by Denovan-Wright et al. (1996), and Nedelcu (1997).

Mitochondrial DNA from *C. smithii* is colinear with that of *C. reinhardtii* except for a 1-kb insertion in the gene encoding cytochrome *b* (Figure 7.7). This insertion contains a mobile intron encoding a site-specific endonuclease (Colleaux et al., 1990; Ma et al., 1992). Boynton et al. (1987) used this polymorphism to demonstrate that in crosses between *C. smithii* and *C. reinhardtii*, the mitochondrial genome shows UP– inheritance.

IV. RNA AND PROTEIN SYNTHESIS IN THE CHLOROPLAST

A. Transcription

In early studies using synchronously grown cells, nuclear genes appeared to be transcribed at specific times during the light–dark cycle (Howell and Walker, 1977; Matsuda and Surzycki, 1980), but some regions of chloroplast

DNA seemed to be transcribed continuously (Matsuda and Surzycki, 1980; Herrin et al., 1986), implying that synthesis of chloroplast proteins is regulated primarily at the translational level. More detailed investigation revealed significant fluctuations in transcript levels, which in most cases were highest early in the light period and lowest at the beginning of the dark period (Leu et al., 1990; Salvador et al., 1993b; Hwang et al., 1996). The fluctuations depend on changes in transcription rate and on RNA degradation (Salvador et al., 1993a; Hwang et al., 1996).

As in land plants, transcription in *Chlamydomonas* chloroplasts conforms to a prokaryotic model, with promoter sequences and an RNA polymerase resembling those of eubacteria. Promoter sequences have been characterized by Thompson and Mosig (1987), Klein et al. (1992, 1994), Klinkert et al. (2005), and Bohne et al. (2006). In contrast to land plants, but similar to other algae, *Chlamydomonas* seems to have only a single chloroplast RNA polymerase, which is chloroplast-encoded (Smith and Purton, 2002). Transcription termination appears to be inefficient (Rott et al., 1996; 1998a, b), with the result that genes may appear both as monocistronic and polycistronic transcripts. Klein (Volume 2, Chapter 25) presents a table of genes that are typically co-transcribed, but comments that this list is not necessarily complete since evidence is lacking for some genes. Additional details on promoters and other aspects of transcription can also be found in Klein's chapter.

B. RNA splicing and processing

The *psbA* gene of *C. reinhardtii* contains four group I introns (Erickson et al., 1984b; Bao and Herrin, 1993) whose splicing has been studied by Herrin and colleagues (Holloway et al., 1999; Odom et al., 2001; Li et al., 2002). An intronless strain, constructed by transformation of a deletion mutant with a cDNA clone of *psbA*, grows normally and is fully photosynthetic (Johanningmeier and Heiss, 1993). Light regulation of *psbA* intron splicing has also been studied. In dark-grown cells, various unspliced precursors accumulate, indicating that order in which the *psbA* introns are spliced is not critical (Deshpande et al., 1997). However, impairing splicing in light-grown cells by site-directed mutagenesis severely affects levels of *psbA* mRNA and consequently photosynthetic capacity (Lee and Herrin, 2003).

The 23S rRNA gene of *C. reinhardtii* contains a single group I intron, within which is an open reading frame encoding a homing endonuclease (I-CreI; Herrin et al., 1990; Thompson et al., 1992; Holloway and Herrin, 1998; Kuo et al., 2006). Introns have also been characterized in the 23S rRNA genes of other *Chlamydomonas* species; the number of introns and their insertion sites are highly variable among species (Gauthier et al., 1991; Marshall and Lemieux, 1991; Turmel et al., 1993; Spiegel et al., 2006).

The *psaA* gene in *C. reinhardtii* is split into three exons that are not contiguous in the genome (Kück et al., 1987; Figure 7.5). These are assembled by a complex *trans*-splicing mechanism, which is reviewed in detail by Goldschmidt-Clermont in Volume 2, Chapter 26. The coding regions of the three parts of *psaA* are flanked by sequences that form group II introns in the precursor mRNA. Splicing of the first intron requires a chloroplast-encoded RNA, *tscA* (Goldschmidt-Clermont et al., 1991). Mutations in several nuclear genes also block *psaA* splicing (Choquet et al., 1988; Herrin and Schmidt, 1988).

Other aspects of processing and stability of chloroplast mRNAs have been reviewed previously by Herrin and Nickelsen (2004) and are discussed at length in Volume 2, Chapter 27.

C. Translation

Chloroplast ribosomes of *Chlamydomonas* resemble those of other photosynthetic organisms in being essentially "prokaryotic" in terms of their RNA and protein composition, and in their functional characteristics. Nearly all the proteins have counterparts in *E. coli*. The reader should be aware however that in the early literature, chloroplast ribosomal proteins were numbered according to their migration by charge and mass in 2-dimensional gel electrophoresis, and these numbers do not correspond to the bacterial counterparts, or to the modern terminology. Approximately one third of the ribosomal proteins are encoded in the chloroplast genome (Table 7.1), the remainder in the nucleus. Yamaguchi et al. (2002, 2003) have done a complete proteomic study.

As discussed earlier, antibiotic resistance mutations in chloroplast ribosomal RNA and in ribosomal proteins, both chloroplast- and nuclear-encoded, have been very important genetic markers in *Chlamydomonas* (see for example Harris et al., 1977; Davidson et al., 1978; Bartlett et al., 1979; McElwain et al., 1993; Bowers et al., 2003). Mutants deficient in chloroplast ribosomes have also been characterized (Goodenough and Levine, 1970; Boynton et al., 1972; Harris et al., 1974; Matsuda et al., 1985b). Cells impaired in chloroplast protein synthesis, whether by ribosome deficiency or antibiotic treatment, show a syndrome of defects, including abnormal chloroplast lamellae and loss of the pyrenoid (Figure 7.8).

Only one ribosomal factor involved in chloroplast protein synthesis, EF-Tu, is encoded in the chloroplast genome, and full characterization of the others began only with the availability of the draft genome sequence. Beligni et al. (2004) and Zerges and Hauser (Volume 2, Chapter 28) have reviewed the current state of our knowledge on this topic.

Zerges and Hauser also provide a thorough discussion of translational regulation in the chloroplast, and only some highlights will be covered here. Most chloroplast genes begin with an AUG codon (Rochaix et al., 1989), and

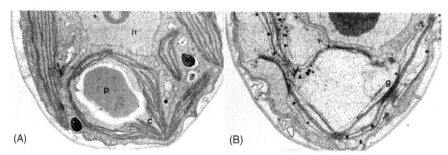

FIGURE 7.8 *Effect of inhibition of chloroplast protein synthesis on chloroplast structure. (A) Median section of wild-type cell grown mixotrophically in liquid culture in the absence of antibiotic. The cells were harvested at mid-log phase. The chloroplast lamellae are paired normally and a large, well-developed pyrenoid (p) is evident. Nucleus is indicated by (n), chloroplast by (c). (B) Similar section of a wild-type cell grown mixotrophically in liquid culture in the presence of 2 µg/ml streptomycin. The chloroplast lamellae are unpaired or stacked in giant grana (g), and no pyrenoid is present. This figure is published here in memory of Mary Feagin Conde (1947–1986), from whose Ph.D. dissertation it is taken (Conde, 1974).*

altering initiation codons by site-directed mutagenesis impairs or even abolishes translation (Chen et al., 1993, 1995; Nickelsen et al., 1999). Shine–Dalgarno sequences are found 5′ to some, but not all, chloroplast genes (Fargo et al., 1998; Hirose and Sugiura, 2004), and can tolerate mutational changes (Sakamoto et al., 1994; Fargo et al., 1998; Nickelsen et al, 1999; Zerges et al., 2003). Sequences capable of forming secondary structures in the 5′ UTR have been implicated in translational regulation (Mayfield et al., 1994; Klinkert et al., 2006), and nuclear-encoded translational activators and their binding sites have also been characterized (Zerges et al., 1997, 2003; Higgs et al., 1999; Ossenbühl and Nickelsen, 2000; Schwarz et al., 2007).

The 3′ UTRs of chloroplast mRNAs are normally not polyadenylated, but do undergo processing that can affect translation, mRNA stability, and degradation. Some key papers include Rott et al. (1996, 1998a, b) Lee et al. (1996), Levy et al. (1997, 1999), Hahn et al. (1998), Katz and Danon (2002), Hicks et al. (2002), Komine et al. (2002), Rymarquis et al. (2006), Zicker et al. (2007), and Goldschmidt-Clermont et al. (2008). Zerges and Hauser (Volume 2, Chapter 28) consider these studies in more detail.

V. COORDINATION OF NUCLEAR AND ORGANELLE GENOMES IN ORGANELLE BIOGENESIS

The evolution of plasmids and mitochondria from endosymbiotic bacteria has fascinated many scientists since the 1960s, when organelle DNA and protein synthesis were first characterized. The ability of *Chlamydomonas* to grow in the absence of photosynthesis made it one of the best systems for studying the interaction of the chloroplast and nuclear genomes in biogenesis of a complex organelle. Unlike higher plants and some other algae, wild-type *Chlamydomonas* cells retain a complete chloroplast structure and pigment content when grown in the dark. However, mutants blocked in the light-independent pathway for conversion of protochlorophyllide to chlorophyllide ("yellow-in-the-dark" or "*y*" mutants) can be maintained indefinitely on acetate medium in the dark, but lose chlorophyll and photosynthetic activity by dilution through cell division. Thylakoid membranes become disorganized and less extensive, although many of the soluble enzymes associated with photosynthesis continue to be synthesized. When light is restored, chlorophyll concentration, thylakoid membranes, and photosynthetic capacity increase in parallel over a 6- to 8-hour period (Ohad et al., 1967a, b). This process was used as the basis for studies on chloroplast biogenesis especially in the laboratories of Ohad and of Hoober (see Harris, 1989, and Hoober et al., 1998, for review). Of particular relevance to the present chapter are papers on the relative contributions of chloroplast and cytosolic protein synthesis to the formation of thylakoid membranes and other components (Hoober and Stegeman, 1973; Bar-Nun and Ohad, 1977).

Several of the classic mutants used by Levine and colleagues to dissect the mechanism of photosynthesis (Levine, 1968; Levine and Goodenough, 1970) received renewed attention when they were discovered to be in nuclear genes that affected expression of genes in the chloroplast genome. For example, Goldschmidt-Clermont et al. (1990) identified 14 complementation groups of nuclear mutants deficient in photosystem I that were involved in *trans*-splicing of the tripartite chloroplast *psaA* gene. Among these were Levine's original *ac9* mutant, and other mutants isolated by Girard et al. (1980). Additional nuclear mutations were characterized that affected expression of other chloroplast genes (Barkan and Goldschmidt-Clermont, 2000; Table 7.4). This line of research illustrates especially well the potential of *Chlamydomonas* for identification of genes controlling organelle biogenesis.

Another dimension to understanding formation of the functional complexes of photosynthesis has been articulated by Choquet and colleagues (Wollman et al., 1999; Choquet and Vallon, 2000; Choquet et al., 2001; Choquet and Wollman, 2002). CES, for "Control by Epistasy of Synthesis," is the name given to assembly-dependent regulation; that is, when a core subunit of a complex is limiting, translation of other chloroplast-encoded subunits of that complex is reduced. This was first discovered in the early characterization of non photosynthetic mutants: mutation in a single protein often resulted in loss of an entire complex. In Volume 2, Chapter 29, Choquet and Wollman discuss the probable origin of this regulatory process in the transition from a cyanobacterial ancestor to a plastid genome controlled by the nucleus, and provide details of its documentation.

Import of proteins synthesized on cytosolic ribosomes into the *Chlamydomonas* chloroplast, and their targeting within the organelle, has been much less extensively investigated than in higher plants. Early studies

Table 7.1 Some nuclear mutations affecting expression of chloroplast genes

Nuclear locus	Chloroplast gene affected	References
RAA1, many others	*psaA*	Goldschmidt-Clermont et al. (1990); Yohn et al. (1996); Balczun et al. (2005a); Merendino et al. (2006)
TBA1	*psbA*	Girard-Bascou et al. (1992); Somanchi et al. (2005)
MBB1	*psbB/psbT/psbH*	Vaistij et al. (2000)
TBC2	*psbC*	Auchincloss et al. (2002)
NAC2	*psbD*	Wu and Kuchka (1995); Boudreau et al. (2000); Nickelsen (2000); Schwarz et al. (2007)
MCA1, TCA1	*petA*	Wostrikoff et al., 2001; Raynaud et al. (2007)
MCD1	*petD*	Murakami et al. (2005)
MDA1	*atpA*	Drapier et al. (2002)
76-5EN	*rbcL*	Hong and Spreitzer (1998)

Note: a consistent nomenclature has been proposed that covers many of these nuclear loci. Three letter names are assigned as follows:
First letter: M for mRNA processing/stability, T for translation, C for cofactor assembly.
Second letter: A for psa genes, B for psb genes, C for pet genes, D for atp genes.
Third letter: the last letter in the target gene name: A for psaA, psbA, petA or atpA, etc.

suggested significant differences in algal and plant transit peptides (Franzén et al., 1990; Franzén, 1994; Su and Boschetti, 1994), and processing enzymes also appear to be distinctive (Su and Boschetti, 1994; Rüfenacht and Boschetti, 2000). Typically chloroplast-targeted proteins have a bipartite transit peptide that first gets the protein into the plastid, and then specifies its location within the organelle. Some mutations in these sequences have been isolated (Smith and Kohorn, 1994; Baillet and Kohorn, 1996; Lawrence and Kindle, 1997; Kindle and Lawrence, 1998). Bernd and Kohorn (1998) identified six additional nuclear loci that suppressed mutations in thylakoid signal sequences. Unfortunately, neither Kindle nor Kohorn has continued this work with *Chlamydomonas*. A system for studying import of proteins into *Chlamydomonas* mitochondria was developed by Nurani et al. (1997), but this work does not seem to have been continued either.

Analysis of the nuclear genomic sequence has revealed signal recognition proteins, and identified targeting sequences in many proteins expected to be found in organelles, although notes made in the genome annotation suggest that prediction algorithms do not always appear to be able to distinguish between chloroplast and mitochondrial proteins of *Chlamydomonas*. Further analysis is expected, and a resurgence of interest in this topic is likely.

CHAPTER 8

Chlamydomonas in the Laboratory

CHAPTER CONTENTS

I. INTRODUCTION

This chapter provides the basic information needed for laboratory culture and genetic and molecular analysis of *Chlamydomonas*. Additional protocols can be found in Harris (1989) and at http://www.chlamy.org. The reader unfamiliar with culturing algae should see Andersen (2005) for fundamental techniques. Information on culture collections, additional web sites, and other useful resources can be found in the Appendix.

II. CULTURE MEDIA

A. Media for *C. reinhardtii*

Three possible growth conditions are customarily defined for algae capable of utilizing organic carbon sources. Phototrophic or photoautotrophic growth implies culture with photosynthetically assimilated CO_2 as sole carbon source; heterotrophic (organotrophic) growth means culture in darkness with an organic carbon source (usually acetate in the case of *Chlamydomonas* species); and mixotrophic (photoheterotrophic) growth is culture in light with added acetate. The changes in metabolism accompanying adaptation to different growth conditions are discussed in Chapter 6.

In nature, *C. reinhardtii* is a soil organism, but it can be grown in the laboratory either in liquid culture or on agar in simple mineral salts, and many of the recipes proposed for algae in general will suffice for its culture (see Bold, 1942, and Watanabe, 2005). Table 8.1 gives the composition of the media most often utilized in current *Chlamydomonas* research, and recipes for their preparation are available online (http://www.chlamy. org/media.html). Of these, probably the most frequently employed is TAP medium (Tris–acetate–phosphate; Gorman and Levine, 1965). The advantages and disadvantages of this and other media will be discussed below.

Preparation of good culture media requires a good water supply. Laboratory water purified by distillation or reverse osmosis systems may need to be passed through an ion exchange column before preparation of media, and users should also be alert to changes in water quality when a building's water system is cleaned, or when ion exchange cartridges are replaced. Some laboratories use glass-distilled water with good results, but in some locations volatile impurities that are toxic or inhibitory to *Chlamydomonas* are present and can be carried over on distillation. Commercially bottled spring water can also be used, providing a brand is selected with low mineral content and no added preservatives. The real test of water quality appears to be whether *Chlamydomonas* cells can not only grow well but also go through their sexual cycle (see Chapter 5). Blankley (1973) discussed potential toxicity to algal cultures of water contaminants and many other laboratory materials.

Table 8.1	Molar composition of culture media					
Component	Sager-Granick	Eversole	Sueoka	Kuhl	Bold	TAP
Major components (mM)						
NH_4^+	3.7	93.5	9.35	6×10^{-5}	–	7.48
K^+	1.88	0.52	22.12	10.0	2.70	1.94
Na^+	5.1	0.027	0.27	5.5	3.37	0.27
Ca^{2+}	0.36	0.09	0.068	0.10	0.17	0.34
Mg^{2+}	1.2	2.03	0.081	1.0	0.30	0.41
Fe^{3+}	0.37	–	–	–	–	–
NO_3^-	3.7	–	–	10.0	2.94	–
Cl^-	1.83	112.9	9.55	0.20	0.78	8.22
SO_4^{2-}	1.2	2.03	0.182	1.03	0.38	0.51
PO_4^{3-}	1.31	0.115	13.6	5.0	1.72	1.00
Tris	–	50.0	–	–	–	20.0
Citrate	1.7	–	–	–	–	–
Acetate	–	–	–	–	–	17.4
Trace components (μM)						
Fe^{2+}	–	1.79	17.9	25	17.9	17.9
Zn^{2+}	3.5	7.65	76.5	1.0	30.7	76.5
Cu^{2+}	0.25	0.63	6.3	0.01	6.3	6.3
Co^{2+}	0.84	0.68	6.8	–	1.7	6.8
Mn^{2+}	2.0	2.56	25.6	1.0	7.3	25.6
Mo^{6+}	0.82	0.62	6.2	0.07	4.9	6.2
BO_3^{3-}	16	18.4	184	1.0	184	184
EDTA	–	13.4	134	25	171.1	134

References are cited in the text; also see text for addition of sodium acetate to minimal media.
Loss of precipitate in Hutner trace elements is disregarded (Sueoka and TAP media).

Wild-type strains of *C. reinhardtii* can be grown in minimal medium, or with acetate added as an alternative carbon source. Nitrogen is usually supplied as ammonium chloride, but most wild-type strains can also assimilate nitrate. However, the Ebersold-Levine 137c strain (see Chapter 1) lacks nitrate reductase activity and must be given a reduced nitrogen source; this is the background strain in which most mutants were isolated. Sager and Granick (1953) tested a wide range of potential carbon and nitrogen sources. Most of the carbon compounds tested, including simple sugars such as glucose, galactose, and arabinose, were found to be non-toxic but did not support growth in the absence of photosynthesis (Chapter 6). Ammonium, nitrate, urea, and glutamine were all good nitrogen sources for the strain tested (21 gr), and arginine can also be used as sole nitrogen source. Sager and Granick reported growth at pH as low as 5.5, but Wang et al. (1975) found in attempting to feed levulinic acid to *C. reinhardtii* for chlorophyll synthesis that pH values below 6.5 were poorly tolerated.

The minimal medium developed by Sager and Granick (1953) is economical to prepare but has several potential disadvantages. It supplies significantly less reduced nitrogen than either TAP or the Sueoka high salt medium (Sueoka, 1960; Table 8.1); so that cells lacking nitrate reductase may become limited by nitrogen deficiency before a high cell density is achieved. In phototrophic cultures the Sager and Granick medium holds a constant pH of about 6.8, but its poor buffering capacity becomes apparent when acetate is added for mixotrophic or heterotrophic growth, and a shift up to pH 8.4 may occur under these conditions. Sager and Granick therefore devised a modified medium for growth on acetate in which the balance of phosphate salts was adjusted to give an initial pH of 6.2. In heterotrophic culture the pH rose but stayed below 8.0. In her later studies Sager (personal communication) used a variation of the original minimal medium in which the concentration of K_2HPO_4 was raised and four times the original molarities of trace elements were used.

The Sueoka high salt medium (HS) is adequately buffered for both phototrophic and acetate-supplemented growth (Sueoka, 1960; Gillham et al., 1970). For some non-photosynthetic mutants, increasing the acetate concentration may prolong culture viability on plates (Shepherd et al., 1979). The Sueoka medium may appear cloudy after autoclaving due to phosphate precipitation, but this does not affect growth of *Chlamydomonas* cells.

TAP medium (Gorman and Levine, 1965) is well buffered and relatively low in phosphate, making it the best medium for ^{32}P labeling and for studies requiring optical clarity of agar, but it is the most expensive of the usual media to prepare. TAP medium contains 17.4 mM acetate as originally formulated. A minimal medium can be made by omitting acetic acid and titrating with HCl to pH 7.0.

Early researchers did not determine precisely what quantities of the various trace elements are required for *C. reinhardtii*. Rather, mixtures developed for other organisms were employed, and since the algae grew satisfactorily, little effort was made to improve them. A rational examination of trace element nutrition came only with availability of EST and genomic sequences, which permitted searches for homologues of metal transporters and other enzymes, and with cumulative increases in knowledge of intermediary metabolism (Merchant et al., 2006).

Most of the trace elements mixtures in current use contain a chelating agent, either EDTA or citrate, to permit a high initial concentration of the required metals without toxicity to the cells, and thus allowing a maximum final cell density to be reached. When iron salts are required in relatively high concentration as in the Sager and Granick medium, citrate is required to prevent precipitation of iron oxides. Note that citrate cannot support dark growth of *C. reinhardtii* and is therefore probably not a significant source of carbon in this medium (Sager and Granick, 1953). TAP and the Sueoka medium use the same trace elements mixture, developed by Hutner et al. (1950), who also discussed general principles of trace element nutrition. Although somewhat complicated to make, and of indeterminate final composition since a precipitate is discarded during the preparation, its use is supported by tradition and satisfactory experimental results. Other trace elements mixtures are given by MaciasR and Eppley (1963), MaciasR (1965), and Bailey and Taub (1980). Theoretical considerations of trace element nutrition are discussed by Sunda et al. (2005), and by Merchant et al. (2006), who concluded that none of the currently used media are optimal for *C. reinhardtii*. Specifically, the Sager and Granick medium is deficient in copper, and Hutner's trace elements mixture as used in TAP and the Sueoka medium contains excess zinc and manganese. The influence of specific elements on growth of *Chlamydomonas*, for the most part in the context of environmental pollution, was discussed by Button and Hostetter (1977), Fennikoh et al. (1978), Macka et al. (1978), and Cain and Allen (1980).

Bold's basal medium (Bischoff and Bold, 1963) is an all-purpose algal medium based on Bristol's solution (see Bold, 1942) and is used in some *Chlamydomonas* work, although not usually for *C. reinhardtii*. It does not contain a reduced nitrogen source and is therefore unsuitable for the Ebersold/Levine strain of *C. reinhardtii* unless ammonium chloride is substituted for sodium nitrate. Eversole's pH 8.3 medium (1956) contains essentially the same components as the Sueoka medium, but in significantly different proportions. It is useful for scoring a few mutants whose phenotype is more clearly expressed at high pH, but otherwise is rarely used in current research. The medium of Kuhl and Lorenzen (1964) is used by the Sammlung von Algenkulturen for maintenance of many of their unicellular green algae, including *Chlamydomonas* species, and was used by Lien and Knutsen (1979) in their studies of cell synchrony. The University of Texas algal collection (UTEX) uses a soil-extract medium for most of their *Chlamydomonas* stocks. Algal media buffered with HEPES were recommended by McFadden and Melkonian (1986), particularly for cultures being grown for electron microscopy. Trainor et al. (1991) tested a simple medium based on a commercially available plant food for growth of a number of different freshwater algal taxa. Their paper does not provide growth data for any *Chlamydomonas* species, but contains a note that *Chlamydomonas* was among the organisms observed in their isolates from nature.

B. Adaptations of the standard media for special purposes

Yeast extract (4 g/liter) is a suitable source of amino acids and vitamins for auxotrophic mutants and is recommended as an addition to all solid media on which stock cultures are maintained because it permits growth of most bacterial contaminants, making them easy to detect. It does not increase the growth rate of wild-type cells (Sager and Granick, 1953) although anecdotal evidence suggests it prolongs the viability of stored cultures. Other supplements that have been used are casein hydrolysate (casamino acids at a suggested concentration of 2.5–5 g/liter), liver extract (1 g/liter), peptone (1 g/liter), and a nucleic acid hydrolysate (Eversole, 1956). Multivitamin supplements can be prepared, or purchased from suppliers of tissue culture media. These are unnecessary for wild-type cells but may be useful when working with auxotrophic strains.

Table 8.2 gives appropriate concentrations in solid media of the compounds required for growth and scoring of the most frequently studied mutants of *C. reinhardtii*. Responses of *Chlamydomonas* cells to various other inhibitors were also tabulated by McBride and Gowans (1970) and by Mottley and Griffiths (1977). In general, lower concentrations of inhibitors will suffice in liquid media than on agar, and these should be determined empirically for the medium and growth conditions in use. In cases where additives are not soluble in water, it is worth noting that *C. reinhardtii* cells will tolerate up to 5% ethanol or 10% acetone in agar media.

Nitrogen-free medium for gametogenesis and zygote maturation is prepared by omitting the ammonium and/or nitrate salts. The low-nitrogen medium developed by Sears et al. (1980), a modification of the Sager and Granick medium, works particular well for gametogenesis. Nitrate medium for testing *nit$^-$* mutants and selecting transformants with the wild-type *NIA1* (*NIT1*) gene (Kindle et al., 1989) is made by substituting equimolar potassium nitrate for ammonium chloride. Low-sulfur medium for ^{35}S labeling or to induce hydrogen production (Melis et al., 2000) can be prepared by substituting an equimolar quantity of magnesium chloride for magnesium sulfate in any of the standard media. The Hutner trace elements recipe contains a significant quantity of sulfate as zinc, copper, and ferrous salts. For short-term ^{35}S labeling experiments, the trace elements can be omitted altogether. For hydrogen production, the recipe can be prepared using chloride salts of these elements instead of sulfates.

C. Culture media for other *Chlamydomonas* species

C. eugametos and *C. moewusii* are obligate photoautotrophs and are unable to grow in the dark on acetate medium. They grow well in the light, however, on any of the aforementioned media for *C. reinhardtii*. Gowans (1960) gave a minimal medium for *C. eugametos* derived from the media used by Levine and Ebersold (1958a) and by Eversole (1956). A few species

Table 8.2 — Additions to agar media

Compound	Final concentration	Notes
3-Acetyl pyridine	22 µg/ml	1, 5
p-Aminobenzoic acid	50 µg/ml	2
Anisomycin	10 µg/ml	
Arginine	50 µg/ml	2, 3, 4
Sodium arsenate	2–3 mM	3, 4
L-Canavanine sulfate	500 µg/ml	
Chloramphenicol for bacterial control for scoring mutants	25 µg/ml 200 µg/ml	1
Cycloheximide	10 µg/ml	1, 6
3(3, 4-Dichlorophenyl)-1, 1-dimethyl urea (DCMU)	3 µM	
Erythromycin	100–300 µg/ml	7
Fusidic acid (as sodium salt)	400 µg/ml	1
Hygromycin	10 µg/ml	
Kanamycin	100 µg/ml	
Methionine sulfoximine	500 µg/ml	2
Neomycin (for bacterial control)	25–50 µg/ml	
Nicotinamide (niacinamide)	2 µg/ml	2, 3, 4
Oxythiamine hydrochloride	5 µg/ml	2
Paromomycin	15 µg/ml	
Pyrithiamine	10 µg/ml	2
Spectinomycin	100 µg/ml	
Streptomycin sulfate	50 100 µg/ml	
Thiamine hydrochloride	1 µg/ml	2, 4
Zeocin	10 µg/ml	

Notes:
1. Compound can be dissolved in ethanol.
2. Metabolic analogs should be prepared in minimal or acetate media, not in media containing yeast extract or peptone.
3. Can be added to culture medium before autoclaving.
4. Stock solutions can be stored refrigerated for several weeks. Nicotinamine solutions should be stored in dark bottles.
5. 3-Acetyl pyridine is solid when removed from the refrigerator but partially liquefies on standing at room temperature. It is most conveniently dispensed by warming the bottle in a water bath and then measuring the desired volume with a micropipette.
6. Cycloheximide has a very narrow range of effective concentrations. Usually 10 µg/ml will kill wild-type cells and allow mutants to grow, but this may need to be adjusted depending on the supplier and the purity.
7. 100 µg/ml erythromycin kills wild-type cells on minimal medium. Higher levels are required on medium containing acetate.

of *Chlamydomonas*, for example *C. chlamydogama*, *C. spreta*, *C. pulsatilla*, and *C. pallens*, require vitamin B_{12}, which must be added (5 µg/ml) to media not containing yeast extract or another vitamin source (Bold, 1949a; Trainor, 1958; Pringsheim, 1962; Turner, 1979).

Most of the other freshwater and soil-dwelling *Chlamydomonas* species also appear to do well on the media listed for *C. reinhardtii*, although a few seem to prefer media without acetate. Marine species are usually grown on either natural or artificial seawater, often supplemented with other compounds (Harrison and Berges, 2005). For species with uncertain nutritional requirements that do not grow well on the usual media, soil extract is the remedy of traditional phycology (Pringsheim, 1946b; Watanabe, 2005). Schlösser (1982) gives instructions for its preparation, as does Watanabe (2005).

III. CULTURE CONDITIONS

A. Temperature and light

The usual laboratory species of *Chlamydomonas* grow well in the range of 20–25°C. Wild types and most mutant strains of *C. reinhardtii* will tolerate temperatures as low as 15°C and as high as 35°C. Gross and Jahn (1962) reported that *C. moewusii* grew at 32.5°C and was stably viable but did not divide at 35°C. *Chlamydomonas* strains native to cooler environments may not do well even at 25°C and should be kept at 15–20°C or lower. Snow algae and other species from very cold habitats may need to be kept below 10°C (see Hoham, 1975).

Light intensities for *Chlamydomonas* cultures are best measured in terms of photosynthetically active radiation (PAR). Cool white fluorescent bulbs are most often used. Two 40-watt bulbs 20 cm above a foil-covered shelf provide about 100 µmol/m²s PAR, a good level for photosynthetically competent cultures on agar. Intensities of 200–400 µmol/m²s are recommended for phototrophic liquid cultures, which should be stirred or shaken not only for aeration but also to overcome shading of the interior of the flask. Wild-type strains of *C. reinhardtii* bleach and die at intensities between 1500 and 2000 µmol/m²s PAR, but mutants resistant to these very high light levels can be isolated (Förster et al., 1999, 2005).

Continuous illumination produces rapid, logarithmic growth in liquid culture over a range of cell densities from about 10^5 to 10^7 cells/ml, with doubling times on the order of 5–8 hours depending on the strain, the light intensity and temperature, and the composition of the medium. The final cell density reached will also depend on these factors. Alternating light: dark cycles (usually 12:12 hour) are used to achieve synchronously dividing cultures in minimal medium and are also sometimes used for stock tube maintenance (see below).

B. Cultures in liquid media

1. Pregrowth cultures

Liquid cultures are conveniently grown in Erlenmeyer flasks, bottles, or carboys. To ensure uniformly growing, logarithmic-phase cells for biochemical work, liquid cultures should not be inoculated directly from petri plates. Rather, a pregrowth liquid culture should be made by inoculating approximately 300 ml of liquid medium with a pea-sized (6–8 mm diameter) pellet of cells taken from agar. This culture is allowed to grow to mid-log phase $(1–5 \times 10^6$ cells/ml). This will require about two days for wild-type cells. An appropriate aliquot is then taken to inoculate the experimental cultures (e.g., 10 ml at 3×10^6 cells/ml will give a starting density of 1×10^5 in a fresh 300 ml culture; the entire 300-ml pregrowth can be poured into a 10–15 liter carboy).

2. Aeration

Cultures should be shaken or stirred continuously for maximum growth. If stirring is used, the culture vessel should be insulated from the stirring motor to prevent overheating. For small flasks, a petri dish lid can be placed on the stirrer beneath the flasks. Large carboys can be set on wooden supports placed on three sides of the magnetic stirrer. The stirring bar can be autoclaved with the medium and centered with a magnet before inserting the stirrer motor into the open side of the support.

Phototrophic cultures should be bubbled with CO_2 (5% in air) for maximum growth. Although this can be purchased already prepared, it may be less expensive in the long run to mix 100% CO_2 with laboratory air using a gas proportioner. The air should pass through a filter to eliminate contaminants. A cartridge-type filter can be used (Snell, 1980), or a glass or a metal drying tube can simply be packed with cotton and autoclaved. Either latex or plastic (Tygon) tubing can be used. Although Tygon is more expensive initially, it lasts much longer, as latex tubing eventually decays under bright light. For multiple cultures, the incoming air–CO_2 mixture can be connected to a row of aquarium pump valves (available from pet stores), from which the gas flow to individual flasks can easily be adjusted. Heterotrophic or mixotrophic cultures will also benefit from bubbling with sterile air, with no added CO_2. An empty buffer flask in the flow line will minimize pressure changes when adding or deleting culture flasks from this setup.

Flasks can be stoppered with foam plugs or polypropylene caps, available from major scientific suppliers, or with homemade cotton and cheesecloth plugs. Bubblers can be prepared by drawing out and fire-polishing one end of pieces of 4-mm ID soft glass tubing. Pasteur pipettes can also be used as bubblers, but they break easily. Most satisfactory in our experience are disposable 1-ml glass pipettes. These can be reused many times, and the calibration markings can be removed by putting the pipettes through a chromic acid wash. Bubblers should be plugged with cotton and can be autoclaved separately in foil-wrapped packages. If they are autoclaved with the media, back-siphoning on cooling may wet the cotton, in which case the bubbler must be replaced, as the wet cotton will not pass air satisfactorily and may allow contaminants to enter by channeling. A further improvement is to prepare a prefilter for each bubbler, consisting of a glass tube stuffed with cotton and having a short length of latex tubing at each end of connection to the bubbler and air lines. Another design for bubblers for culture tubes was described by Thomas and Herbert (2005).

3. Problems with liquid cultures

Excessive foaming of *Chlamydomonas* cultures is not ordinarily a problem. However, some authors (for example, Thacker and Syrett 1972a) recommend adding an antifoaming agent dropwise to cultures as necessary.

As discussed in Chapter 2, *Chlamydomonas* cells in liquid culture sometimes form "palmelloids," which appear to be clumps of non-motile cells. These usually result from failure of cells to hatch from the mother cell wall after mitosis, rather than from aggregation of previously swimming individuals. To eliminate palmelloid formation in laboratory cultures, several possible remedies have been suggested, but none in our experience is effective in all cases. Strain differences are apparent: palmelloids are formed more commonly in the 21 gr strain of *C. reinhardtii* and in *C. smithii* than in the Ebersold-Levine wild-type strain and these differences seem to segregate genetically in crosses among these strains. Surzycki (1971) recommended subcloning for colonies that appear moist or soupy on agar. In general, cells on low phosphate medium (e.g., TAP) form palmelloids less readily than those on high phosphate (e.g., the Sueoka high salt medium). Iwasa and Murakami (1969) reported that palmelloids dissociate more readily in acidic media. Resuspending cells for a few hours in distilled water or one-tenth strength or nitrogen-free culture medium may also help, probably by inducing gametogenesis which in turn may result in some release of the gamete lytic enzyme even in cultures of a single mating type. Keeping a culture in dim light or in the dark for a few hours may also encourage hatching (Spudich and Sager, 1980). Ellis (1972) found that cells of another green alga, *Pediastrum*, formed fewer palmelloids when grown at $300\,\mu M$ $FeCl_3$ (10 times the usual concentration for this alga). This concentration of iron is in the same range as in the Sager and Granick medium (Table 8.1) and is much higher than the iron content of either the Sueoka medium or TAP.

Wall-deficient (*cw*) mutants are more sensitive than wild-type cells to physical disruption and to changes in culture media (pH, etc.). Davies and Plaskitt (1971) recommended adding 0.5% peptone to both liquid and solid media for *cw* mutants and also reducing the agar content of plates from 1.5% to 0.8%. Gresshoff (1976) used 0.1 M mannitol, 0.2% yeast extract and 0.1% casein hydrolysate when growing *cw* strains. R.K. Togasaki (personal communication) suggested including 1% sorbitol in liquid media and noted that cells grown in this medium seemed to withstand centrifugation better than those grown in the usual media. Loppes and Deltour (1975) speculated that loss of essential periplasmic enzymes (phosphatases, sulfatases, carbonic anhydrase; see Chapter 6) into the culture medium was responsible for failure of *cw* mutants to grow in liquid culture, but they found that increasing the phosphate concentration of the medium did not improve growth. Davies and Plaskitt (1971) found no differences in osmotic properties between wild-type cells and *cw* mutants. Cultures of the *cw15* mutant frequently are observed to contain large amounts of cellular debris, which is sometimes mistaken for contamination, especially when small particles are undergoing Brownian motion. Plating aliquots of these cultures on a nutrient-rich medium, either *Chlamydomonas* medium with added yeast extract and acetate, or a general-purpose bacteriological medium, should reveal any true contaminants.

4. Measuring cell density and viability

Chlamydomonas cells are easily counted under a compound microscope (10x ocular, 10x objective) using a standard hemocytometer (Guillard and Sieracki, 2005). Cultures should be kept in dim light after removal from the shaker or stirrer to minimize settling and phototactic responses. A 1-ml sample taken from a thoroughly mixed culture is treated with a drop of tincture of iodine (0.25 g iodine in 100 ml 95% ethanol) or IKI (a dilute Lugol's solution; 1 g iodine, 0.5 g potassium iodide in 100 ml H_2O) to immobilize the cells. A sample from this mixture is then taken with a Pasteur pipette and allowed to flow by capillary action under the hemocytometer cover glass. A convention should be established to avoid duplicate counting of cells that rest on the lines of the hemocytometer grid, for example counting all cells on the top and right margins of a square but excluding those on the bottom and left margins. When palmelloid cells are present, cell counts will be less accurate, and some interpretation will be necessary according to the purpose for which the cells are to be used. For biochemical and physiological studies, each group can be counted as four individuals, and larger groups can be estimated in multiples of four cells. For plating cells on agar, however, a group within a common wall counts as a single colony-forming unit.

Turbidometric measurements standardized to a cell count curve are satisfactory for quick estimates of cell density, but cells need to be thoroughly mixed and measurements taken rapidly to avoid errors due to motility or settling. Phosphate precipitation in the Sueoka high salt medium may interfere with this method. Sager and Granick (1953) found that optical density measured at 750 nm was linear over a range of 2×10^5 to 1×10^7 cells/ml. At this wavelength, chlorophyll absorbance does not interfere. Bernd and Cook (2002) described a method for microscale cultures in 96-well plates, monitored at 750 nm using a commercially available plate reader.

Separate standard curves must be prepared for each strain or genotype of *Chlamydomonas*, as average cell sizes vary greatly. When sampling cells grown on a light:dark cycle, the aliquots should be taken at the same time every day. Spudich and Sager (1980) used laser light-scattering to plot cell size changes in synchronous cultures. *Chlamydomonas* cells can also be counted by flow cytometry (Lien and Knutsen, 1979; Lemaire et al., 1999; Marie et al., 2005). This works best with synchronous cultures, in which all the cells are similar in size. For nonsynchronous cultures, the apparatus must be calibrated to accommodate the broad range of cell sizes.

To obtain a population of cells of uniform size without prior synchronous culture, cells can be starved overnight by placing them in the dark on minimal medium (Conde et al., 1975). Cells growing in continuous light are harvested by centrifugation in sterile containers at room temperature late in the afternoon and are resuspended in minimal medium in flasks wrapped with foil or black electrical tape to exclude all light. These are shaken or stirred (but not bubbled with CO_2) overnight. On return to the light in the morning, the cultures will consist entirely of small cells, which will then increase in size for 6 hours or more before undergoing a more or less synchronous division.

Living cells can be distinguished from dead ones by the autofluorescence of chlorophyll, which is rapidly lost on cell death (Pouneva, 1997). Amano et al. (2003) reported that fluorescein diacetate could be used as a viability probe in plants and also in *Chlamydomonas*. Capasso et al. (2003) suggested a colorimetric method based on reduction of a tetrazolium compound as a viability assay for several algae including *Dunaliella*. Although Capasso et al. did not include *Chlamydomonas* in their report, a similar assay was used by Dorthu et al. (1992) to select respiratory-deficient mutants of *C. reinhardtii* (see below). Evans blue, which is not taken up by intact cells, is also a useful reagent for determining the percent of viable cells in a population (Crutchfield et al., 1999).

5. Synchronous cultures

Lien and Knutsen (1979) summarized earlier literature on synchronous growth of *C. reinhardtii* and *C. moewusii*. Alternation of growth periods in light and darkness on minimal medium (Bernstein, 1960; Kates and Jones 1964) is most often used to achieve synchronous division, although there are reports of synchrony induced by temperature shift (Rooney et al., 1971), by size selection (Knutsen et al. 1973; Tetík and Nečas, 1979), and by alternation of light of different wavelengths (Carroll et al., 1970). A 12:12 light: dark cycle is most commonly used, but 14:10, 12:4, and other cycles have also given good results. The minimum length of the light period for *C. reinhardtii* seems to be about 8 hours, with 12 or 14 hours being preferable.

When grown on a light:dark cycle, *Chlamydomonas* cells divide during the dark phase of the cycle, and may complete two or three successive rounds of mitosis with no intervening growth phase. The number of divisions to be undergone by a single cell is determined by its size at the beginning of the division period (see Chapter 3). Each mitosis is preceded by a round of nuclear DNA synthesis. After each cytokinesis all the daughter progeny remain within the mother cell wall until all the divisions are complete, and then are released simultaneously. Lemaire et al. (1999d) showed that synchrony of the S phase is imperfect, possibly because the timing and/or duration of the division period is different for cells that will ultimately divide once, twice, or three times.

Bernstein (1968) reported that *C. moewusii* could be grown on a 4:20 cycle, doubling once per cycle; on 8:16, to produce a fourfold division; or on 12:12, for an eightfold increase in each cycle. Synchronization by light:dark cycles is effective only for photosynthetically competent strains, because the presence of acetate in the medium permits cell division at irregular times (Kates and Jones, 1964a; Surzycki, 1971). Auxotrophic mutants requiring amino acids or vitamins can be synchronized, however, on minimal medium plus the required supplement.

Surzycki (1971) advised against selection of strains of *C. reinhardtii* that show a high degree of synchronization on the grounds that these are usually found to be yellow mutants unable to synthesize chlorophyll in the dark (*y1* and related mutants). Synchronous cultures of yellow mutants differ from those of wild-type cells in the time of synthesis of chlorophyll, RNA, and other components. Selection of a strain that liberates daughter cells immediately after cytokinesis (i.e., does not form palmelloid colonies) is desirable, however.

Pregrowth cultures grown in continuous light are useful for starting larger volumes of culture to be grown synchronously. Surzycki (1971) recommended a starting density of 5×10^4 cells/ml, with inoculation of the large culture at 6–7 hours before the first dark cycle. Uniform cell division is not expected in the first dark period after inoculation, and the first synchronous cycle is reckoned from the start of the subsequent light period. At 21°C and $100 \mu mol/m^2 s$, on a 12:12 cycle the first two full cycles will produce fourfold divisions to reach a density of approximately 8×10^5 cells/ml. In the next cycle at this light intensity, some cells appear to divide into two and others into four daughters to produce about 2.4×10^6 cells/ml, and in the final cycle a twofold division brings the density to about 4.8×10^6 cells/ml. (Prediction of division number from cell size is discussed in Chapter 3). Increasing the light intensity will give a uniform fourfold division in the third cycle. Cells from the third light period are best for most physiological studies.

Schlösser (1966) observed that cultures on a 12:12 cycle at 30–32°C and $390 \mu mol/m^2 s$ (20,000 lux) sometimes had a fraction of cells that divided during the light phase. Increasing the temperature to 34°C eliminated this effect, as did reducing the light intensity by half. Lien and Knutsen (1979;

Knutsen and Lien, 1981) confirmed that no diurnal division occurred at 35°C and found that the division step was completed in the first 2 hours of the dark period, prompting them to shorten the dark period under these light and temperature conditions to 4 hours (a 12:4 cycle). Lien and Knutsen achieved maximum growth rates with *C. reinhardtii* using this regime at 35°C under 390 μmol/m²s. They observed a 16-fold increase in cell number during the dark phase, equivalent to a 4-hour doubling time. Temperature control in this range should be very precise, since cytokinesis and cell hatching are inhibited above 36°C in wild-type *C. reinhardtii* (Lien and Knutsen, 1979).

6. Chemostats and bioreactors

Either synchronous or asynchronous cultures can be maintained indefinitely, barring contamination, by daily dilution to a predetermined cell density. For example, Lien and Knutsen (1979) maintained synchronous cultures for weeks or even months by dilution to 1.4×10^6 cells/ml at the end of each dark phase. Daily dilutions are most conveniently accomplished with some type of chemostat or turbidostat apparatus. Experiments utilizing cultures grown in such systems were described by Taub and Dollar (1968), Grob et al. (1970), Hudock et al. (1971), Sharaf and Rooney (1982), and Maeda et al. (2006). The theory of continuous culture was explored by Cunningham and collaborators (Cunningham and Maas, 1978; Cunningham and Nisbet, 1980; Cunningham, 1984) and by Tang et al. (1997).

Laboratory research rarely requires growing *Chlamydomonas* in volumes greater than a few liters. For commercial production, however, cost-effective methods for large-scale culture are required. Behrens (2005) reviewed designs and other considerations for photobioreactors and fermentors for algae in general. Although most publications on these types of culture have focused on algae such as *Dunaliella*, *Haematococcus*, and *Chlorella* that are already being grown for commercial purposes, Pottier et al. (2005) used *C. reinhardtii* in a study of light attenuation in a torus photobioreactor, and Janssen et al. (2000) evaluated biomass yield of *C. reinhardtii* in a photobioreactor under different light regimes. Photobioreactor conditions for hydrogen production by *Chlamydomonas* have been described by Kosourov et al. (2007). Outdoor cultivation of several microalgae, although not *Chlamydomonas*, has been reviewed by Borowitzka (2005) and by Del Campo et al. (2007).

C. Cultures on agar

1. Plating on solid substrata

The culture media in Table 8.1 are suitable for plate cultures with addition of 1.5% agar and for slants in test tubes with 1.5% or 2.0% agar. For routine cultures and genetic analysis, ordinary bacteriological-grade agar is satisfactory without further purification. *Chlamydomonas* can also be grown on the agar substitute Gelrite, a gellan gum of bacterial origin. When prepared in TAP medium, this product yields a clear gel and gives a high level of cell viability.

To obtain a uniform lawn of cells, or to isolate colonies each arising from single cells, 0.1–0.5 ml of liquid culture can be spread over the surface of an agar plate using a bent glass rod and a rotating turntable. A light touch is desirable, and spreading should stop before the glass rod meets resistance from the agar. Zygospores are more durable and can be spread more vigorously. Plating efficiency with this technique is rarely more than 80–90% and may be considerably less.

Agar overlays are useful for plating a uniform lawn of cells at relatively high density and are especially recommended when working with wall-deficient or other fragile cells. Cells should be suspended by vortex mixing in 2.5–4.0 ml of melted soft agar (0.7%) at 45°C and immediately poured onto the surface of previously prepared plates. Agar overlays are useful for testing different inhibitors on a single population of cells. The test compounds can be spotted (5–10 μl) as soon as the overlay solidifies.

2. Streak tests, spot tests, and replica plating

Auxotrophs, acetate-requiring mutants, and resistance mutants are usually identified after mutagenesis and in crosses by comparison of growth on selective and non-selective media in agar. For best results, cells should be transferred to test plates at relatively low density, by one of the following methods. Cells can be suspended in a small volume of sterile medium and delivered to the test plates with a micropipette or wire loop to make a spot or streak on each. This method usually produces the cleanest results. Cells can be streaked lightly with the broad end of a sterile toothpick. Best results are obtained when a non-selective plate is streaked first, thereby removing most of the cells from the toothpick, which is then used to streak the test plate.

For large numbers of individual samples, replica plating is convenient (see illustrations at http://www.chlamy.org/replica.html). Filter paper (Whatman No. 1) works much better for replica-plating *Chlamydomonas* than does the velvet commonly used for bacteria and yeast. A wooden or metal cylinder 80 mm in diameter, or a plastic food container with a smooth circular surface of the same size, is used to hold the paper for stamping the plates. Two or more sheets of 15 cm diameter paper should be used together, the number depending on how many replica plates are to be made and on how wet the plates are. Only the top sheet needs to be sterile, and the bottom sheets can be air dried and reused indefinitely. The filter paper should be placed on the cylinder top, pressed into place with a sterile petri dish bottom, and secured with a rubber band. The plates to be replicated are then inverted onto the filter paper; this procedure works better than stamping plates from above. Although replicate plates can be made directly from plates of spread colonies, results are cleaner and easier to interpret if master plates are made by transferring individual colonies with sterile toothpicks to a grid pattern of 64 or 100 patches per plate. Each patch should be spread evenly on the agar surface with the blunt end of the toothpick, without gouging the agar. The master plates should be replicated as soon as patches are well grown, usually 2 days for cells growing at wild-type rates. Dry colonies, which are frequently encountered on older plates, can sometimes be softened by leaving plates overnight in darkness. TAP medium produces wetter colonies than the Sueoka high salt medium and may be useful for strains that form very crusty colonies. A non-selective replica should always be made before the selective replicas to ensure perpetuation of the clones, since the original master plates may not grow back well. A second non-selective replica made at the end of the selective series will demonstrate whether all patches were satisfactorily transferred to each plate. For large numbers of samples, cells can be grown in 96-well plates and spotted onto agar using a 96-pin replicator.

3. Control of contamination

Chlamydomonas cultures are highly susceptible to contamination by fungi and bacteria. On minimal or minimal-acetate media, bacterial contaminants in particular may not be obvious on superficial examination but will be seen on transfer to enriched solid medium. For this reason it is advisable to keep all stock cultures on medium containing yeast extract, peptone, or other supplements. Any contaminants arising will then be readily apparent and can be eliminated before experimental results or valuable stocks are endangered. Wrapping petri plates with parafilm or wide rubber bands reduces both contamination and drying of the agar. Fresh agar plates should be stored unwrapped at room temperature in a clean place for 2 days to allow some drying to occur prior to use or storage. Plates should be inspected for bacterial colonies before transfers of *Chlamydomonas* cells are made.

Irradiation of transfer rooms or work areas with germicidal UV light when not in use and periodic swabbing of bench tops with 70% ethanol will reduce contamination substantially. The use of other disinfectants, and their advantages and disadvantages for phycological work, are discussed by Hamilton (1973) and by Kawachi and Noël (2005). Laminar flow hoods are sometimes used in *Chlamydomonas* laboratories to provide a sterile environment. However, in our experience transfers of cultures and particularly zygospore dissection are best performed in still air in an enclosed room, to minimize the danger of blowing mold spores and other contaminants from the lids and rims of petri plates onto the agar surface.

Fungal contaminants are a chronic problem in humid regions, but fortunately removing fungi is relatively easy, and vigilant daily monitoring of cultures will usually keep them under control. Fungi can be detected by holding plates up to a light to look for spreading hyphae. Isolated fungal colonies can be excised with a sterile spatula, preferably at an early stage of growth before extensive sporulation has occurred. When fungi have overgrown the *Chlamydomonas*, the best approach is to remove the algae from the fungi instead. First, use a sterile toothpick or glass needle to scrape away any mycelium overlying the *Chlamydomonas* cells. With a new toothpick, then streak a sample of the mixed fungus and *Chlamydomonas* across one side of a fresh agar plate of the simplest medium on which the strain in question can grow (e.g., minimal, minimal plus arginine). After one day's incubation, the fungus will have extended hyphae into the agar but will usually not have sporulated. Using a glass loop or needle and a dissecting microscope, one can now tease individual *Chlamydomonas* cells or colonies across the agar away from the hyphae. When several streaks of algae have been made away from the advancing hyphae, invert the plate over a dish of alcohol and cut out the fungus-containing streak with a sterile spatula. After 1–3 days more, *Chlamydomonas* colonies free of fungus should be evident and can be transferred to a fresh plate.

Bowne (1964) suggested 10–30 mM caffeine as an additive to media that would eliminate fungal contaminants. *Chlamydomonas* cells (species uncertain) were not inhibited at this concentration in Bowne's experiments. However, McBride and Gowans (1970) reported that 30 mM caffeine was toxic to both *C. eugametos* and *C. reinhardtii*. Isolated fungal colonies can also be killed by addition of a drop of silver nitrate solution (N.W. Gillham, personal communication).

Mahan et al. (2005), working in an old building in humid Austin, Texas, where even the methods suggested earlier were inadequate, screened several fungicides for ability to kill common contaminants without harming *Chlamydomonas*. Three compounds in the benzimidazole family – carbendazim, thiophanate-methyl, and benomyl – all fulfilled the criterion of effective fungal inhibition at concentrations that did not impair growth of *Chlamydomonas* on TAP agar.

Bacterial contaminants are often more difficult to remove than fungi, especially in *Chlamydomonas* cultures that require an enriched medium. However, one or more of the following procedures will usually be successful:

1. Streaking cells on agar to obtain single-cell colonies is often effective. The chances for success are improved if a transfer can first be made to minimal medium to enrich the proportion of *Chlamydomonas* to bacterial cells. Several rounds of gentle centrifugation and washing may also concentrate the algal cells, leaving most bacteria in the supernatant fraction. Streaks can be made from a solid culture or from a liquid suspension, using sterile toothpicks or a wire loop. After 1–3 days incubation, algal colonies free of bacteria can often be seen with a dissecting microscope and can be picked off or streaked further away from the contaminant. Microscopic observation is strongly

advised, since if plates are left until the *Chlamydomonas* forms macroscopic colonies, often the bacteria will have overgrown them.

2. Lewin (1959) recommended "washing" individual contaminated cells by sequential passage through a series of liquid drops on an agar surface, using a capillary pipette under a dissecting microscope to transfer single cells. Pasteur pipettes can be drawn out in a flame to a suitably fine diameter and topped with a rubber bulb or tube for expelling the cells into fresh medium. The process should be repeated several times; Lewin estimated that passage of 10–20 cells each through three droplets would be sufficient to obtain a few clean cultures.

3. Aerosol plating is another means of producing single-cell clones either from a contaminated culture or from a mixed natural population. The following procedure, adapted from Wiedeman et al. (1964), has been used by P. A. Lefebvre and colleagues to isolate *Chlamydomonas* from soil samples incubated in culture medium. A 1 ml aliquot of cell suspension is put in a 1.5-ml plastic microcentrifuge tube, which is then mounted in a burette clamp in a fume hood. A sterile 20-μl glass pipette is broken in half and the broken end put into the solution. A strong stream of air is then directed over the top of the microliter pipette. Laboratory forced air is satisfactory if filtered through a sterile, cotton-plugged Pasteur pipette. When the air stream and the pipette are held at the correct angle, approximately perpendicular, a fine mist will be pulled out of the pipette and sent "downwind." An agar plate is passed quickly through this mist to capture small droplets. The density of the droplets on the plate can be varied by passing the plate more rapidly through the mist or by moving it farther from the stream. Individual, closely spaced droplets should be deposited on the plate and are often small enough to contain individual *Chlamydomonas* separated from any contaminating cells.

4. Contaminated cultures can be streaked out onto agar containing a broad-spectrum antibiotic to which *Chlamydomonas* is insensitive. These plates should be prepared in enriched medium to encourage bacterial growth and thereby promote killing. Useful antibiotics include penicillin (10^6 units/ml); ampicillin (25–100 μg/ml); chloramphenicol (25 μg/ml; although it is toxic to *Chlamydomonas* in liquid culture, very high concentrations are needed to inhibit algal growth on agar); and neomycin (50–100 μg/ml). If the *Chlamydomonas* strain in question carries an antibiotic resistance marker (e.g., streptomycin, erythromycin, or spectinomycin resistance), that antibiotic can also be used. Hoshaw and Rosowski

(1973) recommended potassium tellurite as a bacteriostatic agent for purification of algal cultures, but we know of no publication of its use on *C. reinhardtii*.

5. The natural phototaxis of *Chlamydomonas* cells can sometimes be used to effect cleaning by allowing the algae to move away from contaminants. R.K. Togasaki (personal communication) suggested wrapping a graduated cylinder with dark cloth or paper, with the top exposed to a high-intensity desk lamp. After a few hours, most contaminants will have sunk to the bottom of the cylinder, and the surface of the culture will be enriched in *Chlamydomonas*, which can then be plated on agar.

6. Occasionally bacteria are encountered that adhere tightly to the surface of the *Chlamydomonas* cells. If not antibiotic-sensitive, these can be very difficult to remove. Manual manipulation of individual algal cells across with agar surface with a glass needle sometimes leaves the bacteria behind. Stubbornly adherent bacteria can also sometimes be removed by crossing the contaminated stock to a wild-type strain. Uncontaminated tetrad products can be recovered from isolated zygospores that had been moved across the agar surface.

Nothing in *Chlamydomonas* research quite compares with the surprise of seeing a small arthropod march across the microscopic field of view on a tetrad dissection plate. Mites are a well-known pestilence of *Neurospora* laboratories and occasionally invade *Chlamydomonas* cultures as well, especially in laboratories located close to rooms where plants are being grown in soil. The principal danger is not from the mites themselves, but from the fungi and bacteria they bring with them. Control consists of several measures:

1. Carefully check all petri plates in use for tell-tale tracks on the agar. The tracks are visible on holding a plate up to the light, although the mites themselves are nearly microscopic. Transfer *Chlamydomonas* colonies to fresh plates, being careful not to transfer the mites as well, and discard the infested plates immediately.

2. Thoroughly scrub all shelves where plates are stored and line with clean aluminum foil.

3. For bad infestations, wipe the shelves with a mitocide.

4. A fumigation procedure effective for closed culture rooms is to expose the room to an open beaker of formaldehyde for several days, then rinse shelves, walls and floor with household ammonia (Jane Aldrich, personal communication).

IV. HARVESTING AND FRACTIONATING CULTURES

Small volumes of *Chlamydomonas* culture are easily harvested by centrifugation at $5000g$ or less in a laboratory centrifuge. For large volumes, such as carboy cultures, repetitive rounds of centrifugation or continuous flow systems can be used. Centrifugation over a 25% Percoll cushion can be used to remove cell debris, which accumulates especially in cultures of wall-deficient cells.

In an attempt to determine quantitatively the yield of packed cells expected from a log-phase culture, we grew wild-type cells on minimal medium in continuous light to a density of 5×10^6/ml, harvested them by centrifugation for 10 minutes at $10,000g$, resuspended them in culture medium, and centrifuged them again under the same conditions. We calculated that 1×10^9 cells treated in this way generated a pellet of 0.5 g wet weight. This quantity of cells contains about 4 mg of chlorophyll and 70 mg of protein, measured by Lowry analysis after extraction of chlorophyll with 80% acetone. DNA content can be estimated to be about 120 μg per 10^9 cells.

Preparation of cellular components and assay of enzymes begins with breakage or lysis of cells and preparation of a crude cell extract. Soluble enzymes can be prepared from cells frozen in liquid nitrogen and then thawed and centrifuged to produce a supernatant fraction (Fernández and Cárdenas, 1982). Acetone powders for enzyme extraction are also easily prepared (Lien et al., 1975). To obtain particulate fractions, more drastic disruption is needed. The most frequently used breakage methods are sonication (Curtis et al., 1975; Yannai et al., 1976), the French press (Milner et al., 1950), the Yeda press (Belknap and Togasaki, 1981; Mendiola-Morgenthaler et al., 1985a), the Bionebulizer (Okpodu et al., 1994), and homogenization with glass or zirconium beads (Tetík, 1976; Maitz et al., 1982) or alumina (Fernández and Cárdenas, 1982; Lang, 1982). For some purposes lysis rather than physical disruption of cells is desirable. Wall-deficient mutants such as *cw15* and *cw92* are readily lysed with 1% NP-40, digitonin, or other detergents. Wild-type cells can be lysed by treatment with gamete lytic enzyme to remove their walls, plus detergent as for *cw* mutants.

V. HISTOLOGICAL TECHNIQUES

General methods for staining and fixation of algal cells are given by Gantt (1980; see also Zhang and Robinson, 1986b). The following suggestions specifically for *Chlamydomonas* are taken from Ettl (1976a). The simplest stain for general visualization of cell structure is iodine-potassium iodide (I-KI: 1 g iodine, 0.5 g KI, in 100 ml H_2O). If this stain is combined with aqueous eosin (1:1, freshly prepared), nuclei will be stained reddish, in good contrast to the blue pyrenoids. The cell wall can be emphasized for light microscopy by staining with ruthenium red, which binds to the outer layer, or with zinc chloride iodide or I-KI plus diluted sulfuric acid. The latter reagents stain the inner wall layer blue. The shape of the papilla, if any, is accentuated by staining with 0.1% aqueous gentian violet, which also stains gelatinous layers found external to the cell wall in some *Chlamydomonas* species. Flagella can be seen easily with phase contrast and do not need staining for most purposes. However, Giemsa stain gives particularly good results, with the flagella and basal bodies colored red, in contrast to a bluish cytoplasm. The pyrenoid can be stained with carmine in propionic acid after fixation in ethanol-sodium hypochlorite (Rosowski and Hoshaw, 1970). Associated starch bodies can be stained with I-KI, but better results are obtained with iodine crystals in chloral hydrate solution. This preparation stains starch bodies dark blue and leaves the cytoplasm clear.

Contractile vacuoles are best observed in living cells but can also be seen on staining with 1% tannin or fixation with osmic acid vapor. Mitochondria can be stained with Janus green or iron hematoxylin. The Golgi apparatus is revealed with 0.5% aqueous trypan red. Cytoplasmic granules and small vacuoles bind neutral red, and lipids are easily seen with alcoholic sudan III solution.

Nuclei can be stained by several classical techniques, including hematoxylin, borax carmine, safranin-light green, and Giemsa. Quick stains include acetocarmine and gentian violet. Schaechter and DeLamater (1956), Eves and Chiang (1982), and Levine and Folsome (1959) all provided methods for fixation and staining of nuclei. Methods for fluorochrome staining of DNA are given by Coleman (1978, 1984), and by Kuroiwa et al. (1982). Immunofluorescence methods for *Chlamydomonas* are given in many papers, for example Sanders and Salisbury (1995), Silflow et al. (2001), Lechtreck et al. (2002), and Mahjoub et al. (2004).

VI. ISOLATION OF *CHLAMYDOMONAS* SPECIES FROM NATURE

Although for most experimental studies use of defined laboratory strains is nearly essential, there are times when new isolates may be needed for comparative purposes or to seek properties not obtainable among the laboratory strains. The time-honored techniques of Bold (1942) and Pringsheim (1946a, 1954) for isolation of algal cultures are still recommended. Lewin (1959) summarized these in an excellent article, from which most of the following suggestions are adapted. Starr (1971, 1973), Guillard (1973), and Hoshaw and Rosowski (1973) offered additional advice. Soil or water samples should be suspended first if possible in filtered water from the collection site, or in medium containing prepared soil extract, at the same pH as the original sample. Enrichment cultures in soil–water should be used to obtain a good growth of the desired species before unialgal cultures are attempted. These should be started from single-cell clones, obtained either by spreading or streaking samples of the enrichment culture on agar or by using sterile micropipettes to transfer individual cells to small quantities of fresh sterile medium. Aerosol plating can also be used (see section III.C.3). Often natural phototaxis can be used to collect *Chlamydomonas* cells in a mixed culture into one portion of the culture dish, from which they can then be removed for subculture (Adams, 1969). The isolation step will probably need to be repeated several times before a true clonal culture is established, and bacterial contaminants may still need to be removed.

Specialized methods for obtaining sexually competent strains of *Chlamydomonas* were used by Smith (1946, 1950) and by Lewin (1951, 1959). Since zygospore walls are very resistant to desiccation, these forms survive in dry soil samples and can be germinated to yield clonal cultures that should be capable of mating with one another. Lewin (1951) found that *Chlamydomonas* zygospores were impervious to acetone and used this property as the basis for direct isolation of zygospores from soil samples. Smith (unpublished correspondence to Lewin and others) simply spread soil samples on agar containing minimal medium and incubated them in light for several weeks until green colonies were formed. Colonies arising from a single zygote should contain cells capable of mating with one another regardless of whether the species is heterothallic or homothallic. Single cells cloned from these colonies will then either show homothallic mating or will be capable of mating with a subset of clones from the same colonies.

VII. STORAGE OF *CHLAMYDOMONAS* STRAINS

A. Low-temperature storage

Maintenance of defined mutant strains in continuous culture demands periodic phenotypic checking to detect reversion or suppression of the mutant phenotype. Even with such screening procedures, there is danger of inadvertent selection of spontaneous mutations or variants on subculturing, such that isolates of a given mutant may acquire different properties from the original strain. Thus, a reliable long-term storage method is highly desirable. Unfortunately, green algae in general, and especially *Chlamydomonas*, have proved notably difficult to preserve by most of the freezing and lyophilization methods that are widely used for bacteria, fungi, and animal cells (see Day and Brand, 2005, for a comprehensive review).

Cellular damage to plant cells on freezing occurs in two ways. When cell suspensions are cooled slowly, formation of extracellular ice crystals occurs. The medium surrounding the cells becomes increasingly hypotonic, and osmotic loss of water with cell shrinkage ensues. In cells surrounded by a rigid cell wall, plasmolysis (shrinkage of the protoplast away from the wall) takes place, often causing eventual collapse of the wall itself and sometimes accompanied by formation of ice between the wall and the cell membrane.

More rapid cooling minimizes the time in which this osmotic dehydration can occur but increases the likelihood of intracellular ice crystal formation. Cell types differ in their permeability and consequent osmotic adaptation. In cells with low water permeability, very rapid cooling can bypass both osmotic damage and ice nucleation, so that the intracellular water reaches low temperatures in an amorphous, noncrystalline glass state that causes minimal cellular damage. Cell survival thus depends on minimizing both osmotic effects (favored by slow cooling) and ice nucleation (favored by rapid cooling). Since cell types differ in their relative susceptibility to each of these types of damage, optimal rates of cooling must be determined experimentally, but there now seems to be consensus that slow freezing is preferable for algae in general.

The best results to date for *C. reinhardtii* have been achieved using a cooling rate of approximately 1°C per minute until −55°C is reached, and then transferring the samples immediately to liquid nitrogen (Crutchfield et al., 1999). Rapid thawing is recommended to minimize recrystallization of intracellular ice.

Cryophilic species of *Chlamydomonas*, for example *C. nivalis*, are much more resistant to damage by freezing than are most temperate zone species, which probably survive cold temperatures in nature as zygospores rather than as vegetative cells (Morris et al., 1979). Clarke and Leeson (1985) demonstrated that these freezing-tolerant species show several consistent differences in cellular ultrastructure compared to sensitive ones. Most notably, vegetative cells of freezing-tolerant species have invaginations in the cell membrane, trough-shaped areas approximately $0.5\,\mu$m long $\times$ $0.05\,\mu$m wide and $0.05\,\mu$m deep, which are randomly distributed over the cell surface (Figure 8.1). These invaginations appear to be filled with cell wall material. Similar invaginations are seen in freezing-tolerant yeast cells, and are also typical of halotolerant algal species (Clarke and Leeson, 1985; Elspeth Leeson, personal communication). Freezing-tolerant species of green algae also have mitochondria that are embedded in the cytoplasm, in contrast to the peripheral mitochondria of freezing-sensitive strains, and either lack pyrenoids or have pyrenoids that contain dispersed starch grains. Freezing-sensitive species such as *C. reinhardtii* have pyrenoids surrounded by starch plates (Figure 8.2).

The growth phase of cells may also make a difference in freezing resistance. Working with *Chlorella*, Morris et al. (1981) found that stationary-phase cells survived better than those from exponentially growing cultures, probably because lipid accumulation and decreased vacuolization in the older cultures provided better adaptation to hypertonic conditions. Algae that naturally lack contractile vacuoles also appeared to be relatively resistant to freezing. Prior acclimatization to low temperature before freezing may result in changes in membrane lipid composition, which in turn determines the temperature at which the membrane undergoes a phase transition from

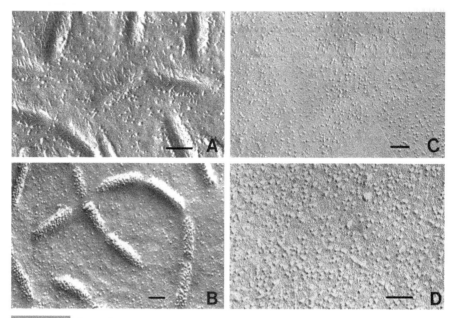

FIGURE 8.1 *Freeze-fracture micrographs of plasma membranes from freezing-tolerant and freezing-sensitive Chlamydomonas species. (A and B) The freezing-tolerant snow alga C. nivalis; P and E faces respectively; (C and D) C. reinhardtii, which is freezing-sensitive; P and E faces respectively. Bars = 0.1 μm. Courtesy of Elspeth Leeson.*

relatively fluid to a more rigid state. However, Morris et al. (1979) reported that although the proportion of unsaturated fatty acids increased in *C. reinhardtii* cultures maintained at 4°C for 3 or 4 weeks and was naturally higher in *C. nivalis*, preadaptation did not increase freezing tolerance in *C. reinhardtii*.

The highest survival rates reported to date for *C. reinhardtii* under have been achieved using 2–10% methanol as cryoprotectant and storage in liquid nitrogen (Crutchfield et al., 1999). Dimethylsulfoxide (DMSO) was reported to be an adequate protectant for short-term storage at −80°C (Johnson and Dutcher, 1993) but viability is lost over time, and this method is not considered to be reliable for storage over a period of years. Glycerol is not suitable for preservation of *C. reinhardtii* (McGrath and Daggett, 1977; Crutchfield et al., 1999).

Protocols for freezing using methanol as cryoprotectant, followed by liquid nitrogen storage, are available online (http://www.chlamy.org/freezing.html).

B. Lyophilization

Attempts at freeze-drying *Chlamydomonas* and other green algae have been relatively unsuccessful. Holm-Hansen (1964) attempted lyophilization of algae from temperature and Antarctic habitats. Most of the Antarctic green algae survived well if milk solids were added to the suspension before freezing,

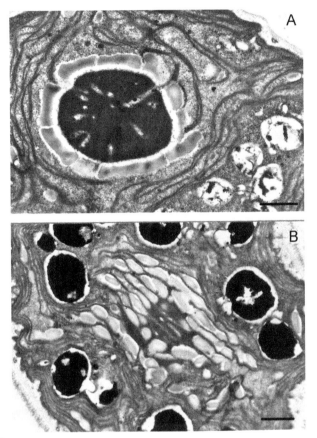

FIGURE 8.2   *Differences in starch grain dispersal around pyrenoids of freezing-tolerant and freezing-sensitive species of Chlamydomonas. (A) C. komma, freezing-sensitive, showing starch plates around pyrenoid. Compare to Figure 2.1, C. reinhardtii. (B) C. gerloffii, freezing-tolerant, showing dispersed starch grains. Courtesy of Elspeth Leeson.*

but survival of the temperate species was much less predictable. Horse serum and dextran were also tested as adjuvants but were found to have little if any protective effect. Tsuru (1973) succeeded in lyophilizing *Chlorella pyrenoidosa* and several other types of algae but did not test *Chlamydomonas*. McGrath and Daggett (1977) were unable to recover any *C. reinhardtii* flagellar mutants but cited a 1954 report by Daily and McGuire that *C. pseudococcum* could be successfully lyophilized. In a subsequent study, McGrath et al. (1978) obtained survival with *C. pseudococcum*, *C. humicola*, and *C. pseudagloë*, but failed with all other *Chlamydomonas* species tested, among them *C. eugametos*, *C. moewusii*, and *C. reinhardtii*.

C. Storage of zygospores

An alternative to freezing vegetative or gametic cells of *Chlamydomonas* is preservation of the thick-walled zygospores, which serve as a natural means by which the alga survives during periods of environmental stress (Coleman, 1983). Bristol (1920) cited a report of algal spores that remained viable for 66 years. In her own studies, Bristol covered natural soil samples with sterile paper and allowed them to dry for 1 month in a warm room before storing them in tin boxes. After 6 weeks, the dried samples were inoculated into sterile medium and left for several months to develop. Blue-green algae and diatoms predominated in these samples, but various Chlorophyceae were found, including two species of *Chlamydomonas*. Starr (1973) gave additional recommendations on recovery of algae from dried soils. Trainor (1985a) reported survival of *Chlamydomonas* species in soil for 10 and 25 years, and Hardy and Curl (1972) reported that "resting cells" [akinetes?] of *C. nivalis* had remained viable after dry, frozen storage for 7 years.

Coleman (1975) demonstrated that zygospores of two species of the colonial green alga *Pandorina* air-dried in cotton-stoppered tubes from soil-water suspensions were still viable after 16 years at temperatures in the range 6–38°C. Germination was induced by addition of water or culture medium to a small quantity of the soil containing zygospores. Zygospores scraped from the walls of dry test tubes did not germinate. Reed et al. (1976) used shredded filter paper as a substratum for preserving cysts of the colorless flagellate *Polytomella*.

Bennoun (1972) published a procedure for freezing zygospores of *C. reinhardtii* on agar, the essential feature of which was to allow maturation of the zygospores at room temperature on agar in a petri dish before placing the dish in a freezer at −70°C for long-term storage. He found that freezing zygospores in liquid suspension was not satisfactory. Germination was achieved by thawing the plate in darkness at room temperature and then transferring the zygospores to fresh medium and placing them in the light. This method was tested by the Chlamydomonas Genetics Center at Duke University and found to work well in short-term experiments, but was not pursued as a long-term storage method. However, several plates of zygospores dating from the mid-1970s and kept at −80°C until 2004 yielded viable cells on thawing.

Coleman's dry storage method also worked well in tests by the Chlamydomonas Genetics Center, using dried clay particles (generic cat litter) as an adsorbent. Mating mixtures of gametes were pipetted onto autoclaved clay and allowed to dry slowly for several weeks under illumination, then transferred to screw-cap vials for long-term storage. Germination was excellent after 6–18 months storage at room temperature but has declined over longer periods.

Zygospore storage of mutant strains poses of a choice of potential difficulties. On the one hand, mutant x mutant crosses may produce relatively few zygotes or give low subsequent viability, especially if auxotrophic or non-photosynthetic strains are involved (see Bennoun, 1972). On the other hand, if mutants are crossed to wild-type strains, mutant clones must be selected from the surviving progeny, where they may be outnumbered by wild-type cells. Nevertheless, the simplicity and low cost of this method may be worth the additional effort required to reselect mutant clones after zygospore storage.

VIII. GENETIC ANALYSIS

A. Gametogenesis and mating

1. Gamete formation

As discussed in Chapter 5, gametogenesis is induced in *C. reinhardtii* in the laboratory by nitrogen deprivation. Gametes prepared from synchronously grown liquid cultures by removing nitrogen at the beginning of the light phase, during early G1 period, will be capable of mating at high efficiency within 6 hours (Abe et al., 2004). Wild-type cells that have been maintained on agar for a week or more may already have undergone gametogenesis and will be able to mate soon after transfer to nitrogen-free medium. Cells from fresh plates or from asynchronous liquid cultures should be left in nitrogen-free medium for several hours, or preferably overnight. Transfer to solid medium with a lower concentration of nitrogen for 1 or 2 days prior to mating often increases the number of zygotes obtained (Sears et al., 1980).

    C. eugametos and *C. moewusii* are customarily induced to form gametes by flooding agar cultures with distilled water and placing the resulting suspensions in darkness for two or more hours (Lewin, 1952a, 1953a). Trainor (1961) reported that *C. chlamydogama* is best induced to gametogenesis by growth at 26°C for 3 days, harvest into nitrogen-free medium, storage at 34°C for 30 hours, and then a shift to 22°C for 12 hours. A similar regime, 12 hours in nitrogen-free medium at 30°C followed by a shift to 22°C in darkness, was effective for *C. eugametos* (Trainor and Roskosky, 1963). These methods should also work for other species of *Chlamydomonas* that use nitrogen deprivation as a gametogenic trigger.

    Distilled water can be used for gametogenesis for most strains but is not advisable for *C. reinhardtii* cultures grown on high salt medium (Sueoka, 1960; Table 8.1), for which the osmotic transition may be damaging. The ordinary culture medium with ammonium and nitrate salts omitted is preferable. For vitamin auxotrophs and acetate-requiring mutants, the needed compound can be included in the gametogenesis medium (although for short periods it is not essential). Honeycutt and Margulies (1972) suggested that arginine should be omitted even for *arg–* mutants, as it can serve as a nitrogen source and thereby repress gametogenesis. However Ursula Goodenough (personal communication) found that arginine did not interfere with gametogenesis of *arg7* mutants.

    Gametogenesis is normally carried out in bright light, although as discussed in Chapter 5, some strains are capable of forming gametes in the dark. Light-sensitive mutants, such as those deficient in chlorophyll or carotenoids, will tolerate light filtered through orange cellophane (Wang et al., 1974). This should be used both for gametogenesis and for subsequent germination of zygospores.

2. Trouble-shooting problems with the mating reaction

As discussed in Chapter 5, sexual agglutination normally begins immediately on mixing suspensions of *plus* and *minus* gametes, and formation of pairs occurs within minutes. The time course may be delayed if one or both partners are impaired in motility, or if gametogenesis was not complete at the time the suspensions were mixed. The gametes soon fuse to form a quadriflagellate zygote, which remains motile for approximately 2 hours before flagellar resorption and zygospore formation begin. Pair formation and appearance of quadriflagellate cells can be monitored by microscopic observation of samples taken at intervals. For genetic analysis, the mating mixture should be plated on agar before the zygospore wall begins to form, so that the young zygospores will not adhere to one another.

Sometimes laboratory strains of *Chlamydomonas* that have not been used in crosses for a long time will not mate well. Trainor (1985b) observed that *C. moewusii* cultures received from the UTEX collection always mated, even though they were not routinely crossed by the UTEX staff. He suggested that the UTEX practice of maintaining stocks on soil-extract medium might be beneficial by contributing some unidentified micronutrient required for gametogenesis, and recommended several rounds of transfer on soil-extract medium to revive sluggish strains. Water quality may also affect gametogenesis. Although this is clearly not the answer when other strains in the same laboratory mate well, it is a prime suspect when wild-type cells fail to mate for a new investigator. Variability in batches of agar has also been implicated. The following suggestions may aid in trouble-shooting problems with gametogenesis and mating:

1. Check each gamete suspension microscopically to verify that cells are motile. If cells are in a "palmelloid" state, mating will be poor. Possible remedies for this condition are discussed in section III. B.3. Paralyzed strains can mate if they have flagella but will be slow to agglutinate since they cannot contact one another as readily as motile cells.

2. Test each partner of the non-mating couple for mating with wild-type strains of known mating type and competence. Often the problem is simply that the mating type of one strain is mislabeled.

3. Try shorter (1–2 hour) or longer (overnight) periods of gametogenesis.

4. Pregrow cultures on different medium, for example in liquid culture, on agar containing yeast extract or soil extract, on nitrogen-free or

low-nitrogen plates, or on plates with reduced sulfur (recommended for certain paralyzed mutants by Susan Dutcher and David Luck, personal communication).

5. Try a longer period of mating, several hours or even overnight. This often helps with palmelloid strains, because the gamete lytic enzyme released by a few mating pairs aids in release of additional cells, which can then mate with others.

6. Increase the culture density.

7. Centrifuge the mating mixture at very low speed to force cell contact and fusion.

8. Treat gametes for 20 minutes prior to mating with 10 mM dibutyryl cyclic AMP and 1 mM isobutylmethylxanthine, combine the gamete suspensions, and allow them to mate in the presence of these compounds for an additional 40-minute period before plating (Pasquale and Goodenough, 1987).

3. Testing mating type

For rapid verification of mating type of a single sample, direct microscopic observation is often satisfactory. Cells should be observed immediately after mixing, as the reaction can be very rapid. The initial clumping phase in particular is easily missed, but quadriflagellate cells of *C. reinhardtii* persist for approximately 2 hours and are readily observed with phase contrast. The vis-à-vis pairs of *C. eugametos* and *C. moewusii* (Chapter 5) are likewise easily seen. To test mating type on a group of samples, one of the following methods will probably prove more convenient:

Cells to be tested can be suspended at 2–5 × 10^6 cells/ml in 1 ml nitrogen-free medium in small test tubes, making duplicate samples. To one set of samples, wild-type *plus* cells in nitrogen-free medium are added; to the other, *minus*. A wild-type *plus* × *minus* control tube is useful. Tubes are left stationary in a rack under light overnight to allow formation of a zygospore pellicle (Figure 8.3A). If mating is good, this will be readily apparent as a reticulate layer on the surface of the liquid and sometimes clinging to the test tube walls beneath the surface. When mating is less complete, the pellicle can still sometimes be seen as a faint meshwork around the meniscus. If the sample cells appear to mate with both *plus* and *minus* testers, the sample should be cloned for single cells and the tests repeated. Mixed colonies are not infrequent in tetrad analysis.

A modification of this technique that is especially well-suited to small sample sizes is to set up replicate samples in multiwell plates (24 wells per plate), using sterile toothpicks to transfer and suspend the cells. Only about 0.5 ml sample volume is needed, to which 0.5 ml wild-type tester cells are

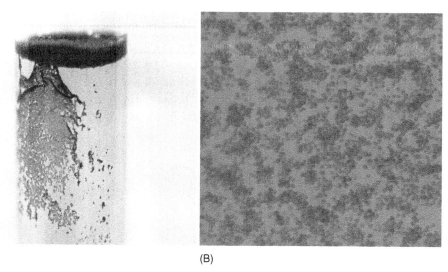

(A)   (B)

FIGURE 8.3   *Mating type tests. Gametes of opposite mating types were mixed and left overnight before photographing.*
(A) Formation of reticulate pellicle on top and sides of tube. (B) Surface of liquid in a multiwell plate, as seen through dissecting
microscope at 25x.

added in parallel rows of alternating mating type. More dilute suspensions can be made than with the tube method, since the pellicle can be observed microscopically. After overnight incubation, the multiwell plates are examined under a dissecting microscope, in each case comparing cells tested with *plus* and *minus*. A reticulate pellicle will be seen only on the surface of wells containing cells of both mating types (Figure 8.3B). The multiwell plates do not need to be kept sterile and can be washed and reused indefinitely. For very small quantities of cells, 96-well plates can be used, but it is harder to score mating in these than in the 24-well plates.

Werner and Mergenhagen (1998) and Zamora et al. (2004) have published methods for determination of mating type using PCR. Both methods depend on amplification of sequences unique to the mating type regions of *plus* and *minus* cells respectively, and can also be used to confirm diploid cells formed by *plus* × *minus* mating, which should have both the *plus* and *minus* mating type loci.

4. Quantitative measurements of mating

Mating efficiency in liquid cultures can be estimated by counting cells before and after gamete mixing. Samples of each mating type are counted, and the starting concentration of cells is calculated from the volumes mixed. After the desired time interval, a count is taken of the mating mixture. The mating efficiency (Chiang et al., 1970) is [2(1 − final number) × 100]/ (starting number of cells).

A more precise measure can be obtained by comparing the numbers of quadriflagellate and biflagellate cells at intervals after mating (Hoffman and Goodenough, 1980). Flagellar adhesiveness can be quantified by using radioactively labeled flagella as a reagent to bind unlabeled gametes (Snell, 1976; Pijst et al., 1984a). Adair et al. (1982) published a quantitative assay for agglutin activity in solution. Collin-Osdoby et al. (1984) developed a method for conjugating agglutinins to agarose beads for use as a probe of adhesion *in vitro*.

Gametes can be "marked" prior to mating by several means to facilitate distinguishing *plus* from *minus* cells at the light microscope level. Sager and Granick (1954) grew one mating type in the light on medium containing 1/10 the usual level of nitrogen, which resulted in accumulation of starch granules. Cells of the opposite mating type, grown on the normal medium, were smaller and darker green. Lewin (1952a) used neutral red, and McLean and Brown (1974) phenol red, to stain one gamete type prior to mating.

5. Zygospore maturation

Light appears to be needed at least in *C. moewusii* at the beginning of zygospore formation (Lewin, 1952a, 1957b) in order for maturation to occur and is traditionally provided to *C. reinhardtii* zygospores also. The standard regime used in many laboratories for both species, which probably dates back to these papers by Lewin, is either 18 hours or 24 hours in light followed by 5–6 days or more in darkness. Lewin (1957b) found that prolonged illumination of *C. moewusii* induced a period of dormancy and formation of a very thick zygospore wall from which subsequent germination was poor, whereas light followed by a dark period produced thinner walls and subsequently good germination. In contrast, if cells of *C. reinhardtii* are left in light for several days on complete medium, premature germination will occur. However, VanWinkle-Swift (1977) found that the entire maturation of *C. reinhardtii* can take place in light if cells are plated on nitrogen free medium. This regime is not satisfactory for *C. moewusii* (Cain, 1980).

Early papers often advised leaving zygospores for long maturation periods, several weeks or even months, before attempting germination. Although this may be necessary for some species, it is obviously a disadvantage in laboratory experiments. With both *C. moewusii* and *C. reinhardtii*, 5–6 days should be sufficient to obtain good germination frequencies. VanWinkle-Swift (1977) and Sears (1980b) reported approximately 80% germination efficiency of *C. reinhardtii* zygospores after only 3 and 2 days, respectively. Many of these young zygospores yielded more than eight progeny, however (VanWinkle-Swift, 1977), rendering them unsuitable for tetrad analysis (see section VIII.C). Bold (1942) and Starr (1949) reviewed methods for inducing zygospore germination in various algae that may provide some helpful suggestions for persons studying unfamiliar *Chlamydomonas* species. Transfer to fresh medium and exposure to light were recognized early on to be most important. For some species, alternately drying and wetting the zygospores during the maturation period was beneficial. Working with *C. chlamydogama*, Starr (1949) found that transfer to fresh agar followed by incubation for 48 hours at 37°C induced approximately 90% germination within the next 48–96 hours. Trainor (1960, 1961), using the same species, recommended mating at 30°C or 34°C followed by 24 hours dark storage at 22°C before transfer to fresh medium in the light. Hommersand (1960) reported that the light requirement for germination of *C. reinhardtii* could be replaced by a combination of cysteine and methionine (10^{-4}M). Homocysteine could be substituted for methionine and any of several oxidizing or reducing agents (e.g., ascorbic acid, cytochrome c, thioglycollic acid) for cysteine. These effects may have some relation to a shift in respiration from CO-sensitive to CO-insensitive which took place during the germination process (Hommersand and Thimann, 1965), but the precise mechanism of action was not reported. Cain (1980) found that germination was impaired by maturation of *C. moewusii* zygospores on medium containing 1.7 mM sodium citrate. Other tricarboxylic acid cycle intermediates at the same concentration did not affect germination, nor did the presence or absence of divalent cations, suggesting that the citrate effect was not attributable to metal chelation.

To prepare zygospores in large quantities for biochemical and molecular analysis, Wegener et al. (1989) recommended preparing gamete suspensions in 2-liter quantities at a density of $4–8 \times 10^6$ cells/ml, and then allowing mating to take place in 100–150 ml batches in Erlenmeyer flasks for 4–6 hours. The zygospore pellicles that formed on the surface of the liquid were transferred to 85-mm TAP plates containing 4% agar, with each such plate containing the pellicles from four or five flasks. Excess liquid was removed with a pipette and the surfaces were allowed to dry under a flow of sterile air. After light exposure for 16–18 hours, the plates were placed in the dark for 5 days to allow maturation. Plates were then exposed to chloroform to kill unmated cells (see section VIII.B.5) and the zygospore pellicles were scraped off with a razor blade and suspended in a small quantity of TAP medium (4 ml for the zygospores from two maturation plates). These suspensions were subjected to 60 seconds sonication at 20–30 W, which destroyed any remaining unmated cells but did not rupture the zygospores. The volume of the tube was brought to 20 ml with TAP, and the suspension was centrifuged at 1500 g for 5 minutes. Two additional washes with TAP medium followed by centrifugation left a preparation of zygospores free of cellular debris. For further analysis, the zygospores were disrupted using a dismembrator with a tungsten carbide ball in Teflon vessels precooled in liquid nitrogen. Disruption was for 30 seconds at a frequency of 50 Hz and maximum amplitude, repeated 20 times with immersion in liquid nitrogen to keep the zygospores frozen.

Umen and Goodenough (2001b) published a method for preparation of DNA from zygospores, using sonication to eliminate unmated gametes, followed by vortexing with zirconium beads in a buffer containing SDS and sarkosyl, then pronase digestion and extraction with phenol/chloroform.

B. Mutagenesis

1. Selection of strains for induction of mutations

As discussed in Chapter 1 (see Figure 1.4), Smith's 1945 Massachusetts isolate of *C. reinhardtii* gave rise to three lines of strains that have been separate since the 1950s. Most of the classic mutant strains were isolated in the Ebersold/Levine 137c background (CC-124 and CC-125 in the Chlamydomonas Resource Center collection; see Table 1.1). The principal exceptions are Sager's collection of antibiotic resistance mutations (Sager and Ramanis, 1970a) and some mutations affecting nitrogen metabolism (Sosa et al., 1978), that were isolated in Sager's 21 gr mt^+ strain (CC-1690) or its derivatives. Strains of the third line, in particular UTEX 89 and UTEX 90, have been reported to produce lethality in crosses (Tan and Hastings, 1977) and are therefore not recommended as a mutant background.

For most purposes, the mating type of the starting strain is unimportant. However, chloroplast mutations should be induced in a *plus* background, and mitochondrial mutations in *minus*, to facilitate later recovery in crosses. Insertional mutagenesis (section IX.D.2) should be done in a cell wall-deficient strain, but this can be either a strain with a *cw* mutation, or one in which the walls have been removed by treatment with the gamete lytic enzyme. Since *cw* mutants often produce incomplete tetrads in crosses, wild-type cells rendered temporarily wall-less may be preferable. Most mutant isolations are carried out on non-synchronized, log-phase cultures. However, synchronized cultures can be used to enrich for chloroplast or nuclear mutations (Lee and Jones, 1973). To ensure that mutations isolated in a given experiment do in fact represent independent mutational events, a "fluctuation test" design can be used (Luria and Delbrück, 1943; see also Harris, 1989).

2. Mutagens

The choice of mutagens to be used is dictated partly by convenience and partly by the type of mutation to be isolated. UV irradiation is effective for inducing nuclear gene mutations and is safer than chemical mutagens. Of the latter, nitrosoguanidine (MNNG) and the alkylating agents methyl methanesulfonate (MMS) and ethylmethanesulfonate (EMS) have been most popular (Table 8.3). The thymidine analog 5-fluorodeoxyuridine is highly effective in facilitating recovery of chloroplast mutations (Wurtz et al., 1977). Treatment of cells with this agent dramatically reduces the chloroplast DNA content with little or no effect on nuclear DNA replication or cell survival. In combination with other mutagenic treatments, 5-fluorodeoxyuridine facilitates recovery of chloroplast mutations in a synergistic manner but also appears to be mutagenic for chloroplast genes in its own

Table 8.3	Efficiency of mutagenic treatments with UV, MNNG, and EMS on isolation of arginine auxotrophs						
	UV			MNNG		EMS	
	3 min	3 min	6 min	30 min	60 min	2 hr	2 hr
Colonies replicated (number)	2820	1181	2299	3381	3002	2546	11,222
Survivors (%)	37	18	0.52	25	14	51	85
Mutants (number)	19	4	7	18	6	9	32
Mutants (%)	0.67	0.34	0.30	0.53	0.20	0.35	0.29

Data are from Loppes (1970a) and represent independent experiments at 25°C with the following regimes: UV treatment with a Hanau Sterilamp, type F2318, at 35 cm distance; MNNG, a dosage of 100 µg/ml N-methyl-N'-nitro-N-nitrosoguanidine in 0.02 M potassium phosphate, pH 6.0; and EMS, a dosage of 0.27 M ethyl methanesulfonate in 0.1 M potassium phosphate, pH 6.0.

right (Shepherd et al., 1979; Wurtz et al., 1979; Myers et al., 1982). Point mutations, deletions, duplications, and rearrangements of chloroplast DNA were all obtained in these experiments. X-rays have been used in relatively few studies but have produced both nuclear and chloroplast mutations, and may be useful if deletion mutations are desired (Myers et al., 1982).

Loppes (1970a) compared mutation rates for UV, MNNG, and EMS in induction of nuclear auxotrophic mutants (Table 8.3). All three agents were effective, and similar types of mutations were recovered. Their data suggested that mutagenic doses giving low survival rates (<20%) were less efficient for recovery of mutations than more moderate doses (20–50%). Also, survivors of high concentrations of mutagens often show undesirable secondary mutations, including product lethality in subsequent crosses. Mutation rates will of course vary not only with the treatment conditions but with the type of mutation being sought. A reasonable expectation for recovery of resistance, auxotrophic and motility mutants from chemical or UV mutagenesis would be in the range of 10^{-6} to 10^{-8}.

3. Recovery of mutants after mutagenesis

In many cases, mutagenized cells can be plated directly on selective medium. However, the problem of phenotypic lag may need to be taken into account. This is the time required before a new mutation is expressed and is a consequence of the retention and expression of wild-type gene products or metabolites within a cell after a mutation has occurred that will block further synthesis of those products. For example, a cell in which a mutation to auxotrophy is induced will continue to grow on minimal medium until the intracellular supply of the needed nutrient is exhausted. An analogous phenomenon is postzygotic lag (Lewin, 1983), in which meiotic products of a cross express wild-type characters for one or more generations despite their mutant genotype. Often a period of growth in non-selective medium is desirable after treatment with a mutagen in order to dilute out wild-type gene products and allow expression of newly induced mutations before challenging the cells with selective conditions. If the fluctuation test design is used, with cells being distributed into individual tube cultures after a mutagenesis treatment, this expression period will take place as the tube cultures grow to final density. If cells are to be plated immediately after mutagenesis, one possible solution to the problem is to plate on non-selective medium and subsequently either to replica-plate to selective medium or to overlay a selective agent (e.g., an antibiotic or other inhibitor) in agar after the plated cells have undergone several divisions (Lee and Jones, 1973).

Success in isolating mutants of particular types has often required development of enrichment schemes that take advantage of particular properties of *Chlamydomonas* biology. The examples that follow are presented partly for their intrinsic value for isolating additional mutants and partly to encourage further creative exploitation of *Chlamydomonas* as a system for genetic analysis.

a. Auxotrophs and conditional lethals

One classic experimental design, for which the prototype is penicillin selection in bacteria (Davis, 1948; Lederberg and Zinder, 1948), employs a toxic agent that kills dividing cells on minimal medium (or under other selective conditions), leaving mutants (e.g., auxotrophs) as survivors that can then be recovered by transfer to supplemented medium in the absence of inhibitor. Penicillin, which interferes with cell wall development in bacteria, is not inhibitory to *Chlamydomonas*. However, nystatin, which has been used in an analogous enrichment technique in yeast (Snow, 1966), is toxic to wild-type *C. reinhardtii* (Lampen and Arnow, 1961) and might be a suitable agent for this purpose. Selection with bromodeoxyuridine and light has been used in mammalian and plant cells (Puck and Kao, 1967; Carlson, 1969). This analog is preferentially incorporated into chloroplast DNA of *C. reinhardtii* and does not kill wild-type cells (Chiang et al., 1975). However, there are reports of the use of bromouracil in a similar fashion to select nuclear mutations in *Chlamydomonas* (Hipkiss, 1968; Sato, 1976). McMahon (1971) showed that canavanine will kill dividing cells of arginine-requiring mutants of *Chlamydomonas* and used this selection method to recover temperature-sensitive mutants with defects in protein synthesis. Nitrogen starvation followed by treatment with 8-azaguanine was used in a similar manner to recover mutants of the colonial green alga *Eudorina* (Toby and Kemp, 1975).

b. Non-photosynthetic mutants

Mutants that require acetate as a carbon source for growth are among the most frequent types recovered after mutagenesis of *C. reinhardtii* cells. In the early studies by Levine and colleagues (see Levine, 1969; Levine and Goodenough, 1970 for review), non-photosynthetic mutants were recovered among acetate-requiring colonies identified by replica plating after mutagenesis. On acetate-containing medium in dim or bright light, these mutants tend to grow more slowly than wild-type cells and will be found to be most common among those isolates that form small colonies (Ladygin, 1977; Spreitzer and Mets, 1981).

Two agents, arsenate (Togasaki and Hudock, 1972; Schneyour and Avron, 1975) and metronidazole (Schmidt et al., 1977; Bennoun et al., 1978), have been used to enhance recovery of non-photosynthetic mutations in a manner analogous to penicillin selection, by killing wild-type photosynthesizing cells, leaving non-photosynthetic mutants as survivors which can then be recovered on transfer to acetate-containing medium (see Harris et al., 1982, for protocols). Spreitzer and Mets (1982) reported no substantial increase in levels of acetate-requiring mutants among cells selected for arsenate resistance, and found that most acetate-requiring cells that were recovered after this treatment were still capable of CO_2 fixation. The method seems to be most efficacious for recovery of mutations with defects in photophosphorylation (Hudock et al., 1979; Shepherd et al., 1979; Woessner et al., 1984). Metronidazole is an effective agent for recovery of mutants blocked in photosynthetic electron transport between water and ferredoxin, and for other mutants whose deficiency leads to a decrease in the electron transfer rate (Bennoun and Delepelaire, 1982).

Both arsenate and metronidazole require prolonged light exposure to effect killing of photosynthetically competent cells and are therefore unsuitable for recovery of mutants that are light-sensitive as a result of pigment deficiency (Tugarinov and Levchenko, 1976; Wang, 1978; Spreitzer and Ogren, 1983a). Spreitzer and Mets (1981) reasoned that mutants capable of photosynthetic electron transport but blocked in CO_2 fixation would also be killed by these enrichment methods because they would accumulate NADPH that could not be oxidized. Using a dark selection procedure based on this rationale, they succeeded in isolating the first chloroplast mutant of *Chlamydomonas* with an altered large subunit of ribulose bisphosphate carboxylase (Spreitzer and Mets, 1980, 1981).

Mutants blocked in photosynthetic electron transport typically show an increase in steady-state levels of chlorophyll fluorescence under constant illumination *in vivo*, because energy absorbed by chlorophyll cannot be utilized. Bennoun and Levine (1967) developed a technique for screening plated colonies for fluorescence that proved to be very useful in recovering mutants with various types of defects in photosynthesis. Mutants blocked in photosystem I, photosystem II, and chloroplast ATP synthase can be distinguished by the kinetics of their response to fluorescence induction (Bennoun and Delepelaire, 1982; Karukstis and Sauer, 1983).

c. Mutants resistant to antibiotics, herbicides and other inhibitors

Recommended concentrations in agar of antibiotics and commonly used metabolic inhibitors are given in Table 8.2. Best results are usually obtained when each test plate contains 1 to 5×10^6 cells, taken either immediately after mutagenesis or after a period of culture in liquid medium. Since wild-type cells of *C. reinhardtii* may undergo one to two divisions in the presence of an inhibitor, a green lawn may appear on the test plate in the early days after plating, but the majority of these cells will ultimately bleach and die, leaving only resistant mutants as survivors.

Gradient plates (Szybalski and Bryson, 1952) are useful for determining the range of concentrations of an inhibitor to which cells are resistant, and they can also be used for isolation of mutants. To prepare these, 10 ml of agar is pipetted into a petri dish and the dish tilted so that the agar solidifies at a slant, with one edge of the plate surface barely covered by agar. The dish is then placed in a level position, and 10 ml of agar containing the highest concentration of inhibitor to be tested is pipetted on top of the other layer. Downward diffusion of the inhibitor through the agar will produce a linear concentration gradient.

Because the chloroplast genome is polyploid, cells expressing chloroplast-mediated antibiotic resistance may still harbor chloroplast DNA molecules containing wild-type (sensitive) alleles of the mutated gene. These alleles can eventually become expressed as a result of somatic segregation (see Chapter 7, section II.F). Antibiotic resistant strains should therefore be cultured for three or more rounds of transfer on antibiotic medium before being considered truly homoplasmic.

d. Motility mutants

Wild-type cells on fresh agar can form flagella and swim short distances. After prolonged culture of cells on agar, the flagella are very short but are able to regenerate immediately on transfer to liquid medium (Lewin, 1952a). The retention of flagellar function on agar accounts for the difference in colony morphology observed between wild-type and non-motile mutants (Lewin, 1954a). Screening after mutagenesis for heaped colonies on agar, especially on 2% or 2.5% agar plates, is an effective way to recover non-motile cells (Lewin, 1954a; Warr et al., 1966; Huang et al., 1977). Mutants lacking flagella also often have difficulty escaping from the mother cell wall after mitosis, and tend to form clumps of cells. Other strains that can form flagella but suffer from hatching defects, perhaps related to cell wall lysis, may also appear crusty or heaped on agar plates.

Jarvik and Rosenbaum (1980) suggested a simple screening method for flagellar mutants. After mutagenesis, cells are spread on agar and allowed to form colonies. Heaped colonies are identified as described above and are picked off with sterile capillary tubes, each tube containing a small volume of liquid medium, the colony, and an agar plug. The tubes are then set in beakers and illuminated from above. Wild-type cells are able to swim toward the light, and they collect at the meniscus. Capillary tubes in which cells do not appear at the meniscus contain putative motility mutants, and these are blown out into tubes of liquid medium to be cultured for further analysis. Adams et al. (1982) constructed an apparatus to enrich for motility mutants in liquid culture by draining off positively phototactic swimming cells from the top of a column and pumping in fresh medium from the bottom. By alternating periods of growth at 20°C and 32°C they were able to isolate a collection of mutants with temperature-conditional defects in flagellar assembly or function.

e. Non-mating mutants

Goodenough et al. (1976) mutagenized wild-type cells prior to gametogenesis. Arginine-requiring (*arg2*) gametes of the opposite mating type, previously demonstrated to have nearly 100% mating efficiency, were then added in excess of the wild-type cells. The mating mixture was left undisturbed overnight to allow formation of a zygospore pellicle, which was lifted off and discarded, and the remaining liquid in the tube was plated on medium without arginine. Since the *arg2* gametes could not grow, the only colonies formed were those of unmated cells of the opposite mating type, which were then tested individually for mating ability.

Forest and Togasaki (1977) used a mutant, *pet-10-1*, deficient in photosynthetic electron transport and thereby relatively resistant to the redox dye methyl viologen, as a source for selection of non-mating mutants. Cells of *pet-10-1* were treated with UV, induced to form gametes, and then added to a gamete suspension of wild-type cells of the opposite mating type. After mating had proceeded for 2 hours or more, methyl viologen was added to kill unmated wild-type cells and newly formed zygotes (which would have received photosynthetic competence from the wild-type parent). The unmated *pet-10-1* cells that survived were presumed to include non-mating mutants. After a second round of gametogenesis, mating and methyl viologen treatment, non-mating mutants in the *pet-10-1* genetic background were recovered and characterized. By carrying out the entire procedure at 35°C, Forest and Togasaki were able to recover temperature-conditional mutants that were unable to mate at high temperature but proved capable of normal mating at 25°C.

f. Respiratory-deficient and other metabolic mutants

Mutants with defects in mitochondrial function are unable to grow on acetate in the dark and can be selected by screening for fluoroacetate resistance followed by replica plating to confirm the dark-dier phenotype (Wiseman et al., 1977). Dorthu et al. (1992) screened for respiratory deficiency by overlaying colonies with soft agar containing triphenyltetrazolium chloride. Wild-type cells turn purple after incubation with this reagent, whereas respiratory mutants remain green.

Colorimetric assays have also been used to identify mutants deficient in phosphatase activity (Loppes and Matagne 1973) and, using the same system, cell wall-deficient mutants, which excrete phosphatases into the medium (Loppes and Deltour, 1977). Chlorate resistance can be used to select for mutants lacking nitrate reductase activity (Prieto and Fernández, 1993; Navarro et al., 2005; see Chapter 6, section III.B.4). Davies et al. (1994) isolated mutants with altered responses to sulfur stress by using a chromogenic substrate for the arylsulfatase enzyme. Mutants deficient in this activity were recovered, as well as mutants with constitutive expression of arylsulfatase. Iodine vapor, which stains starch, is useful for identifying mutants with abnormal starch accumulation. The color developed is indicative of the type of starch that is present (see Volume 2, Chapter 1).

4. Induction of mutations in diploid cells

Mutagenesis of diploid cells (section VIII.E) is an effective way to recover non-photosynthetic mutations in the chloroplast genome, which segregate in mitotic divisions and are therefore expressed when recessive nuclear gene mutations are not (Lee et al., 1973; Myers et al., 1982). Diploid mutagenesis can also be used to select deliberately for dominant nuclear gene mutations. Preble et al. (2001) used diploid cells to obtain mutant alleles at the *BLD2* locus that have defects in flagellar assembly and positioning of the mitotic spindle and cleavage furrow. A null mutant isolated in this way had a lethal phenotype in the haploid state but could be rescued by an extragenic suppressor. Bellafiore et al. (2002) used gamma irradiation of a diploid strain to select new mutations at the *AC29* locus, which is tightly linked to mating type and encodes the Albino3 protein.

5. Determining inheritance pattern and allelism

Zygote streak tests are a quick method for determining whether a given mutation is transmitted by a nuclear (Mendelian) or organelle (uniparentally inherited) gene. Mutant cells are mated with wild-type partners, using standard techniques (see below). After 4–7 days maturation time, zygospores should be scraped off the maturation plate, streaked on a plate of non-selective medium, and exposed to chloroform vapor for 30–45 seconds by inverting the agar plate over a glass petri dish containing chloroform. This treatment will kill any residual unmated cells. When the zygotes have germinated to form a confluent streak of cells, they can be replica-plated to selective media. The 2:2 segregation of nuclear genes will produce both wild-type and mutant cells in every zygote colony, whereas the predominantly uniparental pattern of chloroplast or mitochondrial gene inheritance will result in death of nearly all progeny on the test plate. Table 8.4 shows the way in which these crosses should be made, the nature of the selective test plate, and the possible outcomes from which the inheritance pattern can be inferred. Results obtained with individual mutants should be confirmed by tetrad analysis of the same crosses.

Tetrad analysis should also be used to confirm that no additional mutations have been introduced that would lead to substantial lethality in future

Table 8.4	Expected results of zygote streak tests for nuclear and organelle mutations			
		Growth of streak		
Cross: *plus* × *minus*	**Test plate**	**Nuclear mutation**	**Chloroplast mutation**	**Mitochondrial mutation**
Acetate-requiring × wild type	Minimal	Confluent	Few colonies	
Wild type × antibiotic resistant	Antibiotic	Confluent	Few colonies	
Wild type × dark-dier	Acetate in dark	Confluent		Few colonies

crosses, and if necessary, to recover the desired mutation in the opposite mating type. Additional rounds of backcrossing to a wild-type strain may be needed before the desired mutation is obtained in a background that produces mostly complete tetrads in crosses.

To decide whether two mutations producing the same phenotype affect same or different genes, two criteria have traditionally been used. Recombination analysis will show whether two mutations are linked or unlinked, but cannot be used to establish whether closely linked mutations are in the same gene. Complementation tests in which both mutant genomes are brought together in a diploid cell (see section VIII.E.4) can be used to ascertain whether two recessive mutations are functionally distinct regardless of their linkage.

C. Tetrad analysis

1. Background

Among many algae and fungi, haploid gametic cells fuse to form a diploid zygote that undergoes meiosis, generating four haploid products (or eight, following a mitotic division), which can be recovered as a set. In contrast to genetic analysis of the type done in many animal and plant systems, where single meiotic products are randomly selected from many meiotic events, tetrad analysis gives information about numbers of crossover events and chromatid interference. The mathematical basis of tetrad analysis was reviewed by Gowans (1965), who gave particular emphasis to *Chlamydomonas* in his discussion. The classic papers of Perkins (1949, 1953, 1962) and Whitehouse (1950, 1957) are also recommended, as are treatments by Snow (1979) and by Ma and Mortimer (1983). The notes below summarize the major points of practical use to the *Chlamydomonas* geneticist. Detailed experimental methods for germination of zygotes and separation of the meiotic products are followed by instructions for determination of linkage, calculation of map distances and centromere distances, and estimation of interference.

2. Experimental techniques

The following protocol for tetrad analysis is based on the procedures developed by Levine and Ebersold (1960). Mix *plus* and *minus* gametes and leave in the light for 1 to 2 hours, or longer if needed to achieve good mating efficiency. Ideally, mated cells should be plated before the secondary zygospore wall forms, in order to avoid formation of a zygote pellicle or clumps. Mix the mating suspension by vortexing, to break up any pellicle that is forming, and pipette approximately 0.5 ml each onto several maturation plates containing 4% agar. Leave the plates upright for a few hours to allow the liquid to soak into the agar. If complete medium (containing nitrogen) is used for the maturation plate, the plates should be exposed to light for 18–24 hours and then transferred to a dark incubator or wrapped in foil until germination is desired. An alternative method (VanWinkle-Swift, 1977) is to mature zygospores on 4% agar made with nitrogen-free medium, in which case they can be kept in the light throughout the maturation period, since both light and nitrogen are required for germination to occur. This method of maturation is preferred in crosses of obligate photoautotrophic (dark-dier) mutants and is convenient in any case.

Zygospores can be germinated any time after the first 4–6 days of maturation but will remain viable for several weeks on the maturation plates. Germination plates should be prepared, containing a nitrogen source and whatever nutrients (acetate, arginine, etc.) are required for the strains being studied. Making the germination plates with 2% or 2.5% agar aids in learning to manipulate tetrad products. However, colony growth is somewhat slower on these plates than on the usual 1.5% agar, and most experienced tetrad dissecters prefer fresh (1- or 2-day-old) 1.5% plates. Mark the germination plates by cutting parallel vertical lines 5–8 mm apart with a sterile razor blade. These lanes will be used as a guide for separating the tetrad products. Cain and Cain (1984a, b) described a device that clamps several razor blades together to facilitate cutting the lanes.

To remove unmated gametes from the maturation plate, dip a clean, preferably new, razor blade in 95% alcohol and flame briefly to sterilize. With a firm but gentle motion, draw the razor blade across the surface of the agar. Unmated cells will be scraped into a pile at the side of the plate, leaving the zygospores adherent to the agar (Figure 8.4). The razor blade should be dipped again in alcohol, wiped clean, and dipped and flamed again before further use.

Under a dissecting microscope, collect zygospores together with a sterile microspatula (made by flattening a bacteriological transfer needle; see Figure 8.5) and transfer them to the germination plate in a streak 5–10 mm above

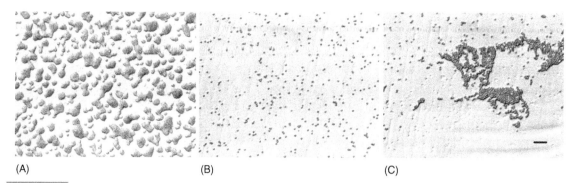

(A) (B) (C)

FIGURE 8.4 *Steps in collecting zygospores for tetrad analysis. (A) Mixed zygospores and unmated gametes on maturation plate (4% agar). (B) Zygospores clinging to agar surface of maturation plate after scraping with razor blade. (C) Zygospores collected in pile (using microspatula; see Figure 8.5), ready for transfer to germination plate. Bar = 100 μm.*

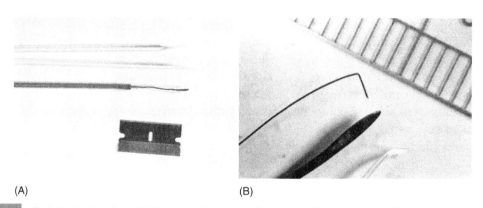

(A) (B)

FIGURE 8.5 *Tools for tetrad analysis. (A) Two glass tools made by drawing out soft glass tubing in a flame, used to separate tetrad products on germination plate, and a microspatula, made by pounding a bacteriological transfer wire. (B) Tungsten wire tool as described by Cain and Cain (1984b), microspatula, and another glass tool, with millimeter scale.*

the top of the cut lanes. A small block of agar can be transferred with the zygospores if necessary and should be placed face down on the germination plate. The zygospores can then be wiped off onto the surface by pushing the agar block across the plate. Using a glass needle (Levine and Ebersold, 1960) or a wire tool (Cain and Cain, 1984b; Figure. 8.5), move a few individual zygospores to the top of each lane. The needle or wire can be sterilized by dipping in alcohol and blotting the excess liquid on the agar. If agar adheres to a glass needle after repeated use, it can be removed by dipping into a small vial of chromic acid cleaning solution and then rinsing in alcohol.

If any unmated cells have been transferred to the germination plate, they can be killed by exposing the inverted plate for 30–45 seconds over a glass dish containing chloroform. This must be done immediately after the zygospores have been transferred to the germination plate, as chloroform treatment is lethal once germination begins and the zygospore wall starts to expand. Occasionally crosses produce chloroform-sensitive zygospores, and for this reason it is usually preferable to rely on visual recognition of zygospores. Chloroform treatment should be done in a fume hood.

Germination takes about 16–20 hours in bright light at 25°C for most strains. At this time the zygospore will be noticeably enlarged and may have already ruptured. Touching a ripe but unruptured zygospore gently with a microneedle will usually release the four (or eight) products. The products can be separated using a glass needle or wire tool under a dissecting microscope at 25–50x. Try to separate the products by at least the width of one microscope field at this magnification, so that colonies will form at least 5 mm apart. When separating eight products, try to keep track of the number so that none are lost. Products that stick together can sometimes be separated by rolling them on the agar surface with the needle, but this process may damage the cells. A better option is to leave the coupled products together, mark the location with a small slash in the agar, and return to the plate several hours later to see if they can then be separated more easily. Products that fall into holes in the agar (a frequent hazard for beginners) can sometimes be salvaged by transferring in a droplet of culture medium to float the lost cells to the surface; buried products rarely survive.

Usually, 5–8 days are required for meiotic products to grow into visible colonies that can be picked off for subsequent analysis. The plates should be monitored daily for contamination, since mold spores falling onto the open plate during tetrad dissection can germinate and rapidly destroy the entire plate. Hyphae can usually be seen if plates are held in front of a light, and can be removed by cutting out a block of agar with a sterile spatula over a dish of alcohol. For additional anti-fungal techniques, see section III.C.3.

The design of glass tetrad-pulling tools is highly personal, and one should take time for experimentation in creating a suitable set of one's own use. These should then be kept in a safe place and vigorously defended from other users. Pasteur pipettes or soft glass rod of about 3 mm diameter can be drawn out by heating the center in a flame until it becomes pliable, then quickly pulling from both sides to taper the new ends. A long, thin thread of glass may be left in the center, and can be broken off and discarded. The tips of the tapered ends can then be smoothed or bent as desired by brief heating (Figure. 8.5). If hollow glass tubing is used, the end of the tool should be closed off so that cells are not drawn into it by capillary action.

The wire tool described by Cain and Cain (1984b) is made from 5.0-mil (0.127 mm) tungsten wire, mounted in a glass rod, and heated in a flame to form a fine pointed tip. The tip is then bent to an angle with watchmaker's forceps. This tool is more durable than glass needles, and it can be flame-sterilized repeatedly.

Rather than moving tetrad products into a linear array as described earlier, some persons simply spread zygospores randomly on agar, draw circles on the bottom of the petri dish to mark off a few well-isolated spores, and after germination separate the individual cells by a few millimeters within the circle. Sager (1955) recommended using a 23-gauge hypodermic needle filed flat to form a tube and bent at a 90° angle to pick up agar plugs containing single zygospores and blow them out onto germination plates.

3. Interpretation of tetrad analysis data

Zygotes of *C. reinhardtii* may germinate into either four or eight products, depending on whether a mitotic division follows meiosis prior to rupture of the zygospore wall. Sager (1972) and Maguire (1976) attributed the production of four vs. eight products primarily to strain differences, but environmental factors such as length of zygospore maturation time, composition of the maturation medium, light intensity, and temperature may also have a considerable effect in determining whether a given zygote forms a tetrad or an octet (VanWinkle-Swift, 1977; Sears, 1980b). In general, older zygotes and those which are relatively nutrient-deprived or otherwise environmentally stressed seem to give a higher proportion of four-product germinations. For tetrad analysis of nuclear genes, germination into four products is highly desirable since it reduces by half the number of individual clones that must subsequently be selected and assayed, but germination into eight products was preferred for study of postmeiotic segregation of chloroplast genes (Forster et al., 1980). Obviously selecting four-product zygotes at the beginning will reduce the subsequent labor of picking and testing product colonies in tetrad analysis. However, sometimes only octets will be formed, and the first step in their analysis then is to reduce them to tetrads by identifying the pairs of identical products formed by the last (mitotic) division.

Tetrads can be classified according to the segregation pattern observed for each pair of markers in the cross (Table 8.5). The relative numbers of parental ditype (PD), nonparental ditype (NPD), and tetratype (T) tetrads obtained are indicative of whether or not the loci in question are linked (on the same chromosome), and how far they are from their respective centromeres. The generation of these three types of tetrads is shown in Table 8.6, and some sample data from typical crosses of *C. reinhardtii* are given in Table 8.7.

Table 8.5   Identification of parental and recombinant tetrad types

Sample cross: *ac17* × *nic13*

Parental ditype (PD)	Nonparental ditype (NPD)	Tetratype (T)
ac+	ac nic	ac+
ac+	ac nic	ac nic
+nic	++	+nic
+nic	++	++

Table 8.6   Generation of PD, NPD, and T tetrads by crossing over

Two loci, linked		Two loci, unlinked	
		Crossovers between loci and	
Crossovers	**Tetrad types**	**centromeres**	**Tetrad types**
None	PD	None	1 PD : 1 NPD
Single	T	One locus and its centromere	T
Double, two-strand	PD		
Double, three-strand	T	Both loci and their centromeres	1 PD : 2 T : 1 NPD
Double, four-strand	NPD		

Adapted from Perkins (1949).

Table 8.7   Sample tetrad data from *Chlamydomonas* crosses

Relationship of loci	Tetrad distribution	Ratio PD:NPD:T	Map distance
Closely linked loci	NPD = 0; PD > T		
act1 × *pf12* (II)		77:0:10	5.7
ac30 × *pf14* (VI)		78:0:12	6.7
Distantly linked loci, same side of centromere	PD > NPD; T > PD; T > 4 NPD		
act2 × *pf14* (VI)		25:0:34	29.8
msr1 × *arg2* (I)		22:1:43	34.1
Distantly linked loci, opposite side of centromere	PD ≥ NPD: T > PD		
ac15 × *sr1* (IX)		65:1:43	20.6
thi9 × *nic2* (II)		6:4:36	47.8
act2 × *thi10* (VI)		4:5:47	50.9
Unlinked, both loci close to centromeres	T < PD or NPD; PD and NPD about equal		
pf27 (XII/XIII) × *ac46* (XVI/XVII)		12:14:0	–
ac17 (III) × *pyr1* (IV)		36:33:14	–
ac17 (III) × *pf2* (XI)		42:34:14	–
Unlinked, both loci distant from centromeres	PD and NPD equal, but T > PD		
act2 (VI) × *nr1* (VIII)		22:20:101	–
act2 (VI) × *msr1* (I)		18:16:78	–

Linkage groups are given in parentheses.

Linkage or non-linkage can be determined by the ratio of PD to NPD tetrads. A cross involving unlinked loci should give statistically equal numbers of PD and NPD tetrads because their respective chromosomes assort independently. In contrast, a cross involving two closely linked loci will show very few if any NPD tetrads, because the double crossovers needed to generate these are rare. Although there may be an appreciable number of NPD tetrads in a cross between two distantly linked loci, they will usually be outnumbered substantially by the PD tetrads. Exceptions to this rule generally involve markers very far from the centromere. A second criterion for assessing linkage in such cases is that the number of T tetrads is substantially greater than four times the number of NPDs. From estimates of chiasma interference, Smyth and Ebersold (1985) derived a ratio of NPD: T > 0.32 as a criterion for non-linkage. Perkins (1953) gave confidence limits for presuming linkage from PD:NPD ratios. Recombination between linked genes 50 map units apart cannot be distinguished from that between unlinked genes.

Map distances are usually calculated on the premise that each T represents two recombinant products, and each NPD represents four. The total percentage of recombination is then $(4NPD + 2T)/[4(PD + NPD + T)] \times 100$, or $(NPD + 0.5T)/(PD + NPD + T) \times 100$. A more complicated analysis (Gowans, 1965) takes into consideration that PDs can also arise from two-strand double exchanges and assumes these to be equal in frequency to four-strand doubles, which generate NPDs. Three-strand vs. single-strand exchanges giving rise to T tetrads are also distinguished. For practical laboratory use, however, the equation above is satisfactory.

4. Interference

Interference refers to deviation of the observed frequency of double crossover events from the number expected from the probability of single crossovers in an adjacent interval. Chiasma interference is seen if the formation of one chiasma affects the probability of forming a second chiasma between the same two chromosomes. Chromatid interference reflects non-random participation of the chromatids of a tetrad in multiple crossover events and is manifest in deviations from the predicted ratio of 1:2:1 for two-, three-, and four-strand double crossovers (see Table 8.6). Early studies of these phenomena in *Chlamydomonas* (summarized by Desborough and Shult, 1961; Smyth and Ebersold, 1985) in general supported a finding of positive chiasma interference but have not been pursued with more extensive work.

D. Genetic maps

1. Linkage groups

Seventeen linkage groups are now recognized in *C. reinhardtii* (Kathir et al., 2003; Rymarquis et al., 2005; Merchant et al., 2007). They are numbered I through XIX, with two pairs (XII + XIII and XVI + XVII) having been consolidated into single linkage groups (Dutcher et al., 1991). Linkage group XIX, previously known as the UNI linkage group, was at one time thought to have a circular map and other anomalous properties (Ramanis and Luck,

1986), and was proposed to be associated with the basal body (Hall et al., 1989). Subsequent studies did not confirm the existence of basal body DNA, however (Johnson and Rosenbaum, 1990; Kuroiwa et al., 1990), and genetic analysis showed that this linkage group has a normal linear organization (Johnson and Dutcher, 1991; Schloss and Croom, 1991; Holmes et al. 1993).

All the linkage groups were defined initially based on tetrad analysis of crosses between mutants with easily scored phenotypes such as acetate requirement, auxotrophy, impaired motility, or resistance to inhibitors (Ebersold et al., 1962; Hastings et al., 1965), and many of these mutants are still useful for traditional mapping. Smyth et al. (1975) created multiply marked strains to facilitate assigning new mutants to linkage groups with a minimal number of crosses. These and other combination strains are available from the Chlamydomonas Resource Center (http://www.chlamycollection.org).

2. Centromere mapping

Since tetratype segregations for a given pair of unlinked markers arise when a crossover occurs at the four-strand state of meiosis between one locus and the centromere of its chromosome, the relative frequency of such second-division exchanges is a measure of the distance from a locus to its centromere. This is a useful piece of information to obtain early in mapping studies, as it aids in selecting appropriate tester strains for future crosses and can sometimes be sufficient in itself to distinguish two similar mutations. In species that produce an ordered tetrad (e.g., *Neurospora crassa*), the second-division exchanges can be determined directly. In an unordered tetrad such as those of *C. reinhardtii*, the second-division exchange frequency must be determined with reference to other genetic markers, for example *ac17*, *pf27*, *y1*, or *y6*, each of which is closely linked to its own centromere. Using segregation data from tetrad analysis for more than 100 molecular markers, Preuss and Mets (2002) were able to identify centromere regions on 15 of the 17 linkage groups, and found them to be notably shorter than centromere sequences in *Arabidopsis*.

3. Molecular markers

Ranum et al. (1988) used restriction fragment length polymorphisms to facilitate mapping, cloned nuclear gene sequences from genes involved in flagellar assembly and function. Using one of the multiply marked *C. reinhardtii* strains created by Smyth et al. (1975) as a source of phenotypic markers in a cross to *C. smithii*, they were able to assign 12 new clones to linkage groups, and created a set of 22 tetrads that could be used as tester strains for further mapping. Gross et al. (1988) described a new isolate (S1 D2, see Chapter 1) that also had extensive polymorphisms compared to the laboratory strain. This isolate was used for subsequent molecular mapping by random spore analysis, in which single progeny from many zygotes were examined by PCR amplification without going through tetrad analysis. Parallel genetic and molecular maps were then constructed for each linkage group, and were connected by anchor markers originally identified by phenotypic mapping and subsequently cloned (Kathir et al., 2003).

Rymarquis et al. (2005) developed additional PCR-based molecular markers to aid in map-based cloning, a technique especially valuable for genes that cannot be identified by complementation of mutants with wild-type sequences introduced by transformation. Many of their markers are suitable for bulked segregant analysis, which is done by pooling DNA from many segregant progeny and looking for deviations from an approximately 1:1 ratio of DNA from the two alleles that would indicate linkage of one allele to the test marker. This technique reduces the number of separate PCR reactions required and streamlines the process of analysis. This mapping kit is available from the Chlamydomonas Resource Center (http://www.chlamy.org/kit.html).

As an increasing number of the old genetic markers are being identified with gene sequences, the molecular and genetic maps are being consolidated and aligned with the genomic sequence (Merchant et al., 2007). Maps of the linkage groups are available at http://www.chlamy.org and are periodically updated to include new data.

E. Working with diploid cells

1. Selection of vegetative diploid cells

As discussed in Chapter 5, as many as 5% of mated gamete pairs of *C. reinhardtii* may divide mitotically as vegetative diploids rather than forming meiotic zygotes (Ebersold, 1963, 1967). Such diploids can be selected deliberately through the use of complementing auxotrophic markers, and can be used to test allelism of other pairs of mutants. To obtain diploids, gamete formation is induced and mating begun as for tetrad analysis, but the mating mixture is then plated on a medium on which only complementing diploids (and meiotic recombinants) can grow. For example, to select diploids from complementing acetate-requiring mutants, a minimal salts medium would be appropriate, while for the arginine-requiring auxotrophs, minimal or acetate medium with no added arginine would be used. Diploid cells begin to divide shortly after mating, whereas meiotic zygotes require several days to mature before germination. The diploid colonies can be identified under a dissecting microscope 3 or 4 days after mating as bright green, hemispherical colonies on a background of ungerminated zygospores and non-growing residual gametes (Figure 8.6). Best results are usually obtained by plating one or more dilutions of the mating mixture to deliver roughly 100–500 potential zygotes per plate. With high mating efficiency the diploid colonies may be so numerous as to merge into a green lawn.

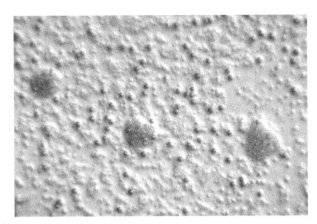

FIGURE 8.6 *Three vegetative diploid colonies on a field of ungerminated zygospores (dark circles) and unmated cells (pale, unable to grow on minimal medium). Photograph by Matt Laudon.*

Individual diploid cells are typically larger than haploids. Biochemical verification involves quantifying DNA content relative to cell number, using diphenylamine, diaminobenzoic acid or ethidium bromide (Cattolico, 1978; Valle et al., 1981; Zachleder, 1984). Genetic proof of diploidy can be obtained by crossing the presumptive diploid to a haploid strain (Gillham, 1969), or by using a PCR-based assay for mating type, as discussed above (section VIII.A.3; Werner and Mergenhagen, 1998; Zamora et al., 2004).

2. Recovery of haploid cells from diploids

Diploid cells formed in sexual crosses have the constitution mt^+/mt^- but mate as *minus* (see Chapter 5). Gametes made from diploid cells will mate readily with wild-type *plus* gametes, but product viability will be poor in the first round of crossing due to formation of aneuploid cells in the triploid meiosis. When the diploid stock was constructed using the complementing *arg2* and *arg7* arginine auxotrophy mutations, zygospores should be germinated on medium containing 10 µg/ml arginine. Progeny clones are selected and crossed again to wild-type cells, and this process should be repeated until a high frequency of complete tetrads is obtained. The arginine marker, if retained, can be removed in a subsequent round of crossing.

Wetherell and Krauss (1956) induced homozygous diploid strains of both *C. reinhardtii* and *C. eugametos* by colchicine treatment. A spontaneous stable diploid homozygous for mating type was identified by Eves and Chiang (1982). Smyth and Ebersold (1985) reported apparent fusion of four nuclei into a tetraploid zygote that yielded four diploid products after meiosis. Partial diploids were obtained by Hourcade (1983), who treated cells with a lethal dose of bleomycin, which induces chromosome breakage, and then mated these cells with untreated ones. Selection for genetic markers from the bleomycin-treated parent resulted in rescue of chromosomal fragments, which were often unstable. Diploids homozygous for mating type can also be obtained using the *iso1* mutation (Campbell et al., 1995; see Chapter 5). Heterozygous diploids containing *iso1* and its wild-type allele can be self-mated to form tetraploid zygotes, which after meiosis produce mt^+/mt^+ and mt^-/mt^- diploid progeny, from which the *iso1* mutation can then be eliminated by backcrossing (Lee et al., 2008).

In contrast to yeast, *Chlamydomonas* diploids do not sporulate naturally, nor can they be readily induced to undergo meiosis. A low frequency of apparent haploid cells was obtained by Davies and Plaskitt (1971) after treatment of diploids with an unspecified concentration of *p*-fluorophenylalanine, which induces mitotic haploidization in *Aspergillus* (Lhoas, 1961). Rare spontaneous haploidization was reported by Lee et al. (1976). Dutcher (1988) suggested an alternative method for haploidizing diploid isolates. Gametes prepared from the diploid strain and a haploid *plus* strain were treated with 6 mM colchicine for 4 hours prior to mating. Colchicine was left in the mating mixture and mating allowed to proceed for 5 hours before plating. Zygotes formed normally, and germinated to produce four haploid progeny derived from the original diploid nucleus; the chromosomes of the haploid gamete were lost.

As discussed in Chapter 5, the *GSP1* and *GSM1* genes encode homeodomain proteins that interact to control zygote formation. Vegetative diploid strains containing ectopic copies of these genes under control of constitutive promoters can be induced to form zygospore walls and undergo meiosis (Goodenough et al., 2007; Lee et al., 2008). These strains should greatly facilitate recovery of recessive mutations and maintenance of otherwise lethal alleles in a heterozygous diploid background.

3. Production of diploid and hybrid cells by somatic fusion

Polyethylene glycol-mediated cell fusion is very effective for production of diploids from *C. reinhardtii* cells (Hermesse and Matagne, 1980). This technique has been exploited for studies on inheritance of chloroplast genes (Matagne and Hermesse, 1980a, b; 1981; Matsuda et al., 1983; Galloway and Holden, 1984; Matagne and Schaus, 1985). Diploids with the constitution mt^+/mt^- or mt^-/mt^- mate as *minus* cells, whereas those containing mt^+/mt^+ mate as *plus*. Matagne and Mathieu (1983) also reported mating between mt^+/mt^+ diploids and either mt^+/mt^- or mt^-/mt^- to form tetraploid zygotes. Fusion diploids containing mt^+/mt^- cannot be distinguished from sexually generated ones with respect to mating behavior, but they differ in transmission of chloroplast genes (Chapter 7).

Somatic fusion is possible only when cell walls are absent. Matagne and colleagues incorporated two complementing cell wall deficiency mutations (*cw15* and *cwd*) into their stocks, permitting fusion and subsequent recognition of diploid colonies able to form normal walls. Galloway and Holden (1984) used wild-type cells but treated them prior to fusion with a filter-sterilized preparation of gamete lytic enzyme.

Nghia et al. (1986) used polyethylene glycol-mediated somatic fusion to achieve uptake of nitrogen-fixing bacteria (*Azotobacter*) by *Chlamydomonas* cells. Electron micrographs showed bacterial cells in membrane-bound vesicles within the algal cytoplasm, and these hybrid cells remained stable and viable for as much as 10 months under phototrophic conditions on medium lacking a nitrogen source. Nghia et al. were unable to confirm with an acetylene reduction assay that nitrogen fixation was occurring, however. Uptake of bacterial cells by *Chlamydomonas* cells in polyethylene glycol was also reported by Vladimirescu and Romanowska-Duda (2003).

Fusion of *Chlamydomonas cw15* cells with carrot protoplasts was reported by Fowke et al. (1979). Several algal cells could be fused with a single carrot cell. Soon after fusion, basal bodies, microtubule rootlets, chloroplasts, and nuclei were seen free in the carrot cytoplasm. After several days of culture, the carrot protoplasts regenerated cell walls and divided, but *Chlamydomonas* chloroplasts could still be seen.

4. Complementation tests

Complementation between recessive nuclear mutations with similar phenotypes is easily assessed in vegetative diploid cells, selected 3 or 4 days after mating as described above. For example, to test two nuclear acetate-requiring mutations, one mutant could be coupled with the arginine auxotrophy marker *arg2* (also known as *arg7-8)* and the other with the closely linked complementing marker *arg7*. Vegetative diploids that will grow on acetate medium in the absence of arginine are selected as fast-growing colonies that appear within the first 3 or 4 days after mating and are then tested for ability to grow on medium without acetate.

For quick complementation tests, mutants can be crossed with one another without being coupled to auxotrophic markers, and the mating mixture plated directly on an appropriate medium under selective conditions (Harris et al., 1974; Wang et al., 1974). The drawback of this method is that failure to form diploids can occur from other causes besides non-complementation, so negative results cannot be interpreted. The method is especially useful for identifying which of several phenotypically similar tetrad products has which genotype. For example, in a cross between two acetate-requiring mutants, tetratype tetrads are formed in which three products are acetate-requiring and one is wild type (Table 8.6). By quick complementation tests against the two parents, one can distinguish which product is the double mutant.

Complementation can also be analyzed in young zygotes. During the transient quadriflagellate stage after mating, the flagella of some paralyzed mutants can be restored to function by wild-type gene products contributed by the mating partner (Lewin, 1954a; Starling and Randall, 1971; Luck et al., 1977; Huang et al., 1981). Use of this technique in conjunction with radioactive labeling to identify polypeptides affected by motility mutations is discussed in Chapter 4. Bennoun and colleagues (Bennoun et al., 1980; Bennoun and Delepelaire, 1982; Woessner et al., 1984; also Baldan et al., 1991) used chlorophyll fluorescence induction kinetics and delayed luminescence patterns to demonstrate complementation between pairs of non-photosynthetic mutants. Cells were mated in a petri dish and exposed to bright light for 4 hours. The newly formed zygotes clumped together and adhered to the bottom of the dish, allowing unmated cells to be removed by washing with buffer. Fluorescence or luminescence measurements were then made directly on the zygote layer in the petri dish bottom. Complementation between nuclear mutations deficient in photosystem I or photosystem II activity was seen 24–48 hours after mating as appearance of the wild-type patterns of fluorescence.

IX. THE MOLECULAR TOOLKIT

A. Overview

"The Molecular Toolkit" has been used informally within the *Chlamydomonas* community since the 1990s to refer to the techniques and materials of transformation and gene manipulation (Fuhrmann, 2002; Grossman et al., 2003). Despite the objections of a reviewer of the prospectus for this book, who said it was "an odd name," it conveys exactly the scope of the section to follow here. Reviews of the current status of transformation and manipulation of the *Chlamydomonas* genome have been published at intervals over the years (Stevens and Purton, 1997; Lumbreras and Purton, 1998; Fuhrmann, 2002; Walker et al., 2005). For the early background, and the frustrations that marked the 1980s as *Chlamydomonas* appeared to be lagging behind other model organisms in joining the molecular age, see Rochaix et al. (1985b), and also Joel Rosenbaum's reminiscences in Volume 3, Chapter 1. The book by León et al. (2007) is especially recommended for a comprehensive review of the literature and a broad perspective on the potential of *Chlamydomonas* and other algae for genetic engineering and commercial applications.

B. History of chloroplast and nuclear transformation

In experiments that foreshadowed the achievement of transformation of plant cells, Lurquin and Behki (1975) showed that bacterial DNA could be taken up by *Chlamydomonas*. Irreversible binding of foreign DNA to the cell wall or membrane (in the case of *cw15* cells) was enhanced by poly-L-lysine or poly-L-ornithine. No integration of bacterial sequences was observed, but intracellular degradation led to reutilization of thymine derivatives in *Chlamydomonas* DNA.

Expression of foreign DNA sequences in *Chlamydomonas* was first reported by Rochaix and Van Dillewijn (1982), who used a mutant at the *ARG7* (argininosuccinate lyase) locus as recipient for a plasmid containing the yeast *ARG4* gene. Cells of a double mutant, *cw15 arg7*, were either pretreated with poly-L-ornithine before addition of the yeast plasmid DNA or were mixed with plasmid DNA and then treated with polyethylene glycol in a procedure derived from methods for transformation of yeast. Prototrophic colonies were obtained at a frequency of roughly 10^{-6} to 10^{-7}, no more than about 10-fold the expected frequency of reversion of the *arg7* allele used, but two putative transformants were shown to contain DNA sequences that hybridized with the transforming plasmid DNA and had slight argininosuccinate lyase activity.

Rochaix et al. (1984, 1986) also constructed plasmids containing *Chlamydomonas* chloroplast sequences and the yeast *ARG4* gene and showed that these were capable of autonomous replication in *Chlamydomonas*, but using these plasmids gave no significant increase in transformation frequency over that observed by Rochaix and Van Dillewijn. In both sets of experiments, the plasmid sequences were lost within about 70–100 generations, and persistence of prototrophy was attributed to reversion of the *arg7* mutation.

Hasnain et al. (1985) and J.C. Cox (abstracts from meetings in 1985, cited by Harris, 1989) both reported transformation of *Chlamydomonas* cells using bacterial genes encoding resistance to kanamycin or G418 and modifications of the poly-L-ornithine protocol of Rochaix and van Dillewijn (1982). Again, plasmid DNA was detected in transformed cells, but in most cases did not appear to be stably integrated into the *Chlamydomonas* genome, and resistance was lost when isolates were transferred to non-selective medium. Bingham et al. (1989) reported a single rare transformant in which the bacterial *nptII* gene was integrated into the *Chlamydomonas* nuclear genome. A few other apparently stable kanamycin-resistant isolates were later shown to have acquired mutations in chloroplast DNA, distinct from the enzymatic resistance conferred by the transforming plasmid (unpublished data from Harris, Gillham, and Boynton, in collaboration with David Luck).

By the late 1980s it was clear that several problems needed to be solved to develop an efficient transformation system for *Chlamydomonas*. A better method was needed to get DNA into the cell and to increase the frequency of transformants recovered. Gene expression and methods for selecting transformants also needed to be improved. It was recognized even at this time, based on the handful of genes that had been sequenced, that codon usage in nuclear, chloroplast and mitochondrial genes was restricted, with the prevalent codons differing in the three genomes (see Harris, 1989, for tabular data), and that this codon bias could present problems in expressing foreign genes. For this reason, and also to achieve stable integration of the transforming DNA, it was apparent that the best hopes lay in transformation with native *Chlamydomonas* DNA rather than yeast or bacterial genes. At that time only a few nuclear genes had been cloned from

Chlamydomonas, and not all of those were marked by mutations that would be suitable recipients for transformation. The chloroplast genome offered more possibilities for an effective combination of transforming DNA and a mutant recipient, and it was with this genome that the first high-frequency transformation was obtained.

John Sanford and colleagues at Cornell University had developed a device that used a gunpowder charge to propel DNA-coated tungsten microparticles into living cells (Klein et al., 1987; Sanford et al., 1987). This method was later given the name "biolistic transformation" (Sanford, 1990). A seminar presented by Sanford at Duke University in 1987 provided the opportunity for collaborative work that led to the first successful transformation of the *Chlamydomonas* chloroplast genome (Boynton et al., 1988) and also of yeast mitochondria (Johnston et al., 1988). Boynton et al. used a cloned fragment of wild-type *Chlamydomonas* chloroplast DNA to transform a mutant having a 2.5-kb deletion in the *atpB* gene. Photosynthetically competent cells were recovered at a frequency in the range of 1–5 per 10^6 bombarded cells, and were shown to have a restored wild-type sequence, suggesting direct replacement of the deleted fragment by its wild-type homologue, without integration of vector DNA. Blowers et al. (1989) showed that foreign DNA also integrates effectively into the chloroplast genome if it is flanked at both ends by chloroplast DNA.

Successful transformation of the nuclear genome was reported by two laboratories in 1989, both using particle bombardment. Debuchy et al. (1989) used the *Chlamydomonas ARG7* gene for argininosuccinate lyase to transform the *arg7* mutant, and Kindle et al. (1989) used the gene encoding nitrate reductase (*NIT1* or *NIA1*) to transform the *nit1* mutant. In both cases, DNA integrated into multiple sites in the nuclear genome, and did not replace the mutant sequence. Shortly after this, Kindle (1990) published an alternative transformation method based on agitation with glass beads in the presence of polyethylene glycol. This offered several advantages: it was inexpensive and required no special equipment, and transformation frequencies were higher than with the biolistic method, yet fewer copies of the transforming DNA integrated into the genome.

C. Methods for delivery of DNA

Glass beads continue to be the simplest and most convenient method for nuclear transformation. A protocol is available at http://www.chlamy.org/methods/beads.html, and other variations on this method have also been published (Kozminski et al., 1993; Nelson et al., 1994). The chief disadvantage of glass bead transformation is that it requires using wall-less cells, either a wall-deficient mutant such as *cw15*, or wild-type cells from which the walls are removed by treatment with gamete lytic enzyme. Wall-deficient mutant cells transform well but show low plating efficiencies. They also often lack flagella and therefore can be difficult to outcross afterwards, in order to get the transformed genotype into a background with a wild-type wall. Preparation of gamete lysin is not difficult (http://www.chlamy.org/methods/autolysin2.html), but anecdotal reports suggest variability in its effectiveness. Simply shaking wild-type cells in nitrogen-free medium for several hours to induce gametogenesis prior to transformation may release enough lytic enzyme to remove most of the cell walls (Nelson et al., 1994).

Dunahay (1993; Dunahay et al., 1997) published a method for transformation by agitation with silicon carbide whiskers. Although effective, even for cells with walls, this has not been widely adopted, presumably because these fibers are somewhat expensive and hard to obtain, and may also be hazardous to handle.

The biolistic method is still used for chloroplast transformation, where integration is homologous and multiple insertions of the transforming DNA are therefore not a problem, and it is also the preferred method to transform the mitochondrial genome (Randolph-Anderson et al., 1993; Remacle et al., 2006). An apparatus is now commercially available, incorporating significant modifications over the original design. Compressed helium has replaced gunpowder for propulsion of the particles (Bio-Rad Laboratories, Hercules, CA), and sub-micron gold particles have been substituted for the tungsten particles first used (see Randolph-Anderson et al., 1995b, for information on the use of gold particles for both nuclear and chloroplast transformation).

Electroporation methods have also been developed (Brown et al., 1991; Tang et al., 1995; Shimogawara et al., 1998; Ladygin, 2004; Azencott et al., 2007). Shimogawara et al. reported efficiencies of 2×10^5 transformants per microgram of DNA, substantially better than what had been achieved previously, but this level of success depends on very accurate control of many different parameters in the procedure.

Transformation of *Chlamydomonas* by co-cultivation with *Agrobacterium tumefaciens* was reported by Kumar et al. (2004; Rajam and Kumar, 2006, Hema et al., 2007), but this method has not yet been adopted widely.

One may generalize that successful transformation by whatever method requires transient opening of the cell membrane without killing the cell. If this is achieved, then entry of DNA apparently occurs readily. Co-transformation experiments with two different vectors generally show integration and expression of both transforming DNAs (Kindle et al., 1989; Diener et al., 1990; Newman et al. 1991; Roffey et al., 1991; Quinn et al., 1993). This is true both for nuclear and for chloroplast transformation and has been exploited as a means of introducing non-selectable genes into cells. Results from the first successful transformations clearly showed however that the mechanisms of incorporation of DNA into the nuclear and chloroplast genomes were very different, and these will be considered separately in the sections that follow.

D. Nuclear transformation

1. Native Chlamydomonas genes as transforming DNA

The *arg7* and *nit1* mutants, first to be transformed in 1989 as described earlier, have remained mainstays of transformation strategy in *Chlamydomonas*. In both these cases the gene had already been identified by mutation and partial sequencing as a structural gene for an enzymatic function. Complete sequences and additional characterization followed for both *ARG7* (Purton and Rochaix, 1995; Auchincloss et al., 1999) and *NIT1* (Fernández et al., 1989; Kalakoutskii and Fernández, 1995). Of the several transformable mutant alleles at the *ARG7* locus, *arg7-8*, also known as *arg2*, is the most stable and least subject to reversion or suppression. Cells die quickly on minimal medium and prototrophic transformants are easily recognized. However, arginine-requiring strains typically grow somewhat less vigorously than wild type. In contrast, *nit1* mutants grow well on the usual media containing ammonium chloride as the nitrogen source, but they can survive for days on nitrate so transformants must be selected as dark green colonies against a yellow background.

Transformation with other cloned genes soon followed. Mayfield and Kindle (1990) rescued a non-photosynthetic mutation with the *OEE1* gene, and Mayfield (1991) showed that insertion of multiple copies of this DNA resulted in over-expression of the protein. Diener et al. (1990) transformed a *nit1 pf14* strain with two plasmids, selecting first for restoration of nitrate reductase activity and then screening those transformants for motility restored by integration of the *RSP3* gene. Smart and Selman (1993) transformed a non-photosynthetic mutant lacking the gamma subunit of chloroplast ATP synthase, encoded by the nuclear *ATPC* gene.

Purton and Rochaix (1994) demonstrated the feasibility of cloning mutants by complementation, using the *arg7-8* mutation as a test recipient for transformation with a cosmid library. Zhang et al. (1994) created an indexed cosmid library that was subsequently used to clone the *CAH3* gene encoding an intracellular carbonic anhydrase previously identified by mutation (Funke et al., 1997). Randolph-Anderson et al. (1998) cloned the gene for protoporphyrinogen oxidase using a cosmid library prepared from an herbicide-resistant mutant strain. A YAC library constructed by Infante et al. (1995) was used by Vashishtha et al. (1996b) to complement flagellar mutants. Chang et al. (2003) constructed a cosmid vector containing the *ble* marker, selectable by resistance to the antibiotic zeocin (Stevens et al., 1996, and see below), expressed under control of the 5′ and 3′ flanking regions of the *RBCS2* gene. This can facilitate cloning of genes for which no selectable phenotype is available.

The gene encoding ribosomal protein S14 can be mutated to confer resistance to emetine and cryptopleurine. This provides a dominant selectable marker (*CRY1*) that can be used to transform wild-type cells

(Nelson et al., 1994). Herbicide resistance resulting from mutations in the gene encoding acetolactate synthase can also be used in this way (Kovar et al., 2002).

2. Insertional mutagenesis

The random insertion of transforming DNA into the nuclear genome, although disconcerting at first, was turned to advantage in creating methods for insertional mutagenesis (Tam and Lefebvre, 1993, 1995; reviewed by Galván et al., 2007). Mutant cells are transformed with the wild-type DNA, and transformants are then screened for other phenotypes. Both *arg7* and *nit1* have been used extensively for this purpose (Adam et al., 1993; Gumpel et al., 1995; Prieto et al., 1996a, b; Zhang et al., 1997; Adam and Loppes, 1998; Van et al., 2001). Genes thus tagged can then be cloned and confirmed by their ability to rescue the insertional mutant.

Insertion can consist of simple integration of the transforming DNA, or can cause deletions of as much as 50 kb. Best results are obtained when the transforming DNA is linearized, since linear DNA is more likely to give transformants in which the transforming phenotype (*ARG* or *NIT*) is linked to the insertional mutation, and is less likely to result in deletion or rearrangement of the vector part of the plasmid, an important consideration if the vector is going to be used for marker rescue to clone the gene (Gumpel and Purton, 1994).

Methods for insertional mutagenesis, and general considerations on the construction of libraries of insertional mutants, are provided by Pazour and Witman (2000) and by Galván et al. (2007).

3. Improvements in promoters

Once transformation was working well for rescue of mutants with cloned genes, attention turned to enhancing gene expression (Table 8.8). Blankenship and Kindle (1992) constructed a fusion of the light-regulated *cabII-1* promoter with the *NIT1* gene, thus permitting *NIT1* expression in medium containing ammonium and in strains carrying both the *nit1* and *nit2* mutations. Davies et al. (1992) used the β2-tubulin promoter to express an arylsulfatase gene; this gene was also expressed under the control of the *NIT1* promoter in experiments by Ohresser et al. (1997). The 5′ and 3′ flanking regions of the *RBCS2* gene, encoding the small subunit of Rubisco, were used by Kozminski et al. (1993) to express epitope-tagged α-tubulin, and this soon became the choice for high-level constitutive expression, being used for example for *CRY1* by Nelson et al. (1994), and in the first constructs made with bacterial genes (see below). *RBCS2*-driven expression improved significantly when the first intron was added back as an enhancer-like element (Kang and Mitchell, 1998; Lumbreras et al., 1998). A further enhancement was a fusion of the *HSP70A* promoter with *RBCS* 5′ (Schroda et al., 2000, 2002; Lodha and Schroda, 2005;

Table 8.8	Representative plasmids used for transformation of the nuclear genome	
Plasmid	**Description**	**References**
pARG7.8	*ARG7* gene	Debuchy et al. (1989)
pMN24	*NIT1* gene	Kindle et al. (1989)
cabII-1	*cabII-1* promoter fused to *NIT1* gene	Blankenship and Kindle (1992)
pJD100	Arylsulfatase reporter with β-tubulin promoter	Davies et al. (1992)
pCRY1-1	*CRY1* (emetine resistance) controlled by *RBCS2*	Nelson et al. (1994)
pSP108	*ble* (zeocin resistance) controlled by *RBCS2*	Stevens et al. (1996)
P-681	*aadA* (spectinomycin resistance) controlled by *RBCS2*	Cerutti et al. (1997a)
pMH213 and others	Promoter trap: *pf14 arg7* transformed with *ARG7* as selectable marker, and a promoterless *RSP3* gene; useful for tagging genes	Haring and Beck (1997)
pSP124	*ble* construct in which introns were inserted into the *RBCS2* 5′ region to improve expression	Lumbreras et al. (1998)
pCrGFP, pKScGFP	Synthetic *GFP* reporter optimized for nuclear codon bias; fused to *ble*, under control of *RBCS2*	Fuhrmann et al. (1999)
pT7ARG7	Shuttle vector with *ARG7* controlled by *RBCS2* and T7 phage promoters; can complement both *Chlamydomonas arg7* and *E. coli argH* mutants	Auchincloss et al. (1999)
pCB745, pCB801, and others	*HSP70A/RBCS2* constructs, light and heat-shock inducible, high expression levels	Schroda et al. (1999, 2000)
pSI103	*aphVIII* (paromomycin resistance) with *HSP70A/RBCS2* promoter	Sizova et al. (2001)
pGenD, pGenD-ble	*PSAD* gene, from which coding region can be removed and replaced with a gene of interest, under control of the *PSAD* flanking sequences, for example, *ble* in pGenD-Ble	Fischer and Rochaix (2001)
p3xHA	Plasmid for epitope tagging; contains three copies of the 9-amino acid influenza hemagglutinin epitope	Silflow et al. (2001)
pCB797	*ble* with *HSP70A/RBCS2* promoter	Schroda et al. (2002)
pJK7, pRP-ALS	Acetolactate synthase with resistance to sulfonylurea herbicides	Kovar et al. (2002)
Maa7/X IR (NE-537)	Construct for post-transcriptional gene silencing, based on tandem inverted repeat sequences from the *MAA7* gene encoding the beta subunit of tryptophan synthase, and the selectable *AphVIII* antibiotic resistance marker under control of *HSP70A/RBCS2*	Rohr et al. (2004)
pRbcRL(Hsp196)	Synthetic luciferase reporter gene, controlled by *HSP70A/RBCS2*	Fuhrmann et al. (2004)
nitRNAi	*NIT1* promoter, followed by inverted repeat of centrin gene, *RBCS2* 3′; induced by ammonium starvation	Koblenz and Lechtreck (2005)
M13-BZ301	Single-stranded DNA for homologous integration, targeted gene disruption	Zorin et al. (2005)
pHsp70A/RbcS2-Chlamy, and others	Modular tandem expression vectors constructed using the Cre/lox-system	Heitzer and Zschoernig (2007)

Lodha et al., 2008). *HSP70A* is inducible both by light and by heat shock, and also functions as an activator for *RBCS2*. In addition, transformants made with this construct appear to be less sensitive to transcriptional silencing than those containing *RBCS2* alone (Schroda et al., 2002). This combination now appears in many of the most frequently used plasmids (Table 8.8). The *PSAD* promoter also gives high-level expression of many genes and is available as a cassette construction (pGenD) to which genes of interest can be added (Fischer and Rochaix, 2001).

4. Expression of foreign genes

Even after nuclear transformation was working well for native *Chlamydomonas* genes, attempts to express foreign genes were unsatisfactory (Blankenship and Kindle, 1992; Hall et al., 1993). Possible explanations included inadequate promoters and other regulatory elements, silencing by methylation or other processes, and inappropriate codon usage compared to the native genes. Stevens et al. (1996) chose the *ble* gene from *Streptoalloteichus hindustanus*, and Sizova et al. (1996) the *aphVIII* gene from *Streptomyces rimosus* as candidates to develop selectable antibiotic resistance markers, based on their having similar codon usage to *Chlamydomonas*. Expression of both was improved by modification of the *Chlamydomonas* promoters (Lumbreras et al. 1998; Fischer and Rochaix, 2001; Sizova et al., 2001). A third bacterial gene, *aadA* from *E. coli*, was expressed in the nuclear genome despite its dissimilar codon bias, but its transcription became inactivated when transformants were not maintained continuously on antibiotic (Cerutti et al., 1997a, b). Bacterial hygromycin resistance has also been developed as a selectable marker for *Chlamydomonas* (Butanaev, 1994; Berthold et al., 2002; Ladygin and Boutanaev, 2002).

The *GFP* (green fluorescent protein) gene from the jellyfish *Aequorea victoria* is a standard reporter in other systems, but was poorly expressed in first attempts with *Chlamydomonas*, and Fuhrmann et al. (1999) resynthesized the gene with appropriate substitutions to adapt it to the *Chlamydomonas* codon bias. A synthetic version of the luciferase gene from the coral *Renilla reniformis* has also been made (Fuhrmann et al., 2004). Heitzer et al. (2007) have reviewed problems of codon bias in *Chlamydomonas* and other algae.

5. Targeted integration of DNA

As discussed earlier, homologous recombination of transforming DNA into the nuclear genome is very rare, and improving its efficiency is a high priority. Sodeinde and Kindle (1993) were able to transform the *nit1* mutant with the 5′ end of the wild-type *NIT1* gene, and Gumpel et al. (1994) did similar transformations with truncated versions of *ARG7* that restored prototrophy to the *arg7-8* mutant. Nelson and Lefebvre (1995a) achieved the first targeted disruption of a nuclear gene, using the *CRY1* selectable marker inserted within the *NIT8* coding region to produce *nit8* mutants with emetine resistance. A more general method for targeted integration was introduced by Zorin et al. (2005), who demonstrated homologous recombination of single-stranded DNA with nuclear genes.

6. Gene silencing

Overcoming the problems of undesirable silencing of introduced genes has had the added benefit of laying the foundation for deliberate inactivation of gene expression. Cerutti et al. (1997a, b) first analyzed silencing of an *RBCS2*-driven *aadA* construct, and concluded that it occurred at both transcriptional and post-transcriptional levels, but was not correlated with methylation. Using the same system, Wu-Scharf et al. (2000) isolated an insertional mutant (*mut6*) in which silencing did not occurred, and cloned the corresponding gene encoding an RNA helicase of the DEAH-box family. They postulated that it had a direct role in degradation of mRNA. Jeong et al. (2002) isolated two additional mutants (*mut9* and *mut11*) in which expression of a previously introduced *aadA* gene was reactivated. These mutations also enhanced mobilization of class I and class II transposons (*TOC1*, Day and Rochaix, 1991a, b; *Gulliver*, Ferris, 1989; see also Chapter 6), and increased sensitivity of cells to agents that induce double-strand breaks in DNA. MUT9 is a serine/threonine protein kinase (Jeong et al., 2002), and MUT11 is a conserved protein with WD40 repeats (Zhang et al., 2002).

Intentional inactivation of gene expression by antisense techniques was reported by Schroda et al. (1999), who used the *HSP70A-RBCS2* promoter coupled to an antisense copy of the *HSP70B* cDNA. Fuhrmann et al. (2001) used a similar construct with the chlamyopsin gene as target, as did Pfannenschmid et al. (2003) for the *DIP3* gene encoding a microtubule-associated protein induced by deflagellation, and Chen and Melis (2004) for the *SULP* gene encoding sulfate permease. Schroda (2006) reviews this topic thoroughly, and includes a table listing constructs, results, and additional data for these and other publications in the 2003–2006 period.

A different approach was taken by Rohr et al. (2004), who developed a generally useful construct consisting of the 3′ UTR of the *MAA7* gene, encoding the beta subunit of tryptophan synthetase, and the 3′ UTR of the target gene, repeated by the same sequence in antisense orientation to form a tandem inverted repeat. Under control of the *RBCS2* promoter, this construct allows selection of transformants in which the *MAA7* gene has been silenced, by their ability to grow in the presence of 5-fluoroindole (see Palombella and Dutcher, 1998, for more on *MAA7*). The target gene is also expected to be silenced in the transformants thus selected.

E. Chloroplast and mitochondrial transformation

The basic strategies and techniques for chloroplast transformation have not changed substantially since the first successful experiments, and the methods presented by Boynton and Gillham (1993) are still useful. For more recent improvements see Ramesh et al. (2004) and Purton (2007). Similar methods are used for delivery of DNA to the 15-kb mitochondrial genome (Boynton and Gillham, 1996; Remacle et al., 2006; Bonnefoy et al., 2007).

As discussed earlier, transformation of the chloroplast genome normally occurs by homologous integration to a target site, provided the transforming DNA is flanked by sequences identical to the target. As little as 100–200 bp of flanking sequence can be sufficient (Cerutti et al., 1995; Dauvillée et al., 2004). Deletion mutants in chloroplast genes such as *atpB* and *psbA* are excellent recipients for transformation, since they require acetate for growth and do not grow on minimal medium, and are incapable of reversion to prototrophy.

Because each *Chlamydomonas* cell typically contains about 80 chloroplast DNA molecules (see Chapter 7), not all DNA copies will be transformed by the initial event, and chloroplast transformants should therefore be purified by subcloning on selective medium for several transfers so that untransformed DNA copies are eliminated. Pretreating cells with 5-fluorodeoxyuridine prior to transformation reduces the number of copies of the chloroplast genome per cell (Wurtz et al., 1979). This increases the transformation frequency and facilitates purification to homoplasmy (Boynton and Gillham, 1993). A few chloroplast genes appear to be essential for cell growth, however, and interruption of these by transformation can result in a persistent heteroplasmic state in which some of the copies of chloroplast DNA in a cell retain the untransformed gene (see Chapter 7 for a general discussion of persistent heteroplasmy, and Purton, 2007, for additional information on its effect on selecting transformants).

If transformation occurs within the inverted repeat region, the introduced DNA will be subjected to the "copy-correction" mechanism (Chapter 7), and either duplicated in the opposite inverted repeat or eliminated by replacement with the corresponding sequence as it existed in the recipient cell prior to transformation.

The entire chloroplast genome is available as a set of cloned BamHI, EcoRI, and PstI fragments, and many additional plasmids have been constructed with various combinations of 5′ and 3′ flanking sequences for chloroplast genes, including *atpA*, *atpB*, *petD*, *rbcL*, and others (for review, see Goldschmidt-Clermont, 1998, and Purton, 2007; also http://www.chlamy.org/strains/plasmids.html for information on available plasmids). Site-directed mutagenesis is easily accomplished (Erickson et al., 1992, Lee et al., 1998, Xiong and Sayre, 2004, and many other papers).

Foreign genes can be introduced downstream of an expressed gene selected by transformation of a deletion mutant, or by co-transformation on a separate plasmid flanked by a chloroplast-homologous sequence. The first such gene expressed in the chloroplast was the bacterial *aadA* gene (Goldschmidt-Clermont, 1991) and this construct and others derived from it continue to be useful. Fischer et al. (1996) developed a method for recycling the *aadA* cassette so that the same antibiotic selection could be used repeatedly on one cell line to make successive modifications in the chloroplast genome. The *aadA* gene can be flanked by direct repeat elements so that it excises readily once selection pressure is removed. An alternative is to introduce it within an essential gene so that it is maintained heteroplasmically during the selection period, and then eliminated.

Other genes that have been expressed in the *Chlamydomonas* chloroplast include bacterial *uidA* (GUS; Sakamoto et al., 1993; Fargo et al., 1998; Ishikura et al., 1999), *aphA6*, a bacterial gene encoding resistance to kanamycin and amikacin (Bateman and Purton, 2000); *GFP* (Komine et al., 2002; Franklin et al., 2002); the native *Renilla* luciferase (Minko et al., 1999), and synthetic bacterial and firefly luciferases adapted to the chloroplast codon usage (Mayfield and Schultz, 2004; Matsuo et al., 2006).

F. The *Chlamydomonas* chloroplast as a protein factory

It seems appropriate to conclude with the potential of *Chlamydomonas* as a system for commercial production of proteins. Purton (2007) discusses why the *Chlamydomonas* chloroplast is particularly suitable for transgene expression, because of its ability to accumulate both soluble and membrane proteins, and the multicopy genome which effectively amplifies expression of introduced genes compared to the nuclear genome. *Chlamydomonas* is non-toxic, classified as "generally regarded as safe" (GRAS), and is not known to harbor any pathogenic viruses, prions, or other agents, nor does it pose any obvious threats to agriculture. It can be grown much more economically than the transgenic mammalian cells used for commercial production of monoclonal antibodies.

Mayfield et al. (2003) synthesized a human antibody to a protein of the herpes simplex virus in *Chlamydomonas*, and showed that it is fully active, and Manuell et al. (2007) reported production of bovine mammary-associated serum amyloid, a protein which stimulates production of mucin in the gut and helps protect infants from viral and bacterial infections. Sun et al. (2003) produced a protein from foot and mouth virus, fused to the cholera toxin B subunit as an adjuvant. This demonstrates the significant potential of *Chlamydomonas* chloroplasts for production of vaccines.

Franklin and Mayfield (2004) reviewed the topic of "molecular farming" in *Chlamydomonas*, and various chapters in the volume edited by León et al. (2007) provide additional perspectives, as do the reviews by Griesbeck et al. (2006) and Mayfield et al. (2007), which includes an excellent annotated bibliography. The successes to date in production of individual proteins with therapeutic value promise a new and exciting future for our favorite microorganism.

Appendix: Resources for the Investigator

THE *CHLAMYDOMONAS* COMMUNITY

http://www.chlamy.org/

This web site is the gateway to general resources for *Chlamydomonas* research. The site includes news of interest to the *Chlamydomonas* community, announcements of meetings and employment opportunities, links to web sites of *Chlamydomonas* researchers, educational material, and methods files. The core of this site is a comprehensive database that interfaces with the genome sequence, gene expression, proteomics, and metabolic projects.

http://www.bio.net/hypermail/chlamydomonas/

The bionet.chlamydomonas newsgroup can be read from this web site, or obtained as an e-mail subscription.

CULTURE COLLECTIONS

http://www.chlamycollection.org/

The Chlamydomonas Genetics Center was created in 1979 in the laboratory of John Boynton and Nicholas Gillham at Duke University, with Elizabeth Harris as director, and received continuous support from the National Science Foundation for 25 years. When the Duke administration refused to support a renewal of this grant in 2004, the cultures were moved to the University of Minnesota, and a new grant was funded with P.A. Lefebvre and C.D. Silflow as principal investigators. The name of the project was changed at that time to the Chlamydomonas Resource Center.

This collection provides wild-type and mutant strains of *C. reinhardtii* and its close relatives, and an archival set of strains of *C. eugametos* and *C. moewusii*. Other resources include plasmids useful for transformation and genetic engineering of the nuclear and organelle genomes (see Table 8.8), a molecular mapping kit (Rymarquis et al., 2005), and kits suitable for high school science projects and undergraduate teaching.

```
http://www.utex.org/
```

```
http://www.ccap.ac.uk/
```

```
http://www.epsag.uni-goettingen.de/
```

```
http://www.nies.go.jp/
```

Wild-type strains of many other *Chlamydomonas* species can be obtained from the University of Texas Algal Collection (UTEX), the Culture Collection of Algae and Protozoa (CCAP; Gachon et al., 2007), the Sammlung von Algenkulturen Göttingen (SAG), the Microbial Culture Collection at the Japanese National Institute for Environmental Studies (NIES), and from numerous smaller collections.

```
http://www.chlamy.org/species.html
```

This page provides a tabulation of equivalent strains in the various collections.

GENETIC AND MOLECULAR MAPS

```
http://www.chlamy.org/nuclear_maps.html
```

```
http://www.chlamy.org/bac.html
```

The first genetic maps for *Chlamydomonas* were based on tetrad analysis of crosses among mutants with phenotypes that were easy to score, including requirement for acetate as a carbon source, impaired motility, resistance to inhibitors, and auxotrophy (Ebersold et al., 1962; Hastings et al., 1965; Levine and Goodenough, 1970; Smyth et al., 1975). Crossing data from the Levine laboratory, the Chlamydomonas Genetics Center, and various publications were summarized by Harris (1989). Nineteen linkage groups were eventually defined, but were reduced to seventeen with the consolidation of two pairs (XII/XIII and XVI/XVII) by Dutcher et al. (1991).

Kathir et al. (2003) constructed maps of molecular markers based on crosses between polymorphic strains, which were anchored to the genetic maps by sequenced genes corresponding to the phenotypic markers. BAC clones were mapped to linkage groups by Silflow, Lefebvre and colleagues, and formed the basis for alignment of linkage groups to the genomic sequence. The supporting on-line material for the publication by Merchant et al. (2007) includes maps for each linkage group based on the version 3 genomic sequence.

cDNA LIBRARIES

```
http://www.chlamy.org/libraries.html
```

```
http://est.kazusa.or.jp/en/plant/chlamy/EST/
```

cDNA libraries were prepared by John Davies and others from the standard laboratory strain 21 gr and the field isolate S1 D2 (Shrager et al., 2003; Jain et al., 2007), and by Asamizu et al. (2000, 2004) from strain C9.

EST sequences of clones from these libraries are available in GenBank. Information on ordering libraries and/or individual clones can be found on the following pages:

```
http://www.chlamycollection.org/libraries.html
```

```
http://www.kazusa.or.jp/clonereq/
```

BAC LIBRARY

```
http://www.genome.clemson.edu/capabilities/bacCenter.shtml
```

A BAC library prepared from CC-503 *cw92 mt+* by P.A. Lefebvre has been made available through the Clemson University BAC/EST Resource Center. Users may purchase the entire library or individual clones.

MICROARRAYS

```
http://www.chlamy.org/micro.html
```

Microarray chips are available from Arthur Grossman and colleagues at the Carnegie Institution, Stanford, CA (Zhang et al., 2004; Eberhard et al., 2006).

ANTIBODIES

```
http://www.agrisera.com/shop/
```

The Swedish company Agrisera sells antibodies to *Chlamydomonas* proteins and provides trouble-shooting assistance with western blots at <inquiry@agrisera.com>.

GENOME SEQUENCE

```
http://genome.jgi-psf.org/Chlre3/Chlre3.home.html
```

Sequencing of the nuclear genome of *C. reinhardtii* has been a project of the DOE Joint Genome Institute (Merchant et al., 2007). Unless otherwise specified, JGI version 3.0 has been the reference genome sequence throughout the three volumes of *The Chlamydomonas Sourcebook*. Future updates should be accessible from the JGI home page (http://genome.jgi-psf.org/).

The DNA that was sequenced was prepared by P.A. Lefebvre from strain CC-503 *cw92 mt+*, which is in the Ebersold/Levine 137c lineage (Chapter 1).

For additional information on analysis of the nuclear genome, see Grossman (2007), Jain et al. (2007), and Misumi et al. (2007).

`http://www.cse.ucsc.edu/~dkulp/cgi-bin/greenGenie`

Green Genie is a gene prediction tool trained on *Chlamydomonas* sequences.

`http://www.chlamy.org/chloro.html`

This site provides maps and sequence for the chloroplast genome.

PROTEOMICS, METABOLIC PATHWAYS, AND OTHER PROJECTS

Stauber and Hippler (2004) reviewed the status of proteomics studies in *Chlamydomonas*. Other web sites and publications include the following:

`http://www.chlamy.org/proteomics.html`
[links to all proteomics projects]

`http://chlamycyc.mpimp-golm.mpg.de/`
[metabolic pathways database]

`http://www.uni-jena.de/Protein_network_of__C__reinhardtii-lang-en.html`
[protein network and eyespot database]

`http://labs.umassmed.edu/chlamyfp/protector_login.php`
Pazour et al. (2005) [flagella]

`http://www.sfu.ca/~leroux/ciliome_home.htm`
Inglis et al. (2006) [cilia, comprehensive multiorganism database]

`http://plntfdb.bio.uni-potsdam.de/`
Riano-Pachon et al. (2007) [plant transcription factor database]

Allmer et al. (2006) [bioenergetic pathways]

Baginsky and Gruissem (2004) [chloroplast proteins]

Cardol et al. (2005) [mitochondrial oxidative phosphorylation]

Förster et al. (2006) [high light stress responses]

Gillet et al. (2006) [soluble proteins, analyzed for cadmium responses]

Keller et al. (2005) [centrioles]

Li et al. (2004) [flagellar and basal body proteome]

Manuell et al. (2005) [cytosolic ribosomes]

Naumann et al. (2007) [bioenergetics, iron deficiency]

Rolland et al. (2003) [chloroplast envelope membranes]

Schmidt et al. (2006) [cycspot]

Turkina et al. (2006) [photosynthetic membranes, phosphoproteome]

Van Wijk (2004) [chloroplast proteins]

Wagner et al. (2008) [phosphoproteome]

Wagner et al. (2005) [circadian rhythm system]

Whitelegge (2003) [thylakoid membranes]

Yamaguchi et al. (2002, 2003) [chloroplast ribosomes]

TAXONOMY

http://www.ncbi.nlm.nih.gov/sites/entrez?db=taxonomy

> Phylum Chlorophyta
> Class Chlorophyceae
> Order Chlamydomonadales or Volvocales (synonymous)
> Family Chlamydomonadaceae

NCBI uses the name Chlamydomonadales in the same sense that Volvocales is used in the present book and in many taxonomic treatments, to designate the order comprising unicellular forms such as *Chlamydomonas*, *Dunaliella*, and *Haematococcus*, and colonial forms such as *Volvox*, *Gonium*, and *Pandorina*. This is consistent with contemporary molecular phylogeny. In the older literature these algae were often divided among separate orders.

NOMENCLATURE

http://www.chlamy.org/nomenclature.html

Guidelines for naming *Chlamydomonas* genes and proteins, summarized on this web page, were outlined by Susan Dutcher in the *Genetic Nomenclature Guide* published by *Trends in Genetics* in 1998. Recommendations for names used in genome annotation are discussed in more detail on the web page cited here.

CHLAMYDOMONAS AS AN EDUCATIONAL TOOL

http://www.chlamy.org/info.html

This page includes links to many other sites with introductory material on *Chlamydomonas* and other algae.

http://www.chlamycollection.org/projects.html

The Chlamydomonas Resource Center supplies kits suitable for high school science projects and undergraduate laboratory teaching. These include hydrogen production, demonstration of phototaxis and circadian rhythms, mating and dikaryon rescue, motility, and uniparental inheritance.

http://www.dartmouth.edu/~cellccli/Laboratory%20Exercises.html

[deflagellation and flagellar regeneration, by Elizabeth Smith, Dartmouth College]

http://www.lifescied.org/cgi/content/full/5/3/239

[laboratory project on analysis of flagellar proteins; Mitchell and Graziano, 2006]

Some additional laboratory exercises are described in the first edition of *The Chlamydomonas Sourcebook* (Harris, 1989, pp. 626–634).

The following DVD includes videos of *Chlamydomonas* cells. It is distributed by Sinauer Associates.

Living Cells: Structure, Function, Diversity
by Jeremy D. Pickett-Heaps and Julianne Pickett-Heaps

http://www.elsevierdirect.com/companions/9780123708731

This is the companion web site for *The Chlamydomonas Sourcebook* and includes excellent videos to accompany chapters in volume 3.

METHODS COMPILATIONS

http://www.chlamy.org/methods.html

This site provides many short protocols contributed by members of the *Chlamydomonas* community and citations for other widely used methods. Volume 47 of *Methods in Cell Biology* (1995) is devoted largely to procedures for *Chlamydomonas*. *Methods in Enzymology* Volume 297 (1998), entitled *Photosynthesis: Molecular Biology of Energy Capture*, includes several articles on *Chlamydomonas*.

SPECIAL JOURNAL ISSUES DEVOTED TO *CHLAMYDOMONAS*

Photosynthesis Res. **82**(3), December, 2004.
Plant Physiology **137**(2), February, 2005.
Genetics **179**(1), May, 2008.

Abe, J., Kubo, T., Takagi, Y., Saito, T., Miura, K., Fukuzawa, H., and Matsuda, Y. (2004). The transcriptional program of synchronous gametogenesis in *Chlamydomonas reinhardtii*. *Curr. Genet.* **46**, 304–315.

Abe, J., Kubo, T., Saito, T., and Matsuda, Y. (2005). The regulatory networks of gene expression during the sexual differentiation of *Chlamydomonas reinhardtii*, as analyzed by mutants for gametogenesis. *Plant Cell Physiol.* **46**, 312–316.

Abeles, F.B. (1964). Cell-free hydrogenase from *Chlamydomonas*. *Plant Physiol.* **39**, 169–176.

Adair, W.S. (1985). Characterization of *Chlamydomonas* sexual agglutinins. *J. Cell Sci. Suppl.* **2**, 233–260.

Adair, W.S. and Appel, H. (1989). Identification of a highly conserved hydroxyproline-rich glycoprotein in the cell walls of *Chlamydomonas reinhardtii* and two other Volvocales. *Planta* **179**, 381–386.

Adair, W.S. and Apt, K.E. (1990). Cell wall regeneration in *Chlamydomonas*. Accumulation of messenger RNAs encoding cell wall hydroxyproline-rich glycoproteins. *Proc. Natl. Acad. Sci. U. S. A.* **87**, 7355–7359.

Adair, W.S. and Snell, W.J. (1990). The *Chlamydomonas reinhardtii* cell wall: structure, biochemistry and molecular biology. In: *Organization and Assembly of Plant and Animal Extracellular Matrix* (R.P. Mecham and W.S. Adair, Eds.), pp. 15–84. Academic Press, Orlando.

Adair, W.S., Monk, B.C., Cohen, R., Hwang, C., and Goodenough, U.W. (1982). Sexual agglutinins from the *Chlamydomonas* flagellar membrane. Partial purification and characterization. *J. Biol. Chem.* **257**, 4593–4602.

Adair, W.S., Hwang, C., and Goodenough, U.W. (1983). Identification and visualization of the sexual agglutinin from the mating-type plus flagellar membrane of *Chlamydomonas*. *Cell* **33**, 183–193.

Adair, W.S., Steinmetz, S.A., Mattson, D.M., Goodenough, U.W., and Heuser, J.E. (1987). Nucleated assembly of *Chlamydomonas* and *Volvox* cell walls. *J. Cell Biol.* **105**, 2373–2382.

Adam, M. and Loppes, R. (1998). Use of the *ARG7* gene as an insertional mutagen to clone *PHON24*, a gene required for derepressible neutral phosphatase activity in *Chlamydomonas reinhardtii*. *Mol. Gen. Genet.* **258**, 123–132.

Adam, M., Lentz, K.E., and Loppes, R. (1993). Insertional mutagenesis to isolate acetate-requiring mutants in *Chlamydomonas reinhardtii*. *FEMS Microbiol. Lett.* **110**, 265–268.

Adams, G.M.W. (1975). An investigation of the mechanism of inheritance of chloroplast genes in *Chlamydomonas reinhardtii*. Ph.D. dissertation, Duke University, Durham NC.

Adams, G.M.W. (1978). Chloroplast gene transmission in *Chlamydomonas reinhardtii*: a random choice model. *Plasmid* **1**, 522–535.

309

Adams, G.M.W. (1982). Chloroplast gene transmission in *Chlamydomonas* somatic fusion products. Effects of pretreatment with isolated flagella. *Curr. Genet.* **5**, 1–3.

Adams, G.M.W., Huang, B., and Luck, D.J.L. (1982). Temperature-sensitive, assembly-defective flagella mutants of *Chlamydomonas reinhardtii*. *Genetics* **100**, 579–586.

Adams, G.M.W., Wright, R.L., and Jarvik, J.W. (1985). Defective temporal and spatial control of flagellar assembly in a mutant of *Chlamydomonas reinhardtii* with variable flagellar number. *J. Cell Biol.* **100**, 955–964.

Adams, K.J. (1969). A simple device for obtaining axenic cultures of phototactic unicells. *J. Gen. Microbiol.* **59**, 427–428.

Aguilar, M., Cárdenas, J., and Fernández, E. (1992a). Quantitation of molybdopterin oxidation product in wild-type and molybdenum cofactor deficient mutants of *Chlamydomonas reinhardtii*. *Biochim. Biophys. Acta* **1160**, 269–274.

Aguilar, M.R., Prieto, R., Cárdenas, J., and Fernández, E. (1992b). *nit 7*: a new locus for molybdopterin cofactor biosynthesis in the green alga *Chlamydomonas reinhardtii*. *Plant Physiol.* **98**, 395–398.

Aguilera, A. and Amils, R. (2005). Tolerance to cadmium in *Chlamydomonas sp.* (Chlorophyta) strains isolated from an extreme acidic environment, the Tinto River (SW, Spain). *Aquat. Toxicol.* **75**, 316–329.

Ahlf, W., Irmer, U., and Weber, A. (1980). Über die Anreicherung von Blei durch Süsswassergrünalgen unter Berücksichtigung verscheidener Aussenfaktoren. *Z. Pflanzenphysiol.* **100**, 197–207.

Ahmed, N.T. and Mitchell, D.R. (2005). ODA16p, a *Chlamydomonas* flagellar protein needed for dynein assembly. *Mol. Biol. Cell* **16**, 5004–5012.

Albertanto, P., Reale, L., Palladino, L., Reale, A., Cotton, R., Bollanti, S., Di-Lazzaro, P., Flora, F., Lisi, N., Nottola, A., Papadaki, K.V., Letardi, T., Batani, D., Conti, A., Moret, M., and Grilli, A. (1997). X-ray contact microscopy using an excimer laser plasma source with different target materials and laser pulse durations. *J. Microsc.* **87**, 96–103.

Aldrich, J., Cherney, B., Merlin, E., Williams, C., and Mets, L. (1985). Recombination within the inverted repeat sequences of the *Chlamydomonas reinhardii* chloroplast genome produces two orientation isomers. *Curr. Genet.* **9**, 233–238.

Alexander, N.J., Gillham, N.W., and Boynton, J.E. (1974). The mitochondrial genome of *Chlamydomonas*. Induction of minute colony mutations by acriflavin and their inheritance. *Mol. Gen. Genet.* **130**, 275–290.

Alizadeh, S., Nechustai, R., Barber, J., and Nixon, P. (1994). Nucleotide sequence of the *psbE*, *psbF* and *trnM* genes from the chloroplast genome of *Chlamydomonas reinhardtii*. *Biochim. Biophys. Acta* **1188**, 439–442.

Allen, C. and Borisy, G.G. (1974). Structural polarity and directional growth of microtubules of *Chlamydomonas* flagella. *J. Mol. Biol.* **90**, 381–402.

Allen, M.D., Kropat, J., Tottey, S., Campo, J.A., and Merchant, S.S. (2007a). Manganese deficiency in *Chlamydomonas* results in loss of PSII and MnSOD function, sensitivity to peroxides, and secondary phosphorus- and iron-deficiency. *Plant Physiol.* **143**, 263–277.

Allen, M.D., Del Campo, J.A., Kropat, J., and Merchant, S.S. (2007b). *FEA1*, *FEA2* and *FRE1*, encoding two homologous secreted proteins and a candidate ferri-reductase, are expressed coordinately with *FOX1* and *FTR1* in iron-deficient *Chlamydomonas reinhardtii*. *Eukaryotic Cell* **6**, 1841–1852.

Allet, B. and Rochaix, J.-D. (1979). Structure analysis at the ends of the intervening DNA sequences in the chloroplast 23S ribosomal genes of *C. reinhardii*. *Cell* **18**, 55–60.

Allmer, J., Naumann, B., Markert, C., Zhang, M., and Hippler, M. (2006). Mass spectrometric genomic data mining: Novel insights into bioenergetic pathways in *Chlamydomonas reinhardtii*. *Proteomics* **6**, 6207–6220.

Amano, T., Hirasawa, K., O'Donohue, M.J., Pernolle, J.-C., and Shioi, Y. (2003). A versatile assay for the accurate, time-resolved determination of cellular viability. *Anal. Biochem.* **314**, 1–7.

Andersen, R.A. (Ed.) (2005). *Algal Culturing Techniques*, Elsevier, Amsterdam. 578 pp.

Angeler, D.G. and Schagerl, M. (1997). Distribution of the xanthophyll loroxanthin in selected members of the Chlamydomonadales and Volvocales (Chlorophyta). *Phyton* **37**, 119–132.

Antal, M., Mougin, A., Kis, M., Boros, E., Steger, G., Jakab, G., Solymosy, F., and Branlant, C. (2000). Molecular characterization at the RNA and gene levels of U3 snoRNA from a unicellular green alga, *Chlamydomonas reinhardtii*. *Nucleic Acids Res.* **28**, 2959–2968.

Antal, T.K., Krendeleva, T.E., and Rubin, A.B. (2004). The photochemical activity of photosystem II in sulfur-deprived *Chlamydomonas reinhardtii* cells depends on the redox state of the quinone pool during the transition to anaerobiosis. *Biofizika* **49**, 499–505.

Antaramian, A., Coria, R., Ramirez, J., and González-Halphen, D. (1996). The deduced primary structure of subunit I from cytochrome c oxidase suggests that the genus *Polytomella* shares a common mitochondrial origin with *Chlamydomonas*. *Biochim. Biophys. Acta* **1273**, 198–202.

Antia, N.J., Berland, B.R., Bonin, D.J., and Maestrini, S.Y. (1975). Comparative evaluation of certain organic and inorganic sources of nitrogen for phototrophic growth of marine microalgae. *J. Mar. Biol. Assoc. U.K.* **55**, 519–539.

Antia, N.J., Berland, B.R., Bonin, D.J., and Maestrini, S.Y. (1977). Effects of urea concentration in supporting growth of certain marine microplanktonic algae. *Phycologia* **16**, 105–112.

Aoyama, H., Hagiwara, Y., Misumi, O., Kuroiwa, T., and Nakamura, S. (2006). Complete elimination of maternal mitochondrial DNA during meiosis resulting in the paternal inheritance of the mitochondrial genome in *Chlamydomonas* species. *Protoplasma* **228**, 231–242.

Aparicio, P.J., Azuara, M.P., Ballesteros, A., and Fernández, V.M. (1985). Effects of light intensity and oxidized nitrogen sources on hydrogen production by *Chlamydomonas reinhardii*. *Plant Physiol.* **78**, 803–806.

Ares, M., Jr. and Howell, S.H. (1982). Cell cycle stage-specific accumulation of mRNAs encoding tubulin and other polypeptides in *Chlamydomonas*. *Proc. Natl. Acad. Sci. U. S. A.* **79**, 5577–5581.

Armbrust, E.V., Ferris, P.J., and Goodenough, U.W. (1993). A mating type-linked gene cluster expressed in *Chlamydomonas* zygotes participates in the uniparental inheritance of the chloroplast genome. *Cell* **74**, 801–811.

Armbrust, E.V., Ibrahim, A., and Goodenough, U.W. (1995). A mating type-linked mutation that disrupts the uniparental inheritance of chloroplast DNA also disrupts cell-size control in *Chlamydomonas*. *Mol. Biol. Cell* **6**, 1807–1818.

Armstrong, W.P. (1987). Watermelon snow. *Environmental Southwest, San Diego Natural History Museum*, spring 1987 issue, pp. 20–23.

Arnold, C.G., Schimmer, O., Schötz, F., and Bathelt, H. (1972). Die Mitochondrien von *Chlamydomonas reinhardii*. *Arch. Mikrobiol.* **81**, 50–67.

Asamizu, E., Miura, K., Kucho, K., Inoue, Y., Fukuzawa, H., Ohyama, K., Nakamura, Y., and Tabata, S. (2000). Generation of expressed sequence tags from low-CO_2 and high-CO_2 adapted cells of *Chlamydomonas reinhardtii*. *DNA Res.* **7**, 305–307.

Asamizu, E., Nakamura, Y., Miura, K., Fukuzawa, H., Fujiwara, S., Hirono, M., Iwamoto, K., Matsuda, Y., Minagawa, J., Shimogawara, K., Takahashi, Y., and Tabata, S. (2004). Establishment of publicly available cDNA material and information resource of *Chlamydomonas reinhardtii* (Chlorophyta) to facilitate gene function analysis. *Phycologia* **43**, 722–726.

Asleson, C.M. and Lefebvre, P.A. (1998). Genetic analysis of flagellar length control in *Chlamydomonas reinhardtii*: a new long-flagella locus and extragenic suppressor mutations. *Genetics* **148**, 693–702.

Auchincloss, A.H., Loroch, A.I., and Rochaix, J.D. (1999). The argininosuccinate lyase gene of *Chlamydomonas reinhardtii*: cloning of the cDNA and its characterization as a selectable shuttle marker. *Mol. Gen. Genet.* **261**, 21–30.

Auchincloss, A.H., Zerges, W., Perron, K., Girard-Bascou, J., and Rochaix, J.D. (2002). Characterization of Tbc2, a nucleus-encoded factor specifically required for translation of the chloroplast *psbC* mRNA in *Chlamydomonas reinhardtii*. *J. Cell Biol.* **157**, 953–962.

Azencott, H.R., Peter, G.F., and Prausnitz, M.R. (2007). Influence of the cell wall on intracellular delivery to algal cells by electroporation and sonication. *Ultrasound Med. Biol.* **33**, 1805–1817.

Azuara, M.P. and Aparicio, P.J. (1983). *In vivo* blue light activation of *Chlamydomonas reinhardii* nitrate reductase. *Plant Physiol.* **71**, 286–290.

Bachir, F. and Loppes, R. (1997). Identification of a new derepressible phosphatase in *Chlamydomonas reinhardtii*. *FEMS Microbiol. Lett.* **149**, 195–200.

Bachir, F., Baise, E., and Loppes, R. (1996). Mutants impaired in derepressible alkaline phosphatase activity in *Chlamydomonas reinhardtii*. *Plant Sci.* **119**, 93–101.

Badger, M.R., Andrews, T.J., Whitney, S.M., Ludwig, M., Yellowlees, D.C., Leggat, W., and Price, G.D. (1998). The diversity and coevolution of Rubisco, plastids, pyrenoids, and chloroplast-based CO_2-concentrating mechanisms in algae. *Can. J. Bot.* **76**, 1052–1071.

Badour, S.S. and Irvine, B.R. (1990). Activities of photosystems I and II in *Chlamydomonas segnis* adapted and adapting to air and air-enriched with carbon dioxide. *Bot. Acta* **103**, 149–154.

Badour, S.S., Tan, C.K., Van Caeseele, L.A., and Isaac, P.K. (1973). Observations on the morphology, reproduction, and fine structure of *Chlamydomonas segnis* from Delta Marsh, Manitoba. *Can. J. Bot.* **51**, 67–72.

Baginsky, S. and Gruissem, W. (2004). Chloroplast proteomics: potentials and challenges. *J. Exp. Bot.* **55**, 1213–1220.

Bailey, K.M. and Taub, F.B. (1980). Effects of hydroxamate siderophores (strong Fe (III) chelators) on the growth of algae. *J. Phycol.* **16**, 334–339.

Baillet, B. and Kohorn, B.D. (1996). Hydrophobic core but not amino-terminal charged residues are required for translocation of an integral thylakoid membrane protein *in vivo*. *J. Biol. Chem.* **271**, 18375–18378.

Balczun, C., Bunse, A., Hahn, D., Bennoun, P., Nickelsen, J., and Kück, U. (2005). Two adjacent nuclear genes are required for functional complementation of a chloroplast *trans*-splicing mutant from *Chlamydomonas reinhardtii*. *Plant J.* **43**, 636–648.

Balczun, C., Bunse, A., Schwarz, C., Piotrowski, M., and Kück, U. (2006). Chloroplast heat shock protein Cpn60 from *Chlamydomonas reinhardtii* exhibits a novel function as a group II intron-specific RNA-binding protein. *FEBS Lett.* **580**, 4527–4532.

Baldan, B., Girard-Bascou, J., Wollman, F.-A., and Olive, J. (1991). Evidence for thylakoid membrane fusion during zygote formation in *Chlamydomonas reinhardtii*. *J. Cell Biol.* **114**, 905–915.

Baldauf, S.L. and Palmer, J.D. (1990). Evolutionary transfer of the chloroplast *tufA* gene to the nucleus. *Nature* **344**, 262–265.

Ball, S., Marianne, T., Dirick, L., Fresnoy, M., Delrue, B., and Decq, A. (1991). A *Chlamydomonas reinhardtii* low-starch mutant is defective for 3-phosphoglycerate activation and orthophosphate inhibition of ADP-glucose pyrophosphorylase. *Planta* **185**, 17–26.

Banerjee, R. and Batschauer, A. (2005). Plant blue-light receptors. *Planta* **220**, 498–502.

Bao, Y. and Herrin, D.L. (1993). Nucleotide sequence and secondary structure of the chloroplast group I intron *Cr.psbA 2*: novel features of this self splicing ribozyme. *Nucleic Acids Res.* **21**, 1667.

Bar-Nun, S. and Ohad, I. (1977). Presence of polypeptides of cytoplasmic and chloroplastic origin in isolated photoactive preparations of photosystems I and II in *Chlamydomonas reinhardi y-1*. *Plant Physiol.* **59**, 161–166.

Barclay, W.R. and Lewin, R.A. (1985). Microalgal polysaccharide production for the conditioning of agricultural soils. *Plant Soil* **88**, 159–169.

Bard, M., Wilson, K.J., and Thompson, R.M. (1978). Isolation of sterol mutants in *Chlamydomonas reinhardi*: chromatographic analyses. *Lipids* **13**, 533–539.

Barkan, A. and Goldschmidt-Clermont, M. (2000). Participation of nuclear genes in chloroplast gene expression. *Biochimie* **82**, 559–572.

Baroli, I., Do, A.D., Yamane, T., and Niyogi, K.K. (2003). Zeaxanthin accumulation in the absence of a functional xanthophyll cycle protects *Chlamydomonas reinhardtii* from photooxidative stress. *Plant Cell* **15**, 992–1008.

Baroli, I., Gutman, B.L., Ledford, H.K., Shin, J.W., Chin, B.L., Havaux, M., and Niyogi, K.K. (2004). Photo-oxidative stress in a xanthophyll-deficient mutant of *Chlamydomonas*. *J. Biol. Chem.* **279**, 6337–6344.

Barsel, S.-E., Wexler, D.E., and Lefebvre, P.A. (1988). Genetic analysis of long-flagella mutants of *Chlamydomonas reinhardtii*. *Genetics* **118**, 637–648.

Bartlett, S.G., Harris, E.H., Grabowy, C.T., Gillham, N.W., and Boynton, J.E. (1979). Ribosomal subunits affected by antibiotic resistance mutations at seven chloroplast loci in *Chlamydomonas reinhardtii*. *Mol. Gen. Genet.* **176**, 199–208.

Bastia, D., Chiang, K.-S., Swift, H., and Siersma, P. (1971). Heterogeneity, complexity, and repetition of the chloroplast DNA of *Chlamydomonas reinhardtii*. *Proc. Natl. Acad. Sci. U. S. A.* **68**, 1157–1161.

Bateman, J.M. and Purton, S. (2000). Tools for chloroplast transformation in *Chlamydomonas*: expression vectors and a new dominant selectable marker. *Mol. Gen. Genet.* **263**, 404–410.

Baymann, F., Zito, F., Kuras, R., Minai, L., Nitschke, W., and Wollman, F.A. (1999). Functional characterization of *Chlamydomonas* mutants defective in cytochrome *f* maturation. *J. Biol. Chem.* **274**, 22957–22967.

Bean, B. (1977). Geotactic behavior of *Chlamydomonas*. *J. Protozool.* **24**, 394–401.

Bean, B. (1984). Microbial geotaxis. In: *Membranes and Sensory Transduction* (G. Colombetti and F. Lenci, Eds.), pp. 163–198. Plenum, New York.

Bean, B., Savitsky, R., and Grossinger, B. (1982). Strontium ion (Sr^{2+}) induces backward swimming of *Chlamydomonas*. *J. Protozool.* **29**, 296.

Beck, C.F. and Acker, A. (1992). Gametic differentiation of *Chlamydomonas reinhardtii*. Control by nitrogen and light. *Plant Physiol.* **98**, 822–826.

Beck, C.F. and Haring, M.A. (1996). Gametic differentiation of *Chlamydomonas*. *Int. Rev. Cytol.* **168**, 259–302.

Becker, D. and Melkonian, M. (1992). N-linked glycoproteins associated with flagellar scales in a flagellate green alga: characterization of interactions. *Eur. J. Cell Biol.* **57**, 109–116.

Becker, B., Marin, B., and Melkonian, M. (1994). Structure, composition and biogenesis of prasinophyte cell coverings. *Protoplasma* **181**, 233–244.

Becker, B., Melkonian, M., and Kamerling, J.P. (1998). The cell wall (theca) of *Tetraselmis striata* (Chlorophyta): macromolecular composition and structural elements of the complex polysaccharides. *J. Phycol.* **34**, 779–787.

Beckers, M.-C., Munaut, C., Minet, A., and Matagne, R.F. (1991). The fate of mitochondrial DNAs of mt^+ and mt^- origin in gametes and zygotes of *Chlamydomonas*. *Curr. Genet.* **20**, 239–243.

Beckmann, M. and Hegemann, P. (1991). *In vitro* identification of rhodopsin in the green alga *Chlamydomonas*. *Biochemistry* **30**, 3692–3697.

Beezley, B.B., Gruber, P.J., Frederick, S.E. (1976). Cytochemical localization of glycolate dehydrogenase in mitochondria of *Chlamydomonas*. *Plant Physiol.* **58**, 315–319.

Behlau, J. (1939). Der Generationswechsel zwischen *Chlamydomonas variabilis* Dangeard und *Carteria ovata* Jacobsen. *Beitr. Biol. Pflanzen* **26**, 221–249.

Behn, W. and Arnold, C.G. (1974). Die Wirkung von Streptomycin und Neamin auf die Chloroplasten- und Mitochondrienstruktur von *Chlamydomonas reinhardii*. *Protoplasma* **82**, 77–89.

Behra, R. (1993). *In vitro* effects of cadmium, zinc and lead on calmodulin-dependent actions in *Oncorhynchus mykiss*, *Mytilus sp.*, and *Chlamydomonas reinhardtii*. *Arch. Environ. Contam. Toxicol.* **24**, 21–27.

Behrens, P.W. (2005). Photobioreactors and fermentors: the light and dark sides of growing algae. In: *Algal Culturing Techniques* (R.A. Andersen, Ed.), pp. 189–203. Elsevier, Amsterdam.

Belar, E. (1926). Der Formwechsel der Protistenkerne. *Ergebn. Fortschr. Zool.* **6**, 1–420.

Beligni, M.V., Yamaguchi, K., and Mayfield, S.P. (2004). The translational apparatus of *Chlamydomonas reinhardtii* chloroplast. *Photosynth. Res.* **82**, 315–325.

Belknap, W.R. and Togasaki, R.K. (1981). *Chlamydomonas reinhardtii* cell preparation with altered permeability toward substrates of organellar reactions. *Proc. Natl. Acad. Sci. U. S. A.* **78**, 2310–2314.

Bell, E.J. (1955). Some effects of colchicine on *Chlamydomonas chlamydogama* Bold. *Exp. Cell Res.* **9**, 350–353.

Bell, G. (1978). The evolution of anisogamy. *J. Theor. Biol.* **73**, 247–270.

Bell, G. (1990a). The ecology and genetics of fitness in *Chlamydomonas*. I. Genotype-by-environment interaction among pure strains. *Proc. R. Soc. Lond. [Biol.]* **240**, 295–321.

Bell, G. (1990b). The ecology and genetics of fitness in *Chlamydomonas*. II. The properties of mixtures of strains. *Proc. R. Soc. Lond. [Biol.]* **240**, 323–350.

Bell, G. (2005). Experimental sexual selection in *Chlamydomonas*. *J. Evol. Biol.* **18**, 722–734.

Bell, R.A. and Cain, J.R. (1983). Sexual reproduction and hybridization in *Chlamydomonas smithii* and *C. reinhardtii* (Chlorophyceae, Volvocales). *Phycologia* **22**, 243–247.

Bellafiore, S., Ferris, P., Naver, H., Gohre, V., and Rochaix, J.D. (2002). Loss of *albino3* leads to the specific depletion of the light-harvesting system. *Plant Cell* **14**, 2303–2314.

Ben-Amotz, A. and Avron, M. (1983). Accumulation of metabolites by halotolerant algae and its industrial potential. *Annu. Rev. Microbiol.* **37**, 95–119.

Ben-Amotz, A. and Gibbs, M. (1975). H_2 metabolism in photosynthetic organisms. II. Light-dependent H_2 evolution by preparations from *Chlamydomonas*, *Scenedesmus*, and spinach. *Biochem. Biophys. Res. Commun.* **64**, 355–359.

Ben-Bassat, D., Shelef, G., Gruner, N., and Shuval, H.I. (1972). Growth of *Chlamydomonas* in a medium containing mercury. *Nature* **240**, 43–44.

Bennett, M.E. and Hobbie, J.E. (1972). The uptake of glucose by *Chlamydomonas sp.. J. Phycol.* **8**, 392–398.

Bennoun, P. (1972). Conservation des souches de *Chlamydomonas reinhardi* à basse temperature. *C. R. Acad. Sci* **275**, 1777–1778.

Bennoun, P. (1982). Evidence for a respiratory chain in the chloroplast. *Proc. Natl. Acad. Sci. U. S. A.* **79**, 4352–4356.

Bennoun, P. (1983). Effects of mutations and of ionophore on chlororespiration in *Chlamydomonas reinhardtii*. *FEBS Lett.* **156**, 363–365.

Bennoun, P. and Delepelaire, P. (1982). Isolation of photosynthesis mutants in *Chlamydomonas*. In: *Methods in Chloroplast Molecular Biology* (M. Edelman, N.-H. Chua, and R.B. Hallick, Eds.), pp. 25–38. Elsevier/North-Holland, Amsterdam.

Bennoun, P. and Levine, R.P. (1967). Detecting mutants that have impaired photosynthesis by their increased level of fluorescence. *Plant Physiol.* **42**, 1284–1287.

Bennoun, P., Masson, A., Piccioni, R., and Chua, N.-H. (1978). Uniparental mutants of *Chlamydomonas reinhardi* defective in photosynthesis. In: *Chloroplast Development* (G. Akoyunoglou and J.H. Argyroudi-Akoyunglou, Eds.), pp. 721–726. Elsevier/North-Holland, Amsterdam.

Bennoun, P., Masson, A., and Delosme, M. (1980). A method for complementation analysis of nuclear and chloroplast mutants of photosynthesis in *Chlamydomonas*. *Genetics* **95**, 39–47.

Bennoun, P., Spierer-Herz, M., Erickson, J., Girard-Bascou, J., Pierre, Y., Delosme, M., and Rochaix, J.-D. (1986). Characterization of photosystem II mutants of *Chlamydomonas reinhardii* lacking the *psbA* gene. *Plant Mol. Biol.* **6**, 151–160.

Bennoun, P., Delosme, M., and Kück, U. (1991). Mitochondrial genetics of *Chlamydomonas reinhardtii*: resistance mutations marking the cytochrome *b* gene. *Genetics* **127**, 335–343.

Bennoun, P., Delosme, M., Godehardt, I., and Kück, U. (1992). New tools for mitochondrial genetics of *Chlamydomonas reinhardtii*: manganese mutagenesis and cytoduction. *Mol. Gen. Genet.* **234**, 147–154.

Bergman, K., Goodenough, U.W., Goodenough, D.A., Jawitz, J., and Martin, H. (1975). Gametic differentiation in *Chlamydomonas reinhardtii*. II. Flagellar membranes and the agglutination reaction. *J. Cell Biol.* **67**, 606–622.

Berman, S.A., Wilson, N.F., Haas, N.A., and Lefebvre, P.A. (2003). A novel MAP kinase regulates flagellar length in *Chlamydomonas*. *Curr. Biol.* **13**, 1145–1149.

Bernd, K.K. and Cook, N. (2002). Microscale assay monitors algal growth characteristics. *BioTechniques* **32**, 1256–1259.

Bernd, K.K. and Kohorn, B.D. (1998). Tip loci: six *Chlamydomonas* nuclear suppressors that permit the translocation of proteins with mutant thylakoid signal sequences. *Genetics* **149**, 1293–1301.

Bernstein, E. (1960). Synchronous division in *Chlamydomonas moewusii*. *Science* **131**, 1528–1529.

Bernstein, E. (1964). Physiology of an obligate photoautotroph (*Chlamydomonas moewusii*). I. Characteristics of synchronously and randomly reproducing cells, and an hypothesis to explain their population curves. *J. Protozool.* **11**, 56–74.

Bernstein, E. (1968). Induction of synchrony in *Chlamydomonas moewusii* as a tool for the study of cell division. *Methods Cell Physiol.* **3**, 119–145.

Bernstein, E. and Jahn, T.L. (1955). Certain aspects of the sexuality of two species of *Chlamydomonas*. *J. Protozool.* **2**, 81–85.

Berry-Lowe, S.L., Johnson, C.H., and Schmidt, G.W. (1992). Nucleotide sequence of the *psbB* gene of *Chlamydomonas reinhardtii* chloroplasts. *Plant Physiol.* **98**, 1541–1543.

Berthold, D.A., Schmidt, C.L., and Malkin, R. (1995). The deletion of *petG* in *Chlamydomonas reinhardtii* disrupts the cytochrome *bf* complex. *J. Biol. Chem.* **270**, 29293–29298.

Berthold, P., Schmitt, R., and Mages, W. (2002). An engineered *Streptomyces hygroscopicus aph* 7 gene mediates dominant resistance against hygromycin B in *Chlamydomonas reinhardtii*. *Protist* **153**, 401–412.

Bertsch, J. and Malkin, R. (1991). Nucleotide sequence of the *petA* (cytochrome *f*) gene from the green alga, *Chlamydomonas reinhardtii*. *Plant Mol. Biol.* **17**, 131–133.

Bessen, M., Fay, R.B., and Witman, G.B. (1980). Calcium control of waveform in isolated flagellar axonemes of *Chlamydomonas*. *J. Cell Biol.* **86**, 446–455.

Bidigare, R.R., Ondrusek, M.E., Kennicutt, M.C., II, Iturriaga, R., Harvey, H.R., Hoham, R.W., and Macko, S.A. (1993). Evidence for a photoprotective function for secondary carotenoids of snow algae. *J. Phycol.* **29**, 427–434.

Bingham, S.E., Cox, J.C., and Strem, M.D. (1989). Expression of foreign DNA in *Chlamydomonas reinhardtii. FEMS Microbiol. Lett.* **65**, 77–82.

Bingham, S.E., Xu, R., and Webber, A.N. (1991). Transformation of chloroplasts with the *psaB* gene encoding a polypeptide of the photosystem I reaction center. *FEBS Lett.* **292**, 137–140.

Birky, C.W., Jr., VanWinkle-Swift, K.P., Sears, B.B., Boynton, J.E., and Gillham, N.W. (1981). Frequency distributions for chloroplast genes in *Chlamydomonas* zygote clones: evidence for random drift. *Plasmid* **6**, 173–192.

Birky, C.W., Jr., Katko, P., and Lorenz, M. (1984). Cytological demonstration of chloroplast DNA behavior during gametogenesis and zygote formation in *Chlamydomonas reinhardtii. Curr. Genet.* **8**, 1–7.

Birmingham, B.C. and Colman, B. (1977). The effect of two organophosphate insecticides on the growth of freshwater algae. *Can. J. Bot.* **55**, 1453–1456.

Bischoff, H.W. (1959). Some observations on *Chlamydomonas microhalophila sp. Nov. Biol. Bull.* **117**, 54–62.

Bischoff, H.W. and Bold, H.C. (1963). Phycological studies. IV. Some soil algae from enchanted rock and related algal species. *Univ. Texas Publ. No. 6318*, 95 pp.

Bishop, C.L., Cain, A.J., Purton, S., and Nugent, J.H.A. (1999). Molecular cloning and sequence analysis of the *Chlamydomonas reinhardtii* nuclear gene encoding the photosystem II subunit PsbW. *Plant Physiol.* **121**, 313.

Bishop, C.L., Purton, S., and Nugent, J.H.A. (2003). Molecular analysis of the *Chlamydomonas* nuclear gene encoding PsbW and demonstration that PsbW is a subunit of photosystem II, but not I. *Plant Mol. Biol.* **52**, 285–289.

Bisova, K., Krylov, D.M., and Umen, J.G. (2005). Genome-wide annotation and expression profiling of cell cycle regulatory genes in *Chlamydomonas reinhardtii. Plant Physiol.* **137**, 475–491.

Blank, R. and Arnold, C.-G. (1980). Variety of mitochondrial shapes, sizes, and volumes in *Chlamydomonas reinhardii. Protoplasma* **104**, 187–191.

Blank, R. and Arnold, C.-G. (1981). Structural changes of mitochondria in *Chlamydomonas reinhardii* after chloramphenicol treatment. *Eur. J. Cell Biol.* **24**, 244–251.

Blank, R., Grobe, B., and Arnold, C.-G. (1978). Time sequence of nuclear and chloroplast fusions in the zygote of *Chlamydomonas reinhardii. Planta* **138**, 63–64.

Blank, R., Hauptmann, E., and Arnold, C.-G. (1980). Variability of mitochondrial population in *Chlamydomonas reinhardii. Planta* **150**, 236–241.

Blankenship, J.E. and Kindle, K.L. (1992). Expression of chimeric genes by the light-regulated *cabII-1* promoter in *Chlamydomonas reinhardtii*: a *cabII-1/nit1* gene functions as a dominant selectable marker in a *nit1⁻ nit2⁻* strain. *Mol. Cell. Biol.* **12**, 5268–5279.

Blankley, W.F. (1973). Toxic and inhibitory materials associated with culturing. In: *Handbook of Phycological Methods. Vol. I. Culture Methods and Growth Measurements* (J.R. Stein, Ed.), pp. 207–228. Cambridge University Press, Cambridge.

Blokker, P., Schouten, S., De Leeuw, J.W., Damste, J.S.S., and van den Ende, H. (1999). Molecular structure of the resistant biopolymer in zygospore cell walls of *Chlamydomonas monoica. Planta* **207**, 539–543.

Bloodgood, R.A. (1977). Motility occurring in association with the surface of the *Chlamydomonas* flagellum. *J. Cell Biol.* **75**, 983–989.

Bloodgood, R.A. (1981a). Flagella-dependent gliding motility in *Chlamydomonas. Protoplasma* **106**, 183–192.

Bloodgood, R.A. (1981b). Flagellum as a model system for studying dynamic cell-surface events. *Cold Spring Harb. Symp. Quant. Biol.* **46**, 683–693.

Bloodgood, R.A. (1987). Glycoprotein dynamics in the *Chlamydomonas* flagellar membrane. *Adv. Cell Biol.* **1**, 97–130.

Bloodgood, R.A. (1990). Gliding motility and flagellar glycoprotein dynamics in *Chlamydomonas*. In: *Ciliary and Flagellar Membranes* (R.A. Bloodgood, Ed.), pp. 91–124. Plenum Press, New York and London.

Bloodgood, R.A. and Salomonsky, N.L. (1994). The transmembrane signaling pathway involved in directed movements of *Chlamydomonas* flagellar membrane glycoproteins involves the dephosphorylation of a 60-kD phosphoprotein that binds to the major flagellar membrane glycoprotein. *J. Cell Biol.* **127**, 803–811.

Bloodgood, R.A. and Salomonsky, N.L. (1998). Microsphere attachment induces glycoprotein redistribution and transmembrane signaling in the *Chlamydomonas* flagellum. *Protoplasma* **202**, 76–83.

Bloodgood, R.A., Leffler, E.M., and Bojczuk, A.T. (1979). Reversible inhibition of *Chlamydomonas* flagellar surface motility. *J. Cell Biol.* **82**, 664–674.

Bloodgood, R.A., Woodward, M.P., and Salomonsky, NL. (1986). Redistribution and shedding of flagellar membrane glycoproteins visualized using an anti-carbohydrate monoclonal antibody and concanavalin A. *J. Cell Biol.* **102**, 1797–1812.

Blowers, A.D., Bogorad, L., Shark, K.B., and Sanford, J.C. (1989). Studies on *Chlamydomonas* transformation. Foreign DNA can be stably maintained in the chromosome. *Plant Cell* **1**, 123–132.

Boer, P.H. and Gray, M.W. (1986a). The *URF 5* gene of *Chlamydomonas reinhardtii* mitochondria: DNA sequence and mode of transcription. *EMBO J.* **5**, 21–28.

Boer, P.H. and Gray, M.W. (1986b). Nucleotide sequence of a protein coding region in *Chlamydomonas reinhardtii* mitochondrial DNA. *Nucleic Acids Res.* **14**, 7506–7507.

Boer, P.H. and Gray, M.W. (1988a). Genes encoding a subunit of respiratory NADH dehydrogenase (*ND1*) and a reverse transcriptase-like protein (*RTL*) are linked to ribosomal RNA gene pieces in *Chlamydomonas reinhardtii* mitochondrial DNA. *EMBO J.* **7**, 3501–3508.

Boer, P.H. and Gray, M.W. (1988b). Scrambled ribosomal RNA gene pieces in *Chlamydomonas reinhardtii* mitochondrial DNA. *Cell* **55**, 399–411.

Boer, P.H. and Gray, M.W. (1988c). Transfer RNA genes and the genetic code in *Chlamydomonas reinhardtii* mitochondria. *Curr. Genet.* **14**, 583–590.

Boer, P.H. and Gray, M.W. (1989). Nucleotide sequence of a region encoding subunit 6 of NADH dehydrogenase (ND6) and tRNATrp in *Chlamydomonas reinhardtii* mitochondrial DNA. *Nucleic Acids Res.* **17**, 3993.

Bohne, A.V., Irihimovitch, V., Weihe, A., and Stern, D.B. (2006). *Chlamydomonas reinhardtii* encodes a single sigma[70]-like factor which likely functions in chloroplast transcription. *Curr. Genet.* **49**, 333–340.

Bohne, F. and Linden, H. (2002). Regulation of carotenoid biosynthesis genes in response to light in *Chlamydomonas reinhardtii*. *Biochim. Biophys. Acta* **1579**, 26–34.

Bold, H.C. (1942). The cultivation of algae. *Bot. Rev.* **8**, 69–138.

Bold, H.C. (1949a). The morphology of *Chlamydomonas chlamydogama*, sp. nov. *Bull. Torrey Bot. Club* **76**, 101–108.

Bold, H.C. (1949b). Some cytological aspects of *Chlamydomonas chlamydogama*. *Am. J. Bot.* **36**, 795.

Boldina, O.N. (2000). The ultrastructure of *Chlamydomonas pseudopertusa* (Chlamydomonadaceae, Chlorophyta). *Bot. Zh. St. Petersburg* **85**, 56–60.

Bolen, P.L., Boynton, J.E., and Gillham, N.W. (1980). Evidence for persistence of chloroplast markers in the heteroplasmic state in *Chlamydomonas reinhardtii*. *Curr. Genet.* **2**, 159–167.

Bolen, P.L., Grant, D.M., Swinton, D., Boynton, J.E., and Gillham, N.W. (1982). Extensive methylation of chloroplast DNA by a nuclear gene mutation does not affect chloroplast gene transmission in *Chlamydomonas*. *Cell* **28**, 335–343.

Bollig, K., Lamshöft, M., Schweimer, K., Marner, F.J., Budzikiewicz, H., and Waffenschmidt, S. (2007). Structural analysis of linear hydroxyproline bound O glycans of *Chlamydomonas reinhardtii* – conservation of the inner core in *Chlamydomonas* and land plants. *Carbohydr. Res.* **342**, 2557–2566.

Bonnefoy, N., Remacle, C., and Fox, T.D. (2007). Genetic transformation of *Saccharomyces cerevisiae* and *Chlamydomonas reinhardtii* mitochondria. *Methods Cell Biol.* **80**, 525–548.

Borkhsenious, O.N., Mason, C.B., and Moroney, J.V. (1998). The intracellular localization of ribulose-1,5-bisphosphate carboxylase/oxygenase in *Chlamydomonas reinhardtii. Plant Physiol.* **116**, 1585–1591.

Borowitzka, M.A. (2005). Culturing microalgae in outdoor ponds. In: *Algal Culturing Techniques* (R.A. Andersen, Ed.), pp. 205–218. Elsevier, Amsterdam.

Boscov, J.S. and Feinleib, M.E. (1979). Phototactic response of *Chlamydomonas* to flashes of light. II. Response of individual cells. *Photochem. Photobiol.* **30**, 499–505.

Boswell, C., Sharma, N.C., and Sahi, S.V. (2002). Copper tolerance and accumulation potential of *Chlamydomonas reinhardtii. Bull. Environ. Contam. Toxicol.* **69**, 546–553.

Bothe, H. (1982). Hydrogen production by algae. *Experientia* **38**, 59–64.

Boudreau, E. and Turmel, M. (1996). Extensive gene rearrangements in the chloroplast DNAs of *Chlamydomonas* species featuring multiple dispersed repeats. *Mol. Biol. Evol.* **13**, 233–243.

Boudreau, E., Turmel, M., Goldschmidt-Clermont, M., Rochaix, J.-D., Sivan, S., Michaels, A., and Leu, S. (1997a). A large open reading frame (orf1995) in the chloroplast DNA of *Chlamydomonas reinhardtii* encodes an essential protein. *Mol. Gen. Genet.* **253**, 649–653.

Boudreau, E., Takahashi, Y., Lemieux, C., Turmel, M., and Rochaix, J.D. (1997b). The chloroplast *ycf3* and *ycf4* open reading frames of *Chlamydomonas reinhardtii* are required for the accumulation of the photosystem I complex. *EMBO J.* **16**, 6095–6104.

Boudreau, E., Nickelsen, J., Lemaire, S.D., Ossenbuhl, F., and Rochaix, J.-D. (2000). The *Nac2* gene of *Chlamydomonas* encodes a chloroplast TRP-like protein involved in *psbD* mRNA stability. *EMBO J.* **19**, 3366–3376.

Bourque, D.P., Boynton, J.E., and Gillham, N.W. (1971). Studies on the structure and cellular location of various ribosome and ribosomal RNA species in the green alga *Chlamydomonas reinhardi. J. Cell Sci.* **8**, 153–183.

Bourrelly, P. (1966). *Les Algues d'Eau Douce. Initiation à la Systematique. Tome I: Les Algues Vertes*. Editions N. Boubée & Cie, Paris, 511 pp.

Boussiba, S. and Vonshak, A. (1991). Astaxanthin accumulation in the green alga *Haematococcus pluvialis. Plant Cell Physiol.* **32**, 1077–1082.

Bowers, A.K., Keller, J.A., and Dutcher, S.K. (2003). Molecular markers for rapidly identifying candidate genes in *Chlamydomonas reinhardtii*: *ERY1* and *ERY2* encode chloroplast ribosomal proteins. *Genetics* **164**, 1345–1353.

Bowne, S.W. (1964). Purification of algal cultures with caffeine. *Nature* **204**, 801.

Bowser, S.S. and Bloodgood, R.A. (1984). Evidence against surf-riding as a general mechanism for surface motility. *Cell Motil.* **4**, 305–314.

Boynton, J.E. and Gillham, N.W. (1993). Chloroplast transformation in *Chlamydomonas. Methods Enzymol.* **217**, 510–536.

Boynton, J.E. and Gillham, N.W. (1996). Genetics and transformation of mitochondria in the green alga *Chlamydomonas. Methods Enzymol.* **264**, 279–296.

Boynton, J.E., Gillham, N.W., and Chabot, J.F. (1972). Chloroplast ribosome deficient mutants in the green alga *Chlamydomonas reinhardi* and the question of chloroplast ribosome function. *J. Cell Sci.* **10**, 267–305.

Boynton, J.E., Harris, E.H., Burkhart, B.D., Lamerson, P.M., and Gillham, N.W. (1987). Transmission of mitochondrial and chloroplast genomes in crosses of *Chlamydomonas*. *Proc. Natl. Acad. Sci. U. S. A.* **84**, 2391–2395.

Boynton, J.E., Gillham, N.W., Harris, E.H., Hosler, J.P., Johnson, A.M., Jones, A.R., Randolph-Anderson, B.L., Robertson, D., Klein, T.M., Shark, K.B., and Sanford, J.C. (1988). Chloroplast transformation in *Chlamydomonas* with high velocity microprojectiles. *Science* **240**, 1534–1538.

Brand, J.J., Lien, S., and Wright, J.N. (1989). Hydrogen production by eukaryotic algae. *Biotechnol. Bioeng.* **33**, 1482–1488.

Bray, D.F. and Nakamura, K. (1986). Rectangular arrays of intramembranous particles on freeze-fractured *Chlamydomonas* plasma membranes. *Micron Microsc. Acta* **17**, 1–10.

Bray, D.F., Nakamura, K., Costerton, J.W., and Wagenaar, E.B. (1974). Ultrastructure of *Chlamydomonas eugametos* as revealed by freeze-etching: cell wall, plasmalemma and chloroplast membrane. *J. Ultrastruct. Res.* **47**, 125–141.

Bray, D.F., Nakamura, K., and Wagenaar, E.B. (1983). Spiral arrays: novel intramembranous particle arrays of the plasma membrane of *Chlamydomonas eugametos*. *Micron Microsc. Acta* **14**, 345–349.

Brazelton, W.J., Amundsen, C.D., Silflow, C.D., and Lefebvre, P.A. (2001). The *bld1* mutation identifies the *Chlamydomonas osm-6* homolog as a gene required for flagellar assembly. *Curr. Biol.* **11**, 1591–1594.

Breton, G. and Kay, S.A. (2006). Circadian rhythms lit up in *Chlamydomonas*. *Genome Biol.* **7**, 215.

Bristol, B.M. (1920). On the alga-flora of some desiccated English soils: an important factor in soil biology. *Ann. Bot.* **34**, 35–80.

Brokaw, C.J. and Kamiya, R. (1987). Bending patterns of *Chlamydomonas* flagella. IV. Mutants with defects in inner and outer dynein arms indicate differences in dynein arm function. *Cell Motil. Cytoskeleton* **8**, 68–75.

Brokaw, C.J. and Luck, D.J.L. (1983). Bending patterns of *Chlamydomonas* flagella. I. Wild-type bending patterns. *Cell Motil.* **3**, 131–150.

Brokaw, C.J. and Luck, D.J.L. (1985). Bending patterns of *Chlamydomonas* flagella. III. A radial spoke head deficient mutant and a central pair deficient mutant. *Cell Motil.* **5**, 195–208.

Brokaw, C.J., Luck, D.J.L., and Huang, B. (1982). Analysis of the movement of *Chlamydomonas* flagella: the function of the radial-spoke system is revealed by comparison of wild-type and mutant flagella. *J. Cell Biol.* **92**, 722–732.

Brown, L.E., Sprecher, S.L., and Keller, L.R. (1991). Introduction of exogenous DNA into *Chlamydomonas reinhardtii* by electroporation. *Mol. Cell. Biol.* **11**, 2328–2332.

Brown, R.M., Jr., Larson, D.A., and Bold, H.C. (1964). Airborne algae: their abundance and heterogeneity. *Science* **143**, 583–585.

Brown, R.M., Jr., Johnson, C., and Bold, H.C. (1968). Electron and phase-contrast microscopy of sexual reproduction in *Chlamydomonas moewusii*. *J. Phycol.* **4**, 100–120.

Bruce, V.G. (1970). The biological clock in *Chlamydomonas reinhardi*. *J. Protozool.* **17**, 328–334.

Bruce, V.G. (1972). Mutants of the biological clock in *Chlamydomonas reinhardi*. *Genetics* **70**, 537–548.

Bruce, V.G. (1973). The role of the clock in controlling phototactic rhythms. In: *Behaviour of Micro-organsims* (A. Perez-Miravete, Ed.), pp. 257–266. Plenum, New York.

Bruce, V.G. (1974). Recombinants between clock mutants of *Chlamydomonas reinhardi*. *Genetics* **77**, 221–230.

Bruce, V.G. and Bruce, N.C. (1981). Circadian clock-controlled growth cycle in *Chlamydomonas reinhardi*. In: *International Cell Biology 1980–1981* (H.-G. Schweiger, Ed.), pp. 823–830. Springer-Verlag, Berlin.

Bryant, P.E. and Parker, J. (1977). Evidence for location of the site of accumulation for sub-lethal damage in *Chlamydomonas*. *Int. J. Radiat. Biol.* **32**, 237–246.

Bryant, P.E. and Parker, J. (1978). Absence of repair of sublethal U.V. light damage in *Chlamydomonas*. *Int. J. Radiat. Biol.* **34**, 273–278.

Buchanan, M.J. and Snell, W.J. (1988). Biochemical studies on lysin, a cell wall degrading enzyme released during fertilization in *Chlamydomonas*. *Exp. Cell Res.* **179**, 181–193.

Buchanan, M.J., Imam, S.H., Eskue, W.A., and Snell, W.J. (1989). Activation of the cell wall degrading protease, lysin, during sexual signalling in *Chlamydomonas*: the enzyme is stored as an inactive, higher relative molecular mass precursor in the periplasm. *J. Cell Biol.* **108**, 199–207.

Buchheim, M.A. and Chapman, R.L. (1991). Phylogeny of the colonial green flagellates: a study of 18S and 26S rRNA sequence data. *BioSystems* **25**, 85–100.

Buchheim, M.A. and Chapman, R.L. (1992). Phylogeny of *Carteria* Chlorophyceae inferred from molecular and organismal data. *J. Phycol.* **28**, 362–374.

Buchheim, M.A., Turmel, M., Zimmer, E.A., and Chapman, R.L. (1990). Phylogeny of *Chlamydomonas* (Chlorophyta): an investigation based on cladistic analysis of nuclear 18S rRNA sequence data. *J. Phycol.* **26**, 689–699.

Buchheim, M.A., Lemieux, C., Otis, C., Gutell, R.R., Chapman, R.L., and Turmel, M. (1996). Phylogeny of the Chlamydomonadales (Chlorophyceae): a comparison of ribosomal RNA gene sequences from the nucleus and the chloroplast. *Mol. Phylogenet. Evol.* **5**, 391–402.

Buchheim, M.A., Buchheim, J.A., and Chapman, R.L. (1997a). Phylogeny of *Chloromonas*: a study of 18S ribosomal RNA gene sequences. *J. Phycol.* **33**, 286–293.

Buchheim, M.A., Buchheim, J.A., and Chapman, R.L. (1997b). Phylogeny of the VLE-14 *Chlamydomonas* (Chlorophyceae) group: a study of 18S rRNA gene sequences. *J. Phycol.* **33**, 1024–1030.

Buder, J. (1919). Zur Kenntnis der phototaktischen Richtungsbewegungen. *Jahrb. Wiss. Bot.* **58**, 105–220.

Buerkle, S., Gloeckner, G., and Beck, C.F. (1993). *Chlamydomonas* mutants affected in the light-dependent step of sexual differentiation. *Proc. Natl. Acad. Sci. U. S. A.* **90**, 6981–6985.

Buffaloe, N.D. (1958). A comparative cytological study of four species of *Chlamydomonas*. *Bull. Torrey Bot. Club* **85**, 157–178.

Buléon, A., Gallant, D.J., Bouchet, B., Mouille, G., D'Hulst, C., Kossmann, J., and Ball, S. (1997). Starches from A to C – *Chlamydomonas reinhardtii* as a model microbial system to investigate the biosynthesis of the plant amylopectin crystal. *Plant Physiol.* **115**, 949–957.

Burrascano, C.G. and VanWinkle-Swift, K.P. (1984). Interspecific matings of five *Chlamydomonas* species. *Genetics* **107**, s15.

Burton, W.G., Grabowy, C.T., and Sager, R. (1979). Role of methylation in the modification and restriction of chloroplast DNA in *Chlamydomonas*. *Proc. Natl. Acad. Sci. U. S. A.* **76**, 1390–1394.

Büschlen, S., Choquet, Y., Kuras, R., and Wollman, F.-A. (1991). Nucleotide sequences of the continuous and separated *petA*, *petB* and *petD* chloroplast genes in *Chlamydomonas reinhardtii*. *FEBS Lett.* **284**, 257–262.

Butanaev, A.M. (1994). Use of the hygromycin phosphotransferase gene as the dominant selective marker for *Chlamydomonas reinhardtii* transformation. *Mol. Biol. (Moscow)* **28**, 5–7.

Button, K.S. and Hostetter, H.P. (1977). Copper sorption and release by *Cyclotella meneghiniana* (Bacillariophyceae) and *Chlamydomonas reinhardtii* (Chlorophyceae). *J. Phycol.* **13**, 198–202.

Byrne, T.E., Wells, M.R., and Johnson, C.H. (1992). Circadian rhythms of chemotaxis to ammonium and of methylammonium uptake in *Chlamydomonas*. *Plant Physiol.* **98**, 879–886.

Cain, J. (1965). Nitrogen utilization in 38 freshwater Chlamydomonad algae. *Can. J. Bot.* **43**, 1367–1378.

Cain, J.R. (1979). Survival and mating behaviour of progeny and germination of zygotes from intra- and interspecific crosses of *Chlamydomonas eugametos* and *C. moewusii* (Chlorophyceae, Volvocales). *Phycologia* **18**, 24–29.

Cain, J.R. (1980). Inhibition of zygote germination in *Chlamydomonas moewusii* (Chlorophyceae, Volvocales) by nitrogen deficiency and sodium citrate. *Phycologia* **19**, 184–189.

Cain, J.R. and Allen, R.K. (1980). Use of a cell wall-less mutant strain to assess the role of the cell wall in cadmium and mercury tolerance by *Chlamydomonas reinhardtii*. *Bull. Environ. Contam. Toxicol.* **25**, 797–801.

Cain, J.R. and Cain, R.K. (1984a). Effects of five insecticides on zygospore germination and growth of the green alga *Chlamydomonas moewusii*. *Bull. Environ. Contam. Toxicol.* **33**, 571–574.

Cain, J.R. and Cain, R.K. (1984b). Tools to facilitate isolation of *Chlamydomonas* zygospores and recombinant progeny. *Ann. Bot.* **54**, 445–446.

Camargo, A., Llamas, A., Schnell, R.A., Higuera, J.J., González-Ballester, D., Lefebvre, P.A., Fernández, E., and Galván, A. (2007). Nitrate signaling by the regulatory gene *NIT2* in *Chlamydomonas*. *Plant Cell* **19**, 3491–3503.

Campbell, A.M., Rayala, H.J., and Goodenough, U.W. (1995). The *iso1* gene of *Chlamydomonas* is involved in sex determination. *Mol. Biol. Cell* **6**, 87–95.

Cann, J.P. and Pennick, N.C. (1982). The fine structure of *Chlamydomonas bullosa* Butcher. *Arch. Protistenk.* **125**, 241–248.

Capasso, J.M., Cossio, B.R., Berl, T., Rivard, C.J., and Jimenez, C. (2003). A colorimetric assay for determination of cell viability in algal cultures. *Biomol. Eng.* **20**, 133–138.

Cardol, P., Matagne, R.F., and Remacle, C. (2002). Impact of mutations affecting ND mitochondria-encoded subunits on the activity and assembly of complex I in *Chlamydomonas*. Implication for the structural organization of the enzyme. *J. Mol. Biol.* **319**, 1211–1221.

Cardol, P., González-Halphen, D., Reyes-Prieto, A., Baurain, D., Matagne, R.F., and Remacle, C. (2005). The mitochondrial oxidative phosphorylation proteome of *Chlamydomonas reinhardtii* deduced from the genome sequencing project. *Plant Physiol.* **137**, 447–459.

Carlson, P.S. (1969). Production of auxotrophic mutants in ferns. *Genet. Res.* **14**, 337–339.

Carpita, N.C. (1985). Tensile strength of cell walls of living cells. *Plant Physiol.* **79**, 485–488.

Carroll, J.W., Thomas, J., Dunaway, C., and O'Kelley, J.C. (1970). Light induced synchronization of algal species that divide preferentially in darkness. *Photochem. Photobiol.* **12**, 91–98.

Carter, M.L., Smith, A.C., Kobayashi, H., Purton, S., and Herrin, D.L. (2004). Structure, circadian regulation and bioinformatic analysis of the unique sigma factor gene in *Chlamydomonas reinhardtii*. *Photosynth. Res.* **82**, 339–349.

Casas-Mollano, J.A., van Dijk, K., Eisenhart, J., and Cerutti, H. (2007). SET3p monomethylates histone H3 on lysine 9 and is required for the silencing of tandemly repeated transgenes in *Chlamydomonas*. *Nucleic Acids Res.* **35**, 939–950.

Casey, D.M., Inaba, K., Pazour, G.J., Takada, S., Wakabayashi, K., Wilkerson, C.G., Kamiya, R., and Witman, G.B. (2003a). DC3, the 21-kD subunit of the outer dynein arm-docking complex (ODA-DC), is a novel EF-hand protein important for assembly of both the outer arm and the ODA-DC. *Mol. Biol. Cell* **14**, 3650–3663.

Casey, D.M., Yagi, T., Kamiya, R., and Witman, G.B. (2003b). DC3, the smallest subunit of the *Chlamydomonas* flagellar outer dynein arm-docking complex, is a redox-sensitive calcium-binding protein. *J. Biol. Chem.* **278**, 42652–42659.

Cassin, P.E. (1974). Isolation, growth, and physiology of acidophilic Chlamydomonads. *J. Phycol.* **10**, 439–447.

Catt, J.W. (1979). The isolation and chemical composition of the zygospore cell wall of *Chlamydomonas reinhardii*. *Plant Sci. Lett.* **15**, 69–74.

Catt, J.W., Hills, G.J., and Roberts, K. (1976). A structural glycoprotein, containing hydroxyproline, isolated from the cell wall of *Chlamydomonas reinhardii*. *Planta* **131**, 165–171.

Catt, J.W., Hills, G.J., and Roberts, K. (1978). Cell wall glycoproteins from *Chlamydomonas reinhardii*, and their self-assembly. *Planta* **138**, 91–98.

Catto, P. and Le Gal, Y. (1972). Etude de la replication de l'ADN au cours du cycle végétatif de *Chlamydomonas reinhardtii*. *C. R. Acad. Sci.* **275**, 2081–2084.

Cattolico, R.A. (1978). Nucleic acids. In: *Handbook of Phycological Methods, Vol. II. Physiological and Biochemical Methods* (J.A. Hellebust and J.S. Craigie, Eds.), pp. 81–90. Cambridge University Press, Cambridge.

Caunter, T. and Weinberger, P. (1988). Effects of algae on the aquatic persistence of fenitrothion. *Water Poll. Res. J. Canada* **23**, 388–395.

Cavalier-Smith, T. (1970). Electron microscopic evidence for chloroplast fusion in zygotes of *Chlamydomonas reinhardii*. *Nature* **228**, 333–335.

Cavalier-Smith, T. (1974). Basal body and flagellar development during the vegetative cell cycle and the sexual cycle of *Chlamydomonas reinhardii*. *J. Cell Sci.* **16**, 529–556.

Cavalier-Smith, T. (1975). Electron and light microscopy of gametogenesis and gamete fusion in *Chlamydomonas reinhardii*. *Protoplasma* **86**, 1–18.

Cavalier-Smith, T. (1976). Electron microscopy of zygospore formation in *Chlamydomonas reinhardii*. *Protoplasma* **87**, 297–315.

Cenkci, B., Petersen, J.L., and Small, G.D. (2003). *REX1*, a novel gene required for DNA repair. *J. Biol. Chem.* **278**, 22574–22577.

Cepák, V., Zobacová, M., and Zachleder, V. (2002). The effect of cadmium ions on the cell cycle of the green flagellate *Chlamydomonas noctigama*. *Arch. Hydrobiol. Suppl* **144**, 117–129.

Cerutti, H. and Casas-Mollano, J.A. (2006). On the origin and functions of RNA-mediated silencing: from protists to man. *Curr. Genet.* **50**, 81–99.

Cerutti, H., Johnson, A.M., Boynton, J.E., and Gillham, N.W. (1995). Inhibition of chloroplast DNA recombination and repair by dominant negative mutants of *Escherichia coli* RecA. *Mol. Cell. Biol.* **15**, 3003–3011.

Cerutti, H., Johnson, A.M., Gillham, N.W., and Boynton, J.E. (1997a). A eubacterial gene conferring spectinomycin resistance on *Chlamydomonas reinhardtii*: integration into the nuclear genome and gene expression. *Genetics* **145**, 97–110.

Cerutti, H., Johnson, A.M., Gillham, N.W., and Boynton, J.E. (1997b). Epigenetic silencing of a foreign gene in nuclear transformants of *Chlamydomonas*. *Plant Cell* **9**, 925–945.

Chang, C.H. and Wu, M. (2000). The effects of transcription and RNA processing on the initiation of chloroplast DNA replication in *Chlamydomonas reinhardtii*. *Mol. Gen. Genet.* **263**, 320–327.

Chang, C.W., Moseley, J.L., Wykoff, D., and Grossman, A.R. (2005). The *LPB1* gene is important for acclimation of *Chlamydomonas reinhardtii* to phosphorus and sulfur deprivation. *Plant Physiol.* **138**, 319–329.

Chang, M., Li, F., Odom, O.W., Lee, J., and Herrin, D.L. (2003). A cosmid vector containing a dominant selectable marker for cloning *Chlamydomonas* genes by complementation. *Plasmid* **49**, 75–78.

Chankova, S.G., Dimova, E., Dimitrova, M., and Bryant, P.E. (2007). Induction of DNA double-strand breaks by zeocin in *Chlamydomonas reinhardtii* and the role of increased DNA double-strand breaks rejoining in the formation of an adaptive response. *Radiat. Environ. Biophys.* **46**, 409–416.

Chapman, R.L. and Buchheim, M.A. (1992). Green algae and the evolution of land plants: inferences from nuclear-encoded rRNA gene sequences. *BioSystems* **28**, 127–137.

Charlesworth, B. (1983). Mating types and uniparental transmission of chloroplast genes. *Nature* **304**, 211.

Chekounova, E., Voronetskaya, V., Papenbrock, J., Grimm, B., and Beck, C.F. (2001). Characterization of *Chlamydomonas* mutants defective in the H subunit of Mg-chelatase. *Mol. Gen. Genet.* **266**, 363–373.

Chen, H.C. and Melis, A. (2004). Localization and function of SulP, a nuclear-encoded chloroplast sulfate permease in *Chlamydomonas reinhardtii*. *Planta* **220**, 198–210.

Chen, X., Kindle, K., and Stern, D. (1993). Initiation codon mutations in the *Chlamydomonas* chloroplast *petD* gene result in temperature-sensitive photosynthetic growth. *EMBO J.* **12**, 3627–3635.

Chen, X.M., Kindle, K.L., and Stern, D.B. (1995). The initiation codon determines the efficiency but not the site of translation initiation in *Chlamydomonas* chloroplasts. *Plant Cell* **7**, 1295–1305.

Cheshire, J.L., Evans, J.H., and Keller, L.R. (1994). Ca^{2+} signaling in the *Chlamydomonas* flagellar regeneration system. *J. Cell Sci.* **107**, 2491–2498.

Chiang, K.-S. (1968). Physical conservation of parental cytoplasmic DNA through meiosis in *Chlamydomonas reinhardi*. *Proc. Natl. Acad. Sci. U. S. A.* **60**, 194–200.

Chiang, K.-S. (1971). Replication, transmission and recombination of cytoplasmic DNAs in *Chlamydomonas reinhardi*. In: *Autonomy and Biogenesis of Mitochondria and Chloroplasts* (N.K. Boardman, A.W. Linnane, and R.M. Smillie, Eds.), pp. 235–249. Elsevier/North-Holland, Amsterdam.

Chiang, K.-S. and Sueoka, N. (1967a). Replication of chloroplast DNA in *Chlamydomonas reinhardi* during vegetative cell cycle: its mode and regulation. *Proc. Natl. Acad. Sci. U. S. A.* **57**, 1506–1513.

Chiang, K.-S. and Sueoka, N. (1967b). Replication of chromosomal and cytoplasmic DNA during mitosis and meiosis in the eucaryote *Chlamydomonas reinhardi*. *J. Cell. Physiol.* **70**, 89–112.

Chiang, K.-S., Kates, J.R., Jones, R.F., and Sueoka, N. (1970). On the formation of a homogeneous zygotic population in *Chlamydomonas reinhardti*. *Dev. Biol.* **22**, 655–669.

Chiang, K.-S., Eves, E., and Swinton, D. (1975). Variation of thymidine incorporation patterns in the alternating vegetative and sexual life cycles of *Chlamydomonas reinhardtii*. *Dev. Biol.* **42**, 53–63.

Chiang, K.-S., Friedman, E., Malavasic, M.J., Jr., Feng, M.L., Eves, E.M., Feng, T.Y., and Swinton, D.C. (1981). On the folding and organization of chloroplast DNA in *Chlamydomonas reinhardtii*. *Ann. N. Y. Acad. Sci.* **361**, 219–247.

Chiu, S.M. and Hastings, P.J. (1973). Pre-meiotic DNA synthesis and recombination in *Chlamydomonas reinhardi*. *Genetics* **73**, 29–43.

Choquet, Y. and Vallon, O. (2000). Synthesis, assembly and degradation of thylakoid membrane proteins. *Biochimie* **82**, 615–634.

Choquet, Y. and Wollman, F.A. (2002). Translational regulations as specific traits of chloroplast gene expression. *FEBS Lett.* **529**, 39–42.

Choquet, Y., Goldschmidt-Clermont, M., Girard-Bascou, J., Kück, U., Bennoun, P., and Rochaix, J.-D. (1988). Mutant phenotypes support a *trans*-splicing mechanism for the expression of the tripartite *psaA* gene in the *C. reinhardtii* chloroplast. *Cell* **52**, 903–913.

Choquet, Y., Rahire, M., Girard-Bascou, J., Erickson, J., and Rochaix, J.-D. (1992). A chloroplast gene is required for the light-independent accumulation of chlorophyll in *Chlamydomonas reinhardtii*. *EMBO J.* **11**, 1697–1704.

Choquet, Y., Wostrikoff, K., Rimbault, B., Zito, F., Girard-Bascou, J., Drapier, D., and Wollman, F.A. (2001). Assembly-controlled regulation of chloroplast gene translation. *Biochem. Soc. Trans.* **29**, 421–426.

Chorin-Kirsh, I. and Mayer, A.M. (1964a). Induction of phototaxis in *Chlamydomonas snowiae* by indolyl-3-acetic acid and ethylene-diamine tetraacetic acid. *Nature* **203**, 1085–1086.

Chorin-Kirsh, I. and Mayer, A.M. (1964b). ATP-ase activity in isolated flagella of *Chlamydomonas snowiae*. *Plant Cell Physiol.* **5**, 441–445.

Christensen, E.R. and Zielski, P.A. (1980). Toxicity of arsenic and PCB to a green alga (*Chlamydomonas*). *Bull. Environ. Contam. Toxicol.* **25**, 43–48.

Chua, N.-H. (1976). A uniparental mutant of *Chlamydomonas reinhardtii* with a variant thylakoid membrane polypeptide. In: *Genetics and Biogenesis of Chloroplasts and Mitochondria* (Th. Bücher, W. Neupert, W. Sebald, and S. Werner, Eds.), pp. 323–330. Elsevier/North-Holland, Amsterdam.

Chua, N.-H. and Gillham, N.W. (1977). The sites of synthesis of the principal thylakoid membrane polypeptides in *Chlamydomonas reinhardtii*. *J. Cell Biol.* **74**, 441–452.

Chua, N.-H., Blobel, G., and Siekevitz, P. (1973a). Isolation of cytoplasmic and chloroplast ribosomes and their dissociation into active subunits from *Chlamydomonas reinhardtii*. *J. Cell Biol.* **57**, 798–814.

Chua, N.-H., Blobel, G., Siekevitz, P., and Palade, G.E. (1973b). Attachment of chloroplast polysomes to thylakoid membranes in *Chlamydomonas reinhardtii*. *Proc. Natl. Acad. Sci. U. S. A.* **70**, 1554–1558.

Chun, E.H.L., Vaughan, M.H., Jr., and Rich, A. (1963). The isolation and characterization of DNA associated with chloroplast preparations. *J. Mol. Biol.* **7**, 130–141.

Chunayev, A.S., Ladygin, V.G., Kornyushenko, G.A., Gaevskii, N.A., and Mirnaya, O.N. (1987). Chlorophyll *b*-less mutants in *Chlamydomonas reinhardtii*. *Photosynthetica* **21**, 301–307.

Cinco, R.M., MacInnis, J.M., and Greenbaum, E. (1993). The role of carbon dioxide in light-activated hydrogen production by *Chlamydomonas reinhardtii*. *Photosynth. Res.* **38**, 27–33.

Claes, H. (1971). Autolyse der Zellwand bei den Gameten von *Chlamydomonas reinhardii*. *Arch. Mikrobiol.* **78**, 180–188.

Claes, H. (1977). Non-specific stimulation of the autolytic system in gametes from *Chlamydomonas reinhardii*. *Exp. Cell Res.* **108**, 221–229.

Clarke, K.J. and Leeson, E.A. (1985). Plasmalemma structure in freezing tolerant unicellular algae. *Protoplasma* **129**, 120–126.

Cohen, A., Yohn, C.B., and Mayfield, S.P. (2001). Translation of the chloroplast-encoded *psbD* mRNA is arrested post-initiation in a nuclear mutant of *Chlamydomonas reinhardtii*. *J. Plant Physiol.* **158**, 1069–1075.

Cohen, I., Knopf, J.A., Irihimovitch, V., and Shapira, M. (2005). A proposed mechanism for the inhibitory effects of oxidative stress on Rubisco assembly and its subunit expression. *Plant Physiol.* **137**, 738–746.

Cole, D.G., Diener, D.R., Himelblau, A.L., Beech, P.L., Fuster, J.C., and Rosenbaum, J.L. (1988). *Chlamydomonas* kinesin-II-dependent intraflagellar transport (IFT): IFT particles contain proteins required for ciliary assembly in *Caenorhabditis elegans* sensory neurons. *J. Cell Biol.* **141**, 993–1008.

Coleman, A.W. (1962). Sexuality. In: *Biochemistry and Physiology of Algae* (R.A. Lewin, Ed.), pp. 711–729. Academic Press, New York.

Coleman, A.W. (1975). Long-term maintenance of fertile algal clones: experience with *Pandorina* (Chlorophyccac). *J. Phycol.* **11**, 282–286.

Coleman, A.W. (1978). Visualization of chloroplast DNA with two fluorochromes. *Exp. Cell Res.* **114**, 95–100.

Coleman, A.W. (1982a). The nuclear cell cycle in *Chlamydomonas* (Chlorophyceae). *J. Phycol.* **18**, 192–195.

Coleman, A.W. (1982b). Sex is dangerous in a world of potential symbionts or the basis of selection for uniparental inheritance. *J. Theor. Biol.* **97**, 367–369.

Coleman, A.W. (1983). The roles of resting spores and akinetes in chlorophyte survival. In: *Survival Strategies of the Algae* (G.A. Fryxell, Ed.), pp. 1–21. Cambridge University Press, Cambridge.

Coleman, A.W. (1984). The fate of chloroplast DNA during cell fusion, zygote maturation and zygote germination in *Chlamydomonas reinhardi* as revealed by DAPI staining. *Exp. Cell Res.* **152**, 528–540.

Coleman, A.W. and Maguire, M.J. (1983). Cytological detection of the basis of uniparental inheritance of plastid DNA in *Chlamydomonas moewusii. Curr. Genet.* **7**, 211–218.

Coleman, A.W. and Mai, J.C. (1997). Ribosomal DNA ITS-1 and ITS-2 sequence comparisons as a tool for predicting genetic relatedness. *J. Mol. Evol.* **45**, 168–177.

Coleman, A.W. and Pröschold, T. (2005). Control of sexual reproduction in algae in culture. In: *Algal Culturing Techniques* (R.A. Andersen, Ed.), pp. 389–397. Elsevier, Amsterdam.

Coleman, A.W., Preparata, R.M., Mehrotra, B., and Mai, J.C. (1998). Derivation of the secondary structure of the ITS-1 transcript in Volvocales and its taxonomic correlations. *Protist* **149**, 135–146.

Coleman, A.W., Jaenicke, L., and Starr, R.C. (2001). Genetics and sexual behavior of the pheromone producer *Chlamydomonas allensworthii* (Chlorophyceae). *J. Phycol.* **37**, 345–349.

Colin, M., Dorthu, M.P., Duby, F., Remacle, C., Dinant, M., Wolwertz, M.R., Duyckaerts, C., Sluse, F., and Matagne, R.F. (1995). Mutations affecting the mitochondrial genes encoding the cytochrome oxidase subunit I and apocytochrome *b* of *Chlamydomonas reinhardtii. Mol. Gen. Genet.* **249**, 179–184.

Collard, J.-M. and Matagne, R.F. (1990). Isolation and genetic analysis of *Chlamydomonas reinhardtii* strains resistant to cadmium. *Appl. Environ. Microbiol.* **56**, 2051–2055.

Collcaux, L., Michcl-Wolwcrtz, M.-R., Matagnc, R.F., and Dujon, B. (1990). Thc apocytochrome *b* gene of *Chlamydomonas smithii* contains a mobile intron related to both *Saccharomyces* and *Neurospora* introns. *Mol. Gen. Genet.* **223**, 288–296.

Colleoni, C., Dauvillée, D., Mouille, G., Buleón, A., Gallant, D., Bouchet, B., Morell, M., Samucl, M., Delrue, B., D'Hulst, C., Bliard, C., Nuzillard, J.M., and Ball, S. (1999a). Genetic and biochemical evidence for the involvement of a-1,4 glucanotransferases in amylopectin synthesis. *Plant Physiol.* **120**, 993–1004.

Colleoni, C., Dauvillée, D., Mouille, G., Morell, M., Samuel, M., Slomiany, M.C., Liénard, L., Wattebled, F., D'Hulst, C., and Ball, S. (1999b). Biochemical characterization of the *Chlamydomonas reinhardtii* a-1,4 glucanotransferase supports a direct function in amylopectin biosynthesis. *Plant Physiol.* **120**, 1005–1013.

Collin-Osdoby, P., Adair, W.S., and Goodenough, U.W. (1984). *Chlamydomonas* agglutinin conjugated to agarose beads as an *in vitro* probe of adhesion. *Exp. Cell Res.* **150**, 282–291.

Collins, S., Sültemeyer, D., and Bell, G. (2006a). Changes in C uptake in populations of *Chlamydomonas reinhardtii* selected at high CO_2. *Plant Cell Environ.* **29**, 1812–1819.

Collins, S., Sültemeyer, D., and Bell, G. (2006b). Rewinding the tape: selection of algae adapted to high CO_2 at current and pleistocene levels of CO_2. *Evolution* **60**, 1392–1401.

Collins, S., Sültemeyer, D., and Bell, G. (2007). Changes in C uptake in populations of *Chlamydomonas reinhardtii* selected at high CO_2. *Plant Cell Environ.* **30**, 1812–1819.

Colón-Ramos, D.A., Salisbury, J.L., Sanders, M.A., Shenoy, S.M., Singer, R.H., and García-Blanco, M.A. (2003). Asymmetric distribution of nuclear pore complexes and the cytoplasmic localization of beta2-tubulin mRNA in *Chlamydomonas reinhardtii. Dev. Cell* **4**, 941–952.

Conde, M.A.F. (1974). Cellular sites of resistance in two uniparentally-inherited streptomcyin-resistant mutants of *Chlamydomonas reinhardtii*. Ph.D. dissertation, Duke University, Durham NC.

Conde, M.F., Boynton, J.E., Gillham, N.W., Harris, E.H., Tingle, C.L., and Wang, W.L. (1975). Chloroplast genes in *Chlamydomonas* affecting organelle ribosomes. Genetic and biochemical analysis of antibiotic-resistant mutants at several gene loci. *Mol. Gen. Genet.* **140**, 183–220.

Conner, A.J. (1981). The differential sensitivity of phytoplankton to polychlorinated biphenyls when cultured heterotrophically and photoautotrophically. *Environ. Exp. Bot.* **21**, 241–247.

Cooper, J.B., Adair, W.S., Mecham, R.P., Heuser, J.E., and Goodenough, U.W. (1983). *Chlamydomonas* agglutinin is a hydroxyproline-rich glycoprotein. *Proc. Natl. Acad. Sci. U. S. A.* **80**, 5898–5901.

Coss, R.A. (1974). Mitosis in *Chlamydomonas reinhardtii*. Basal bodies and the mitotic apparatus. *J. Cell Biol.* **63**, 325–329.

Coughlan, S. (1977). The effect of organic substrates on the growth, photosynthesis and dark survival of marine algae. *Br. Phycol. J.* **12**, 155–162.

Cournac, L., Latouche, G., Cerovic, Z., Redding, K., Ravenel, J., and Peltier, G. (2002). *In vivo* interactions between photosynthesis, mitorespiration, and chlororespiration in *Chlamydomonas reinhardtii. Plant Physiol.* **129**, 1921–1928.

Cox, J.L. and Small, G.D. (1985). Isolation of a photoreactivation-deficient mutant of *Chlamydomonas. Mutat. Res.* **146**, 249–255.

Coyne, B. and Rosenbaum, J.L. (1970). Flagellar elongation and shortening in *Chlamydomonas*. II. Re-utilization of flagellar proteins. *J. Cell Biol.* **47**, 777–781.

Crabbendam, K.J., Nanninga, N., Musgrave, A., and van den Ende, H. (1984). Flagellar tip activation in *vis-à-vis* pairs of *Chlamydomonas eugametos. Arch. Microbiol.* **138**, 220–223.

Crabbendam, K.J., Klis, F.M., Musgrave, A., and van den Ende, H. (1986). Ultrastructure of the plus and minus mating-type sexual agglutinins of *Chlamydomonas eugametos*, as visualized by negative staining. *J. Ultrastruct. Mol. Struct. Res.* **96**, 151–159.

Craigie, R.A. and Cavalier-Smith, T. (1982). Cell volume and the control of the *Chlamydomonas* cell cycle. *J. Cell Sci.* **54**, 173–191.

Cramer, W.A., Yan, J., Zhang, H., Kurisu, G., and Smith, J.L. (2005). Structure of the cytochrome b_6f complex: new prosthetic groups, Q-space, and the 'hors d'oeuvres hypothesis' for assembly of the complex. *Photosynth. Res.* **85**, 133–143.

Crayton, M.A. (1982). A comparative cytochemical study of Volvocacean matrix polysaccharides. *J. Phycol.* **18**, 336–344.

Crescitelli, F., James, T.W., Erickson, J.M., Loew, E.R., and McFarland, W.N. (1992). The eyespot of *Chlamydomonas reinhardtii*. A comparative microspectrophotometric study. *Vision Res.* **32**, 1593–1600.

Croft, M.T., Warren, M.J., and Smith, A.G. (2006). Algae need their vitamins. *Eukaryotic Cell* **5**, 1175–1183.

Croft, M.T., Moulin, M., Webb, M.E., and Smith, A.G. (2007). Thiamine biosynthesis in algae is regulated by riboswitches. *Proc. Natl. Acad. Sci. U. S. A.* **104**, 20770–20775.

Crutchfield, A.L.M., Diller, K.R., and Brand, J.J. (1999). Cryopreservation of *Chlamydomonas reinhardtii* (Chlorophyta). *Eur. J. Phycol.* **34**, 43–52.

Cruz, J.A., Salbilla, B.A., Kanazawa, A., and Kramer, D.M. (2001). Inhibition of plastocyanin to P700+ electron transfer in *Chlamydomonas reinhardtii* by hyperosmotic stress. *Plant Physiol.* **127**, 1167–1179.

Cullen, M., Ray, N., Husain, S., Nugent, J., Nield, J., and Purton, S. (2007). A highly active histidine-tagged *Chlamydomonas reinhardtii* Photosystem II preparation for structural and biophysical analysis. *Photochem. Photobiol. Sci.* **6**, 1177–1183.

Cunningham, A. (1984). The impulse response of *Chlamydomonas reinhardii* in nitrite-limited chemostat culture. *Biotechnol. Bioeng.* **26**, 1430–1435.

Cunningham, A. and Maas, P. (1978). Time lag and nutrient storage effects in the transient growth response of *Chlamydomonas reinhardtii* in nitrogen-limited batch and continuous culture. *J. Gen. Microbiol.* **104**, 227–231.

Cunningham, A. and Nisbet, R.M. (1980). Time lag and co-operativity in the transient growth dynamics of microalgae. *J. Theor. Biol.* **84**, 189–203.

Curry, A.M., Williams, B.D., and Rosenbaum, J.L. (1992). Sequence analysis reveals homology between two proteins of the flagellar radial spoke. *Mol. Cell. Biol.* **12**, 3967–3977.

Curtis, V.A., Brand, J.J., and Togasaki, R.K. (1975). Partial reactions of photosynthesis in briefly sonicated *Chlamydomonas*. I. Cell breakage and electron transport activities. *Plant Physiol.* **55**, 183–186.

Czirók, A., Jánosi, I.M., and Kessler, J.O. (2000). Bioconvective dynamics: dependence on organism behaviour. *J. Exp. Biol.* **203**, 3345–3354.

Czurda, V. (1935). Über die Variabilität von *Chlamydomonas eugametos* Moewus. *Beih. Bot. Zentralbl.* **53/A**, 133–157.

Czygan, F.-C. (1970). Blutregen und Blutschnee: Stickstoffmangel-Zellen von *Haematococcus pluvialis* und *Chlamydomonas nivalis*. *Arch. Mikrobiol.* **74**, 69–76.

Dacks, J.B. and Doolittle, W.F. (2002). Novel syntaxin gene sequences from *Giardia*, *Trypanosoma* and algae: implications for the ancient evolution of the eukaryotic endomembrane system. *J. Cell Sci.* **115**, 1635–1642.

Dame, G., Gloeckner, G., and Beck, C.F. (2002). Knock-out of a putative transporter results in altered blue-light signalling in *Chlamydomonas*. *Plant J.* **31**, 577–587.

Dangeard, P.-A. (1888). Recherches sur les algues inferieures. *Ann. Sci. Nat., 7th ser., Bot.* **4**, 105–175.

Dangeard, P.-A. (1899). Memoire sur les Chlamydomonadinees ou l'histoire d'une cellule. *Le Botaniste* **6**, 65–292.

Daniel, P., Henley, J., and VanWinkle-Swift, K. (2007). Altered zygospore wall ultrastructure correlates with reduced abiotic stress resistance in a mutant strain of *Chlamydomonas monoica* (Chlorophyta). *J. Phycol.* **43**, 112–119.

Darlix, J.-L. and Rochaix, J.-D. (1981). Nucleotide sequence and structure of cytoplasmic 5S RNA and 5.8S RNA of *Chlamydomonas reinhardtii*. *Nucleic Acids Res.* **9**, 1291–1299.

Dauvillée, D., Colleoni, C., Shaw, E., Mouille, G., D'Hulst, C., Morell, M., Samuel, M.S., Bouchet, B., Gallant, D.J., Sinskey, A., and Ball, S. (1999). Novel, starch-like polysaccharides are synthesized by an unbound form of granule-bound starch synthase in glycogen-accumulating mutants of *Chlamydomonas reinhardtii*. *Plant Physiol.* **119**, 321–329.

Dauvillée, D., Colleoni, C., Mouille, G., Buléon, A., Gallant, D.J., Bouchet, B., Morell, M.K., D'Hulst, C., Myers, A.M., and Ball, S.G. (2001a). Two loci control phytoglycogen production in the monocellular green alga *Chlamydomonas reinhardtii*. *Plant Physiol.* **125**, 1710–1722.

Dauvillée, D., Colleoni, C., Mouille, G., Morell, M.K., D'Hulst, C., Wattebled, F., Liénard, L., Delvalle, D., Ral, J.P., Myers, A.M., and Ball, S.G. (2001b). Biochemical characterization of wild-type and mutant isoamylases of *Chlamydomonas reinhardtii* supports a function of the multimeric enzyme organization in amylopectin maturation. *Plant Physiol.* **125**, 1723–1731.

Dauvillée, D., Hilbig, L., Preiss, S., and Johanningmeier, U. (2004). Minimal extent of sequence homology required for homologous recombination at the *psbA* locus in *Chlamydomonas reinhardtii* chloroplasts using PCR-generated DNA fragments. *Photosynth. Res.* **79**, 219–224.

Davidson, J.N., Hanson, M.R., and Bogorad, L. (1978). Erythromycin resistance and the chloroplast ribosome in *Chlamydomonas reinhardi. Genetics* **89**, 281–297.

Davies, D.R. (1967a). UV-sensitive mutants of *Chlamydomonas reinhardi. Mutat. Res.* **4**, 765–770.

Davies, D.R. (1967b). The control of dark repair mechanisms in meiotic cells. *Mol. Gen. Genet.* **100**, 140–149.

Davies, D.R. (1972). Electrophoretic analyses of wall glycoproteins in normal and mutant cells. *Exp. Cell Res.* **73**, 512–516.

Davies, D.R. and Lawrence, C.W. (1967). The mechanism of recombination in *Chlamydomonas reinhardi.* II. The influence of inhibitors of DNA synthesis on intergenic recombination. *Mutat. Res.* **4**, 147–154.

Davies, D.R. and Lyall, V. (1973). The assembly of a highly ordered component of the cell wall: the role of heritable factors and of physical structure. *Mol. Gen. Genet.* **124**, 21–34.

Davies, D.R. and Plaskitt, A. (1971). Genetical and structural analyses of cell-wall formation in *Chlamydomonas reinhardi. Genet. Res.* **17**, 33–43.

Davies, D.R., Holt, P.D., and Papworth, D.G. (1969). The survival curves of haploid and diploid *Chlamydomonas reinhardtii* exposed to radiations of different LET. *Int. J. Radiat. Biol.* **15**, 75–87.

Davies, J.P., Weeks, D.P., and Grossman, A.R. (1992). Expression of the arylsulfatase gene from the beta2-tubulin promoter in *Chlamydomonas reinhardtii. Nucleic Acids Res.* **20**, 2959–2965.

Davies, J.P., Yildiz, F., and Grossman, AR. (1994). Mutants of *Chlamydomonas* with aberrant responses to sulfur deprivation. *Plant Cell* **6**, 53–63.

Davies, J.P., Yildiz, F.H., and Grossman, A.R. (1996). Sac1, a putative regulator that is critical for survival of *Chlamydomonas reinhardtii* during sulfur deprivation. *EMBO J.* **15**, 2150–2159.

Davis, B.D. (1948). Isolation of biochemically deficient mutants of bacteria by penicillin. *J. Am. Chem. Soc.* **70**, 4267.

Day, A. (1995). A transposon-like sequence with short terminal inverted repeats in the nuclear genome of *Chlamydomonas reinhardtii. Plant Mol. Biol.* **28**, 437–442.

Day, A. and Rochaix, J.-D. (1989). Characterization of transcribed dispersed repetitive DNAs in the nuclear genome of the green alga *Chlamydomonas reinhardtii. Curr. Genet.* **16**, 165–176.

Day, A. and Rochaix, J.-D. (1991a). A transposon with an unusual LTR arrangement from *Chlamydomonas reinhardtii* contains an internal tandem array of 76 bp repeats. *Nucleic Acids Res.* **19**, 1259–1266.

Day, A. and Rochaix, J.-D. (1991b). Conservation in structure of *TOC1* transposons from *Chlamydomonas reinhardtii. Gene* **104**, 235–239.

Day, A. and Rochaix, J.-D. (1991c). Structure and inheritance of sense and anti-sense transcripts from a transposon in the green alga *Chlamydomonas reinhardtii. J. Mol. Biol.* **218**, 273–291.

Day, A., Schirmer-Rahire, M., Kuchka, M.R., Mayfield, S.P., and Rochaix, J.-D. (1988). A transposon with an unusual arrangement of long terminal repeats in the green alga *Chlamydomonas reinhardtii. EMBO J.* **7**, 1917–1927.

Day, J.G. and Brand, J.J. (2005). Cryopreservation methods for maintaining microalgal cultures. In: *Algal Culturing Techniques* (R.A. Andersen, Ed.), pp. 165–187. Elsevier, Amsterdam.

de Hostos, E.L., Togasaki, R.K., and Grossman, A. (1988). Purification and biosynthesis of a derepressible periplasmic arylsulfatase from *Chlamydomonas reinhardtii*. *J. Cell Biol.* **106**, 29–37.

de Hostos, E.L., Schilling, J., and Grossman, A.R. (1989). Structure and expression of the gene encoding the periplasmic arylsulfatase of *Chlamydomonas reinhardtii*. *Mol. Gen. Genet.* **218**, 229–239.

de Vitry, C. (1994). Characterization of the gene of the chloroplast Rieske iron-sulfur protein in *Chlamydomonas reinhardtii*. Indications for an uncleaved lumen targeting sequence. *J. Biol. Chem.* **269**, 7603–7609.

de Vitry, C., Olive, J., Drapier, D., Recouvreur, M., and Wollman, F.-A. (1989). Post-translational events leading to the assembly of photosystem II protein complex: a study using photosynthesis mutants from *Chlamydomonas reinhardtii*. *J. Cell Biol.* **109**, 991–1006.

de Vitry, C., Breyton, C., Pierre, Y., and Popot, J.-L. (1996). The 4-kDa nuclear-encoded PetM polypeptide of the chloroplast cytochrome $b_6 f$ complex: nucleic acid and protein sequences, targeting signals, transmembrane topology. *J. Biol. Chem.* **271**, 10667–10671.

de Vitry, C., Finazzi, G., Baymann, F., and Kallas, T. (1999). Analysis of the nucleus-encoded and chloroplast-targeted Rieske protein by classic and site-directed mutagenesis of *Chlamydomonas*. *Plant Cell* **11**, 2031–2044.

Deane, J.A., Cole, D.G., Seeley, E.S., Diener, D.R., and Rosenbaum, J.L. (2001). Localization of intraflagellar transport protein IFT52 identifies basal body transitional fibers as the docking site for IFT particles. *Curr. Biol.* **11**, 1586–1590.

Deason, T.R. (1967). *Chlamydomonas gymnogama*, a new homothallic species with naked gametes. *J. Phycol.* **3**, 109–112.

Deason, T.R. and Bold, H.C. (1960). Phycological studies. I. Exploratory studies of Texas soil algae. *University of Texas publication No. 6022.* 72 pp.

Deason, T.R. and Ratnasabapathy, M. (1976). A new homothallic variety of *Chlamydomonas moewusii* (Chlorophyceae). *J. Phycol.* **12**, 82–85.

Debuchy, R., Purton, S., and Rochaix, J.-D. (1989). The argininosuccinate lyase gene of *Chlamydomonas reinhardtii*: an important tool for nuclear transformation and for correlating the genetic and molecular maps of the *ARG7* locus. *EMBO J.* **8**, 2803–2809.

Dechacin, C., Weinberger, P., and Czuba, M. (1991). The interaction between cells in different phases of the cell cycle and aminocarb. *Ecotoxicol. Environ. Saf.* **21**, 25–31.

Decottignies, P., Le Marechal, P., Jacquot, J.P., Schmitter, J.M., and Gadal, P. (1995). Primary structure and post-translational modification of ferredoxin-NADP reductase from *Chlamydomonas reinhardtii*. *Arch. Biochem. Biophys.* **316**, 249–259.

Deininger, W., Kröger, P., Hegemann, U., Lottspeich, F., and Hegemann, P. (1995). Chlamyrhodopsin represents a new type of sensory photoreceptor. *EMBO J.* **14**, 5849–5858.

Del Campo, J.A., García-González, M., and Guerrero, M.G. (2007). Outdoor cultivation of microalgae for carotenoid production: current state and perspectives. *Appl. Microbiol. Biotechnol.* **74**, 1163–1174.

Delcourt, A. and Mestre, J.C. (1978). The effects of phenylmercuric acetate on the growth of *Chlamydomonas variabilis* Dang. *Bull. Environ. Contam. Toxicol.* **20**, 145–148.

Delrue, B., Fontaine, T., Routier, F., Decq, A., Wieruszeski, J.M., van den Koornhuyse, N., Maddelein, M.-L., Fournet, B., and Ball, S. (1992). Waxy *Chlamydomonas reinhardtii*: monocellular algal mutants defective in amylose biosynthesis and granule-bound starch synthase activity accumulate a structurally modified amylopectin. *J. Bacteriol.* **174**, 3612–3620.

Denning, G.M. and Fulton, A.B. (1989a). Electron microscopy of a contractile-vacuole mutant of *Chlamydomonas moewusii* (Chlorophyta) defective in the late stages of diastole. *J. Phycol.* **25**, 667–672.

Denning, G.M. and Fulton, A.B. (1989b). Purification and characterization of clathrin-coated vesicles from *Chlamydomonas*. *J. Protozool.* **36**, 334–340.

Denovan-Wright, E.M. and Lee, R.W. (1994). Comparative structure and genomic organization of the discontinuous mitochondrial ribosomal RNA genes of *Chlamydomonas eugametos* and *Chlamydomonas reinhardtii*. *J. Mol. Biol.* **241**, 298–311.

Denovan-Wright, E.M., Sankoff, D., Spencer, D.F., and Lee, R.W. (1996). Evolution of fragmented mitochondrial ribosomal RNA genes in *Chlamydomonas*. *J. Mol. Evol.* **42**, 382–391.

Dent, R.M., Haglund, C.M., Chin, B.L., Kobayashi, M.C., and Niyogi, K.K. (2005). Functional genomics of eukaryotic photosynthesis using insertional mutagenesis of *Chlamydomonas reinhardtii*. *Plant Physiol.* **137**, 545–556.

Dentler, W.L. (1980). Structures linking the tips of ciliary and flagellar microtubules to the membrane. *J. Cell Sci.* **42**, 207–220.

Dentler, W.L. (1990). Linkages between microtubules and membranes in cilia and flagella. In: *Ciliary and Flagellar Membranes* (R.A. Bloodgood, Ed.), pp. 31–64. Plenum Press, New York.

Dentler, W.L. and Rosenbaum, J.L. (1977). Flagellar elongation and shortening in *Chlamydomonas*. III. Structures attached to the tips of flagellar microtubules and their relationship to the directionality of flagellar microtubule assembly. *J. Cell Biol.* **74**, 747–759.

Derguini, F., Mazur, P., Nakanishi, K., Starace, D.M., Saranak, J., and Foster, K.W. (1991). All-*trans* retinal is the chromophore bound to the photoreceptor of the alga *Chlamydomonas reinhardtii*. *Photochem. Photobiol.* **5**, 1017–1022.

Desborough, S. and Shult, E.E. (1961). Tetrad analysis in *Chlamydomonas*. *Can. J. Genet. Cytol.* **3**, 325–339.

Deshpande, N.N., Hollingsworth, M., and Herrin, D.L. (1997). Evidence for light/redox-regulated splicing of *psbA* pre-RNAs in *Chlamydomonas* chloroplasts. *RNA* **3**, 37–48.

Desroche, P. (1912). *Reactions des Chlamydomonas aux Agents Physiques*. Schulz, Paris, 158 pp.

Detmers, P.A., Goodenough, U.W., and Condeelis, J. (1983). Elongation of the fertilization tubule in *Chlamydomonas*: new observations on the core microfilaments and the effect of transient intracellular signals on their structural integrity. *J. Cell Biol.* **97**, 522–532.

Detmers, P.A., Carboni, J.M., and Condeelis, J. (1985). Localization of actin in *Chlamydomonas* using antiactin and NBD-phallacidin. *Cell Motility* **5**, 415–430.

Devos, N., Ingouff, M., Loppes, R., and Matagne, R.F. (1998). Rubisco adaptation to low temperatures: a comparative study in psychrophilic and mesophilic unicellular algae. *J. Phycol.* **34**, 655–660.

Díaz-Troya, S., Florencio, F.J., and Crespo, J.L. (2008). Target of rapamycin and LST8 proteins associate with membranes from the endoplasmic reticulum in the unicellular green alga *Chlamydomonas reinhardtii*. *Eukaryotic Cell* **7**, 212–222.

DiBella, L.M., Gorbatyuk, O., Sakato, M., Wakabayashi, K.I., Patel-King, R.S., Pazour, G.J., Witman, G.B., and King, S.M. (2005). Differential light chain assembly influences outer arm dynein motor function. *Mol. Biol. Cell* **16**, 5661–5674.

Dieckmann, C.L. (2003). Eyespot placement and assembly in the green alga *Chlamydomonas*. *BioEssays* **25**, 410–416.

Diener, D.R., Curry, A.M., Johnson, K.A., Williams, B.D., Lefebvre, P.A., Kindle, K.L., and Rosenbaum, J.L. (1990). Rescue of a paralyzed-flagella mutant of *Chlamydomonas* by transformation. *Proc. Natl. Acad. Sci. U. S. A.* **87**, 5739–5743.

Diener, D.R., Ang, L.H., and Rosenbaum, J.L. (1993). Assembly of flagellar radial spoke proteins in *Chlamydomonas*: identification of the axoneme binding domain of radial spoke protein 3. *J. Cell Biol.* **123**, 183–190.

Dill, O. (1895). Die Gattung *Chlamydomonas* und ihre nächsten Verwandten. *Jahrb. Wiss. Bot.* **28**, 323–358.

Ding, Y., Miao, J.L., Li, G.Y., Wang, Q.F., Kan, G.F., and Wang, G.D. (2005). Effect of Cd on GSH and GSH-related enzymes of *Chlamydomonas* sp. ICE-L existing in Antarctic ice. *J. Environ. Sci.* **17**, 667–671.

Dionisio, M.L., Tsuzuki, M., and Miyachi, S. (1989). Blue light induction of carbonic anhydrase activity in *Chlamydomonas reinhardtii*. *Plant Cell Physiol.* **30**, 215–219.

Dionisio-Sese, M.L., Fukuzawa, H., and Miyachi, S. (1990). Light-induced carbonic anhydrase expression in *Chlamydomonas reinhardtii*. *Plant Physiol.* **94**, 1103–1110.

Domozych, D.S. (1989). The endomembrane system and mechanism of membrane flow in the green alga *Gloeomonas kupfferi* Volvocales Chlorophyta. II. A cytochemical analysis. *Protoplasma* **149**, 109–119.

Domozych, D.S. and Dairman, M. (1993). Synthesis of the inner cell wall layer of the Chlamydomonad flagellate, *Gloeomonas kupfferi*. *Protoplasma* **176**, 1–13.

Domozych, D.S. and Nimmons, T.T. (1992). The contractile vacuole as an endocytic organelle of the Chlamydomonad flagellate *Gloeomonas kupfferi* (Volvocales, Chlorophyta). *J. Phycol.* **28**, 809–816.

Domozych, D.S., Stewart, K.D., and Mattox, K.R. (1980). The comparative aspects of cell wall chemistry in the green algae (Chlorophyta). *J. Mol. Evol.* **15**, 1–12.

Domozych, D.S., Wells, B., and Shaw, P.J. (1992). The cell wall of the Chlamydomonad flagellate *Gloeomonas kupfferi* (Volvocales, Chlorophyta). *Protoplasma* **168**, 95–106.

Donnan, L. and John, P.C.L. (1983). Cell cycle control by timer and sizer in *Chlamydomonas*. *Nature* **304**, 630–633.

Donnan, L., Carvill, E.P., Gilliland, T.J., and John, P.C.L. (1985). The cell cycles of *Chlamydomonas* and *Chlorella*. *New Phytol.* **99**, 1–40.

Doonan, J.H. and Grief, C. (1987). Microtubule cycle in *Chlamydomonas reinhardtii*: an immunofluorescence study. *Cell Motil. Cytoskeleton* **7**, 381–392.

Dorthu, M.-P., Remy, S., Michel-Wolwertz, M.-R., Colleaux, L., Breyer, D., Beckers, M.-C., Englebert, S., Duyckaerts, C., Sluse, F.E., and Matagne, R.F. (1992). Biochemical, genetic and molecular characterization of new respiratory-deficient mutants in *Chlamydomonas reinhardtii*. *Plant Mol. Biol.* **18**, 759–772.

Drapier, D., Girard-Bascou, J., and Wollman, F.-A. (1992). Evidence for nuclear control of the expression of the *atpA* and *atpB* chloroplast genes in *Chlamydomonas*. *Plant Cell* **4**, 283–295.

Drapier, D., Girard-Bascou, J., Stern, D.B., and Wollman, F.A. (2002). A dominant nuclear mutation in *Chlamydomonas* identifies a factor controlling chloroplast mRNA stability by acting on the coding region of the *atpA* transcript. *Plant J.* **31**, 687–697.

Drapier, D., Rimbault, B., Vallon, O., Wollman, F.A., and Choquet, Y. (2007). Intertwined translational regulations set uneven stoichiometry of chloroplast ATP synthase subunits. *EMBO J.* **26**, 3581–3591.

Dreyfuss, B.W. and Merchant, S. (1999). *CCS5*, a new locus required for *c*-type cytochrome synthesis. In: *Photosynthesis: Mechanisms and Effects*, Vol. IV (G. Garab, Ed.), pp. 3221–3226. Kluwer Academic, Dordrecht.

Dreyfuss, B.W., Hamel, P.P., Nakamoto, S.S., and Merchant, S. (2003). Functional analysis of a divergent system II protein, Ccs1, involved in c-type cytochrome biogenesis. *J. Biol. Chem.* **278**, 2604–2613.

Dron, M., Robreau, G., and Le Gal, Y. (1979). Isolation of a chromoid from the chloroplast of *Chlamydomonas reinhardi. Exp. Cell Res.* **119**, 301–305.

Dron, M., Rahire, M., and Rochaix, J.-D. (1982a). Sequence of the chloroplast DNA region of *Chlamydomonas reinhardii* containing the gene of the large subunit of ribulose bisphosphate carboxylase and parts of its flanking genes. *J. Mol. Biol.* **162**, 775–793.

Dron, M., Rahire, M., and Rochaix, J.-D. (1982b). Sequence of the chloroplast 16S rRNA gene and its surrounding regions of *Chlamydomonas reinhardii. Nucleic Acids Res.* **10**, 7609–7620.

Dron, M., Rahire, M., Rochaix, J.-D., and Mets, L. (1983). First DNA sequence of a chloroplast mutation: a missense alteration in the ribulosebisphosphate carboxylase large subunit gene. *Plasmid* **9**, 321–324.

Droop, M.R. (1961). *Haematococcus pluvialis* and its allies. III. Organic nutrition. *Rev. Algol.* **5**, 247–259.

Droop, M.R. (1962). Organic micronutrients. In: *Biochemistry and Physiology of Algae* (R.A. Lewin, Ed.), pp. 141–149. Academic Press, New York.

Droop, M.R. (1974). Heterotrophy of carbon. In: *Algal Physiology and Biochemistry* (W.D.P. Stewart, Ed.), pp. 530–559. Blackwell, Oxford.

Droop, M.R. and McGill, S. (1966). The carbon nutrition of some algae: the inability to utilize glycollic acid for growth. *J. Mar. Biol. Assoc. U.K.* **46**, 679–684.

Duby, F. and Matagne, R.F. (1999). Alteration of dark respiration and reduction of phototrophic growth in a mitochondrial DNA deletion mutant of *Chlamydomonas* lacking *cob, nd4*, and the 3′ end of *nd5. Plant Cell* **11**, 115–125.

Duby, F., Cardol, P., Matagne, R.F., and Remacle, C. (2001). Structure of the telomeric ends of mt DNA, transcriptional analysis and complex I assembly in the *dum24* mitochondrial mutant of *Chlamydomonas reinhardtii. Mol. Gen. Genet.* **266**, 109–114.

Dujon, B., Slonimski, P.P., and Weill, L. (1974). Mitochondrial genetics IX. A model for recombination and segregation of mitochondrial genomes in *Saccharomyces cerevisiae. Genetics* **78**, 415–437.

Dumont, F., Loppes, R., and Kremers, P. (1990). New polypeptides and *in vitro* translatable messenger RNAs are produced by phosphate-starved cells of the unicellular alga *Chlamydomonas reinhardtii. Planta* **182**, 610–616.

Dunahay, T.G. (1993). Transformation of *Chlamydomonas reinhardtii* with silicon carbide whiskers. *BioTechniques* **15**, 452–460.

Dunahay, T.G., Adler, S.A., and Jarvik, J.W. (1997). Transformation of microalgae using silicon carbide whiskers. *Methods Mol. Biol.* **62**, 503–509.

Dusek, D.M. and Preston, J.F., III (1988). Isolation of amatoxin-resistant lines of *Chlamydomonas reinhardtii. Plant Physiol.* **87**, 286–290.

Dutcher, S.K. (1988). Nuclear fusion-defective phenocopies in *Chlamydomonas reinhardtii*: mating-type functions for meiosis can act through the cytoplasm. *Proc. Natl. Acad. Sci. U. S. A.* **85**, 3946–3950.

Dutcher, S.K. (2003). Long-lost relatives reappear: identification of new members of the tubulin superfamily. *Curr. Opin. Microbiol.* **6**, 634–640.

Dutcher, S.K. and Harris, E.H. (1998). *Chlamydomonas reinhardtii. Trends in Genetics, Genetic Nomenclature Guide*, pp. S18–S19.

Dutcher, S.K. and Trabuco, E.C. (1998). The *UNI3* gene is required for assembly of basal bodies of *Chlamydomonas* and encodes delta-tubulin, a new member of the tubulin superfamily. *Mol. Biol. Cell* **9**, 1293–1308.

Dutcher, S.K., Huang, B., and Luck, D.J.L. (1984). Genetic dissection of the central pair microtubules of the flagella of *Chlamydomonas reinhardtii. J. Cell Biol.* **98**, 229–236.

Dutcher, S.K., Gibbons, W., and Inwood, W.B. (1988). A genetic analysis of suppressors of the *PF10* mutation in *Chlamydomonas reinhardtii. Genetics* **120**, 965–976.

Dutcher, S.K., Power, J., Galloway, R.E., and Porter, M.E. (1991). Reappraisal of the genetic map of *Chlamydomonas reinhardtii*. *J. Hered.* **82**, 295–301.

Dutcher, S.K., Morrissette, N.S., Preble, A.M., Rackley, C., and Stanga, J. (2002). Epsilon-tubulin is an essential component of the centriole. *Mol. Biol. Cell* **13**, 3859–3869.

Duval, B., Shetty, K., and Thomas, W.H. (1999). Phenolic compounds and antioxidant properties in the snow alga *Chlamydomonas nivalis* after exposure to UV light. *J. Appl. Phycol.* **11**, 559–566.

Dyer, T.A. (1982). Methylation of chloroplast DNA in *Chlamydomonas*. *Nature* **298**, 422–423.

Dymek, E.E., Lefebvre, P.A., and Smith, E.F. (2004). PF15p is the *Chlamydomonas* homologue of the katanin p80 subunit and is required for assembly of flagellar central microtubules. *Eukaryotic Cell* **3**, 870–879.

Dymek, E.E., Goduti, D., Kramer, T., and Smith, E.F. (2006). A kinesin-like calmodulin binding protein in *Chlamydomonas*: evidence for a role in cell division and flagellar function. *J. Cell Sci.* **119**, 3107–3116.

Eberhard, S., Jain, M., Im, C.S., Pollock, S., Shrager, J., Lin, Y., Peek, A.S., and Grossman, A.R. (2006). Generation of an oligonucleotide array for analysis of gene expression in *Chlamydomonas reinhardtii*. *Curr. Genet.* **49**, 106–124.

Ebersold, W.T. (1956). Crossing over in *Chlamydomonas reinhardi*. *Am. J. Bot.* **43**, 408–410.

Ebersold, W.T. (1962). Biochemical genetics. In: *Biochemistry and Physiology of Algae* (R.A. Lewin, Ed.), pp. 731–739. Academic Press, New York.

Ebersold, W.T. (1963). Heterozygous diploid strains of *Chlamydomonas reinhardi*. *Genetics* **48**, 888.

Ebersold, W.T. (1967). *Chlamydomonas reinhardi*: heterozygous diploid strains. *Science* **157**, 447–449.

Ebersold, W.T. and Levine, R.P. (1959). A genetic analysis of linkage group I of *Chlamydomonas reinhardi*. *Z. Vererbs.* **90**, 74–82.

Ebersold, W.T., Levine, R.P., Levine, E.E., and Olmsted, M.A. (1962). Linkage maps in *Chlamydomonas reinhardi*. *Genetics* **47**, 531–543.

Ebnet, E., Fischer, M., Deininger, W., and Hegemann, P. (1999). Volvoxrhodopsin, a light-regulated sensory photoreceptor of the spheroidal green alga *Volvox carteri*. *Plant Cell* **11**, 1473–1484.

Egashira, T., Takahama, U., and Nakamura, K. (1989). A reduced activity of catalase as a basis for light dependent methionine sensitivity of a *Chlamydomonas reinhardtii* mutant. *Plant Cell Physiol.* **30**, 1171–1175.

Ehara, M., Inagaki, Y., Watanabe, K.I., and Ohama, T. (2000). Phylogenetic analysis of diatom *coxI* genes and implications of a fluctuating GC content on mitochondrial genetic code evolution. *Curr. Genet.* **37**, 29–33.

Ehara, T., Osafune, T., and Hase, E. (1995). Behavior of mitochondria in synchronized cells of *Chlamydomonas reinhardtii* (Chlorophyta). *J. Cell Sci.* **108**, 499–507.

Ehlenbeck, S., Gradmann, D., Braun, F.J., and Hegemann, P. (2002). Evidence for a light-induced H$^+$ conductance in the eye of the green alga *Chlamydomonas reinhardtii*. *Biophys. J.* **82**, 740–751.

Ehler, L.L. and Dutcher, S.K. (1998). Pharmacological and genetic evidence for a role of rootlet and phycoplast microtubules in the positioning and assembly of cleavage furrows in *Chlamydomonas reinhardtii*. *Cell Motil. Cytoskeleton* **40**, 193–207.

Ehler, L.L., Holmes, J.A., and Dutcher, S.K. (1995). Loss of spatial control of the mitotic spindle apparatus in a *Chlamydomonas reinhardtii* mutant strain lacking basal bodies. *Genetics* **141**, 945–960.

Ehrenberg, C.G. (1833). Über die Entwicklung und Lebensdauer der Infusionsthiere. *Abhandl. Akad. Wiss. Berlin* **1833**, 288.

Ehrenberg, C.G. (1838). Die Infusionsthierchen als vollkommene Organismen. Leipzig, 547 pp.

Eichenberger, W. (1976). Lipids of *Chlamydomonas reinhardi* under different growth conditions. *Phytochemistry* **15**, 459–463.

Eichenberger, W. (1982). Distribution of diacylglyceryl-O-4′-(*N,N,N*-trimethyl) homoserine in different algae. *Plant Sci. Lett.* **24**, 91–95.

Eichenberger, W. and Boschetti, A. (1978). Occurrence of 1(3),2-diacylglyceryl-(3)-O-4′-(*N,N,N*-trimethyl)-homoserine in *Chlamydomonas reinhardi*. *FEBS Lett.* **88**, 201–204.

Eichenberger, W., Schaffner, J.-C., and Boschetti, A. (1977). Characterization of proteins and lipids of photosystem I and II particles from *Chlamydomonas reinhardi*. *FEBS Lett.* **84**, 144–148.

Eisenberg-Domovich, Y., Oelmuller, R., Hermmann, R.G., and Ohad, I. (1994). Reversible membrane association of heat-shock protein 22 in *Chlamydomonas reinhardtii* during heat shock and recovery. *Eur. J. Biochem.* **222**, 1041–1046.

El-Naggar, A.H. (1998). Toxic effects of nickel on photosystem II of *Chlamydomonas reinhardtii*. *Cytobios* **93**, 93–101.

Ellis, R.J. (1972). Control of palmelloid formation in the green alga *Pediastrum*. *Plant Cell Physiol.* **13**, 663–672.

Elrad, D. and Grossman, A.R. (2004). A genome's-eye view of the light-harvesting polypeptides of *Chlamydomonas reinhardtii*. *Curr. Genet.* **45**, 61–75.

Ender, F., Godl, K., Wenzl, S., and Sumper, M. (2002). Evidence for autocatalytic cross-linking of hydroxyproline-rich glycoproteins during extracellular matrix assembly in *Volvox*. *Plant Cell* **14**, 1147–1160.

Eom, H., Lee, C.G., and Jin, E. (2006). Gene expression profile analysis in astaxanthin-induced *Haematococcus pluvialis* using a cDNA microarray. *Planta* **223**, 1231–1242.

Erickson, J.M., Rahire, M., Bennoun, P., Delepelaire, P., Diner, B., and Rochaix, J.-D. (1984a). Herbicide sistance in *Chlamydomonas reinhardtii* results from a mutation in the chloroplast gene for the 32-kilodalton protein of photosystem II. *Proc. Natl. Acad. Sci. U. S. A.* **81**, 3617–3621.

Erickson, J.M., Rahire, M., and Rochaix, J.-D. (1984b). *Chlamydomonas reinhardii* gene for the 32 000 mol.wt. protein of photosystem II contains four large introns and is located entirely within the chloroplast inverted repeat. *EMBO J.* **3**, 2753–2762.

Erickson, J.M., Rahire, M., Rochaix, J.-D., and Mets, L. (1985). Herbicide resistance and cross-resistance: changes at three distinct sites in the herbicide-binding protein. *Science* **228**, 204–207.

Erickson, J.M., Rahire, M., Malnoë, P., Girard-Bascou, J., Pierre, Y., Bennoun, P., and Rochaix, J.-D. (1986). Lack of the D2 protein in a *Chlamydomonas reinhardtii psbD* mutant affects photosystem II stability and D1 expression. *EMBO J.* **5**, 1745–1754.

Erickson, J.M., Pfister, K., Rahire, M., Togasaki, R.K., Mets, L., and Rochaix, J.-D. (1989). Molecular and biophysical analysis of herbicide-resistant mutants of *Chlamydomonas reinhardtii*: structure-function relationship of the Photosystem II D1 polypeptide. *Plant Cell* **1**, 361–372.

Erickson, J.M., Whitelegge, J., Koo, D., and Boyd, K. (1992). Site-directed mutagenesis of the Photosystem II D1 gene and transformation of the *Chlamydomonas* chloroplast genome. In: *Research in Photosynthesis*, Vol. 3 (N. Murata, Ed.), pp. 421–424. Kluwer Academic, Dordrecht.

Eriksson, M., Gardestrom, P., and Samuelsson, G. (1995). Isolation, purification, and characterization of mitochondria from *Chlamydomonas reinhardtii*. *Plant Physiol.* **107**, 479–483.

Erlbaum, P.A. (1968). Heterotrophic capabilities of 2 acidophilic Chlamydomonads. *J. Protozool.* **15**, 20–21.

Ermilova, E.V., Zalutskaya, Z.M., and Gromov, B.V. (1993). Chemotaxis towards sugars in *Chlamydomonas reinhardtii*. *Curr. Microbiol.* **27**, 47–50.

Ermilova, E.V., Chekunova, E.M., Zalutskaya, Z.M., Krupnov, K.R., and Gromov, B.V. (1996). Isolation and characterization of chemotaxis mutants of *Chlamydomonas reinhardtii*. *Curr. Microbiol.* **32**, 357–359.

Ermilova, E.V., Krupnoz, K.R., Sotnikov, A.G., and Gromov, B.V. (1999). Production and characterization of an insertion mutant of *Chlamydomonas reinhardtii* with impaired chemotaxis towards sucrose. *Fiziol. Rast.* **46**, 82–86.

Ermilova, E.V., Zalutskaya, Z.M., Gromov, B.V., Häder, D.P., and Purton, S. (2000). Isolation and characterisation of chemotactic mutants of *Chlamydomonas reinhardtii* obtained by insertional mutagenesis. *Protist* **151**, 127–137.

Ermilova, E.V., Zalutskaya, Z.M., Lapina, T.V., and Nikitin, M.M. (2003). Chemotactic behavior of *Chlamydomonas reinhardtii* is altered during gametogenesis. *Curr. Microbiol.* **46**, 261–264.

Ermilova, E.V., Zalutskaya, Z.M., Huang, K., and Beck, C.F. (2004). Phototropin plays a crucial role in controlling changes in chemotaxis during the initial phase of the sexual life cycle in *Chlamydomonas*. *Planta* **219**, 420–427.

Ermilova, E.V., Nikitin, M.M., and Fernández, E. (2007). Chemotaxis to ammonium/ methylammonium in *Chlamydomonas reinhardtii*: the role of transport systems for ammonium/methylammonium. *Planta* **226**, 1323–1332.

Erwin, J.A. (1973). Comparative biochemistry of fatty acids in eukaryotic microorganisms. In: *Lipids and Biomembranes of Eukaryotic Microorganisms* (J.A. Erwin, Ed.), pp. 41–143. Academic Press, New York.

Ettl, H. (1965a). *Gloeomonas* oder *Chlamydomonas*? *Nova Hedwigia* **9**, 291–297.

Ettl, H. (1965b). Beitrag zur Kenntnis der Morphologie der Gattung *Chlamydomonas* Ehrenberg. *Arch. Protistenk.* **108**, 271–430.

Ettl, H. (1965c). Über systematische Probleme in der Gattung *Chlamydomonas*. *Phycologia* **5**, 61–70.

Ettl, H. (1967). *Chlamydomonas parkeae*, eine neue marine Chlamydomonade. *Int. Rev. Ges. Hydrobiol.* **52**, 437–440.

Ettl, H. (1969). Über einen gelappten Chromatophor bie *Chlamydomonas geitleri* nova spec., seine Entwicklung und Vereinfachung während der Fortpflanzung *Oesterr. Bot. Z.* **116**, 127–144.

Ettl, H. (1970). Die Gattung *Chloromonas* Gobi emend. Wille. *Beihefte zur Nova Hedwigia* **34**, 1–283.

Ettl, H. (1976a). Die Gattung *Chlamydomonas* Ehrenberg. *Beihefte zur Nova Hedwigia* **49**, 1–1122.

Ettl, H. (1976b). Über den Teilungsverlauf des Chloroplasten bei *Chlamydomonas*. *Protoplasma* **88**, 75–84.

Ettl, H. (1979). Die Protoplastenteilung und der Verlauf der Chloroplastenteilung während der asexuellen Fortpflanzung von *Chlamydomonas* (Volvocales, Chlorophyceae). *Arch. Hydrobiol. Suppl.* **56**, 40–113.

Ettl, H. (1981). Die neue Klasse Chlamydophyceae, eine natürliche Gruppe der Grünalgen (Chlorophyta). *Plant Syst. Evol.* **137**, 107–126.

Ettl, H. (1985). Probleme der Variabilitäts-Untersuchungen bei Phytomonadina. *Arch. Hydrobiol. Suppl.* **71**, 603–610.

Ettl, H. (1988). Unterschiedliche Teilungsverläufe bei den Phytomonaden (Chlorophyta). *Arch. Protistenk.* **135**, 85–101.

Ettl, H. and Green, J.C. (1973). *Chlamydomonas reginae sp. nov.* (Chlorophyceae), a new marine flagellate with unusual chloroplast differentiation. *J. Mar. Biol. Assoc. U.K.* **53**, 975–985.

Ettl, H. and Schlösser, U.G. (1992). Towards a revision of the systematics of the genus *Chlamydomonas* Chlorophyta. 1. *Chlamydomonas applanata* Pringsheim. *Bot. Acta* **105**, 323–330.

Evans, J.H. and Keller, L.R. (1997a). Calcium influx signals normal flagellar RNA induction following acid shock of *Chlamydomonas reinhardtii*. *Plant Mol. Biol.* **33**, 467–481.

Evans, J.H. and Keller, L.R. (1997b). Receptor-mediated calcium influx in *Chlamydomonas reinhardtii*. *J. Euk. Microbiol.* **44**, 237–245.

Evans, M.C.W., Purton, S., Patel, V., Wright, D., Heathcote, P., and Rigby, S.E.J. (1999). Modification of electron transfer from the quinone electron carrier, A1, of Photosystem 1 in a site directed mutant D576 L within the Fe-S$_x$ binding site of PsaA and in second site suppressors of the mutation in *Chlamydomonas reinhardtii*. *Photosynth. Res.* **61**, 33–42.

Eversole, R.A. (1956). Biochemical mutants of *Chlamydomonas reinhardi*. *Am. J. Bot.* **43**, 404–407.

Eves, E.M. and Chiang, K.-S. (1982). Genetics of *Chlamydomonas reinhardtii* diploids. I. Isolation and characterization and meiotic segregation pattern of a homozygous diploid. *Genetics* **100**, 35–60.

Eves, E.M. and Chiang, K.-S. (1984). Genetics of *Chlamydomonas reinhardtii* diploids. II. The effects of diploidy and aneuploidy on the transmission of non-Mendelian markers. *Genetics* **107**, 563–576.

Fabry, S., Mueller, K., Lindauer, A., Park, P.B., Cornelius, T., and Schmitt, R. (1995). The organization structure and regulatory elements of *Chlamydomonas* histone genes reveal features linking plant and animal genes. *Curr. Genet.* **28**, 333–345.

Fan, B., Li, Y.Z., and Zhai, Z.H. (1994). The keratin-like intermediate filament network exists in the *Chlamydomonas sp.* pyrenoid. *Acta Bot. Sin.* **36**, 518–521.

Fan, L., Vonshak, A., Zarka, A., and Boussiba, S. (1998). Does astaxanthin protect *Haematococcus* against light damage? *Z. Naturforsch.* **53c**, 93–100.

Fan, J.S., Schnare, M.N., and Lee, R.W. (2003). Characterization of fragmented mitochondrial ribosomal RNAs of the colorless green alga *Polytomella parva*. *Nucleic Acids Res.* **31**, 769–778.

Fan, W.H., Woelfle, M.A., and Mosig, G. (1995). Two copies of a DNA element, 'Wendy', in the chloroplast chromosome of *Chlamydomonas reinhardtii* between rearranged gene clusters. *Plant Mol. Biol.* **29**, 63–80.

Fang, S.C., Reyes, C.D., and Umen, J.G. (2006). Cell size checkpoint control by the retinoblastoma tumor suppressor pathway. *PLoS Genet.* **2**, e167.

Faragó, A. and Dénes, G. (1968). Acetate activation in *Chlamydomonas reinhardti*. I. Partial purification and characterization of acetate kinase. *Acta Biochim. Biophys. Acad. Sci. Hung.* **3**, 3–12.

Farah, J.A., Rappaport, F., Choquet, Y., Joliot, P., and Rochaix, J.-D. (1995a). Isolation of a *psaF*-deficient mutant of *Chlamydomonas reinhardtii*: efficient interaction of plastocyanin with the photosystem I reaction center is mediated by the PsaF subunit. *EMBO J.* **14**, 4976–4984.

Farah, J.A., Frank, G., Zuber, H., and Rochaix, J.-D. (1995b). Cloning and sequencing of a cDNA clone encoding the photosystem I PsaD subunit from *Chlamydomonas reinhardtii*. *Plant Physiol.* **107**, 1485–1486.

Fargo, D.C., Zhang, M., Gillham, N.W., and Boynton, J.E. (1998). Shine-Dalgarno-like sequences are not required for translation of chloroplast mRNAs in *Chlamydomonas reinhardtii* chloroplasts or in *Escherichia coli*. *Mol. Gen. Genet.* **257**, 271–282.

Farooqui, P.B. (1974). Taxonomy of *Chlamydomonas eugametos* Moewus and *Chlamydomonas hydra* Ettl. *Arch. Protistenk.* **116**, 185–191.

Fawley, M.W. and Buchheim, M.A. (1995). Loroxanthin, a phylogenetically useful character in *Chlamydomonas* and other Chlorophycean flagellates. *J. Phycol.* **31**, 664–667.

Fedtke, C., Depka, B., Schallner, O., Tietjen, K., Trebst, A., Wollweber, D., and Wroblowsky, H.J. (2001). Mode of action of new diethylamines in lycopene cyclase inhibition and in photosystem II turnover. *Pest Manag. Sci.* **57**, 278–282.

Fei, X. and Deng, X. (2007). A novel Fe deficiency responsive element (FeRE) regulates the expression of *atx1* in *Chlamydomonas reinhardtii*. *Plant Cell Physiol.* **48**, 1496–1503.

Feinleib, M.E. (1984). Behavioral studies of free-swimming photoresponsive organisms. In: *Sensory Perception and Transduction in Aneural Organisms* (G. Colombetti, F. Lenci, and P. S. Song, Eds.), pp. 119–146. *NATO Adv. Study Inst. Ser. A.* Vol. 89. Plenum Press, London.

Feinleib, M.E.H. and Curry, G.M. (1971). The relationship between stimulus intensity and oriented phototactic response (topotaxis) in *Chlamydomonas*. *Physiol. Plant.* **25**, 346–352.

Feng, T.-Y. and Chiang, K.-S. (1984). The persistence of maternal inheritance in *Chlamydomonas* despite hypomethylation of chloroplast DNA induced by inhibitors. *Proc. Natl. Acad. Sci. U. S. A.* **81**, 3438–3442.

Fennikoh, K.B., Hirshfield, H.I., and Kneip, T.J. (1978). Cadmium toxicity in planktonic organisms of a freshwater food web. *Environ. Res.* **15**, 357–367.

Fernández, E. and Aguilar, M. (1987). Molybdate repair of molybdopterin deficient mutants from *Chlamydomonas reinhardtii*. *Curr. Genet.* **12**, 349–355.

Fernández, E. and Cárdenas, J. (1981). Occurrence of xanthine dehydrogenase in *Chlamydomonas reinhardii*: a common cofactor shared by xanthine dehydrogenase and nitrate reductase. *Planta* **153**, 254–257.

Fernández, E. and Cárdenas, J. (1982). Regulation of the nitrate-reducing system enzymes in wild-type and mutant strains of *Chlamydomonas reinhardii*. *Mol. Gen. Genet.* **186**, 164–169.

Fernández, E. and Galván, A. (2007). Inorganic nitrogen assimilation in *Chlamydomonas*. *J. Exp. Bot.* **58**, 2279–2287.

Fernández, E. and Matagne, R.F. (1984). Genetic analysis of nitrate reductase-deficient mutants in *Chlamydomonas reinhardii*. *Curr. Genet.* **8**, 635–640.

Fernández, E., Schnell, R., Ranum, L.P.W., Hussey, S.C., Silflow, C.D., and Lefebvre, P.A. (1989). Isolation and characterization of the nitrate reductase structural gene of *Chlamydomonas reinhardtii*. *Proc. Natl. Acad. Sci. U. S. A.* **86**, 6449–6453.

Ferris, P.J. (1989). Characterization of a *Chlamydomonas* transposon, *Gulliver*, resembling those in higher plants. *Genetics* **122**, 363–377.

Ferris, P.J. and Goodenough, U.W. (1987). Transcription of novel genes, including a gene linked to the mating-type locus, induced by *Chlamydomonas* fertilization. *Mol. Cell. Biol.* **7**, 2360–2366.

Ferris, P.J. and Goodenough, U.W. (1994). The mating-type locus of *Chlamydomonas reinhardtii* contains highly rearranged DNA sequences. *Cell* **76**, 1135–1145.

Ferris, P.J. and Goodenough, U.W. (1997). Mating type in *Chlamydomonas* is specified by *mid*, the minus-dominance gene. *Genetics* **146**, 859–869.

Ferris, P.J., Woessner, J.P., and Goodenough, U.W. (1996). A sex recognition glycoprotein is encoded by the plus mating-type gene *fus1* of *Chlamydomonas reinhardtii*. *Mol. Biol. Cell* **7**, 1235–1248.

Ferris, P.J., Pavlovic, C., Fabry, S., and Goodenough, U.W. (1997). Rapid evolution of sex-related genes in *Chlamydomonas*. *Proc. Natl. Acad. Sci. U. S. A.* **94**, 8634–8639.

Ferris, P.J., Woessner, J.P., Waffenschmidt, S., Kilz, S., Drees, J., and Goodenough, U. W. (2001). Glycosylated polyproline II rods with kinks as a structural motif in plant hydroxyproline-rich glycoproteins. *Biochemistry* **40**, 2978–2987.

Ferris, P.J., Armbrust, E.V., and Goodenough, U.W. (2002). Genetic structure of the mating-type locus of *Chlamydomonas reinhardtii*. *Genetics* **160**, 181–200.

Ferris, P.J., Waffenschmidt, S., Umen, J.G., Lin, H., Lee, J.H., Ishida, K., Kubo, T., Lau, J., and Goodenough, U.W. (2005). *Plus* and *minus* sexual agglutinins from *Chlamydomonas reinhardtii*. *Plant Cell* **17**, 597–615.

Fiedler, H.R., Schmid, R., Leu, S., Shavit, N., and Strotmann, H. (1995). Isolation of CF_0CF_1 from *Chlamydomonas reinhardtii cw15* and the N-terminal amino acid sequences of the CF_0CF_1 subunits. *FEBS Lett.* **377**, 163–166.

Fiedler, H.R., Schlesinger, J., Strotmann, H., Shavit, N., and Leu, S. (1997). Characterization of *atpA* and *atpB* deletion mutants produced in *Chlamydomonas reinhardtii cw15*: electron transport and photophosphorylation activities of isolated thylakoids. *Biochim. Biophys. Acta* **1319**, 109–118.

Finazzi, G., Büschlen, S., de Vitry, C., Rappaport, F., Joliot, P., and Wollman, F.A. (1997). Function-directed mutagenesis of the cytochrome b_6f complex in *Chlamydomonas reinhardtii*: involvement of the cd loop of cytochrome b_6 in quinol binding to the Q_0 site. *Biochemistry* **36**, 2867–2874.

Finst, R.J., Kim, P.J., and Quarmby, L.M. (1998). Genetics of the deflagellation pathway in *Chlamydomonas*. *Genetics* **149**, 927–936.

Finst, R.J., Kim, P.J., Griffis, E.R., and Quarmby, L.M. (2000). Fa1p is a 171 kDa protein essential for axonemal microtubule severing in *Chlamydomonas*. *J. Cell Sci.* **113**, 1963–1971.

Fiore, D.C., McKee, D.D., and Janiga, M.A. (1997). Red snow: is it safe to eat? A pilot study. *Wilderness Environ. Med.* **8**, 94–95.

Fischer, B.B., Krieger-Liszkay, A., and Eggen, R.I.L. (2005). Oxidative stress induced by the photosensitizers neutral red (type I) or rose bengal (type II) in the light causes different molecular responses in *Chlamydomonas reinhardtii*. *Plant Sci.* **168**, 747–759.

Fischer, B.B., Wiesendanger, M., and Eggen, R.I.L. (2006). Growth-condition dependent sensitivity, photodamage and stress response of *Chlamydomonas reinhardtii* exposed to high light condition. *Plant Cell Physiol.* **47**, 1135–1145.

Fischer, B.B., Dayer, R., Wiesendanger, M., and Eggen, R.I.L. (2007). Independent regulation of the *GPXH* gene expression by primary and secondary effects of high light stress in *Chlamydomonas reinhardtii*. *Physiol. Plant.* **130**, 195–206.

Fischer, N. and Rochaix, J.D. (2001). The flanking regions of PsaD drive efficient gene expression in the nucleus of the green alga *Chlamydomonas reinhardtii*. *Mol. Gen. Genet.* **265**, 888–894.

Fischer, N., Stampacchia, O., Redding, K., and Rochaix, J.-D. (1996). Selectable marker recycling in the chloroplast. *Mol. Gen. Genet.* **251**, 373–380.

Fischer, N., Boudreau, E., Hippler, M., Drepper, F., Haehnel, W., and Rochaix, J.-D. (1999). A large fraction of PsaF is nonfunctional in photosystem I complexes lacking the PsaJ subunit. *Biochemistry* **38**, 5546–5552.

Fischer, N., Setif, P., and Rochaix, J.-D. (1999). Site-directed mutagenesis of the PsaC subunit of photosystem I-FB is the cluster interacting with soluble ferredoxin. *J. Biol. Chem.* **274**, 23333–23340.

Fjerdingstad, E., Kemp, K., Fjerdingstad, E., and Vanggaard, L. (1974). Chemical analyses of red "snow" from East-Greenland with remarks on *Chlamydomonas nivalis* (Bau.) Wille. *Arch. Hydrobiol.* **73**, 70–83.

Fleming, G.H., Boynton, J.E., and Gillham, N.W. (1987a). Cytoplasmic ribosomal proteins from *Chlamydomonas reinhardtii*: characterization and immunological comparisons. *Mol. Gen. Genet.* **206**, 226–237.

Fleming, G.H., Boynton, J.E., and Gillham, N.W. (1987b). The cytoplasmic ribosomes of *Chlamydomonas reinhardtii*: characterization of antibiotic sensitivity and cycloheximide resistant mutants. *Mol. Gen. Genet.* **210**, 419–428.

Fong, S.E. and Surzycki, S.J. (1992a). Chloroplast RNA polymerase genes of *Chlamydomonas reinhardtii* exhibit an unusual structure and arrangement. *Curr. Genet.* **21**, 485–497.

Fong, S.E. and Surzycki, S.J. (1992b). Organization and structure of plastome *psbF*, *psbL*, *petG* and *ORF712* genes in *Chlamydomonas reinhardtii*. *Curr. Genet.* **21**, 527–530.

Ford, C. and Wang, W.-Y. (1980a). Three new yellow loci in *Chlamydomonas reinhardtii*. *Mol. Gen. Genet.* **179**, 259–263.

Ford, C. and Wang, W.-Y. (1980b). Temperature-sensitive yellow mutants of *Chlamydomonas reinhardtii*. *Mol. Gen. Genet.* **180**, 5–10.

Ford, C., Mitchell, S., and Wang, W.-Y. (1981). Protochlorophyllide photoconversion mutants of *Chlamydomonas reinhardtii*. *Mol. Gen. Genet.* **184**, 460–464.

Ford, C., Mitchell, S., and Wang, W.-Y. (1983). Characterization of NADPH: protochlorophyllide oxidoreductase in the *y-7* and *pc-1 y-7* mutants of *Chlamydomonas reinhardtii*. *Mol. Gen. Genet.* **192**, 290–292.

Forest, C.L. (1982). The relationship of flagellar length to sexual signalling in *Chlamydomonas*. *Exp. Cell Res.* **139**, 427–431.

Forest, C.L. (1983a). Specific contact between mating structure membranes observed in conditional fusion-defective *Chlamydomonas* mutants. *Exp. Cell Res.* **148**, 143–154.

Forest, C.L. (1983b). Mutational disruption of the 9 + 2 structure of the axoneme of *Chlamydomonas* flagella. *J. Cell Sci.* **61**, 423–436.

Forest, C.L. (1985). -SH and S-S involvement in *Chlamydomonas* mating. *Gamete Res.* **12**, 139–149.

Forest, C.L. (1987). Genetic control of plasma membrane adhesion and fusion in *Chlamydomonas* gametes. *J. Cell Sci.* **88**, 613–621.

Forest, C.L. and Ojakian, G.K. (1989). Mating structure differences demonstrated by freeze-fracture analysis of fusion-defective *Chlamydomonas* gametes. *J. Protozool.* **36**, 548–556.

Forest, C.L. and Togasaki, R.K. (1975). Selection for conditional gametogenesis in *Chlamydomonas reinhardi*. *Proc. Natl. Acad. Sci. U. S. A.* **72**, 3652–3655.

Forest, C.L. and Togasaki, R.K. (1977). A selection procedure for obtaining conditional gametogenic mutants using a photosynthetically incompetent strain of *Chlamydomonas reinhardi*. *Mol. Gen. Genet.* **153**, 227–230.

Forest, C.L., Goodenough, D.A., and Goodenough, U.W. (1978). Flagellar membrane agglutination and sexual signaling in the conditional *gam-1* mutant of *Chlamydomonas*. *J. Cell Biol.* **79**, 74–84.

Forestier, M., King, P., Zhang, L.P., Posewitz, M., Schwarzer, S., Happe, T., Ghirardi, M.L., and Seibert, M. (2003). Expression of two [Fe]-hydrogenases in *Chlamydomonas reinhardtii* under anaerobic conditions. *Eur. J. Biochem.* **270**, 2750–2758.

Förster, B., Osmond, C.B., Boynton, J.E., and Gillham, N.W. (1999). Mutants of *Chlamydomonas reinhardtii* resistant to very high light. *J. Photochem. Photobiol.* **48**, 127–135.

Förster, B., Osmond, C.B., and Pogson, B.J. (2005). Improved survival of very high light and oxidative stress is conferred by spontaneous gain-of-function mutations in *Chlamydomonas*. *Biochim. Biophys. Acta* **1709**, 45–57.

Förster, B., Mathesius, U., and Pogson, B.J. (2006). Comparative proteomics of high light stress in the model alga *Chlamydomonas reinhardtii*. *Proteomics* **6**, 4309–4320.

Forster, J.L., Grabowy, C.T., Harris, E.H., Boynton, J.E., and Gillham, N.W. (1980). Behavior of chloroplast genes during the early zygotic divisions of *Chlamydomonas reinhardtii*. *Curr. Genet.* **1**, 137–153.

Foster, K.W. and Smyth, R.D. (1980). Light antennas in phototactic algae. *Microbiol. Rev.* **44**, 572–630.

Foster, K.W., Saranak, J., Patel, N., Zarrilli, G., Okabe, M., Kline, T., and Nakanishi, K. (1984). A rhodopsin is the functional photoreceptor for phototaxis in the unicellular eukaryote *Chlamydomonas*. *Nature* **311**, 756–759.

Foster, K.W., Saranak, J., and Zarrilli, G. (1988). Autoregulation of rhodopsin synthesis in *Chlamydomonas reinhardtii*. *Proc. Natl. Acad. Sci. U. S. A.* **85**, 6379–6383.

Fott, B. (1974). Concerning the genus *Chlamydomonas* Ehr. *sensu lato.* – a review of Ettl's treatise of the genus *Chloromonas* Gobi with new transfers and descriptions. *Arch. Protistenk.* **116**, 304–316.

Fott, B. and McCarthy, A.J. (1964). Three acidophilic volvocine flagellates in pure culture. *J. Protozool.* **11**, 116–120.

Fott, B. and Novakova, M. (1971). Taxonomy of the palmelloid genera *Gloeocystis* Nageli and *Palmogloea* Kutzing (Chlorophyceae). *Arch. Protistenk.* **113**, 322–333.

Fowke, L.C., Cresshoff, P.M., and Marchant, H.J. (1979). Transfer of organelles of the alga *Chlamydomonas reinhardii* into carrot cells by protoplast fusion. *Planta* **144**, 341–347.

Franco, A.R., Cárdenas, J., and Fernández, E. (1988a). Regulation by ammonium of nitrate and nitrite assimilation in *Chlamydomonas reinhardtii*. *Biochim. Biophys. Acta* **951**, 98–103.

Franco, A.R., Cárdenas, J., and Fernández, E. (1988b). Two different carriers transport both ammonium and methylammonium in *Chlamydomonas reinhardtii*. *J. Biol. Chem.* **263**, 14039–14043.

François, D.L. and Robinson, G.G.C. (1988). Indices of triazine toxicity in *Chlamydomonas geitleri* Ettl. *Aquat. Toxicol.* **16**, 205–227.

François, L., Fortin, C., and Campbell, P.G. (2007). pH modulates transport rates of manganese and cadmium in the green alga *Chlamydomonas reinhardtii* through non-competitive interactions: implications for an algal BLM. *Aquat. Toxicol.* **84**, 123–132.

Franklin, S.E. and Mayfield, S.P. (2004). Prospects for molecular farming in the green alga *Chlamydomonas reinhardtii*. *Curr. Opin. Plant Biol.* **7**, 159–165.

Franklin, S., Ngo, B., Efuet, E., and Mayfield, S.P. (2002). Development of a GFP reporter gene for *Chlamydomonas reinhardtii* chloroplast. *Plant J.* **30**, 733–744.

Franzén, L.-G. (1994). Analysis of chloroplast and mitochondrial targeting sequences from the green alga *Chlamydomonas reinhardtii*. *Biol. Membrany (Moscow)* **11**, 304–309.

Franzén, L.-G., Frank, G., Zuber, H., and Rochaix, J.-D. (1989a). Isolation and characterization of cDNA clones encoding Photosystem I subunits with molecular masses 11.0, 10.0 and 8.4 kDa from *Chlamydomonas reinhardtii*. *Mol. Gen. Genet.* **219**, 137–144.

Franzén, L.-G., Frank, G., Zuber, H., and Rochaix, J.-D. (1989b). Isolation and characterization of cDNA clones encoding the 17.9 and 8.1 kDa subunits of Photosystem I from *Chlamydomonas reinhardtii*. *Plant Mol. Biol.* **12**, 463–474.

Franzén, L.-G., Rochaix, J.-D., and von Heijne, G. (1990). Chloroplast transit peptides from the green alga *Chlamydomonas reinhardtii* share features with both mitochondrial and higher plant chloroplast presequences. *FEBS Lett.* **260**, 165–168.

Frenkel, A. (1949). A study of the hydrogenase systems of green and blue-green algae. *Biol. Bull.* **97**, 261–262.

Frenkel, A.W. (1951). Hydrogen evolution by the flagellate green alga, *Chlamydomonas moewusii*. *Arch. Biochem. Biophys.* **38**, 219–230.

Frenkel, A.W. and Rieger, C. (1951). Photoreduction in algae. *Nature* **167**, 1030.

Freshour, J., Yokoyama, R., and Mitchell, D.R. (2007). *Chlamydomonas* flagellar outer row dynein assembly protein Oda7 interacts with both outer row and I1 inner row dyneins. *J. Biol. Chem.* **282**, 5404–5412.

Friedmann, I., Colwin, A.L., and Colwin, L.H. (1968). Fine-structural aspects of fertilization in *Chlamydomonas reinhardi*. *J. Cell Sci.* **3**, 115–128.

Fromherz, S., Giddings, T.H., Jr., Gomez-Ospina, N., and Dutcher, S.K. (2004). Mutations in alpha-tubulin promote basal body maturation and flagellar assembly in the absence of delta-tubulin. *J. Cell. Sci.* **117(Pt2)**, 303–314.

Frost, B.F. and Small, G.D. (1987). The apparent lack of O^6-methylguanine in nuclear DNA of *Chlamydomonas reinhardtii*. *Mutat. Res.* **181**, 37–44.

Fu, L.H., Wang, X.F., Eyal, Y., She, Y.M., Donald, L.J., Standing, K.G., and Ben-Hayyim, G. (2002). A selenoprotein in the plant kingdom. Mass spectrometry confirms that an opal codon (UGA) encodes selenocysteine in *Chlamydomonas reinhardtii* glutathione peroxidase. *J. Biol. Chem.* **277**, 25983–25991.

Fuentes, C. and VanWinkle-Swift, K. (2003). Isolation and characterization of a cell wall-defective mutant of *Chlamydomonas monoica* (Chlorophyta). *J. Phycol.* **39**, 1261–1267.

Fuhrmann, M. (2002). Expanding the molecular toolkit for *Chlamydomonas reinhardtii*: from history to new frontiers. *Protist* **153**, 357–364.

Fuhrmann, M., Oertel, W., and Hegemann, P. (1999). A synthetic gene coding for the green fluorescent protein (GFP) is a versatile reporter in *Chlamydomonas reinhardtii*. *Plant J.* **19**, 353–361.

Fuhrmann, M., Stahlberg, A., Govorunova, E., Rank, S., and Hegemann, P. (2001). The abundant retinal protein of the *Chlamydomonas* eye is not the photoreceptor for phototaxis and photophobic responses. *J. Cell Sci.* **114**, 3857–3863.

Fuhrmann, M., Deininger, W., Kateriya, S., and Hegemann, P. (2003). Rhodopsin-related proteins, cop1, cop2 and chop1, in *Chlamydomonas reinhardtii*. In: *Photoreceptors and Light Signalling* (A. Batschauer, Ed.), pp. 124–135. Royal Society of Chemistry, Cambridge, UK.

Fuhrmann, M., Hausherr, A., Ferbitz, L., Schödl, T., Heitzer, M., and Hegemann, P. (2004). Monitoring dynamic expression of nuclear genes in *Chlamydomonas reinhardtii* by using a synthetic luciferase reporter gene. *Plant Mol. Biol.* **55**, 869–881.

Fukuzawa, H., Ishizaki, K., Miura, K., Matsueda, S., Ino-ue, T., Kucho, K., and Ohyama, K. (1998). Isolation and characterization of high-CO_2 requiring mutants from *Chlamydomonas reinhardtii* by gene tagging. *Can. J. Bot.* **76**, 1092–1097.

Fukuzawa, H., Miura, K., Ishizaki, K., Kucho, K., Saito, T., Kohinata, T., and Ohyama, K. (2001). Ccm1, a regulatory gene controlling the induction of a carbon-concentrating mechanism in *Chlamydomonas reinhardtii* by sensing CO_2 availability. *Proc. Natl. Acad. Sci. U. S. A.* **98**, 5347–5352.

Funes, S., Franzén, L.G., and González-Halphen, D. (2007). *Chlamydomonas reinhardtii*: the model of choice to study mitochondria from unicellular photosynthetic organisms. *Methods Mol. Biol.* **372**, 137–149.

Funke, R.P., Kovar, J.L., and Weeks, D.P. (1997). Intracellular carbonic anhydrase is essential to photosynthesis in *Chlamydomonas reinhardtii* at atmospheric levels of CO_2. Demonstration via genomic complementation of the high-CO_2-requiring mutant *ca-1*. *Plant Physiol.* **114**, 237–244.

Gachon, C.M., Day, J.G., Campbell, C.N., Pröschold, T., Saxon, R.J., and Kupper, F.C. (2007). The Culture Collection of Algae and Protozoa (CCAP): a biological resource for protistan genomics. *Gene* **406**, 51–57.

Gaffal, K.P. (1987). Mitosis-specific oscillations of mitochondrial morphology in *Chlamydomonas reinhardii*. *Endocytobiosis Cell Res.* **4**, 41–62.

Gaffal, K.P. (1988). The basal body-root complex of *Chlamydomonas reinhardtii* during mitosis. *Protoplasma* **143**, 118–129.

Gaffal, K.P. and el-Gammal, E.S. (1990). Elucidation of the enigma of the "metaphase band" of *Chlamydomonas reinhardtii*. *Protoplasma* **156**, 139–148.

Gaffal, K.P. and Schneider, G.J. (1978). The changes in ultrastructure during fertilization of the colourless flagellate *Polytoma papillatum* with special reference to the configural changes of their mitochondria. *Cytobiologie* **18**, 161–173.

Gaffal, K.P., el-Gammal, S., and Friedrichs, G.J. (1992). Computer-aided 3D reconstruction of the eyespot-flagellar/basal apparatus – contractile vacuoles – nucleus associations during mitosis of *Chlamydomonas reinhardtii*. *Endocytobiosis Cell Res.* **9**, 177–208.

Gaffal, K.P., Arnold, C.G., Friedrichs, G.J., and Gemple, W. (1995). Morphodynamical changes of the chloroplast of *Chlamydomonas reinhardtii* during the 1st round of division. *Arch. Protistenk.* **145**, 10–23.

Gaillard, A.R., Diener, D.R., Rosenbaum, J.L., and Sale, W.S. (2001). Flagellar radial spoke protein 3 is an A-kinase anchoring protein (AKAP). *J. Cell Biol.* **153**, 443–448.

Gaillard, A.R., Fox, L.A., Rhea, J.M., Craige, B., and Sale, W.S. (2006). Disruption of the A-kinase-anchoring domain in flagellar radial spoke protein 3 results in unregulated axonemal PKA activity and abnormal flagellar motility. *Mol. Biol. Cell* **17**, 2626–2635.

Galloway, R.E. and Goodenough, U.W. (1985). Genetic analysis of mating locus linked mutations in *Chlamydomonas reinhardii*. *Genetics* **111**, 447–461.

Galloway, R.E. and Holden, L.R. (1984). Transmission and recombination of chloroplast genes in asexual crosses of *Chlamydomonas reinhardtii*. I. Flagellar agglutination prior to fusion does not promote uniparental inheritance or affect recombinant frequencies. *Curr. Genet.* **8**, 399–405.

Galloway, R.E. and Holden, L.R. (1985). Transmission and recombination of chloroplast genes in asexual crosses of *Chlamydomonas reinhardii*. II. Comparisons with observations of sexual diploids. *Curr. Genet.* **10**, 221–228.

Galloway, R.E. and Mets, L. (1984). Atrazine, bromacil and diuron resistance in *Chlamydomonas*. A single non-Mendelian genetic locus controls the structure of the thylakoid binding site. *Plant Physiol.* **74**, 469–474.

Galván, A., González-Ballester, D., and Fernández, E. (2007). Insertional mutagenesis as a tool to study genes/functions in *Chlamydomonas*. *Adv. Exp. Med. Biol.* **616**, 77–89.

Gandhi, S.R., Kulkarni, S.B., and Netrawali, M.S. (1988). Comparative effects of synthetic insecticides – endosulfan, phosalone and permethrin – on *Chlamydomonas reinhardtii* algal cells. *Acta Microbiol. Hung.* **35**, 93–99.

Gantt, E. (Ed.) (1980). *Handbook of Phycological Methods, Volume III. Developmental and Cytological Methods*. Cambridge University Press, Cambridge.

Gardner, L.C., O'Toole, E., Perrone, C.A., Giddings, T., and Porter, M.E. (1994). Components of a "dynein regulatory complex" are located at the junction between the radial spokes and the dynein arms in *Chlamydomonas* flagella. *J. Cell Biol.* **127**, 1311–1325.

Garvey, J.E., Owen, H.A., and Winner, R.W. (1991). Toxicity of copper to the green alga *Chlamydomonas reinhardtii* (Chlorophyceae) as affected by humic substances of terrestrial and freshwater origin. *Aquat. Toxicol.* **19**, 89–96.

Gauthier, A., Turmel, M., and Lemieux, C. (1991). A group I intron in the chloroplast large subunit rRNA gene of *Chlamydomonas eugametos* encodes a double-strand endonuclease that cleaves the homing site of this intron. *Curr. Genet.* **19**, 43–47.

Gealt, M.A., Adler, J.H., and Nes, W.R. (1981). The sterols and fatty acids from purified flagella of *Chlamydomonas reinhardi*. *Lipids* **16**, 133–136.

Geimer, S. and Melkonian, M. (2004). The ultrastructure of the *Chlamydomonas reinhardtii* basal apparatus: identification of an early marker of radial asymmetry inherent in the basal body. *J. Cell Sci.* **117**, 2663–2674.

Geimer, S. and Melkonian, M. (2005). Centrin scaffold in *Chlamydomonas reinhardtii* revealed by immunoelectron microscopy. *Eukaryotic Cell* **4**, 1253–1263.

Geoffroy, L., Gilbin, R., Simon, O., Floriani, M., Adam, C., Pradines, C., Cournac, L., and Garnier-Laplace, J. (2007). Effect of selenate on growth and photosynthesis of *Chlamydomonas reinhardtii*. *Aquat. Toxicol.* **83**, 149–158.

Geraghty, A.M. and Spalding, M.H. (1996). Molecular and structural changes in *Chlamydomonas* under limiting CO_2 – A possible mitochondrial role in adaptation. *Plant Physiol.* **111**, 1339–1347.

Gercke, H. and Arnold, C.G. (1981a). Die Wirkung von Nalidixinsaure auf die Feinstruktur der Plastiden und Mitochondrien von *Chlamydomonas reinhardii* und *Polytoma papillatum*. *Biol. Zbl.* **100**, 33–39.

Gercke, H. and Arnold, C.G. (1981b). Die Wirkung von 4′-6-Diamidino-2-phenylindol (DAPI) auf die Feinstruktur der Mitochondrien und Plastiden von *Chlamydomonas reinhardii* und *Polytoma papillatum*. *Biol. Zbl.* **100**, 209–215.

Gerloff, J. (1940). Beiträge zur Kenntnis der Variabilität und Systematik der Gattung *Chlamydomonas*. *Arch. Protistenk.* **94**, 311–502.

Gerloff, J. (1962). Beiträge zur Kenntnis einiger Volvocales. *Nova Hedwigia* **4**, 1–20.

Gerloff-Elias, A., Spijkerman, E., and Pröschold, T. (2005). Effect of external pH on the growth, photosynthesis and photosynthetic electron transport of *Chlamydomonas acidophila* Negoro, isolated from an extremely acidic lake (pH 2.6). *Plant Cell Environ.* **28**, 1218–1229.

Gfeller, R.P. and Gibbs, M. (1984). Fermentative metabolism of *Chlamydomonas reinhardtii*. I. Analysis of fermentative products from starch in dark and light. *Plant Physiol.* **75**, 212–218.

Gfeller, R.P. and Gibbs, M. (1985). Fermentative metabolism of *Chlamydomonas reinhardtii*. II. Role of plastoquinone. *Plant Physiol.* **77**, 509–511.

Gherman, A., Davis, E.E., and Katsanis, N. (2006). The ciliary proteome database: an integrated community resource for the genetic and functional dissection of cilia. *Nat. Genet.* **38**, 961–962.

Ghirardi, M.L. (2006). Hydrogen production by photosynthetic green algae. *Indian J. Biochem. Biophys.* **43**, 201–210.

Ghirardi, M.L., Zhang, J.P., Lee, J.W., Flynn, T., Seibert, M., Greenbaum, E., and Melis, A. (2000). Microalgae: a green source of renewable H_2. *Trends Biotechnol.* **18**, 506–511.

Ghirardi, M.L., King, P.W., Posewitz, M.C., Maness, P.C., Fedorov, A., Kim, K., Cohen, J., Schulten, K., and Seibert, M. (2005). Approaches to developing biological H_2-photoproducing organisms and processes. *Biochem. Soc. Trans.* **33**, 70–72.

Gibbs, M., Gfeller, R.P., and Chen, C. (1986). Fermentative metabolism of *Chlamydomonas reinhardii*. III. Photoassimilation of acetate. *Plant Physiol.* **82**, 160–166.

Gibbs, S.P., Lewin, R.A., and Philpott, D.E. (1958). The fine structure of the flagellar apparatus of *Chlamydomonas moewusii*. *Exp. Cell Res.* **15**, 619–622.

Gillet, S., Decottignies, P., Chardonnet, S., and Le Marechal, P. (2006). Cadmium response and redoxin targets in *Chlamydomonas reinhardtii*: a proteomic approach. *Photosynth. Res.* **89**, 201–211.

Gillham, N.W. (1963). Transmission and segregation of a non-chromosomal factor controlling streptomycin resistance in diploid *Chlamydomonas*. *Nature* **200**, 294.

Gillham, N.W. (1965a). Induction of chromosomal and nonchromosomal mutations in *Chlamydomonas reinhardi* with *N*-methyl-*N'*-nitro-*N*-nitrosoguanidine. *Genetics* **52**, 529–537.

Gillham, N.W. (1965b). Linkage and recombination between non-chromosomal mutations in *Chlamydomonas reinhardi*. *Proc. Natl. Acad. Sci. U. S. A.* **54**, 1560–1567.

Gillham, N.W. (1969). Uniparental inheritance in *Chlamydomonas reinhardi*. *Am. Nat.* **103**, 355–388.

Gillham, N.W. (1974). Genetic analysis of the chloroplast and mitochondrial genomes. *Annu. Rev. Genet.* **8**, 347–391.

Gillham, N.W. (1978). *Organelle Heredity*. Raven Press, New York. 602 pp.

Gillham, N.W., Boynton, J.E., and Burkholder, B. (1970). Mutations altering chloroplast ribosome phenotype in *Chlamydomonas*. I. Non-Mendelian mutations. *Proc. Natl. Acad. Sci. U. S. A.* **67**, 1026–1033.

Gillham, N.W., Boynton, J.E., and Lee, R.W. (1974). Segregation and recombination of non-Mendelian genes in *Chlamydomonas*. *Genetics* **78**, 439–457.

Gillham, N.W., Boynton, J.E., and Harris, E.H. (1987a). Specific elimination of mitochondrial DNA from *Chlamydomonas* by intercalating dyes. *Curr. Genet.* **12**, 41–48.

Gillham, N.W., Boynton, J.E., Johnson, A.M., and Burkhart, B.D. (1987b). Mating type linked mutations which disrupt the uniparental transmission of chloroplast genes in *Chlamydomonas*. *Genetics* **115**, 677–684.

Girard, J., Chua, N.-H., Bennoun, P., Schmidt, G., and Delosme, M. (1980). Studies on mutants deficient in the photosystem I reaction centers in *Chlamydomonas reinhardtii. Curr. Genet.* **2**, 215-221.

Girard-Bascou, J., Choquet, Y., Schneider, M., Delosme, M., and Dron, M. (1987). Characterization of a chloroplast mutation in the *psaA2* gene of *Chlamydomonas reinhardtii. Curr. Genet.* **12**, 489-495.

Girard-Bascou, J., Pierre, Y., and Drapier, D. (1992). A nuclear mutation affects the synthesis of the chloroplast *psbA* gene product in *Chlamydomonas reinhardtii. Curr. Genet.* **22**, 47-52.

Giraud, G. and Czaninski, Y. (1971). Localisation ultrastructurale d'activités oxydasiques chez le *Chlamydomonas reinhardi. C. R. Acad. Sci.* **273**, 2500-2503.

Glöckner, G. and Beck, C.F. (1995). Genes involved in light control of sexual differentiation in *Chlamydomonas reinhardtii. Genetics* **141**, 937-943.

Glöckner, G. and Beck, C.F. (1997). Cloning and characterization of *LRG5*, a gene involved in blue light signaling in *Chlamydomonas* gametogenesis. *Plant J.* **12**, 677-683.

Godde, D. (1982). Evidence for a membrane bound NADH-plastoquinone- oxidoreductase in *Chlamydomonas reinhardii CW-15. Arch. Microbiol.* **131**, 197-202.

Goldschmidt-Clermont, M. (1986). The two genes for the small subunit of RuBP carboxylase/oxygenase are closely linked in *Chlamydomonas reinhardtii. Plant Mol. Biol.* **6**, 13-21.

Goldschmidt-Clermont, M. (1991). Transgenic expression of aminoglycoside adenine transferase in the chloroplast: a selectable marker for site-directed transformation of *Chlamydomonas. Nucleic Acids Res.* **19**, 4083-4089.

Goldschmidt-Clermont, M. (1998). Chloroplast transformation and reverse genetics. In: *The Molecular Biology of Chloroplasts and Mitochondria in Chlamydomonas* (J.-D. Rochaix, M. Goldschmidt-Clermont, and S. Merchant, Eds.), pp. 139-149. Kluwer Adademic, Dordrecht.

Goldschmidt-Clermont, M. and Rahire, M. (1986). Sequence, evolution and differential expression of the two genes encoding variant small subunits of ribulose bisphosphate carboxylase/oxygenase in *Chlamydomonas reinhardtii. J. Mol. Biol.* **191**, 421-432.

Goldschmidt-Clermont, M., Girard-Bascou, J., Choquet, Y., and Rochaix, J.-D. (1990). *Trans*-splicing mutants of *Chlamydomonas reinhardtii. Mol. Gen. Genet.* **223**, 417-425.

Goldschmidt-Clermont, M., Choquet, Y., Girard-Bascou, J., Michel, F., Schirmer-Rahire, M., and Rochaix, J.-D. (1991). A small chloroplast RNA may be required for *trans*-splicing in *Chlamydomonas reinhardtii. Cell* **65**, 135-143.

Goldschmidt-Clermont, M., Rahire, M., and Rochaix, J.D. (2008). Redundant *cis*-acting determinants of 3' processing and RNA stability in the chloroplast *rbcL* mRNA of *Chlamydomonas. Plant J.* **53**, 566-577.

González, M.A., Coleman, A.W., Gómez, P.I., and Montoya, R. (2001). Phylogenetic relationship among various strains of *Dunaliella* (Chlorophyceae) based on nuclear ITS rDNA sequences. *J. Phycol.* **37**, 604-611.

González-Ballester, D., Camargo, A., and Fernández, E. (2004). Ammonium transporter genes in *Chlamydomonas*: the nitrate-specific regulatory gene *Nit2* is involved in *Amt1;1* expression. *Plant Mol. Biol.* **56**, 863-878.

González-Ballester, D., de Montaigu, A., Higuera, J.J., Galván, A., and Fernández, E. (2005). Functional genomics of the regulation of the nitrate assimilation pathway in *Chlamydomonas. Plant Physiol.* **137**, 522-533.

Goodenough, J.E. and Bruce, V.G. (1980). The effects of caffeine and theophylline on the phototactic rhythm of *Chlamydomonas reinhardii. Biol. Bull.* **159**, 649-655.

References **345**

Goodenough, J.E., Bruce, V.G., and Carter, A. (1981). The effects of inhibitors affecting protein synthesis and membrane activity on the *Chlamydomonas reinhardii* phototactic rhythm. *Biol. Bull.* **161**, 371–381.

Goodenough, U.W. (1970). Chloroplast division and pyrenoid formation in *Chlamydomonas reinhardi*. *J. Phycol.* **6**, 1–6.

Goodenough, U.W. (1985). An essay on the origins and evolution of eukaryotic sex. In: *The Origin and Evolution of Sex. MBL Lectures in Biology*, Vol. 7 (H.O. Halvorson and A. Monroy, Eds.), pp. 123–140. Alan R. Liss, New York.

Goodenough, U.W. and Heuser, J.E. (1985a). Substructure of inner dynein arms, radial spokes, and the central pair/projection complex of cilia and flagella. *J. Cell Biol.* **100**, 2008–2018.

Goodenough, U.W. and Heuser, J.E. (1985b). The *Chlamydomonas* cell wall and its constituent glycoproteins analyzed by the quick-freeze, deep-etch technique. *J. Cell Biol.* **101**, 1550–1568.

Goodenough, U.W. and Heuser, J.E. (1988a). Molecular organization of cell-wall crystals from *Chlamydomonas eugametos*. *J. Cell Sci.* **90**, 735–750.

Goodenough, U.W. and Heuser, J.E. (1988b). Molecular organization of cell-wall crystals from *Chlamydomonas reinhardtii* and *Volvox carteri*. *J. Cell Sci.* **90**, 717–733.

Goodenough, U.W. and Jurivich, D. (1978). Tipping and mating-structure activation induced in *Chlamydomonas* gametes by flagellar membrane antisera. *J. Cell Biol.* **79**, 680–693.

Goodenough, U.W. and Levine, R.P. (1969). Chloroplast ultrastructure in mutant strains of *Chlamydomonas reinhardi* lacking components of the photosynthetic apparatus. *Plant Physiol.* **44**, 990–1000.

Goodenough, U.W. and Levine, R.P. (1970). Chloroplast structure and function in *ac-20*, a mutant strain of *Chlamydomonas reinhardi*. III. Chloroplast ribosomes and membrane organization. *J. Cell Biol.* **44**, 547–562.

Goodenough, U.W. and St. Clair, H.S. (1975). *Bald-2*: a mutation affecting the formation of doublet and triplet sets of microtubules in *Chlamydomonas reinhardi*. *J. Cell Biol.* **66**, 480–491.

Goodenough, U.W. and Staehelin, L.A. (1971). Structural differentiation of stacked and unstacked chloroplast membranes. Freeze-etch electron microscopy of wild-type and mutant strains of *Chlamydomonas*. *J. Cell Biol.* **48**, 594–619.

Goodenough, U.W. and Weiss, R.L. (1975). Gametic differentiation in *Chlamydomonas reinhardi*. III. Cell wall lysis and microfilament-associated mating structure activation in wild type and mutant strains. *J. Cell Biol.* **67**, 623–637.

Goodenough, U.W. and Weiss, R.L. (1978). Interrelationships between microtubules, a striated fiber, and the gametic mating structure of *Chlamydomonas reinhardi*. *J. Cell Biol.* **76**, 430–438.

Goodenough, U.W., Hwang, C., and Martin, H. (1976). Isolation and genetic analysis of mutant strains of *Chlamydomonas reinhardi* defective in gametic differentiation. *Genetics* **82**, 169–186.

Goodenough, U.W., Hwang, C., and Warren, A.J. (1978). Sex-limited expression of gene loci controlling flagellar membrane agglutination in the *Chlamydomonas* mating reaction. *Genetics* **89**, 235–243.

Goodenough, U.W., Adair, W.S., Caligor, E., Forest, C.L., Hoffman, J.L., Mesland, D. A., and Spath, S. (1980). Membrane–membrane and membrane–ligand interactions in *Chlamydomonas* mating. *Soc. Gen. Physiol. Ser.* **34**, 131–152.

Goodenough, U.W., Detmers, P.A., and Hwang, C. (1982). Activation for cell fusion in *Chlamydomonas*: analysis of wild-type gametes and nonfusing mutants. *J. Cell Biol.* **92**, 378–386.

Goodenough, U.W., Adair, W.S., Collin-Osdoby, P., and Heuser, J.E. (1985). Structure of the *Chlamydomonas* agglutinin and related flagellar surface proteins *in vitro* and *in situ*. *J. Cell Biol.* **101**, 924–941.

Goodenough, U.W., Gebhart, B., Mecham, R.P., and Heuser, J.E. (1986). Crystals of the *Chlamydomonas reinhardtii* cell wall: polymerization, depolymerization, and purification of glycoprotein monomers. *J. Cell Biol.* **103**, 405–417.

Goodenough, U.W., Armbrust, E.V., Campbell, A.M., and Ferris, P.J. (1995). Molecular genetics of sexuality in *Chlamydomonas*. *Annu. Rev. Plant Physiol. Plant Mol. Biol.* **46**, 21–44.

Goodenough, U., Lin, H., and Lee, J.H. (2007). Sex determination in *Chlamydomonas*. *Semin. Cell Dev. Biol.* **18**, 350–361.

Goodwin, T.J., Butler, M.I., and amd Poulter, R.T. (2006). Multiple, non-allelic, intein-coding sequences in eukaryotic RNA polymerase genes. *BMC Biol.* **4**, 38.

Gorman, D.S. and Levine, R.P. (1965). Cytochrome *f* and plastocyanin: their sequence in the photosynthetic electron transport chain of *Chlamydomonas reinhardi*. *Proc. Natl. Acad. Sci. U. S. A.* **54**, 1665–1669.

Goroschankin, J.N. (1891). Beiträge zur Kenntnis der Morphologie und Systematik der Chlamydomonaden. II. *Ch. reinhardi* Dang. und sein Verwandten. *Bull. Soc. Imp. Natur. Moscou, N.S.* **5**, 101–142.

Gorton, H.L. and Vogelmann, T.C. (2003). Ultraviolet radiation and the snow alga *Chlamydomonas nivalis* (Bauer) Wille. *Photochem. Photobiol.* **77**, 608–615.

Gorton, H.L., Williams, W.E., and Vogelmann, T.C. (2001). The light environment and cellular optics of the snow alga *Chlamydomonas nivalis* (Bauer) Wille. *Photochem. Photobiol.* **73**, 611–620.

Goto, K. and Johnson, C.H. (1995). Is the cell division cycle gated by a circadian clock? The case of *Chlamydomonas reinhardtii*. *J. Cell Biol.* **129**, 1061–1069.

Govindjee, Eggenberg, P., Pfister, K., and Strasser, R.J. (1992). Chlorophyll *a* fluorescence decay in herbicide-resistant D1 mutants of *Chlamydomonas reinhardtii* and the formate effect. *Biochim. Biophys. Acta* **1101**, 353–358.

Govorunova, E.G. and Sineshchekov, O.A. (2003). Integration of photo- and chemosensory signaling pathways in *Chlamydomonas*. *Planta* **216**, 535–540.

Govorunova, E.G. and Sineshchekov, O.A. (2005). Chemotaxis in the green flagellate alga *Chlamydomonas*. *Biochemistry (Moscow)* **70**, 717–725.

Govorunova, E.G., Voytsekh, O.O., and Sineshchekov, O.A. (2007). Changes in photoreceptor currents and their sensitivity to the chemoeffector tryptone during gamete mating in *Chlamydomonas reinhardtii*. *Planta* **225**, 441–449.

Gowans, C.S. (1960). Some genetic investigations on *Chlamydomonas eugametos*. *Z. Vererbs.* **91**, 63–73.

Gowans, C.S. (1963). The conspecificity of *Chlamydomonas eugametos* and *Chlamydomonas moewusii*: an experimental approach. *Phycologia* **3**, 37–44.

Gowans, C.S. (1965). Tetrad analysis. *Taiwania* **11**, 1–19.

Gowans, C.S. (1976a). Genetics of *Chlamydomonas moewusii* and *Chlamydomonas eugametos*. In: *The Genetics of Algae* (R. A. Lewin, Ed.), pp. 145–173. *Botanical Monographs, Vol. 12*, University of California Press, Berkeley.

Gowans, C.S. (1976b). Publications by Franz Moewus on the genetics of algae. In: *The Genetics of Algae* (R. A. Lewin, Ed.), pp. 310–332. *Botanical Monographs, Vol. 12*, University of California Press, Berkeley.

Goyal, A. and Tolbert, N.E. (1989). Variations in the alternative oxidase in *Chlamydomonas* grown in air or high CO_2. *Plant Physiol.* **89**, 958–962.

Gradmann, D., Ehlenbeck, S., and Hegemann, P. (2002). Modeling light-induced currents in the eye of *Chlamydomonas reinhardtii*. *J. Membr. Biol.* **189**, 93–104.

Graham, J.E., Spanier, J.G., and Jarvik, J.W. (1995). Isolation and characterization of Pioneer1, a novel *Chlamydomonas* transposable element. *Curr. Genet.* **28**, 429–436.

Grant, D. and Chiang, K.S. (1980). Physical mapping and characterization of *Chlamydomonas* mitochondrial DNA molecules: their unique ends, sequence homogeneity, and conservation. *Plasmid* **4**, 82–96.

Grant, D., Swinton, D.C., and Chiang, K.-S. (1978). Differential patterns of mitochondrial, chloroplastic and nuclear DNA synthesis in the synchronous cell cycle of *Chlamydomonas reinhardtii*. *Planta* **141**, 259–267.

Grant, D.M., Gillham, N.W., and Boynton, J.E. (1980). Inheritance of chloroplast DNA in *Chlamydomonas reinhardtii*. *Proc. Natl. Acad. Sci. U. S. A.* **77**, 6067–6071.

Gray, M.W. and Boer, P.H. (1988). Organization and expression of algal (*Chlamydomonas reinhardtii*) mitochondrial DNA. *Philos. Trans. R. Soc. Lond., B, Biol. Sci.* **319**, 135–147.

Greenbaum, E., Guillard, R.R.L., and Sunda, W.G. (1983a). Hydrogen and oxygen photoproduction by marine algae. *Photochem. Photobiol.* **37**, 649–655.

Greenbaum, E., Guillard, R.R.L., and Sunda, W.G. (1983b). Biological solar energy production with marine algae. *Bioscience* **33**, 584–585.

Gresshoff, P.M. (1976). Culture of *Chlamydomonas reinhardi* protoplasts in defined media. *Aust. J. Plant Physiol.* **3**, 457–464.

Gresshoff, P.M. (1981). Amide metabolism of *Chlamydomonas reinhardi*. *Arch. Microbiol.* **128**, 303–306.

Gresshoff, P.M., Mahanty, H.K., and Gartner, E. (1977). Fate of polychlorinated biphenyl (Aroclor 1242) in an experimental study and its significance to the natural environment. *Bull. Environ. Contam. Toxicol.* **17**, 686–691.

Grief, C. and Shaw, P.J. (1987). Assembly of cell-wall glycoproteins of *Chlamydomonas reinhardii*: oligosaccharides are added in medial and trans Golgi compartments. *Planta* **171**, 302–312.

Grief, C. and Shaw, P.J. (1990). Cell wall glycoproteins of *Chlamydomonas reinhardii*: negative stain electron microscopy and epitope mapping of the molecules. *Cell Biol. Int. Rep.* **14**, 47–58.

Grief, C., O'Neill, M.A., and Shaw, P.J. (1987). The zygote cell wall of *Chlamydomonas reinhardii*: a structural, chemical and immunological approach. *Planta* **170**, 433–445.

Griesbeck, C., Kobl, I., and Heitzer, M. (2006). *Chlamydomonas reinhardtii*: a protein expression system for pharmaceutical and biotechnological proteins. *Mol. Biotechnol.* **34**, 213–223.

Grob, E.C., Boschetti, A., and Morgenthaler, J.-J. (1970). Automatische Apparatur zur kontinuierlichen Herstellung synchroner Algenkulturen. *Experientia* **26**, 1040–1042.

Grobe, B. and Arnold, C.G. (1975). Evidence of a large, ramified mitochondrium in *Chlamydomonas reinhardii*. *Protoplasma* **86**, 291–294.

Grobe, B. and Arnold, C.G. (1977). The behaviour of mitochondria in the zygote of *Chlamydomonas reinhardii*. *Protoplasma* **93**, 357–361.

Gross, J.A. and Jahn, T.L. (1962). Cellular responses to thermal and photo stress. I. *Euglena* and *Chlamydomonas*. *J. Protozool.* **9**, 340–346.

Gross, C.H., Ranum, L.P.W., and Lefebvre, P.A. (1988). Extensive restriction fragment length polymorphisms in a new isolate of *Chlamydomonas reinhardtii*. *Curr. Genet.* **13**, 503–508.

Grossman, A.R. (2000a). Acclimation of *Chlamydomonas reinhardtii* to its nutrient environment. *Protist* **151**, 201–224.

Grossman, A.R. (2000b). *Chlamydomonas reinhardtii* and photosynthesis: genetics to genomics. *Curr. Opin. Plant Biol.* **3**, 132–137.

Grossman, A.R. (2007). In the grip of algal genomics. *Adv. Exp. Med. Biol.* **616**, 54–76.

Grossman, A.R., Harris, E.H., Hauser, C., Lefebvre, P.A., Martinez, D., Rokhsar, D., Shrager, J., Silflow, C.D., Stern, D., Vallon, O., and Zhang, Z. (2003). *Chlamydomonas reinhardtii* at the crossroads of genomics. *Eukaryotic Cell* **2**, 1137–1150.

Grossman, A.R., Croft, M., Gladyshev, V.N., Merchant, S.S., Posewitz, M.C., Prochnik, S., and Spalding, M.H. (2007). Novel metabolism in *Chlamydomonas* through the lens of genomics. *Curr. Opin. Plant Biol.* **10**, 190–198.

Gruber, H.E. and Rosario, B. (1974). Variation in eyespot ultrastructure in *Chlamydomonas reinhardi (ac-31)*. *J. Cell Sci.* **15**, 481–494.

Gruber, H.E. and Rosario, B. (1979). Ultrastructure of the Golgi apparatus and contractile vacuole in *Chlamydomonas reinhardi*. *Cytologia* **44**, 505–526.

Grunow, A. and Lechtreck, K.F. (2001). Mitosis in *Dunaliella bioculata* (Chlorophyta): centrin but not basal bodies are at the spindle poles. *J. Phycol.* **37**, 1030–1043.

Gudynaite-Savitch, L., Gretes, M., Morgan-Kiss, R.M., Savitch, L.V., Simmonds, J., Kohalmi, S.E., and Huner, N.P. (2006). Cytochrome *f* from the Antarctic psychrophile, *Chlamydomonas raudensis* UWO 241: structure, sequence, and complementation in the mesophile, *Chlamydomonas reinhardtii*. *Mol. Genet. Genomics* **275**, 387–398.

Guillard, R.R.L. (1960). A mutant of *Chlamydomonas moewusii* lacking contractile vacuoles. *J. Protozool.* **7**, 262–268.

Guillard, R.R.L. (1973). Methods for microflagellates and nannoplankton. In *Handbook of Phycological Methods. Vol. I. Culture Methods and Growth Measurements* (J.R. Stein, Ed.), pp. 69–85. Cambridge University Press, Cambridge.

Guillard, R.R.L. and Sieracki, M.S. (2005). Counting cells in cultures with the light microscope. In *Algal Culturing Techniques* (R.A. Andersen, Ed.), pp. 239–252. Elsevier, Amsterdam.

Gumpel, N.J. and Purton, S. (1994). Playing tag with *Chlamydomonas*. *Trends Cell Biol.* **4**, 299–301.

Gumpel, N.J., Rochaix, J.-D., and Purton, S. (1994). Studies on homologous recombination in the green alga *Chlamydomonas reinhardtii*. *Curr. Genet.* **26**, 438–442.

Gumpel, N.J., Ralley, L., Girard-Bascou, J., Wollman, F.A., Nugent, J.H.A., and Purton, S. (1995). Nuclear mutants of *Chlamydomonas reinhardtii* defective in the biogenesis of the cytochrome b_6f complex. *Plant Mol. Biol.* **29**, 921–932.

Haag, E.S. (2007). Why two sexes? Sex determination in multicellular organisms and protistan mating types. *Semin. Cell Dev. Biol.* **18**, 348–349.

Hagen, C., Siegmund, S., and Braune, W. (2002). Ultrastructural and chemical changes in the cell wall of *Haematococcus pluvialis* (Volvocales, Chlorophyta) during aplanospore formation. *Eur. J. Phycol.* **37**, 217–226.

Hagen-Seyfferth, M. (1959). Zur Kenntnis der Geisseln und der Chemotaxis von *Chlamydomonas eugametos* Moewus (*Chl. moewusii* Gerloff). *Planta* **53**, 376–401.

Hahn, D., Bennoun, P., and Kück, U. (1996). Altered expression of nuclear genes encoding chloroplast polypeptides in non-photosynthetic mutants of *Chlamydomonas reinhardtii*: evidence for post transcriptional regulation. *Mol. Gen. Genet.* **252**, 362–370.

Hahn, D., Nickelsen, J., Hackert, A., and Kück, U. (1998). A single nuclear locus is involved in both chloroplast RNA *trans*-splicing and 3′ end processing. *Plant J.* **15**, 575–581.

Hall, J.L., Ramanis, Z., and Luck, D.J.L. (1989). Basal body/centriolar DNA: molecular genetic studies in *Chlamydomonas*. *Cell* **59**, 121–132.

Hall, L.M., Taylor, K.B., and Jones, D.D. (1993). Expression of a foreign gene in *Chlamydomonas reinhardtii*. *Gene* **124**, 75–81.

Hallmann, A. (2003). Extracellular matrix and sex-inducing pheromone in *Volvox*. *Int. Rev. Cytol.* **227**, 131–182.

Hallmann, A. (2006). The pherophorins: common, versatile building blocks in the evolution of extracellular matrix architecture in Volvocales. *Plant J.* **45**, 292–307.

Hallmann, A., Amon, P., Godl, K., Heitzer, M., and Sumper, M. (2001). Transcriptional activation by the sexual pheromone and wounding: a new gene family from *Volvox* encoding modular proteins with (hydroxy) proline-rich and metalloproteinase homology domains. *Plant J.* **26**, 583–593.

Hamaji, T., Ferris, P.J., Coleman, A.W., Waffenschmidt, S., Takahashi, F., Nishii, I., and Nozaki, H. (2008). Identification of the minus-dominance gene ortholog in the mating-type locus of *Gonium pectorale*. *Genetics* **178**, 192–283.

Hamel, P., Olive, J., Pierre, Y., Wollman, F.A., and de Vitry, C. (2000). A new subunit of cytochrome b_6f complex undergoes reversible phosphorylation upon state transition. *J. Biol. Chem.* **275**, 17072–17079.

Hamel, P.P., Dreyfuss, B.W., Xie, Z., Gabilly, S.T., and Merchant, S. (2003). Essential histidine and tryptophan residues in CcsA, a system II polytopic cytochrome c biogenesis protein. *J. Biol. Chem.* **278**, 2593–2603.

Hamilton, R.D. (1973). Sterilization. In: *Handbook of Phycological Methods. Vol. I. Culture Methods and Growth Measurements* (J.R. Stein, Ed.), pp. 181–193. Cambridge University Press, Cambridge.

Hanikenne, M., Matagne, R.F., and Loppes, R. (2001). Pleiotropic mutants hypersensitive to heavy metals and to oxidative stress in *Chlamydomonas reinhardtii*. *FEMS Microbiol. Lett.* **196**, 107–111.

Hanikenne, M., Motte, P., Wu, M.C.S., Wang, T., Loppes, R., and Matagne, R.F. (2005). A mitochondrial half-size ABC transporter is involved in cadmium tolerance in *Chlamydomonas reinhardtii*. *Plant Cell Environ.* **28**, 863–873.

Hanson, D.T., Franklin, L.A., Samuelsson, G., and Badger, M.R. (2003). The *Chlamydomonas reinhardtii cia3* mutant lacking a thylakoid lumen-localized carbonic anhydrase is limited by CO_2 supply to rubisco and not photosystem II function *in vivo*. *Plant Physiol.* **132**, 2267–2275.

Happe, T., Mosler, B., and Naber, J.D. (1994). Induction, localization and metal content of hydrogenase in the green alga *Chlamydomonas reinhardtii*. *Eur. J. Biochem.* **222**, 769–774.

Hardy, J.T. and Curl, H., Jr. (1972). The candy-colored, snow-flaked alpine biome. *Nat. Hist.* **81(9)**, 74–78.

Haring, M.A. and Beck, C.F. (1997). A promoter trap for *Chlamydomonas reinhardtii*: development of a gene cloning method using 5′ RACE-based probes. *Plant J.* **11**, 1341–1348.

Harper, J.D.I. (1999). *Chlamydomonas* cell cycle mutants. *Int. Rev. Cytol.* **189**, 131–176.

Harper, J.D.I. and John, P.C.L. (1986). Coordination of division events in the *Chlamydomonas* cell cycle. *Protoplasma* **131**, 118–130.

Harper, J.D.I., Rao, P.N., and John, P.C.L. (1990). The mitosis-specific monoclonal antibody MPM-2 recognizes phosphoproteins associated with the nuclear envelope in *Chlamydomonas reinhardtii* cells. *Eur. J. Cell Biol.* **51**, 272–278.

Harper, J.D.I., Wu, L., Sakuanrungsirikul, S., and John, P.C.L. (1995). Isolation and partial characterization of conditional cell division cycle mutants in *Chlamydomonas*. *Protoplasma* **186**, 149–162.

Harper, J.D.I., Salisbury, J.L., John, P.C.L., and Koutoulis, A. (2004). Changes in the centrin and microtubule cytoskeletons after metaphase arrest of the *Chlamydomonas reinhardtii met1* mutant. *Protoplasma* **224**, 159–165.

Harris, E.H. (1989). *The Chlamydomonas Sourcebook. A Comprehensive Guide to Biology and Laboratory Use*. Academic Press, San Diego. 780 pp.

Harris, E.H. (1998). Introduction to *Chlamydomonas*. In: *The Molecular Biology of Chloroplasts and Mitochondria in Chlamydomonas* (J.-D. Rochaix, M. Goldschmidt-Clermont, and S. Merchant, Eds.), pp. 1–11. Kluwer Adademic, Dordrecht.

Harris, E.H., Boynton, J.E., and Gillham, N.W. (1974). Chloroplast ribosome biogenesis in *Chlamydomonas*. Selection and characterization of mutants blocked in ribosome formation. *J. Cell Biol.* **63**, 160–179.

Harris, E.H., Boynton, J.E., Gillham, N.W., Tingle, C.L., and Fox, S.B. (1977). Mapping of chloroplast genes involved in chloroplast ribosome biogenesis in *Chlamydomonas reinhardtii*. *Mol. Gen. Genet.* **155**, 249–265.

Harris, E.H., Boynton, J.E., and Gillham, N.W. (1982). Induction of nuclear and chloroplast mutations which affect the chloroplast in *Chlamydomonas reinhardtii*. In: *Methods in Chloroplast Molecular Biology* (M. Edelman, R.B. Hallick, and N.-H. Chua, Eds.), pp. 3–23. Elsevier/North-Holland, Amsterdam.

Harris, E.H., Burkhart, B.D., Gillham, N.W., and Boynton, J.E. (1989). Antibiotic resistance mutations in the chloroplast 16S and 23S rRNA genes of *Chlamydomonas reinhardtii*: correlation of genetic and physical maps of the chloroplast genome. *Genetics* **123**, 281–292.

Harris, E.H., Boynton, J.E., Gillham, N.W., Burkhart, B.D., and Newman, S.M. (1991). Chloroplast genome organization in *Chlamydomonas*. *Arch. Protistenk.* **139**, 183–192.

Harrison, P.J. and Berges, J.A. (2005). Marine culture media. In: *Algal Culturing Techniques* (R.A. Andersen, Ed.), pp. 21–33. Elsevier, Amsterdam.

Hartfiel, G. and Amrhein, N. (1976). The action of methylxanthines on motility and growth of *Chlamydomonas reinhardtii* and other flagellated algae. Is cyclic AMP involved? *Biochem. Physiol. Pflanz.* **169**, 531–556.

Hartmann, M. (1955). Sex problems in algae, fungi, and protozoa. A critical account following the review of R.A. Lewin. *Am. Nat.* **89**, 321–346.

Hartnett, M.E., Newcomb, J.R., and Hodson, R.C. (1987). Mutations in *Chlamydomonas reinhardtii* conferring resistance to the herbicide sulfometuron methyl. *Plant Physiol.* **85**, 898–901.

Hartshorne, J.N. (1953). The function of the eyespot in *Chlamydomonas*. *New Phytol.* **52**, 292–297.

Hartshorne, J.N. (1955). Multiple mutation in *Chlamydomonas reinhardi*. *Heredity* **9**, 239–248.

Harz, H. and Hegemann, P. (1991). Rhodopsin-regulated calcium currents in *Chlamydomonas*. *Nature* **351**, 489–491.

Harz, H., Nonnengässer, C., and Hegemann, P. (1992). The photoreceptor current of the green alga *Chlamydomonas*. *Phil. Trans. R. Soc. Lond. B* **338**, 39–52.

Hasnain, S.E. and Upadhyaya, K.C. (1982). On the transport of 2-aminoisobutyric acid in *Chlamydomonas* protoplasts. *Indian J. Exp. Biol.* **20**, 175–176.

Hasnain, S.E., Manavathu, E.K., and Leung, W.C. (1985). DNA-mediated transformation of *Chlamydomonas reinhardi* cells: use of aminoglycoside 3′-phosphotransferase as a selectable marker. *Mol. Cell. Biol.* **5**, 3647–3650.

Hastings, P.J., Levine, E.E., Cosbey, E., Hudock, M.O., Gillham, N.W., Surzycki, S.J., Loppes, R., and Levine, R.P. (1965). The linkage groups of *Chlamydomonas reinhardi*. *Microb. Genet. Bull.* **23**, 17–19.

Hayashi-Ishimaru, Y., Ehara, M., Inagaki, Y., and Ohama, T. (1997). A deviant mitochondrial genetic code in prymnesiophytes (yellow-algae): UGA codon for tryptophan. *Curr. Genet.* **32**, 296–299.

He, P., Zhang, D., Chen, G., Liu, Q., and Wu, W. (2003). Gold immunolocalization of Rubisco and Rubisco activase in pyrenoid of *Chlamydomonas reinhardtii*. *Algae* **18**, 121–127.

Healey, F.P. (1970a). Hydrogen evolution by several algae. *Planta* **91**, 220–226.

Healey, F.P. (1970b). The mechanism of hydrogen evolution by *Chlamydomonas moewusii*. *Plant Physiol.* **45**, 153–159.

Hegemann, P. and Bruck, B. (1989). Light-induced stop response in *Chlamydomonas reinhardtii*: occurrence and adaptation phenomena. *Cell Motil. Cytoskeleton* **14**, 501–515.

Hegemann, P., Gärtner, W., and Uhl, R. (1991). All-*trans* retinal constitutes the functional chromophore in *Chlamydomonas* rhodopsin. *Biophys. J.* **60**, 1477–1489.

Hegemann, P., Fuhrmann, M., and Kateriya, S. (2001). Algal sensory photoreceptors. *J. Phycol.* **37**, 668–676.

Heifetz, P.B., Lers, A., Turpin, D.H., Gillham, N.W., Boynton, J.E., and Osmond, C.B. (1997). *dr* and *spr/sr* mutations of *Chlamydomonas reinhardtii* affecting D1 protein function and synthesis define two independent steps leading to chronic photoinhibition and confer differential fitness. *Plant Cell Environ.* **20**, 1145–1157.

Heimke, J.W. and Starr, R.C. (1979). The sexual process in several heterogamous *Chlamydomonas* strains in the subgenus Pleiochloris. *Arch. Protistenk.* **122**, 20–42.

Heitzer, M. and Zschoernig, B. (2007). Construction of modular tandem expression vectors for the green alga *Chlamydomonas reinhardtii* using the Cre/lox-system. *BioTechniques* **43**, 324–326, 328.

Heitzer, M., Eckert, A., Fuhrmann, M., and Griesbeck, C. (2007). Influence of codon bias on the expression of foreign genes in microalgae. *Adv. Exp. Med. Biol.* **616**, 46–53.

Hellebust, J.A. and Le Gresley, S.M.L. (1985). Growth characteristics of the marine rock pool flagellate *Chlamydomonas pulsatilla* Wollenweber (Chlorophyta). *Phycologia* **24**, 225–229.

Hellebust, J.A. and Lin, Y.-H. (1989). Regulation of glycerol and starch metabolism in *Chlamydomonas pulsatilla* in response to changes in salinity. *Plant Cell Environ.* **12**, 621–627.

Hellebust, J.A., Soto, C., and Hutchinson, T.C. (1982). Effect of naphthalene and aqueous crude oil extracts on the green flagellate *Chlamydomonas angulosa*. V. Heterotrophic responses. *Can. J. Bot.* **60**, 1495–1502.

Hellebust, J.A., Soto, C., and Hutchinson, T.C. (1985). Effect of naphthalene and aqueous crude oil extracts on the green flagellate *Chlamydomonas angulosa*. VII. Nitrate and methylamine uptake and retention. *Can. J. Bot.* **63**, 834–840.

Hellebust, J.A., Merida, T., and Ahmad, I. (1989). Operation of contractile vacuoles in the euryhaline green flagellate *Chlamydomonas pulsatilla* Chlorophyceae as a function of salinity. *Mar. Biol. (Berlin)* **100**, 373–380.

Hellio, C., Veron, B., and Le Gal, Y. (2004). Amino acid utilization by *Chlamydomonas reinhardtii*: specific study of histidine. *Plant Physiol. Biochem.* **42**, 257–261.

Hema, R., Senthil-Kumar, M., Shivakumar, S., Reddy, P.C., and Udayakumar, M. (2007). *Chlamydomonas reinhardtii*, a model system for functional validation of abiotic stress responsive genes. *Planta* **226**, 655–670.

Hendrickson, T.W., Perrone, C.A., Griffin, P., Wuichet, K., Mueller, J., Yang, P., Porter, M.E., and Sale, W.S. (2004). IC138 is a WD-repeat dynein intermediate chain required for light chain assembly and regulation of flagellar bending. *Mol. Biol. Cell* **15**, 5431–5442.

Hendrychová, J., Vítová, M., Bišová, K., Wiche, G., and Zachleder, V. (2002). Plectin-like proteins are present in cells of *Chlamydomonas eugametos* (Volvocales). *Folia Microbiol.* **47**, 535–539.

Hermesse, M.-P. and Matagne, R.F. (1980). Conditions for induced somatic fusions between protoplasts of *Chlamydomonas reinhardii*. *Arch. Int. Physiol. Biochim.* **88**, B282–B283.

Herrin, D.L. and Nickelsen, J. (2004). Chloroplast RNA processing and stability. *Photosynth. Res.* **82**, 301–314.

Herrin, D.L. and Schmidt, G.W. (1988). *trans*-Splicing of transcripts for the chloroplast *psaA1* gene. *In vivo* requirement for nuclear gene products. *J. Biol. Chem.* **263**, 14601–14604.

Herrin, D.L., Michaels, A.S., and Paul, A.L. (1986). Regulation of genes encoding the large subunit of ribulose-1,5-bisphosphate carboxylase and the photosystem II polypeptides D 1 and D 2 during the cell cycle of *Chlamydomonas reinhardtii*. *J. Cell Biol.* **103**, 1837–1845.

Herrin, D.L., Chen, Y.-F., and Schmidt, G.W. (1990). RNA splicing in *Chlamydomonas* chloroplasts. Self-splicing of 23S pre-RNA. *J. Biol. Chem.* **265**, 21134–21140.

Hicks, A., Drager, R.G., Higgs, D.C., and Stern, D.B. (2002). An mRNA 3′ processing site targets downstream sequences for rapid degradation in *Chlamydomonas* chloroplasts. *J. Biol. Chem.* **277**, 3325–3333.

Higgs, D.C., Kuras, R., Kindle, K.L., Wollman, F.A., and Stern, D.B. (1998). Inversions in the *Chlamydomonas* chloroplast genome suppress a *petD* 5′ untranslated region deletion by creating functional chimeric mRNAs. *Plant J.* **14**, 663–671.

Higgs, D.C., Shapiro, R.S., Kindle, K.L., and Stern, D.B. (1999). Small *cis*-acting sequences that specify secondary structures in a chloroplast mRNA are essential for RNA stability and translation. *Mol. Cell. Biol.* **19**, 8479–8491.

Hills, G.J. (1973). Cell wall assembly *in vitro* from *Chlamydomonas reinhardi*. *Planta* **115**, 17–23.

Hills, G.J., Gurney-Smith, M., and Roberts, K. (1973). Structure, composition, and morphogenesis of the cell wall of *Chlamydomonas reinhardi*. II. Electron microscopy and optical diffraction analysis. *J. Ultrastruct. Res.* **43**, 179–192.

Hills, G.J., Phillips, J.M., Gay, M.R., and Roberts, K. (1975). Self-assembly of a plant cell wall *in vitro*. *J. Mol. Biol.* **96**, 431–441.

Hipkiss, A.R. (1967). The effect of streptomycin on gametogenesis in *Chlamydomonas reinhardi*. *Life Sci.* **6**, 669–672.

Hipkiss, A.R. (1968). Bromouracil-resistant mutant of *Chlamydomonas*. *Can. J. Biochem.* **46**, 621–623.

Hippler, M., Drepper, F., Farah, J., and Rochaix, J.-D. (1997). Fast electron transfer from cytochrome c_6 and plastocyanin to photosystem I of *Chlamydomonas reinhardtii* requires PsaF. *Biochemistry* **36**, 6343–6349.

Hippler, M., Biehler, K., Krieger-Liszkay, A., van Dillewijn, J., and Rochaix, J.-D. (2000). Limitation in electron transfer in photosystem I donor side mutants of *Chlamydomonas reinhardtii*. Lethal photo-oxidative damage in high light is overcome in a suppressor strain deficient in the assembly of the light harvesting complex. *J. Biol. Chem.* **275**, 5852–5859.

Hippler, M., Klein, J., Fink, A., Allinger, T., and Hoerth, P. (2001). Towards functional proteomics of membrane protein complexes: analysis of thylakoid membranes from *Chlamydomonas reinhardtii*. *Plant J.* **28**, 595–606.

Hiramatsu, T., Nakamura, S., Misumi, O., and Kuroiwa, T. (2006). Morphological changes in mitochondrial and chloroplast nucleoids and mitochondria during the *Chlamydomonas reinhardtii* (Chlorophyceae) cell cycle. *J. Phycol.* **42**, 1048–1058.

Hirayama, K. and Hirano, R. (1970). Influences of high temperature and residual chlorine on marine phytoplankton. *Mar. Biol.* **7**, 205–213.

Hirono, M. and Yoda, A. (1997). Isolation and phenotypic characterization of *Chlamydomonas* mutants defective in cytokinesis. *Cell Struct. Funct.* **22**, 1–5.

Hirose, T. and Sugiura, M. (2004). Multiple elements required for translation of plastid *atpB* mRNA lacking the Shine-Dalgarno sequence. *Nucleic Acids Res.* **32**, 3503–3510.

Hirschberg, R. and Rodgers, S. (1978). Chemoresponses of *Chlamydomonas reinhardtii*. *J. Bacteriol.* **134**, 671–673.

Hodson, R.C., Williams, S.K., II, and Davidson, W.R., Jr. (1975). Metabolic control of urea catabolism in *Chlamydomonas reinhardi* and *Chlorella pyrenoidosa*. *J. Bacteriol.* **121**, 1022–1035.

Hoffman, J.L. and Goodenough, U.W. (1980). Experimental dissection of flagellar surface motility in *Chlamydomonas*. *J. Cell Biol.* **86**, 656–665.

Hoffman, N. and Grossman, A.R. (1999). Trafficking of proteins from chloroplast to vacuoles; perspectives and directions. *J. Phycol.* **35**, 443–445.

Hoffman, X.K. and Beck, C.F. (2005). Mating-induced shedding of cell walls, removal of walls from vegetative cells, and osmotic stress induce presumed cell wall genes in *Chlamydomonas*. *Plant Physiol.* **139**, 999–1014.

Hoffmans-Hohn, M., Martin, W., and Brinkmann, K. (1984). Multiple periodicities in the circadian system of unicellular algae. *Z. Naturforsch.* **39c**, 791–800.

Hoham, R.W. (1975). Optimum temperatures and temperature ranges for growth of snow algae. *Arct. Alp. Res.* **7**, 13–24.

Hoham, R.W., Bonome, T.A., Martin, C.W., and Leebens-Mack, J.H. (2002). A combined 18S rDNA and *rbcL* phylogenetic analysis of *Chloromonas* and *Chlamydomonas* (Chlorophyceae, Volvocales) emphasizing snow and other cold-temperature habitats. *J. Phycol.* **38**, 1051–1064.

Holland, E.M., Harz, H., Uhl, R., and Hegemann, P. (1997). Control of phobic behavioral responses by rhodopsin-induced photocurrents in *Chlamydomonas*. *Biophys. J.* **73**, 1395–1401.

Holloway, S.P. and Herrin, D.L. (1998). Processing of a composite large subunit rRNA: studies with *Chlamydomonas* mutants deficient in maturation of the 23S-like rRNA. *Plant Cell* **10**, 1193–1206.

Holloway, S.P., Deshpande, N.N., and Herrin, D.L. (1999). The catalytic group-I introns of the *psbA* gene of *Chlamydomonas reinhardtii*: core structures, ORFs and evolutionary implications. *Curr. Genet.* **36**, 69–78.

Holm-Hansen, O. (1964). Viability of lyophilized algae. *Can. J. Bot.* **42**, 127–137.

Holmes, J.A. and Dutcher, S.K. (1989). Cellular asymmetry in *Chlamydomonas reinhardtii*. *J. Cell Sci.* **94**, 273–285.

Holmes, J.A., Johnson, D.E., and Dutcher, S.K. (1993). Linkage group XIX of *Chlamydomonas reinhardtii* has a linear map. *Genetics* **133**, 865–874.

Homan, W.L., van Kalshoven, H., Kolk, A.H.J., Musgrave, A., Schuring, F., and van den Ende, H. (1987). Monoclonal antibodies to surface glycoconjugates in *Chlamydomonas eugametos* recognize strain-specific O-methyl sugars. *Planta* **170**, 328–335.

Homann, P.H. (2003). Hydrogen metabolism of green algae: discovery and early research – a tribute to Hans Gaffron and his coworkers. *Photosynth. Res.* **76**, 93–103.

Homer, R.B. and Roberts, K. (1979). Glycoprotein conformation in plant cell walls. Circular dichroism reveals a polyproline II structure. *Planta* **146**, 217–222.

Hommersand, M. (1960). On the nature of the light requirement for zygospore germination in a unicellular green alga, *Chlamydomonas reinhardi*. *J. Elisha Mitchell Soc.* **76**, 184–185.

Hommersand, M.H. and Thimann, K.V. (1965). Terminal respiration of vegetative cells and zygospores in *Chlamydomonas reinhardi*. *Plant Physiol.* **40**, 1220–1227.

Honeycutt, R.C. and Margulies, M.M. (1972). Control of ribosome content, gamete formation and amino acid uptake in wild-type and *arg-1 Chlamydomonas reinhardti*. *Biochim. Biophys. Acta* **281**, 399–405.

Hong, S. and Spreitzer, R.J. (1998). Nuclear-gene mutations suppress a defect in the expression of the chloroplast-encoded large subunit of ribulose-1,5-bisphosphate carboxylase/oxygenase. *Plant Physiol.* **116**, 1387–1392.

Hoober, J.K. and Stegeman, W.J. (1973). Control of the synthesis of a major polypeptide of chloroplast membranes in *Chlamydomonas reinhardi*. *J. Cell Biol.* **56**, 1–12.

Hoober, J.K., Park, H., Wolfe, G.R., Komine, Y., and Eggink, L.L. (1998). Assembly of light-harvesting systems. In: *The Molecular Biology of Chloroplasts and*

Mitochondria in Chlamydomonas (J.-D. Rochaix, M. Goldschmidt-Clermont, and S. Merchant, Eds.), pp. 363–376. Kluwer Adademic, Dordrecht.

Hoops, H.J. and Witman, G.B. (1983). Outer doublet heterogeneity reveals structural polarity related to beat direction in *Chlamydomonas* flagella. *J. Cell Biol.* **97**, 902–908.

Horne, R.W., Davies, D.R., Norton, K., and Gurney-Smith, M. (1971). Electron microscope and optical diffraction studies on isolated cell walls from *Chlamydomonas*. *Nature* **232**, 493–495.

Horst, C.J. and Witman, G.B. (1993). *ptx1*, a nonphototactic mutant of *Chlamydomonas*, lacks control of flagellar dominance. *J. Cell Biol.* **120**, 733–741.

Horst, C.J., Fishkind, D.J., Pazour, G.J., and Witman, G.B. (1999). An insertional mutant of *Chlamydomonas reinhardtii* with defective microtubule positioning. *Cell Motil. Cytoskeleton* **44**, 143–154.

Hoshaw, R.W. (1965). Mating types of *Chlamydomonas* from the collection of Gilbert M. Smith. *J. Phycol.* **1**, 194–196.

Hoshaw, R.W. and Ettl, H. (1966). *Chlamydomonas smithii sp. nov.* – A Chlamydomonad interfertile with *Chlamydomonas reinhardtii*. *J. Phycol.* **2**, 93–96.

Hoshaw, R.W. and Rosowski, J.R. (1973). Methods for microscopic algae. In: *Handbook of Phycological Methods. Vol. I. Culture Methods and Growth Measurements* (J.R. Stein, Ed.), pp. 53–68. Cambridge University Press, Cambridge.

Houba, S. and Loppes, R. (1985). The chloroplast DNA of *Chlamydomonas reinhardii* contains at least nine autonomous replication sequences active in yeast. *Arch. Int. Physiol. Biochim.* **93**, B92.

Hourcade, D.E. (1983). Marker rescue from bleomycin-treated *Chlamydomonas reinhardi*. *Genetics* **104**, 391–404.

Howe, G. and Merchant, S. (1992). The biosynthesis of membrane and soluble plastidic *c*-type cytochromes of *Chlamydomonas reinhardtii* is dependent on multiple common gene products. *EMBO J.* **11**, 2789–2801.

Howell, S.H. (1972). The differential synthesis and degradation of ribosomal DNA during the vegetative cell cycle in *Chlamydomonas reinhardi*. *Nature New Biol.* **240**, 264–267.

Howell, S.H. and Naliboff, J.A. (1973). Conditional mutants in *Chlamydomonas reinhardtii* blocked in the vegetative cell cycle. I. An analysis of cell cycle block points. *J. Cell Biol.* **57**, 760–772.

Howell, S.H. and Walker, L.L. (1976). Informational complexity of the nuclear and chloroplast genomes of *Chlamydomonas reinhardi*. *Biochim. Biophys. Acta* **418**, 249–256.

Howell, S.H. and Walker, L.L. (1977). Transcription of the nuclear and chloroplast genomes during the vegetative cell cycle in *Chlamydomonas reinhardi*. *Dev. Biol.* **56**, 11–23.

Hsiao, S.I.C. (1978). Effects of crude oils on the growth of arctic marine phytoplankton. *Environ. Pollut.* **17**, 93–107.

Hsieh, C.-H., Wu, M., and Yang, J. (1991). The sequence-directed bent DNA detected in the replication origin of *Chlamydomonas reinhardtii* chloroplast DNA is important for the replication function. *Mol. Gen. Genet.* **225**, 25–32.

Hu, S.X., Lau, K.W.K., and Wu, M. (2001). Cadmium sequestration in *Chlamydomonas reinhardtii*. *Plant Sci.* **161**, 987–996.

Huang, B. (1986). *Chlamydomonas reinhardtii*: a model system for the genetic analysis of flagellar structure and motility. *Int. Rev. Cytol.* **99**, 181–216.

Huang, B., Rifkin, M.R., and Luck, D.J.L. (1977). Temperature-sensitive mutations affecting flagellar assembly and function in *Chlamydomonas reinhardtii*. *J. Cell Biol.* **72**, 67–85.

Huang, B., Piperno, G., and Luck, D.J.L. (1979). Paralyzed flagella mutants of *Chlamydomonas reinhardtii* defective for axonemal doublet microtubule arms. *J. Biol. Chem.* **254**, 3091–3099.

Huang, B., Piperno, G., Ramanis, Z., and Luck, D.J.L. (1981). Radial spokes of *Chlamydomonas* flagella: genetic analysis of assembly and function. *J. Cell Biol.* **88**, 80–88.

Huang, B., Ramanis, Z., Dutcher, S.K., and Luck, D.J.L. (1982). Uniflagellar mutants of *Chlamydomonas*. Evidence for the role of basal bodies in transmission of positional information. *Cell* **29**, 745–753.

Huang, C. and Liu, X.-Q. (1992). Nucleotide sequence of the *frxC*, *petB* and *trnL* genes in the chloroplast genome of *Chlamydomonas reinhardtii*. *Plant Mol. Biol.* **18**, 985–988.

Huang, C., Wang, S., Chen, L., Lemieux, C., Otis, C., Turmel, M., and Liu, X.-Q. (1994). The *Chlamydomonas* chloroplast *clpP* gene contains translated large insertion sequences and is essential for cell growth. *Mol. Gen. Genet.* **244**, 151–159.

Huang, K.Y. and Beck, C.F. (2003). Phototropin is the blue-light receptor that controls multiple steps in the sexual life cycle of the green alga *Chlamydomonas reinhardtii*. *Proc. Natl. Acad. Sci. U. S. A.* **100**, 6269–6274.

Huang, K.Y., Merkle, T., and Beck, C.F. (2002). Isolation and characterization of a *Chlamydomonas* gene that encodes a putative blue-light photoreceptor of the phototropin family. *Physiol. Plant.* **115**, 613–622.

Huang, K.Y., Kunkel, T., and Beck, C.F. (2004). Localization of the blue-light receptor phototropin to the flagella of the green alga *Chlamydomonas reinhardtii*. *Mol. Biol. Cell* **15**, 3605–3614.

Hudock, G.A. (1962). The pathway of arginine biosynthesis in *Chlamydomonas reinhardi*. *Biochem. Biophys. Res. Commun.* **9**, 551–555.

Hudock, G.A. (1963). Repression of arginosuccinase in *Chlamydomonas reinhardi*. *Biochem. Biophys. Res. Commun.* **10**, 133–138.

Hudock, G.A., Gring, D.M., and Bart, C. (1971). Responses of *Chlamydomonas reinhardti* to specific nutritional limitation in continuous culture. *J. Protozool.* **18**, 128–131.

Hudock, M.O., Togasaki, R.K., Lien, S., Hosek, M., and San Pietro, A. (1979). A uniparentally inherited mutation affecting photophosphorylation in *Chlamydomonas reinhardi*. *Biochem. Biophys. Res. Commun.* **87**, 66–71.

Hummel, E., Schmickl, R., Hinz, G., Hillmer, S., and Robinson, D.G. (2007). Brefeldin A action and recovery in *Chlamydomonas* are rapid and involve fusion and fission of Golgi cisternae. *Plant Biol.* **9**, 489–501.

Hunnicutt, G.R., Kosfiszer, M.G., and Snell, W.J. (1990). Cell body and flagellar agglutinins in *Chlamydomonas reinhardtii*: the cell body plasma membrane is a reservoir for agglutinins whose migration to the flagella is regulated by a functional barrier. *J. Cell Biol.* **111**, 1605–1616.

Hutchinson, T.C., Hellebust, J.A., and Soto, C. (1981). Effect of naphthalene and aqueous crude oil extracts on the green flagellate *Chlamydomonas angulosa*. IV. Decreases in cellular manganese and potassium. *Can. J. Bot.* **59**, 742–749.

Hutchinson, T.C., Soto, C., and Hellebust, J.A. (1985). Effect of naphthalene and aqueous crude oil extracts on the green flagellate *Chlamydomonas angulosa*. VI. Phosphate uptake and retention. *Can. J. Bot.* **63**, 829–833.

Hutner, S.H. and Provasoli, L. (1951). The phytoflagellates. In: *Biochemistry and Physiology of Protozoa*, Vol. 1 (A. Lwoff, Ed.), pp. 27–128. Academic Press, New York.

Hutner, S.H., Provasoli, L., Schatz, A., and Haskins, C.P. (1950). Some approaches to the study of the role of metals in the metabolism of microorganisms. *Proc. Am. Philos. Soc.* **94**, 152–170.

Hwang, C.J., Monk, B.C., and Goodenough, U.W. (1981). Linkage of mutations affecting minus flagellar membrane agglutinability to the *mt*⁻ mating-type locus of *Chlamydomonas*. *Genetics* **99**, 41–47.

Hwang, S., Kawazoe, R., and Herrin, D.L. (1996). Transcription of *tufA* and other chloroplast-encoded genes is controlled by a circadian clock in *Chlamydomonas*. *Proc. Natl. Acad. Sci. U. S. A.* **93**, 996–1000.

Hyams, J. and Davies, D.R. (1972). The induction and characterisation of cell wall mutants of *Chlamydomonas reinhardi. Mutat. Res.* **14**, 381–389.

Iglesias, A.A., Charng, Y., Ball, S., and Preiss, J. (1994). Characterization of the kinetic, regulatory, and structural properties of ADP-glucose pyrophosphorylase from *Chlamydomonas reinhardtii. Plant Physiol.* **104**, 1287–1294.

Iliev, D., Voytsekh, O., Schmidt, E.M., Fiedler, M., Nykytenko, A., and Mittag, M. (2006a). A heteromeric RNA-binding protein is involved in maintaining acrophase and period of the circadian clock. *Plant Physiol.* **142**, 797–806.

Iliev, D., Voystekh, O., and Mittag, M. (2006b). The circadian system of *Chlamydomonas reinhardtii. Biol. Rhythm Res.* **37**, 323–333.

Im, C.S. and Grossman, A.R. (2002). Identification and regulation of high light-induced genes in *Chlamydomonas reinhardtii. Plant J.* **30**, 301–313.

Im, C.S., Matters, G.L., and Beale, S.I. (1996). Calcium and calmodulin are involved in blue light induction of the *gsa* gene for an early chlorophyll biosynthetic step in *Chlamydomonas. Plant Cell* **8**, 2245–2253.

Im, C.S., Eberhard, S., Huang, K., Beck, C.F., and Grossman, A.R. (2006). Phototropin involvement in the expression of genes encoding chlorophyll and carotenoid biosynthesis enzymes and LHC apoproteins in *Chlamydomonas reinhardtii. Plant J.* **48**, 1–16.

Imam, S.H. and Snell, W.J. (1988). The *Chlamydomonas* cell wall degrading enzyme, lysin, acts on two substrates within the framework of the wall. *J. Cell Biol.* **106**, 2211–2221.

Imam, S.H., Buchanan, M.J., Shin, H.C., and Snell, W.J. (1985). The *Chlamydomonas* cell wall: characterization of the wall framework. *J. Cell Biol.* **101**, 1599–1607.

Immeln, D., Schlesinger, R., Heberle, J., and Kottke, T. (2007). Blue light induces radical formation and autophosphorylation in the light-sensitive domain of *Chlamydomonas* cryptochrome. *J. Biol. Chem.* **282**, 21720–21728.

Inagaki, Y., Ehara, M., Watanabe, K.I., Hayashi-Ishimaru, Y., and Ohama, T. (1998). Directionally evolving genetic code: the UGA codon from stop to tryptophan in mitochondria. *J. Mol. Evol.* **47**, 378–384.

Infante, A., Lo, S., and Hall, J.L. (1995). A *Chlamydomonas* genomic library in yeast artificial chromosomes. *Genetics* **141**, 87–93.

Inglis, P.N., Boroevich, K.A., and Leroux, M.R. (2006). Piecing together a ciliome. *Trends Genet.* **22**, 491–500.

Ingram-Smith, C., Martin, S.R., and Smith, K.S. (2006). Acetate kinase: not just a bacterial enzyme. *Trends Microbiol.* **14**, 249–253.

Inoue, K., Dreyfuss, B.W., Kindle, K.L., Stern, D.B., Merchant, S., and Sodeinde, O.A. (1997). *Ccs1*, a nuclear gene required for the post-translational assembly of chloroplast *c*-type cytochromes. *J. Biol. Chem.* **272**, 31747–31754.

Iomini, C., Li, L., Mo, W., Dutcher, S.K., and Piperno, G. (2006). Two flagellar genes, *AGG2* and *AGG3*, mediate orientation to light in *Chlamydomonas. Curr. Biol.* **16**, 1147–1153.

Irihimovitch, V. and Stern, D.B. (2006). The sulfur acclimation SAC3 kinase is required for chloroplast transcriptional repression under sulfur limitation in *Chlamydomonas reinhardtii. Proc. Natl. Acad. Sci. U. S. A.* **103**, 7911–7916.

Irmer, U., Wachholz, I., Schaefer, H., and Lorch, D.W. (1986). Influence of lead on *Chlamydomonas reinhardii* Dangeard (Volvocales, Chlorophyta): accumulation, toxicity and ultrastructural changes. *Environ. Exp. Bot.* **26**, 97–105.

Ishikura, K., Takaoka, Y., Kato, K., Sekine, M., Yoshida, K., and Shinmyo, A. (1999). Expression of a foreign gene in *Chlamydomonas reinhardtii* chloroplast. *J. Biosci. Bioeng.* **87**, 307–314.

Isogai, N., Kamiya, R., and Yoshimura, K. (2000). Dominance between the two flagella during phototactic turning in *Chlamydomonas. Zool. Sci.* **17**, 1261–1266.

Iwasa, K. and Murakami, S. (1968). Palmelloid formation of *Chlamydomonas*. I. Palmelloid induction by organic acids. *Physiol. Plant.* **21**, 1224–1233.

Iwasa, K. and Murakami, S. (1969). Palmelloid formation of *Chlamydomonas*. II. Mechanism of palmelloid formation by organic acids. *Physiol. Plant.* **22**, 43–50.

Jabusch, T.W. and Swackhamer, D.L. (2004). Subcellular accumulation of polychlorinated biphenyls in the green alga *Chlamydomonas reinhardtii. Environ. Toxicol. Chem.* **23**, 2823–2830.

Jacobshagen, S., Kindle, K.L., and Johnson, C.H. (1996). Transcription of CABII is regulated by the biological clock in *Chlamydomonas reinhardtii. Plant Mol. Biol.* **31**, 1173–1184.

Jacobshagen, S., Whetstine, J.R., and Boling, J.M. (2001). Many but not all genes in *Chlamydomonas reinhardtii* are regulated by the circadian clock. *Plant Biol.* **3**, 592–597.

Jaenicke, L. and Starr, R.C. (1996). The lurlenes, a new class of plastoquinone-related mating pheromones from *Chlamydomonas allensworthii* (Chlorophyceae). *Eur. J. Biochem.* **241**, 581–585.

Jaenicke, L. and Waffenschmidt, S. (1981). Liberation of reproductive units in *Volvox* and *Chlamydomonas*: proteolytic processes. *Ber. Deutsch. Bot. Ges.* **94**, 375–386.

Jaenicke, L., Kuhne, W., Spessert, R., Wahle, U., and Waffenschmidt, S. (1987). Cell-wall lytic enzymes (autolysins) of *Chlamydomonas reinhardtii* are (hydroxy) proline-specific proteases. *Eur. J. Biochem.* **170**, 485–491.

Jain, M., Shrager, J., Harris, E.H., Halbrook, R., Grossman, A.R., Hauser, C., and Vallon, O. (2007). EST assembly supported by a draft genome sequence: an analysis of the *Chlamydomonas reinhardtii* transcriptome. *Nucleic Acids Res.* **35**, 2074–2083.

Jakubiec, H. (1984). Studies on the Polish Chlamydomonadaceae. *Nova Hedwigia* **39**, 269–292.

Jamers, A., van der Ven., K., Moens, L., Robbens, J., Potters, G., Guisez, Y., Blust, R., and de Coen, W. (2006). Effect of copper exposure on gene expression profiles in *Chlamydomonas reinhardtii* based on microarray analysis. *Aquat. Toxicol.* **80**, 249–260.

James, S.W. and Lefebvre, P.A. (1992). Genetic interactions among *Chlamydomonas reinhardtii* mutations that confer resistance to anti-microtubule herbicides. *Genetics* **130**, 305–314.

James, S.W., Ranum, L.P.W., Silflow, C.D., and Lefebvre, P.A. (1988). Mutants resistant to anti-microtubule herbicides map to a locus on the UNI linkage group in *Chlamydomonas reinhardtii. Genetics* **118**, 141–147.

James, S.W., Silflow, C.D., Thompson, M.D., Ranum, L.P.W., and Lefebvre, P.A. (1989). Extragenic suppression and synthetic lethality among *Chlamydomonas reinhardtii* mutants resistant to anti-microtubule drugs. *Genetics* **122**, 567–577.

James, S.W., Silflow, C.D., Stroom, P., and Lefebvre, P.A. (1993). A mutation in the alpha1-tubulin gene of *Chlamydomonas reinhardtii* confers resistance to anti-microtubule herbicides. *J. Cell Sci.* **106**, 209–218.

Janero, D.R. and Barrnett, R. (1981a). Cellular and thylakoid-membrane glycolipids of *Chlamydomonas reinhardtii* 137+. *J. Lipid Res.* **22**, 1119–1125.

Janero, D.R. and Barrnett, R. (1981b). Cellular and thylakoid-membrane phospholipids of *Chlamydomonas reinhardtii* 137+. *J. Lipid Res.* **22**, 1126–1130.

Janero, D.R. and Barrnett, R. (1982a). Comparative analysis of diacylglyceryltrimethylhomoserine in *Ochromonas danica* and in *Chlamydomonas reinhardtii* 137+. *Phytochemistry* **21**, 47–50.

Janero, D.R. and Barrnett, R. (1982b). Isolation and characterization of an ether-linked homoserine lipid from the thylakoid membrane of *Chlamydomonas reinhardtii* 137+. *J. Lipid Res.* **23**, 307–316.

Janssen, M., de Bresser, L., Baijens, T., Tramper, J., Mur, L.R., Snel, J.F.H., and Wijffels, R.H. (2000). Scale-up aspects of photobioreactors: effects of mixing-induced light/dark cycles. *J. Appl. Phycol.* **12**, 225–237.

Jarvik, J.W. and Rosenbaum, J. (1980). Oversized flagellar membrane protein in para-lyzed mutants of *Chlamydomonas reinhardtii*. *J. Cell Biol.* **85**, 258–272.

Jarvik, J.W., Reinhart, F.D., Kuchka, M.R., and Adler, S.A. (1984). Altered flagellar size-control in *shf-1* short-flagella mutants of *Chlamydomonas reinhardtii*. *J. Protozool.* **31**, 199–204.

Jasper, F., Quednau, B., Kortenjann, M., and Johanningmeier, U. (1991). Control of *cab* gene expression in synchronized *Chlamydomonas reinhardtii* cells. *J. Photochem. Photobiol.* **11**, 139–150.

Jeong, B.R., Wu-Scharf, D., Zhang, C.M., and Cerutti, H. (2002). Suppressors of transcriptional transgenic silencing in *Chlamydomonas* are sensitive to DNA-damaging agents and reactivate transposable elements. *Proc. Natl. Acad. Sci. U. S. A.* **99**, 1076–1081.

Jiang, K.-S. and Barber, G.A. (1975). Polysaccharide from cell walls of *Chlamydomonas reinhardtii*. *Phytochemistry* **14**, 2459–2461.

Jiao, H.S., Hicks, A., Simpson, C., and Stern, D.B. (2004). Short dispersed repeats in the *Chlamydomonas* chloroplast genome are collocated with sites for mRNA 3′ end formation. *Curr. Genet.* **45**, 311–322.

Johanningmeier, U. and Heiss, S. (1993). Construction of a *Chlamydomonas reinhardtii* mutant with an intronless *psbA* gene. *Plant Mol. Biol.* **22**, 91–99.

Johanningmeier, U., Bodner, U., and Wildner, G.F. (1987). A new mutation in the gene coding for the herbicide-binding protein in *Chlamydomonas*. *FEBS Lett.* **211**, 221–224.

John, P.C.L. (1984). Control of the cell division cycle in *Chlamydomonas*. *Microbiol. Sci.* **1**, 96–101.

John, P.C.L. (1987). Control points in the *Chlamydomonas* cell cycle. In: *Molecular and Cellular Aspects of Algal Development* (W. Wiessner and D.G. Robinson, Eds.), pp. 9–16. Springer-Verlag, Berlin.

Johnson, C.H. (2001). Endogenous timekeepers in photosynthetic organisms. *Annu. Rev. Physiol.* **63**, 695–728.

Johnson, C.H. and Schmidt, G.W. (1993). The *psbB* gene cluster of the *Chlamydomonas reinhardtii* chloroplast: sequence and transcriptional analyses of *psbN* and *psbH*. *Plant Mol. Biol.* **22**, 645–658.

Johnson, C.H., Kondo, T., and Hastings, J.W. (1991). Action spectrum for resetting the circadian phototaxis rhythm in the *CW15* strain of *Chlamydomonas*. II. Illuminated cells. *Plant Physiol.* **97**, 1122–1129.

Johnson, D.E. and Dutcher, S.K. (1991). Molecular studies of linkage group XIX of *Chlamydomonas reinhardtii*: evidence against a basal body location. *J. Cell Biol.* **113**, 339–346.

Johnson, D.E. and Dutcher, S.K. (1993). A simple, reliable method for prolonged frozen storage of *Chlamydomonas*. *Trends Genet.* **9**, 194–195.

Johnson, E. and Melis, A. (2004). Functional characterization of *Chlamydomonas reinhardtii* with alterations in the *atpE* gene. *Photosynth. Res.* **82**, 131–140.

Johnson, K.A. and Rosenbaum, J.L. (1990). The basal bodies of *Chlamydomonas reinhardtii* do not contain immunologically detectable DNA. *Cell* **62**, 339–346.

Johnson, K.A., Porter, M.E., and Shimizu, T. (1984). Mechanism of force production for microtubule-dependent movements. *J. Cell Biol.* **99**, 132s–136s.

Johnson, U.G. and Porter, K.R. (1968). Fine structure of cell division in *Chlamydomonas reinhardi*. Basal bodies and microtubules. *J. Cell Biol.* **38**, 403–425.

Johnston, S.A., Anziano, P.Q., Shark, K., Sanford, J.C., and Butow, R.A. (1988). Mitochondrial transformation in yeast by bombardment with microprojectiles. *Science* **240**, 1538–1541.

Jones, R.F. (1970). Physiological and biochemical aspects of growth and gametogenesis in *Chlamydomonas reinhardtii*. *Ann. N. Y. Acad. Sci.* **175**, 648–659.

Josef, K., Saranak, J., and Foster, K.W. (2005a). An electro-optic monitor of the behavior of *Chlamydomonas reinhardtii* cilia. *Cell Motil. Cytoskeleton* **61**, 83–96.

Josef, K., Saranak, J., and Foster, K.W. (2005b). Ciliary behavior of a negatively phototactic *Chlamydomonas reinhardtii*. *Cell Motil. Cytoskeleton* **61**, 97–111.

Josef, K., Saranak, J., and Foster, K.W. (2006). Linear systems analysis of the ciliary steering behavior associated with negative-phototaxis in *Chlamydomonas reinhardtii*. *Cell Motil. Cytoskeleton* **63**, 758–777.

Jupe, E.R., Chapman, R.L., and Zimmer, E.A. (1988). Nuclear ribosomal RNA genes and algal phylogeny – the *Chlamydomonas* example. *BioSystems* **21**, 223–230.

Kaise, T., Fujiwara, S., Tsuzuki, M., Sakurai, T., Saitoh, T., and Mastubara, C. (1999). Accumulation of arsenic in a unicellular alga *Chlamydomonas reinhardtii*. *Appl. Organometal. Chem.* **13**, 107–111.

Kaiser, J. (2000). Power from pond scum. *Science* **287**, 1580–1583.

Kalakoutskii, K.L. and Fernández, E. (1995). *Chlamydomonas reinhardtii* nitrate reductase complex has 105 kDa subunits in the wild-type strain and a structural mutant. *Plant Sci.* **105**, 195–206.

Kalina, T. and Stefanova, I. (1992). Contribution to the fine structure of the eyespot of *Chlamydomonas geitleri* Ettl (Chlamydophyceae Chlorophyta). *Arch. Protistenk.* **141**, 153–157.

Kalmus, H. (1932). Über den Erhaltungswert der phänotypischen (morphologischen) Anisogamie und die Entstehung der ersten Geschlechtsunterschiede. *Biol. Zentralblatt* **52**, 716–726.

Kam, V., Moseyko, N., Nemson, E., and Feldman, L.J. (1999). Gravitaxis in *Chlamydomonas reinhardtii*: characterization using video microscopy and computer analysis. *Int. J. Plant Sci.* **160**, 1093–1098.

Kamiya, R. and Witman, G.B. (1984). Submicromolar levels of calcium control the balance of beating between the two flagella in demembranated models of *Chlamydomonas*. *J. Cell Biol.* **98**, 97–107.

Kamiya, R., Kurimoto, E., and Muto, E. (1991). Two types of *Chlamydomonas* flagellar mutants missing different components of inner-arm dynein. *J. Cell Biol.* **112**, 441–447.

Kan, G.F., Miao, J.L., Shi, C.J., and Li, G.Y. (2006). Proteomic alterations of antarctic ice microalga *Chlamydomonas sp.* under low-temperature stress. *J. Integrative Plant Biol.* **48**, 965–970.

Kang, Y. and Mitchell, D.R. (1998). An intronic enhancer is required for deflagellation-induced transcriptional regulation of a *Chlamydomonas reinhardtii* dynein gene. *Mol. Biol. Cell* **9**, 3085–3094.

Kapraun, D.F. (2007). Nuclear DNA content estimates in green algal lineages: Chlorophyta and Streptophyta. *Ann. Bot.* **99**, 677–701.

Karlsson, J., Hiltonen, T., Husic, H.D., Ramazanov, Z., and Samuelsson, G. (1995). Intracellular carbonic anhydrase of *Chlamydomonas reinhardtii*. *Plant Physiol.* **109**, 533–539.

Karlsson, J., Clarke, A.K., Chen, Z.Y., Hugghins, S.Y., Park, Y.I., Husic, H.D., Moroney, J.V., and Samuelsson, G. (1998). A novel alpha-type carbonic anhydrase associated with the thylakoid membrane in *Chlamydomonas reinhardtii* is required for growth at ambient CO_2. *EMBO J.* **17**, 1208–1216.

Karukstis, K.K. and Sauer, K. (1983). Fluorescence decay kinetics of chlorophyll in photosynthetic membranes. *J. Cell. Biochem.* **23**, 131–158.

Kaska, D.D. and Gibor, A. (1982). Initiation of cell wall lysis in gametes of *Chlamydomonas reinhardi* by isolated flagella of the complementary mating type. *Exp. Cell Res.* **138**, 121–125.

Kater, J.M. (1929). Morphology and division of *Chlamydomonas* with reference to the phylogeny of the flagellate neuromotor system. *Univ. Calif. Publ. Zool.* **33**, 125 169.

Kateriya, S., Nagel, G., Bamberg, E., and Hegemann, P. (2004). "Vision" in single-celled algae. *News Physiol. Sci.* **19**, 133–137.

Kates, J.R. and Jones, R.F. (1964). The control of gametic differentiation in liquid cultures of *Chlamydomonas*. *J. Cell Comp. Physiol.* **63**, 157–164.

Kates, J.R. and Jones, R.F. (1967). Periodic increases in enzyme activity in synchronized cultures of *Chlamydomonas reinhardtii*. *Biochim. Biophys. Acta* **145**, 153–158.

Kates, J.R., Chiang, K.-S., and Jones, R.F. (1968). Studies on DNA replication during synchronized vegetative growth and gametic differentiation in *Chlamydomonas reinhardtii*. *Exp. Cell Res.* **49**, 121–135.

Kathir, P., LaVoie, M., Brazelton, W.J., Haas, N.A., Lefebvre, P.A., and Silflow, C.D. (2003). A molecular map of the *Chlamydomonas reinhardtii* nuclear genome. *Eukaryotic Cell* **2**, 362–379.

Kato, J., Yamahara, T., Tanaka, K., Takio, S., and Satoh, T. (1997). Characterization of catalase from green algae *Chlamydomonas reinhardtii*. *J. Plant Physiol.* **151**, 262–268.

Kato, K., Marui, T., Kasai, S., and Shinmyo, A. (2007). Artificial control of transgene expression in *Chlamydomonas reinhardtii* chloroplast using the *lac* regulation system from *Escherichia coli*. *J. Biosci. Bioeng.* **104**, 207–213.

Kato-Minoura, T., Hirono, M., and Kamiya, R. (1997). *Chlamydomonas* inner-arm dynein mutant, *ida5*, has a mutation in an actin-encoding gene. *J. Cell Biol.* **137**, 649–656.

Katoh, S. (2003). Early research on the role of plastocyanin in photosynthesis. *Photosynth. Res.* **76**, 255–261.

Katz, K.R. and McLean, R.J. (1979). Rhizoplast and rootlet system of the flagellar apparatus of *Chlamydomonas moewusii*. *J. Cell Sci.* **39**, 373–381.

Katz, Y.S. and Danon, A. (2002). The 3'-untranslated region of chloroplast *psbA* mRNA stabilizes binding of regulatory proteins to the leader of the message. *J. Biol. Chem.* **277**, 18665–18669.

Kawachi, M. and Noël, M.-H. (2005). Sterilization and sterile technique. In: *Algal Culturing Techniques* (R.A. Andersen, Ed.), pp. 65–81. Elsevier, Amsterdam.

Kawecka, B. (1981). Biology and ecology of snow algae. 2. Formation of aplanospores in *Chlamydomonas nivalis* (Bauer) Wille (Chlorophyta, Volvocales). *Acta Hydrobiol.* **23**, 211–215.

Kawecka, B. and Drake, B.G. (1978). Biology and ecology of snow algae. l. The sexual reproduction of *Chlamydomonas nivalis* (Bauer) Wille (Chlorophyta, Volvocales). *Acta Hydrobiol.* **20**, 111–116.

Keller, L.C. and Marshall, W.F. (2008). Isolation and proteomic analysis of *Chlamydomonas* centrioles. *Methods Mol. Biol.* **432**, 289–300.

Keller, L.C., Romijn, E.P., Zamora, I., Yates, J.R., III, and Marshall, W.F. (2005). Proteomic analysis of isolated *Chlamydomonas* centrioles reveals orthologs of ciliary-disease genes. *Curr. Biol.* **15**, 1090–1098.

Keller, S.J. and Ho, C. (1981). Chloroplast DNA replication in *Chlamydomonas reinhardtii*. *Int. Rev. Cytol.* **69**, 157–190.

Kent, R.A. and Caux, P.Y. (1995). Sublethal effects of the insecticide fenitrothion on freshwater phytoplankton. *Can. J. Bot.* **73**, 45–53.

Kent, R.A. and Currie, D. (1995). Predicting algal sensitivity to a pesticide stress. *Environ. Toxicol. Chem.* **14**, 983–991.

Kent, R.A. and Weinberger, P. (1991). Multibiological-level responses of freshwater phytoplankton to pesticide stress. *Environ. Toxicol. Chem.* **10**, 209–216.

Kern, J. and Renger, G. (2007). Photosystem II: structure and mechanism of the water:plastoquinone oxidoreductase. *Photosynth. Res.* **94**, 183–202.

Keskiaho, K., Hieta, R., Sormunen, R., and Myllyharju, J. (2007). *Chlamydomonas reinhardtii* has multiple prolyl 4-hydroxylases, one of which is essential for proper cell wall assembly. *Plant Cell* **19**, 256–269.

Kessler, E. (1974). Hydrogenase, photoreduction and anaerobic growth. In: *Algal Physiology and Biochemsitry* (W.D.P. Stewart, Ed.), pp. 456–473. Blackwell, Oxford.

Kessler, J.O. (1985). Hydrodynamic focusing of motile algal cells. *Nature* **313**, 218–220.

Kessler, J.O. (1991). The dynamics of unicellular swimming organisms. *ASGSB Bull.* **4**, 97–105.

Kessler, J.O., Hill, N.A., and Häder, D.-P. (1992). Orientation of swimming flagellates by simultaneously acting external factors. *J. Phycol.* **28**, 816–822.

Ketchner, S.L. and Malkin, R. (1996). Nucleotide sequence of the *PetM* gene encoding a 4 kDa subunit of the cytochrome b_6f complex from *Chlamydomonas reinhardtii*. *Biochim. Biophys. Acta* **1273**, 195–197.

Ketchner, S.L., Drapier, D., Olive, J., Gaudriault, S., Girard-Bascou, J., and Wollman, F.-A. (1995). Chloroplasts can accommodate inclusion bodies. Evidence from a mutant of *Chlamydomonas reinhardtii* defective in the assembly of the chloroplast ATP synthase. *J. Biol. Chem.* **270**, 15299–15306.

Khrebtukova, I. and Spreitzer, R.J. (1996a). Nucleotide sequences of the chloroplast *trnS-GCU* and *ycf12* genes (Accession No. U40346) of *Chlamydomonas reinhardtii* (PGR95-117). *Plant Physiol.* **110**, 336.

Khrebtukova, I. and Spreitzer, R.J. (1996b). Elimination of the *Chlamydomonas* gene family that encodes the small subunit of ribulose-1,5-bisphosphate carboxylase oxygenase. *Proc. Natl. Acad. Sci. U. S. A.* **93**, 13689–13693.

Kiaulehn, S., Voytsekh, O., Fuhrmann, M., and Mittag, M. (2007). The presence of UG-repeat sequences in the 3′-UTRs of reporter luciferase mRNAs mediates circadian expression and can determine acrophase in *Chlamydomonas reinhardtii*. *J. Biol. Rhythms* **22**, 275–277.

Kilz, S., Waffenschmidt, S., and Budzikiewicz, H. (2000). Mass spectrometric analysis of hydroxyproline glycans. *J. Mass Spectrom.* **35**, 689–697.

Kilz, S., Budzikiewicz, H., and Waffenschmidt, S. (2002). In-gel deglycosylation of sodiumdodecyl sulfate polyacrylamide gel electrophoresis-separated glycoproteins for carbohydrate estimation by matrix-assisted laser desorption/ionization time-of-flight mass spectrometry. *J. Mass Spectrom.* **37**, 331–335.

Kim, H.Y., Fomenko, D.E., Yoon, Y.E., and Gladyshev, V.N. (2006). Catalytic advantages provided by selenocysteine in methionine-*S*-sulfoxide reductases. *Biochemistry* **45**, 13697–13704.

Kim, K.S., Feild, E., King, N., Yaoi, T., Kustu, S., and Inwood, W. (2005). Spontaneous mutations in the ammonium transport gene *AMT4* of *Chlamydomonas reinhardtii*. *Genetics* **170**, 631–644.

Kim, K.S., Kustu, S., and Inwood, W. (2006). Natural history of transposition in the green alga *Chlamydomonas reinhardtii*: use of the *AMT4* locus as an experimental system. *Genetics* **173**, 2005–2019.

Kim, Y.S., Oyaizu, H., Matsumoto, S., Watanabe, M.M., and Nozaki, H. (1994). Chloroplast small subunit ribosomal RNA gene sequence from *Chlamydomonas parkeae* (Chlorophyta): molecular phylogeny of a green alga with a peculiar pigment composition. *Eur. J. Phycol.* **29**, 213–217.

Kindle, K.L. (1990). High-frequency nuclear transformation of *Chlamydomonas reinhardtii*. *Proc. Natl. Acad. Sci. U. S. A.* **87**, 1228–1232.

Kindle, K.L. and Lawrence, S.D. (1998). Transit peptide mutations that impair *in vitro* and *in vivo* chloroplast protein import do not affect accumulation of the gamma-subunit of chloroplast ATPase. *Plant Physiol.* **116**, 1179–1190.

Kindle, K.L., Schnell, R.A., Fernández, E., and Lefebvre, P.A. (1989). Stable nuclear transformation of *Chlamydomonas* using the *Chlamydomonas* gene for nitrate reductase. *J. Cell Biol.* **109**, 2589–2601.

King, S.J. and Dutcher, S.K. (1997). Phosphoregulation of an inner dynein arm complex in *Chlamydomonas reinhardtii* is altered in phototactic mutant strains. *J. Cell Biol.* **136**, 177–191.

King, S.M. and Patel-King, R.S. (1995). The Mr = 8,000 and 11,000 outer arm dynein light chains from *Chlamydomonas* flagella have cytoplasmic homologues. *J. Biol. Chem.* **270**, 11445–11452.

Kinoshita, T., Fukuzawa, H., Shimada, T., Saito, T., and Matsuda, Y. (1992). Primary structure and expression of a gamete lytic enzyme in *Chlamydomonas reinhardtii*: similarity of functional domains to matrix metalloproteases. *Proc. Natl. Acad. Sci. U. S. A.* **89**, 4693–4697.

Kirk, D.L. (1998). *Volvox. A Search for the Molecular and Genetic Origins of Multicellularity and Cellular Differentiation*. Cambridge University Press, Cambridge. Developmental and Cell Biology Series, No. 33. 381 pp.

Kirk, D.L. (2006). Oogamy: inventing the sexes. *Curr. Biol.* **16**, R1028–R1030.

Kirk, D.L. and Kirk, M.M. (1978a). Amino acid and urea uptake in ten species of Chlorophyta. *J. Phycol.* **14**, 198–203.

Kirk, D.L. and Kirk, M.M. (1978b). Carrier-mediated uptake of arginine and urea by *Chlamydomonas reinhardtii*. *Plant Physiol.* **61**, 556–560.

Kitayama, M., Kitayama, K., and Togasaki, R.K. (1994). A cDNA clone encoding a ferredoxin-NADP$^+$ reductase from *Chlamydomonas reinhardtii*. *Plant Physiol.* **106**, 1715–1716.

Klein, T.M., Wolf, E.D., Wu, R., and Sanford, J.C. (1987). High-velocity microprojectiles for delivering nucleic acids into living cells. *Nature* **327**, 70–73.

Klein, U. (1987). Intracellular carbon partitioning in *Chlamydomonas reinhardtii*. *Plant Physiol.* **85**, 892–897.

Klein, U. and Betz, A. (1978a). Induced protein synthesis during the adaptation to H_2 production in *Chlamydomonas moewusii*. *Physiol. Plant.* **42**, 1–4.

Klein, U. and Betz, A. (1978b). Fermentative metabolism of hydrogen-evolving *Chlamydomonas moewusii*. *Plant Physiol.* **61**, 953–956.

Klein, U., De Camp, J.D., and Bogorad, L. (1992). Two types of chloroplast gene promoters in *Chlamydomonas reinhardtii*. *Proc. Natl. Acad. Sci. U. S. A.* **89**, 3453–3457.

Klein, U., Salvador, M.L., and Bogorad, L. (1994). Activity of the *Chlamydomonas* chloroplast *rbcL* gene promoter is enhanced by a remote sequence element. *Proc. Natl. Acad. Sci. U. S. A.* **91**, 10819–10823.

Klinkert, B., Schwarz, C., Pohlmann, S., Pierre, Y., Girard-Bascou, J., and Nickelsen, J. (2005). Relationship between mRNA levels and protein accumulation in a chloroplast promoter-mutant of *Chlamydomonas reinhardtii*. *Mol. Genet. Genomics* **274**, 637–643.

Klinkert, B., Elles, I., and Nickelsen, J. (2006). Translation of chloroplast *psbD* mRNA in *Chlamydomonas* is controlled by a secondary RNA structure blocking the AUG start codon. *Nucleic Acids Res.* **34**, 386–394.

Klis, F.M., Samson, M.R., Touw, E., Musgrave, A., and van den Ende, H. (1985). Sexual agglutination in the unicellular green alga *Chlamydomonas eugametos*. Identification and properties of the mating type plus agglutination factor. *Plant Physiol.* **79**, 740–745.

Klis, F.M., Crabbendam, K., van Egmond, P., and van den Ende, H. (1989). Ultrastructure and properties of the sexual agglutinins of the biflagellate green alga *Chlamydomonas moewusii*. *Sex. Plant Reprod.* **2**, 213–218.

Kloppstech, K., Meyer, G., Schuster, G., and Ohad, I. (1985). Synthesis, transport and localization of a nuclear coded 22-kd heat-shock protein in the chloroplast membranes of peas and *Chlamydomonas reinhardi*. *EMBO J.* **4**, 1901–1909.

Knutsen, G. and Lien, T. (1981). Properties of synchronous cultures of *Chlamydomonas reinhardti* under optimal conditions, and some factors influencing them. *Ber. Deutsch. Bot. Ges.* **94**, 599–611.

Knutsen, G., Lien, T., Schreiner, Ø., and Vaage, R. (1973). Selection synchrony of *Chlamydomonas* using the Rastgeldi Threshold centrifuge. *Exp. Cell Res.* **81**, 26–30.

Kobayashi, I., Fujiwara, S., Sacgusa, H., Inouhc, M., Matsumoto, H., and Tsuzuki, M. (2005). Relief of arsenate toxicity by Cd-stimulated phytochelatin synthesis in the green alga *Chlamydomonas reinhardtii. Mar. Biotechnol.* **8**, 94–101.

Koblenz, B. and Lechtreck, K.-F. (2005). The *NIT1* promoter allows inducible and reversible silencing of centrin in *Chlamydomonas reinhardtii. Eukaryotic Cell* **4**, 1959–1962.

Koblenz, B., Schoppmeier, J., Grunow, A., and Lechtreck, K.F. (2003). Centrin deficiency in *Chlamydomonas* causes defects in basal body replication, segregation and maturation. *J. Cell Sci.* **116**, 2635–2646.

Kochert, G. (1978). Sexual pheromones in algae and fungi. *Annu. Rev. Plant Physiol.* **29**, 461–486.

Kohinata, T., Nishino, H., and Fukuzawa, H. (2008). Significance of zinc in a regulatory protein, CCM1, which regulates the carbon-concentrating mechanism in *Chlamydomonas reinhardtii. Plant Cell Physiol.* **49**, 273–283.

Kojima, K.K. and Fujiwara, H. (2005). An extraordinary retrotransposon family encoding dual endonucleases. *Genome Res.* **15**, 1106–1117.

Kol, E. (1941). The green snow of Yellowstone National Park. *Am. J. Bot.* **28**, 185–191.

Kol, E. and Flint, E.A. (1968). Algae in green ice from the Balleny Islands, Antarctica. *N.Z. J. Bot.* **6**, 249–261.

Komine, Y., Eggink, L.L., Park, H.S., and Hoober, J.K. (2000). Vacuolar granules in *Chlamydomonas reinhardtii*: polyphosphate and a 70-kDa polypeptide as major components. *Planta* **210**, 897–905.

Komine, Y., Kikis, E., Schuster, G., and Stern, D. (2002). Evidence for *in vivo* modulation of chloroplast RNA stability by 3′-UTR homopolymeric tails in *Chlamydomonas reinhardtii. Proc. Natl. Acad. Sci. U. S. A.* **99**, 4085–4090.

Kondo, T., Johnson, C.H., and Hastings, J.W. (1991). Action spectrum for resetting the circadian phototaxis rhythm in the *CW15* strain of *Chlamydomonas*. I. Cells in darkness. *Plant Physiol.* **95**, 197–205.

Kosourov, S., Tsygankov, A., Seibert, M., and Ghirardi, M.L. (2002). Sustained hydrogen photoproduction by *Chlamydomonas reinhardtii*: effects of culture parameters. *Biotechnol. Bioeng.* **78**, 731–740.

Kosourov, S., Makarova, V., Fedorov, A.S., Tsygankov, A., Seibert, M., and Ghirardi, M.L. (2005). The effect of sulfur re-addition on H_2 photoproduction by sulfur-deprived green algae. *Photosynth. Res.* **85**, 295–305.

Kosourov, S., Patrusheva, E., Ghirardi, M.L., Seibert, M., and Tsygankov, A. (2007). A comparison of hydrogen photoproduction by sulfur-deprived *Chlamydomonas reinhardtii* under different growth conditions. *Biotechnol. J.* **128**, 776–787.

Koutoulis, A., Pazour, G.J., Wilkerson, C.G., Inaba, K., Sheng, H., Takada, S., and Witman, G.B. (1997). The *Chlamydomonas reinhardtii ODA3* gene encodes a protein of the outer dynein arm docking complex. *J. Cell Biol.* **137**, 1069–1080.

Kovar, J.L., Zhang, J., Funke, R.P., and Weeks, D.P. (2002). Molecular analysis of the acetolactate synthase gene of *Chlamydomonas reinhardtii* and development of a genetically engineered gene as a dominant selectable marker for genetic transformation. *Plant J.* **29**, 109–117.

Kozminski, K.G., Johnson, K.A., Forscher, P., and Rosenbaum, J.L. (1993). A motility in the eukaryotic flagellum unrelated to flagellar beating. *Proc. Natl. Acad. Sci. U. S. A.* **90**, 5519–5523.

Kozminski, K.G., Johnson, K.A., Forscher, P., and Rosenbaum, J.L. (1995). The *Chlamydomonas* kinesin-like protein FLA10 is involved in motility associated with the flagellar membrane. *J. Cell Biol.* **131**, 1517–1527.

Kramzar, L.M., Mueller, T., Erickson, B., and Higgs, D.C. (2006). Regulatory sequences of orthologous *petD* chloroplast mRNAs are highly specific among *Chlamydomonas* species. *Plant Mol. Biol.* **60**, 105–122.

Kreimer, G. (1994). Cell biology of phototaxis in flagellate algae. *Int. Rev. Cytol.* **148**, 229–310.

Kreimer, G., Overländer, C., Sineshchekov, O.A., Stolzis, H., Nultsch, W., and Melkonian, M. (1992). Functional analysis of the eyespot in *Chlamydomonas reinhardtii* mutant *ey 627, mt⁻*. *Planta* **188**, 513–521.

Kretzer, F. (1973). Molecular architecture of the chloroplast membranes of *Chlamydomonas reinhardi* as revelated by high resolution electron microscopy. *J. Ultrastruct. Res.* **44**, 146–178.

Kreuzberg, K. (1984a). Starch fermentation via a formate producing pathway in *Chlamydomonas reinhardii, Chlorogonium elongatum* and *Chlorella fusca*. *Physiol. Plant.* **61**, 87–94.

Kreuzberg, K. (1984b). Evidences for the role of the chloroplast in algal fermentation. *Adv. Photosynth. Res.* **3**, 437–440.

Kroen, W.K. (1984). Growth and polysaccharide production by the green alga *Chlamydomonas mexicana* (Chlorophyceae) on soil. *J. Phycol.* **20**, 616–618.

Kropat, J. and Beck, C.F. (1998). Characterization of photoreceptor and signaling pathway for light induction of the *Chlamydomonas* heat-shock gene *HSP70A*. *Photochem. Photobiol.* **68**, 414–419.

Kropat, J., von Gromoff, E.D., Mueller, F.W., and Beck, C.F. (1995). Heat shock and light activation of a *Chlamydomonas HSP70* gene are mediated by independent regulatory pathways. *Mol. Gen. Genet.* **248**, 727–734.

Kropat, J., Oster, U., Rüdiger, W., and Beck, C.F. (2000). Chloroplast signalling in the light induction of nuclear *HSP70* genes requires the accumulation of chlorophyll precursors and their accessibility to cytoplasm/nucleus. *Plant J.* **24**, 523–531.

Kruse, O., Rupprecht, J., Bader, K.P., Thomas-Hall, S., Schenk, P.M., Finazzi, G., and Hankamer, B. (2005). Improved photobiological H₂ production in engineered green algal cells. *J. Biol. Chem.* **280**, 34170–34177.

Kubo, T., Saito, T., Fukuzawa, H., and Matsuda, Y. (2001). Two tandemly-located matrix metalloprotease genes with different expression patterns in the *Chlamydomonas* sexual cell cycle. *Curr. Genet.* **40**, 136–143.

Kubo, T., Abe, J., Saito, T., and Matsuda, Y. (2002). Genealogical relationships among laboratory strains of *Chlamydomonas reinhardtii* as inferred from matrix metalloprotease genes. *Curr. Genet.* **41**, 115–122.

Kubo, Y., Ikeda, T., Yang, S.Y., and Tsuboi, M. (2000). Orientation of carotenoid molecules in the eyespot of alga: *in situ* polarized resonance Raman spectroscopy. *Appl. Spectrosci.* **54**, 1114–1119.

Kuchitsu, K., Tsuzuki, M., and Miyachi, S. (1988). Characterization of the pyrenoid isolated from unicellular green alga *Chlamydomonas reinhardtii*: particulate form of RuBisCO protein. *Protoplasma* **144**, 17–24.

Kuchitsu, K., Tsuzuki, M., and Miyachi, S. (1991). Polypeptide composition and enzyme activities of the pyrenoid and its regulation by CO₂ concentration in unicellular green algae. *Can. J. Bot.* **69**, 1062–1069.

Kuchka, M.R. and Jarvik, J.W. (1982). Analysis of flagellar size control using a mutant of *Chlamydomonas reinhardtii* with a variable number of flagella. *J. Cell Biol.* **92**, 170–175.

Kuchka, M.R. and Jarvik, J.W. (1987). Short-flagella mutants of *Chlamydomonas reinhardtii*. *Genetics* **115**, 685–691.

Kuchka, M.R., Mayfield, S.P., and Rochaix, J.-D. (1988). Nuclear mutations specifically affect the synthesis and/or degradation of the chloroplast-encoded D2 polypeptide of photosystem II in *Chlamydomonas reinhardtii*. *EMBO J.* **7**, 319–324.

Kuchka, M.R., Goldschmidt-Clermont, M., van Dillewijn, J., and Rochaix, J.-D. (1989). Mutation at the *Chlamydomonas* nuclear *NAC2* locus specifically affects stability of the chloroplast *psbD* transcript encoding polypeptide D2 of PS II. *Cell* **58**, 869–876.

Kucho, K., Ohyama, K., and Fukuzawa, H. (2005). Identification of novel clock-controlled genes by cDNA macroarray analysis in *Chlamydomonas reinhardtii*. *Plant Mol. Biol.* **57**, 889–906.

Kück, U. and Neuhaus, H. (1986). Universal genetic code evidenced in mitochondria of *Chlamydomonas reinhardii*. *Appl. Microbiol. Biotechnol.* **23**, 462–469.

Kück, U., Choquet, Y., Schneider, M., Dron, M., and Bennoun, P. (1987). Structural and transcription analysis of two homologous genes for the P700 chlorophyll *a*-apoproteins in *Chlamydomonas reinhardii*: evidence for *in vivo trans*-splicing. *EMBO J.* **6**, 2185–2195.

Künstner, P., Guardiola, A., Takahashi, Y., and Rochaix, J.-D. (1995). A mutant strain of *Chlamydomonas reinhardtii* lacking the chloroplast photosystem II *psbI* gene grows photoautotrophically. *J. Biol. Chem.* **270**, 9651–9654.

Kuhl, A. (1972). Zur Physiologie der Speicherung kondensierter anorganischer Phosphate in *Chlorella*. *Betr. Physiol. Morphol. Algen, Vortr. Gesamteb. Bot., Deutsch. Bot. Ges., Neue Folge* **No. 1.** 157–166.

Kuhl, A. and Lorenzen, H. (1964). Handling and culturing of *Chlorella*. In: *Methods in Cell Physiology*, Vol. 1 (D.M. Prescott, Ed.), pp. 159–187. Academic Press, New York.

Kumar, S.V., Misquitta, R.W., Reddy, V.S., Rao, B.J., and Rajam, M.V. (2004). Genetic transformation of the green alga *Chlamydomonas reinhardtii* by *Agrobacterium tumefaciens*. *Plant Sci.* **166**, 731–738.

Kumekawa, N., Ohtsubo, E., and Ohtsubo, H. (1999). Identification and phylogenetic analysis of *gypsy*-type retrotransposons in the plant kingdom. *Genes Genet. Syst.* **74**, 299–307.

Kuo, T.-C., Odom, O.W., and Herrin, D.L. (2006). Unusual metal specificity and structure of the group I ribozyme from *Chlamydomonas reinhardtii* 23S rRNA. *FEBS J.* **273**, 2631–2644.

Kuras, R. and Wollman, F.-A. (1994). The assembly of cytochrome b_6/f complexes: an approach using genetic transformation of the green alga *Chlamydomonas reinhardtii*. *EMBO J.* **13**, 1019–1027.

Kuroiwa, T., Suzuki, T., Ogawa, K., and Kawano, S. (1981). The chloroplast nucleus: distribution, number, size, and shape, and a model for the multiplication of the chloroplast genome during chloroplast development. *Plant Cell Physiol.* **22**, 381–396.

Kuroiwa, T., Kawano, S., Nishibayashi, S., and Sato, C. (1982). Epifluorescent microscopic evidence for maternal inheritance of chloroplast DNA. *Nature* **298**, 481–483.

Kuroiwa, T., Kawano, S., and Sato, C. (1983a). Mechanisms of maternal inheritance. I. Protein synthesis involved in preferential destruction of chloroplast DNA of male origin. *Proc. Jpn. Acad.* **59B**, 177–181.

Kuroiwa, T., Kawano, S., and Sato, C. (1983b). Mechanisms of maternal inheritance. II. RNA synthesis involved in preferential destruction of chloroplast DNA of male origin. *Proc. Jpn. Acad.* **59B**, 182–185.

Kuroiwa, T., Nakamura, S., Sato, C., and Tsubo, Y. (1985). Epifluorescent microscopic studies on the mechanism of preferential destruction of chloroplast nucleoids of male origin in young zygotes of *Chlamydomonas reinhardtii*. *Protoplasma* **125**, 43–52.

Kurvari, V. (1997). Cell wall biogenesis in *Chlamydomonas*: molecular characterization of a novel protein whose expression is up-regulated during matrix formation. *Mol. Gen. Genet.* **256**, 572–580.

Kurvari, V., Grishin, N.V., and Snell, W.J. (1998). A gamete-specific, sex-limited homeodomain protein in *Chlamydomonas*. *J. Cell Biol.* **143**, 1971–1980.

Kustu, S. and Inwood, W. (2006). Biological gas channels for NH_3 and CO_2: evidence that Rh (Rhesus) proteins are CO_2 channels. *Transfus. Clin. Biol.* **13**, 103–110.

L'Hernault, S.W. and Rosenbaum, J.L. (1983). *Chlamydomonas* alpha-tubulin is post-translationally modified in the flagella during flagellar assembly. *J. Cell Biol.* **97**, 258–263.

Lacoste-Royal, G. and Gibbs, S.P. (1987). Immunocytochemical localization of ribulose-1,5-bisphosphate carboxylase in the pyrenoid and thylakoid region of the chloroplast of *Chlamydomonas reinhardtii*. *Plant Physiol.* **83**, 602–606.

Ladygin, V.G. (1977). Method of isolating nonphotosynthetic mutants of algae with dwarf colonies. *Sov. Genet.* **13**, 620–623. [Russian original *Genetika* **13**, 905–909]

Ladygin, V.G. (2004). Efficient transformation of mutant cells of *Chlamydomonas reinhardtii* by electroporation. *Process Biochem.* **39**, 1685–1691.

Ladygin, V.G. and Boutanaev, A.M. (2002). Transformation of *Chlamydomonas reinhardtii CW-15* with the hygromycin phosphotransferase gene as a selectable marker. *Genetika* **38**, 1196–1202.

Laliberte, G. and De La Noue, J. (1993). Auto-, hetero-, and mixotrophic growth of *Chlamydomonas humicola* (Chlorophyceae) on acetate. *J. Phycol.* **29**, 612–620.

Lamb, M.R., Dutcher, S.K., Worley, C.K., and Dieckmann, C.L. (1999). Eyespot-assembly mutants in *Chlamydomonas reinhardtii*. *Genetics* **153**, 721–729.

Lampen, J.O. and Arnow, P. (1961). Inhibition of algae by nystatin. *J. Bacteriol.* **82**, 247–251.

Lang, W.C. (1982). Glycoprotein biosynthesis in *Chlamydomonas*. I. *In vitro* incorporation of galactose from UDP-[^{14}C]galactose into membrane-bound protein. *Plant Physiol.* **69**, 678–681.

Lang, W.C. and Chrispeels, M. (1976). Biosynthesis and release of cell wall-like glycoproteins during the vegetative cell cycle of *Chlamydomonas reinhardii*. *Planta* **129**, 183–189.

Lardans, A., Förster, B., Prasil, O., Falkowski, P.G., Sobolev, V., Edelman, M., Osmond, C.B., Gillham, N.W., and Boynton, J.E. (1998). Biophysical, biochemical, and physiological characterization of *Chlamydomonas reinhardtii* mutants with amino acid substitutions at the Ala251 residue in the D1 protein that result in varying levels of photosynthetic competence. *J. Biol. Chem.* **273**, 11082–11091.

Larson, A., Kirk, M.M., and Kirk, D.L. (1992). Molecular phylogeny of the volvocine flagellates. *Mol. Biol. Evol.* **9**, 85–105.

Lau, K.W., Ren, J., and Wu, M. (2000). Redox modulation of chloroplast DNA replication in *Chlamydomonas reinhardtii*. *Antioxid. Redox Signal.* **2**, 529–535.

Lawrence, C.W. (1965a). Influence of non-lethal doses of radiation on recombination in *Chlamydomonas reinhardi*. *Nature* **206**, 789–791.

Lawrence, C.W. (1965b). The effect of dose duration in the influence of irradiation on recombination in *Chlamydomonas*. *Mutat. Res.* **2**, 487–493.

Lawrence, C.W. (1970). Dose dependence for radiation-induced allelic recombination in *Chlamydomonas reinhardi*. *Mutat. Res.* **10**, 557–566.

Lawrence, C.W. and Davies, D.R. (1967). The mechanism of recombination in *Chlamydomonas reinhardi*. I. The influence of inhibitors of protein synthesis on recombination. *Mutat. Res.* **4**, 137–146.

Lawrence, C.W. and Holt, P.D. (1970). Effect of gamma radiation and alpha particles on gene recombination in *Chlamydomonas reinhardi*. *Mutat. Res.* **10**, 545–555.

Lawrence, S.D. and Kindle, K.L. (1997). Alterations in the *Chlamydomonas* plastocyanin transit peptide have distinct effects on *in vitro* import and *in vivo* protein accumulation. *J. Biol. Chem.* **272**, 20357–20363.

Lawrence, S.G., Holoka, M.H., and Hamilton, R.D. (1989). Effects of cadmium on a microbial food chain, *Chlamydomonas reinhardii* and *Tetrahymena vorax*. *Sci. Total Environ.* **87–88**, 381–395.

Lawson, M.A. and Satir, P. (1994). Characterization of the eyespot regions of 'blind' *Chlamydomonas* mutants after restoration of photophobic responses. *J. Euk. Microbiol.* **41**, 593–601.

Lawson, M.A., Zacks, D.N., Derguini, F., Nakanishi, K., and Spudich, J.L. (1991). Retinal analog restoration of photophobic responses in a blind *Chlamydomonas reinhardtii* mutant. Evidence for an archaebacterial like chromophore in a eukaryotic rhodopsin. *Biophys. J.* **60**, 1490–1498.

Lechtreck, K.F. (1998). Analysis of striated fiber formation by recombinant SF-assemblin *in vitro*. *J. Mol. Biol.* **279**, 423–438.

Lechtreck, K.F. and Melkonian, M. (1998). SF-assemblin, striated fibers, and segmented coiled coil proteins. *Cell Motil. Cytoskeleton* **41**, 289–296.

Lechtreck, K.F. and Silflow, C.D. (1997). SF-assemblin in *Chlamydomonas*: sequence conservation and localization during the cell cycle. *Cell Motil. Cytoskeleton* **36**, 190–201.

Lechtreck, K.F., Rostmann, J., and Grunow, A. (2002). Analysis of *Chlamydomonas* SF-assemblin by GFP tagging and expression of antisense constructs. *J. Cell Sci.* **115**, 1511–1522.

Lederberg, J. and Zinder, N. (1948). Concentration of biochemical mutants of bacteria with penicillin. *J. Am. Chem. Soc.* **70**, 4267–4268.

Ledford, H.K. and Niyogi, K.K. (2005). Singlet oxygen and photo-oxidative stress management in plants and algae. *Plant Cell Environ.* **28**, 1037–1045.

Ledford, H.K., Chin, B.L., and Niyogi, K.K. (2007). Acclimation to singlet oxygen stress in *Chlamydomonas reinhardtii*. *Eukaryotic Cell* **6**, 919–930.

LeDizet, M. and Piperno, G. (1986). Cytoplasmic microtubules containing acetylated alpha-tubulin in *Chlamydomonas reinhardtii*: spatial arrangement and properties. *J. Cell Biol.* **103**, 13–22.

LeDizet, M. and Piperno, G. (1995). *ida4-1*, *ida4-2*, and *ida4-3* are intron splicing mutations affecting the locus encoding p28, a light chain of *Chlamydomonas* axonemal inner dynein arms. *Mol. Biol. Cell* **6**, 713–723.

Lee, H., Bingham, S.E., and Webber, A.N. (1996). Function of 3′ non-coding sequences and stop codon usage in expression of the chloroplast *psaB* gene in *Chlamydomonas reinhardtii*. *Plant Mol. Biol.* **31**, 337–354.

Lee, H., Bingham, S.E., and Webber, A.N. (1998). Specific mutagenesis of reaction center proteins by chloroplast transformation of *Chlamydomonas reinhardtii*. *Methods Enzymol.* **297**, 310–320.

Lee, J. and Herrin, D.L. (2003). Mutagenesis of a light-regulated *psbA* intron reveals the importance of efficient splicing for photosynthetic growth. *Nucleic Acids Res.* **31**, 4361–4372.

Lee, J.-H., Lin, H., Joo, S., and Goodenough, U.G. (2008). Early sexual origins of homeoprotein heterodimers and their role in the evolution of algae and land plants. *Cell* **133**, 829–840.

Lee, J.J. and McEnery, M.M. (1983). Symbiosis in foraminifera. In: *Algal Symbiosis* (L.J. Goff, Ed.), pp. 37–68. Cambridge University Press, Cambridge.

Lee, J.J., Crockett, L.J., Hagen, J., and Stone, R.J. (1974). The taxonomic identity and physiological ecology of *Chlamydomonas hedleyi sp. nov.*, algal flagellate symbiont from the foraminifer *Archaias angulatus*. *Br. Phycol. J.* **9**, 407–422.

Lee, J.W. and Greenbaum, E. (2003). A new oxygen sensitivity and its potential application in photosynthetic H_2 production. *Appl. Biochem. Biotechnol.* **105**, 303–313.

Lee, R.W. and Jones, R.F. (1973). Induction of Mendelian and non-Mendelian streptomycin resistant mutants during the synchronous cell cycle of *Chlamydomonas reinhardtii*. *Mol. Gen. Genet.* **121**, 99–108.

Lee, R.W. and Lemieux, C. (1986). Biparental inheritance of non-Mendelian gene markers in *Chlamydomonas moewusii*. *Genetics* **113**, 589–600.

Lee, R.W., Gillham, N.W., Van Winkle, K.P., and Boynton, J.E. (1973). Preferential recovery of uniparental streptomycin resistant mutants from diploid *Chlamydomonas reinhardtii*. *Mol. Gen. Genet.* **121**, 109–116.

Lee, R.W., Whiteway, M.S., and Yorke, M.A. (1976). Recovery of sexually viable non-diploids from diploid *Chlamydomonas reinhardtii*. *Genetics* **83**, s44.

Lee, R.W., Langille, B., Lemieux, C., and Boer, P.H. (1990). Inheritance of mito-chondrial and chloroplast genome markers in backcrosses of *Chlamydomonas eugametos* X *Chlamydomonas moewusii* hybrids. *Curr. Genet.* **17**, 73–76.

Lefebvre, P.A., Nordstrom, S.A., Moulder, J.E., and Rosenbaum, J.L. (1978). Flagellar elongation and shortening in *Chlamydomonas*. IV. Effects of flagellar detach-ment, regeneration, and resorption on the induction of flagellar protein synthe-sis. *J. Cell Biol.* **78**, 8–27.

Lefebvre, P.A., Silflow, C.D., Wieben, E.D., and Rosenbaum, J.L. (1980). Increased levels of mRNAs for tubulin and other flagellar proteins after amputation or shortening of *Chlamydomonas* flagella. *Cell* **20**, 469–477.

Leisinger, U., Rüfenacht, K., Fischer, B., Pesaro, M., Spengler, A., Zehnder, A.J.B., and Eggen, R.I.L. (2001). The glutathione peroxidase homologous gene from *Chlamy-domonas reinhardtii* is transcriptionally up-regulated by singlet oxygen. *Plant Mol. Biol.* **46**, 395–408.

Lemaire, C. and Wollman, F.-A. (1989). The chloroplast ATP synthase in *Chlamydo-monas reinhardtii*. II. Biochemical studies on its biogenesis using mutants defec-tive in photophosphorylation. *J. Biol. Chem.* **264**, 10235–10242.

Lemaire, C., Girard-Bascou, J., Wollman, F.-A., and Bennoun, P. (1986). Studies on the cytochrome b_6/f complex. I. Characterization of the complex subunits in *Chlamydomonas reinhardtii*. *Biochim. Biophys. Acta* **851**, 229–238.

Lemaire, C., Wollman, F.-A., and Bennoun, P. (1988). Restoration of phototrophic growth in a mutant of *Chlamydomonas reinhardtii* in which the chloroplast *atpB* gene of the ATP synthase has a deletion: an example of mitochondria-dependent photosynthesis. *Proc. Natl. Acad. Sci. U. S. A.* **85**, 1344–1348.

Lemaire, S., Stein, M., Issakidis, E., Keryer, E., Benoit, V., Gerard-Hirne, C., Miginiac-Maslow, M., Jacquot, J.P., Lemaire, S.D., Hours, M., Trouabal, A., and Roche, O. (1999). Analysis of light/dark synchronization of cell-wall-less *Chlamydo-monas reinhardtii* (Chlorophyta) cells by flow cytometry. *Eur. J. Phycol.* **34**, 279–286.

Lembi, C.A. and Lang, N.J. (1965). Electron microscopy of *Carteria* and *Chlamydo-monas*. *Am. J. Bot.* **52**, 464–477.

Lemieux, B. and Lemieux, C. (1985). Extensive sequence rearrangements in the chlo-roplast genomes of the green algae *Chlamydomonas eugametos* and *Chlamydo-monas reinhardtii*. *Curr. Genet.* **10**, 213–219.

Lemieux, B., Turmel, M., and Lemieux, C. (1985). Chloroplast DNA variation in *Chlamydomonas* and its potential application to the systematics of this genus. *BioSystems* **18**, 293–298.

Lemieux, B., Turmel, M., and Lemieux, C. (1988). Unidirectional gene conversions in the chloroplast of *Chlamydomonas* interspecific hybrids. *Mol. Gen. Genet.* **212**, 48–55.

Lemieux, B., Turmel, M., and Lemieux, C. (1990). Recombination of *Chlamydo-monas* chloroplast DNA occurs more frequently in the large inverted repeat sequence than in the single-copy regions. *Theor. Appl. Genet.* **79**, 17–27.

Lemieux, C., Turmel, M., and Lee, R.W. (1980). Characterization of chloroplast DNA in *Chlamydomonas eugametos* and *C. moewusii* and its inheritance in hybrid progeny. *Curr. Genet.* **2**, 139–147.

Lemieux, C., Turmel, M., and Lee, R.W. (1981). Physical evidence for recombina-tion of chloroplast DNA in hybrid progeny of *Chlamydomonas eugametos* and *C. moewusii*. *Curr. Genet.* **3**, 97–103.

Lemieux, C., Turmel, M., Seligy, V.L., and Lee, R.W. (1984a). Chloroplast DNA recombination in interspecific hybrids of *Chlamydomonas*: linkage between a nonmendelian locus for streptomycin resistance and restriction fragments coding for 16S rRNA. *Proc. Natl. Acad. Sci. U. S. A.* **81**, 1164–1168.

Lemieux, C., Turmel, M., Seligy, V.L., and Lee, R.W. (1984b). A genetical approach to the physical mapping of chloroplast genes in *Chlamydomonas*. *Can. J. Biochem. Cell Biol.* **62**, 225–229.

Lemieux, C., Boulanger, J., Otis, C., and Turmel, M. (1989). Nucleotide sequence of the chloroplast large subunit rRNA gene from *Chlamydomonas reinhardtii*. *Nucleic Acids Res.* **17**, 7997.

León, R. and Fernández, E. (2007). Nuclear transformation of eukaryotic microalgae: historical overview, achievements and problems. *Adv. Exp. Med. Biol.* **616**, 1–11.

León, R., Galván, A., and Fernández, E. (Eds.) (2007). *Transgenic Microalgae as Green Cell Factories. Advances in Experimental Medicine and Biology, Volume 616.* Landes Bioscience and Springer Science+Business Media, LLC, New York. 131 pp.

Leu, S. (1998). Extraordinary features in the *Chlamydomonas reinhardtii* chloroplast genome: (1) *rps2* as part of a large open reading frame; (2) a *C. reinhardtii* specific repeat sequence. *Biochim. Biophys. Acta* **1365**, 541–544.

Leu, S., White, D., and Michaels, A. (1990). Cell cycle-dependent transcriptional and post-transciptional regulation of chloroplast gene expression in *Chlamydomonas reinhardtii*. *Biochim. Biophys. Acta* **1049**, 311–317.

Leu, S., Schlesinger, J., Michaels, A., and Shavit, N. (1992). Complete DNA sequence of the *Chlamydomonas reinhardtii* chloroplast *atpA* gene. *Plant Mol. Biol.* **18**, 613–616.

Levine, R.P. (1960). Genetic control of photosynthesis in *Chlamydomonas reinhardi*. *Proc. Natl. Acad. Sci. U. S. A.* **46**, 972–977.

Levine, R.P. (1968). Genetic dissection of photosynthesis. *Science* **162**, 768–771.

Levine, R.P. (1969). The analysis of photosynthesis using mutant strains of algae and higher plants. *Annu. Rev. Plant Physiol.* **20**, 523–540.

Levine, R.P. and Ebersold, W.T. (1958a). The relation of calcium and magnesium to crossing over in *Chlamydomonas reinhardi*. *Z. Vererbs.* **89**, 631–635.

Levine, R.P. and Ebersold, W.T. (1958b). Gene recombination in *Chlamydomonas reinhardi*. *Cold Spring Harb. Symp. Quant. Biol.* **23**, 101–109.

Levine, R.P. and Ebersold, W.T. (1960). The genetics and cytology of *Chlamydomonas*. *Annu. Rev. Microbiol.* **14**, 197–216.

Levine, R.P. and Folsome, C.E. (1959). The nuclear cycle in *Chlamydomonas reinhardi*. *Z. Vererbs.* **90**, 215–222.

Levine, R.P. and Goodenough, U.W. (1970). The genetics of photosynthesis and the chloroplast in *Chlamydomonas reinhardi*. *Annu. Rev. Genet.* **4**, 397–408.

Levy, E.M. (1974). Flagellar elongation: an example of controlled growth. *J. Theor. Biol.* **43**, 133–149.

Levy, H., Kindle, K.L., and Stern, D.B. (1997). A nuclear mutation that affects the 3′ processing of several mRNAs in *Chlamydomonas* chloroplasts. *Plant Cell* **9**, 825–836.

Levy, H., Kindle, K.L., and Stern, D.B. (1999). Target and specificity of a nuclear gene product that participates in mRNA 3′-end formation in *Chlamydomonas* chloroplasts. *J. Biol. Chem.* **274**, 35955–35962.

Lewin, J.C. (1950). Obligate autotrophy in *Chlamydomonas moewusii* Gerloff. *Science* **112**, 652–653.

Lewin, R.A. (1949). Genetics of *Chlamydomonas* – paving the way. *Biol. Bull.* **97**, 243–244.

Lewin, R.A. (1951). Isolation of sexual strains of *Chlamydomonas*. *J. Gen. Microbiol.* **5**, 926–929.

Lewin, R.A. (1952a). Studies on the flagella of algae. I. General observations on *Chlamydomonas moewusii* Gerloff. *Biol. Bull.* **103**, 74–79.

Lewin, R.A. (1952b). Ultraviolet induced mutations in *Chlamydomonas moewusii* Gerloff. *J. Gen. Microbiol.* **6**, 233–248.

Lewin, R.A. (1952c). The primary zygote membrane in *Chlamydomonas moewusii*. *J. Gen. Microbiol.* **6**, 249–250.

Lewin, R.A. (1953a). The genetics of *Chlamydomonas moewusii* Gerloff. *J. Genet.* **51**, 543–560.

Lewin, R.A. (1954a). Mutants of *Chlamydomonas moewusii* with impaired motility. *J. Gen. Microbiol.* **11**, 358–363.

Lewin, R.A. (1954b). The utilization of acetate by wild-type and mutant *Chlamydomonas dysosmos*. *J. Gen. Microbiol.* **11**, 459–471.

Lewin, R.A. (1954c). Sex in unicellular algae. In: *Sex in Microorganisms* (D.H. Wenrich, I.F. Lewis, and J.R. Raper, Eds.), pp. 100–133. AAAS, Washington DC.

Lewin, R.A. (1956). Extracellular polysaccharides of green algae. *Can. J. Microbiol.* **2**, 665–672.

Lewin, R.A. (1957a). Four new species of *Chlamydomonas*. *Can. J. Bot.* **35**, 321–326.

Lewin, R.A. (1957b). The zygote of *Chlamydomonas moewusii*. *Can. J. Bot.* **35**, 795–804.

Lewin, R.A. (1959). The isolation of algae. *Rev. Algol.* **3**, 181–198.

Lewin, R.A. (1975). A *Chlamydomonas* with black zygospores. *Phycologia* **14**, 71–74.

Lewin, R.A., (Ed.) (1976). *The Genetics of Algae*. Blackwell Scientific Publications, Oxford, and University of California Press, Berkeley, 360 pp.

Lewin, R.A. (1977). The use of algae as soil conditioners. *CIBCASIO Trans. (La Jolla, CA)* **3**, 31–35.

Lewin, R.A. (1982). A new kind of motility mutant (non-gliding) in *Chlamydomonas*. *Experientia* **38**, 348–349.

Lewin, R.A. (1983). Evidence for post-zygotic lag in *Chlamydomonas moewusii* (Chlorophyta, Volvocales). *Experientia* **39**, 612–613.

Lewin, R.A. (1984). *Chlamydomonas sajao nov. sp.* (Chlorophyta, Volvocales). *Chin. J. Oceanol. Limnol.* **2**, 92–96.

Lewin, R.A. (2000). Experimental phycology: some historic diversions and aberrations. *J. Appl. Phycol.* **12**, 197–200.

Lewin, R.A. and Burrascano, C. (1983). Another new kind of *Chlamydomonas* mutant, with impaired flagellar autotomy. *Experientia* **39**, 1397–1398.

Lewin, R.A. and Lee, K.W. (1985). Autotomy of algal flagella: electron microscope studies of *Chlamydomonas* (Chlorophyceae) and *Tetraselmis* (Prasinophyceae). *Phycologia* **24**, 311–316.

Lewin, R.A. and Meinhart, J.O. (1953). Studies on the flagella of algae. III. Electron micrographs of *Chlamydomonas moewusii*. *Can. J. Bot.* **31**, 711–717.

Lewis, L.A. and McCourt, R.M. (2004). Green algae and the origin of land plants. *Am. J. Bot.* **91**, 1535–1556.

Lhoas, P. (1961). Mitotic haploidization by treatment of *Aspergillus niger* diploids with para-fluorophenylalanine. *Nature* **190**, 744.

Li, F., Holloway, S.P., Lee, J., and Herrin, D.L. (2002). Nuclear genes that promote splicing of group I introns in the chloroplast 23S rRNA and *psbA* genes in *Chlamydomonas reinhardtii*. *Plant J.* **32**, 467–480.

Li, H.H. and Merchant, S. (1995). Degradation of plastocyanin in copper-deficient *Chlamydomonas reinhardtii*. Evidence for a protease-susceptible conformation of the apoprotein and regulated proteolysis. *J. Biol. Chem.* **270**, 23504–23510.

Li, H.H., Quinn, J., Culler, D., Girard-Bascou, J., and Merchant, S. (1996). Molecular genetic analysis of plastocyanin biosynthesis in *Chlamydomonas reinhardtii*. *J. Biol. Chem.* **271**, 31283–31289.

Li, J., Goldschmidt-Clermont, M., and Timko, M.P. (1993). Chloroplast-encoded *chlB* is required for light-independent protochlorophyllide reductase activity in *Chlamydomonas reinhardtii*. *Plant Cell* **5**, 1817–1829.

Li, J.B., Gerdes, J.M., Haycraft, C.J., Fan, Y., Teslovich, T.M., May-Simera, H., Li, H., Blacque, O.E., Li, L., Leitch, C.C., Lewis, R.A., Green, J.S., Parfrey, P.S., Leroux, M.R., Davidson, W.S., Beales, P.L., Guay-Woodford, L.M., Yoder, B.K., Stormo, G.D., Katsanis, N., and Dutcher, S.K. (2004). Comparative genomics identifies a flagellar and basal body proteome that includes the *BBS5* human disease gene. *Cell* **117**, 541–552.

Li, J.M. and Timko, M.P. (1996). The *pc-1* phenotype of *Chlamydomonas reinhardtii* results from a deletion mutation in the nuclear gene for NADPH:protochlorophyllide oxidoreductase. *Plant Mol. Biol.* **30**, 15–37.

Libessart, N., Maddelein, M.L., van den Koornhuyse, N., Decq, A., Delrue, B., Mouille, G., D'Hulst, C., and Ball, S. (1995). Storage, photosynthesis, and growth: the conditional nature of mutations affecting starch synthesis and structure in *Chlamydomonas*. *Plant Cell* **7**, 1117–1127.

Lien, T. and Knutsen, G. (1973). Synchronous cultures of *Chlamydomonas reinhardti*: properties and regulation of repressible phosphatases. *Physiol. Plant.* **28**, 291–298.

Lien, T. and Knutsen, G. (1979). Synchronous growth of *Chlamydomonas reinhardtii* (Chlorophyceae): a review of optimal conditions. *J. Phycol.* **15**, 191–200.

Lien, T., Schreiner, O., and Steine, M. (1975). Purification of a derepressible arylsulfatase from *Chlamydomonas reinhardti*. Properties of the enzyme in intact cells and in purified state. *Biochim. Biophys. Acta* **384**, 168–179.

Lin, H. and Goodenough, U.W. (2007). Gametogenesis in the *Chlamydomonas reinhardtii minus* mating type is controlled by two genes, *MID* and *MTD1*. *Genetics* **176**, 913–925.

Lindauer, A., Müller, K., and Schmitt, R. (1993). Two histone H1-encoding genes of the green alga *Volvox carteri* with features intermediate between plant and animal genes. *Gene* **129**, 59–68.

Lindemann, C.B. and Mitchell, D.R. (2007). Evidence for axonemal distortion during the flagellar beat of *Chlamydomonas*. *Cell Motil. Cytoskeleton* **64**, 580–589.

Liss, M., Kirk, D.L., Beyser, K., and Fabry, S. (1997). Intron sequences provide a tool for high-resolution phylogenetic analysis of volvocine algae. *Curr. Genet.* **31**, 214–227.

Liu, C., Wang, X., and Huang, X. (2006). Phylogenetic studies on two strains of Antarctic ice algae based on morphological and molecular characteristics. *Phycologia* **45**, 190–198.

Liu, X.-Q., Gillham, N.W., and Boynton, J.E. (1989). Chloroplast ribosomal protein gene *rps12* of *Chlamydomonas reinhardtii*. Wild-type sequence, mutation to streptomycin resistance and dependence, and function in *Escherichia coli*. *J. Biol. Chem.* **264**, 16100–16108.

Liu, X.-Q., Huang, C., and Xu, H. (1993). The unusual *rps3*-like *orf712* is functionally essential and structurally conserved in *Chlamydomonas*. *FEBS Lett.* **336**, 225–230.

Llamas, A., Kalakoutskii, K.L., and Fernández, E. (2000). Molybdenum cofactor amounts in *Chlamydomonas reinhardtii* depend on the *Nit5* gene function related to molybdate transport. *Plant Cell Environ.* **23**, 1247–1255.

Llamas, A., Igeño, M.I., Galván, A., and Fernández, E. (2002). Nitrate signalling on the nitrate reductase gene promoter depends directly on the activity of the nitrate transport systems in *Chlamydomonas*. *Plant J.* **30**, 261–271.

Llamas, A., Tejada-Jiménez, M., González-Ballester, D., Higuera, J.J., Schwarz, G., Galván, A., and Fernández, E. (2007). *Chlamydomonas* CNX1E reconstitutes molybdenum cofactor biosynthesis in *E. coli* mutants. *Eukaryotic Cell* **6**, 1063–1067.

Lloyd, D. and Cantor, M.H. (1979). Subcellular structure and function in acetate flagellates. In: *Biochemistry and Physiology of Protozoa 2nd edition*, Vol. 2 (M. Levandowsky and S.H. Hutner, Eds.), pp. 9–65. Academic Press, New York.

Lodha, M. and Schroda, M. (2005). Analysis of chromatin structure in the control regions of the *Chlamydomonas HSP70A* and *RBCS2* genes. *Plant Mol. Biol.* **59**, 501–513.

Lodha, M., Schulz-Raffelt, M., and Schroda, M. (2008). A new assay for promoter analysis in *Chlamydomonas* reveals a role for heat shock elements and TATA-box in *HSP70A* promoter mediated activation of transgene expression. *Eukaryotic Cell* **7**, 172–176.

Lohr, M., Im, C.S., and Grossman, A.R. (2005). Genome-based examination of chlorophyll and carotenoid biosynthesis in *Chlamydomonas reinhardtii*. *Plant Physiol.* **138**, 490–515.

Lohret, T.A., McNally, F.J., and Quarmby, L.M. (1998). A role for katanin-mediated axonemal severing during *Chlamydomonas* deflagellation. *Mol. Biol. Cell* **9**, 1195–1207.

Lohret, T.A., Zhao, L.F., and Quarmby, L.M. (1999). Cloning of *Chlamydomonas* p60 katanin and localization to the site of outer doublet severing during deflagellation. *Cell Motil. Cytoskeleton* **43**, 221–231.

López-Ruiz, A., Verbelen, J.P., Roldan, J.M., and Diez, J. (1985). Nitrate reductase of green algae is located in the pyrenoid. *Plant Physiol.* **79**, 1006–1010.

López-Ruiz, A., Verbelen, J.-P., Bocanegra, J.A., and Diez, J. (1991). Immunocytochemical localization of nitrite reductase in green algae. *Plant Physiol.* **96**, 699–704.

Loppes, R. (1969). A new class of arginine-requiring mutants in *Chlamydomonas reinhardi*. *Mol. Gen. Genet.* **104**, 172–177.

Loppes, R. (1970). Selection of arginine-requiring mutants in *Chlamydomonas reinhardi* after treatment with three mutagens. *Experientia* **26**, 660–661.

Loppes, R. (1976a). Release of enzymes by normal and wall-free cells of *Chlamydomonas*. *J. Bacteriol.* **128**, 114–116.

Loppes, R. (1976b). Genes involved in the regulation of the neutral phosphatase in *Chlamydomonas reinhardi*. *Mol. Gen. Genet.* **148**, 315–321.

Loppes, R. and Deltour, R. (1975). Changes in phosphatase activity associated with cell wall defects in *Chlamydomonas reinhardi*. *Arch. Microbiol.* **103**, 247–250.

Loppes, R. and Deltour, R. (1977). A biochemical assay for detecting cell-wall mutants of *Chlamydomonas reinhardi*. *Plant Sci. Lett.* **8**, 261–265.

Loppes, R. and Deltour, R. (1978). A temperature-conditional protoplast of *Chlamydomonas reinhardi*. *Exp. Cell Res.* **117**, 439–441.

Loppes, R. and Deltour, R. (1981). A pleiotropic mutant of *Chlamydomonas reinhardi* showing cell wall abnormalities and altered phosphatase activities. *Plant Sci. Lett.* **21**, 193–197.

Loppes, R. and Denis, C. (1983). Chloroplast and nuclear DNA fragments from *Chlamydomonas* promoting high frequency transformation of yeast. *Curr. Genet.* **7**, 473–480.

Loppes, R. and Heindricks, R. (1986). New arginine-requiring mutants in *Chlamydomonas reinhardtii*. *Arch. Microbiol.* **143**, 348–352.

Loppes, R. and Matagne, R. (1972). Allelic complementation between *arg-7* mutants in *Chlamydomonas reinhardi*. *Genetica* **43**, 422–430.

Loppes, R. and Matagne, R.F. (1973). Acid phosphatase mutants in *Chlamydomonas*: isolation and characterization by biochemical, electrophoretic and genetic analysis. *Genetics* **75**, 593–604.

Loppes, R., Braipson, J., Matagne, R.F., Sassen, A., and Ledoux, L. (1977). Regulation of the neutral phosphatase in *Chlamydomonas reinhardi*: an immunogenetic study of wild-type and mutant strains. *Biochem. Genet.* **15**, 1147–1157.

Loppes, R., Devos, N., Willem, S., Barthelemy, P., and Matagne, R.F. (1996). Effect of temperature on two enzymes from a psychrophilic *Chloromonas* (Chlorophyta). *J. Phycol.* **32**, 276–278.

Lorenz, R.T. and Cysewski, G.R. (2000). Commercial potential for *Haematococcus* microalgae as a natural source of astaxanthin. *Trends Biotechnol.* **18**, 160–167.

Lou, J.K., Wu, M., Chang, C.H., and Cuticchia, A.J. (1987). Localization of a r-protein gene within the chloroplast DNA replication origin of *Chlamydomonas*. *Curr. Genet.* **11**, 537–541.

Lou, J.K., Cruz, F.D., and Wu, M. (1989). Nucleotide sequence of the chloroplast ribosomal protein gene L14 in *Chlamydomonas reinhardtii*. *Nucleic Acids Res.* **17**, 3587.

Luck, D.J.L., Piperno, G., Ramanis, Z., and Huang, B. (1977). Flagellar mutants of *Chlamydomonas*: studies of radial spoke-defective strains by dikaryon and revertant analysis. *Proc. Natl. Acad. Sci. U. S. A.* **74**, 3456–3460.

Lucksch, I. (1932). Ernährungsphysiologische Untersuchungen an Chlamydomonadeen. *Beih. Bot. Zentralbl.* **50**, 64–94.

Luis, P., Behnke, K., Toepel, J., and Wilhelm, C. (2006). Parallel analysis of transcript levels and physiological key parameters allows the identification of stress phase gene markers in *Chlamydomonas reinhardtii* under copper excess. *Plant Cell Environ.* **29**, 2043–2054.

Lumbreras, V. and Purton, S. (1998). Recent advances in *Chlamydomonas* transgenics. *Protist* **149**, 23–27.

Lumbreras, V., Stevens, D.R., and Purton, S. (1998). Efficient foreign gene expression in *Chlamydomonas reinhardtii* mediated by an endogenous intron. *Plant J.* **14**, 441–447.

Luria, S.E. and Delbrück, M. (1943). Mutations of bacteria from virus sensitivity to virus resistance. *Genetics* **28**, 491–511.

Lurling, M. and Beekman, W. (2006). Palmelloids formation in *Chlamydomonas reinhardtii*: defence against rotifer predators. *Ann. Limnol.* **42**, 65–72.

Lurquin, P.F. and Behki, R.M. (1975). Uptake of bacterial DNA by *Chlamydomonas reinhardi*. *Mutat. Res.* **29**, 35–51.

Lustigman, B., Lee, L.H., and Weiss-Magasic, C. (1995). Effects of cobalt and pH on the growth of *Chlamydomonas reinhardtii*. *Bull. Environ. Contam. Toxicol.* **55**, 65–72.

Lütz-Meindl, U. and Lütz, C. (2006). Analysis of element accumulation in cell wall attached and intracellular particles of snow algae by EELS and ESI. *Micron* **37**, 452–458.

Lux, F.G., III and Dutcher, S.K. (1991). Genetic interactions at the *FLA10* locus: suppressors and synthetic phenotypes that affect the cell cycle and flagellar function in *Chlamydomonas reinhardtii*. *Genetics* **128**, 549–561.

Luykx, P. (2000). Contractile vacuoles. In: *Vacuolar Compartments* (D.G. Robinson and J.C. Rogers, Eds.), pp. 43–70. Sheffield Academic Press, Sheffield.

Luykx, P., Hoppenrath, M., and Robinson, D.G. (1997a). Structure and behavior of contractile vacuoles in *Chlamydomonas reinhardtii*. *Protoplasma* **198**, 73–84.

Luykx, P., Hoppenrath, M., and Robinson, D.G. (1997b). Osmoregulatory mutants that affect the function of the contractile vacuole in *Chlamydomonas reinhardtii*. *Protoplasma* **200**, 99–111.

Ma, C. and Mortimer, R.K. (1983). Empirical equation that can be used to determine genetic map distances from tetrad data. *Mol. Cell. Biol.* **3**, 1886–1887.

Ma, D.-P., Yang, Y.-W., and Hasnain, S.E. (1988). Nucleotide sequence of *Chlamydomonas reinhardtii* mitochondrial genes coding for subunit 6 of NADH dehydrogenase and tRNATrp. *Nucleic Acids Res.* **16**, 11373.

Ma, D.-P., Yang, Y.-W., and Hasnain, S.E. (1989a). Nucleotide sequence of *Chlamydomonas reinhardtii* mitochondrial genes coding for tRNAGln (UUG) and tRNA Met (CAU). *Nucleic Acids Res.* **17**, 1256.

Ma, D.-P., Yang, Y.-W., King, Y.-T., and Hasnain, S.E. (1989b). Nucleotide sequence of cloned *nad4* (*urf4*) gene from *Chlamydomonas reinhardtii* mitochondrial DNA. *Gene* **85**, 363–370.

Ma, D.-P., Yang, Y.W., King, T.Y., and Hasnain, S.E. (1990). The mitochondrial apocytochrome *b* gene from *Chlamydomonas reinhardtii*. *Plant Mol. Biol.* **15**, 357–359.

Ma, D.-P., King, Y.-T., Kim, Y., and Luckett, W.S. (1992). The group I intron of apocytochrome *b* gene from *Chlamydomonas smithii* encodes a site-specific endonuclease. *Plant Mol. Biol.* **18**, 1001–1004.

Macfie, S.M. and Welbourn, P.M. (2000). The cell wall as a barrier to uptake of metal ions in the unicellular green alga *Chlamydomonas reinhardtii* (Chlorophyceae). *Arch. Environ. Contam. Toxicol.* **39**, 413–419.

Macfie, S.M., Tarmohamed, Y., and Welbourn, P.M. (1995). Effects of cadmium, cobalt, copper, and nickel on growth of the green alga *Chlamydomonas reinhardtii*: the influences of the cell wall and pH. *Arch. Environ. Contam. Toxicol.* **27**, 454–458.

MaciasR, F.M. (1965). Effect of pH of the medium on the availability of chelated iron for *Chlamydomonas mundana*. *J. Protozool.* **12**, 500–504.

MaciasR, F.M. and Eppley, R.W. (1963). Development of EDTA media for the rapid growth of *Chlamydomonas mundana*. *J. Protozool.* **10**, 243–246.

MacIntyre, H.L. and Cullen, J.J. (2005). Using cultures to investigate the physiological ecology of microalgae. In: *Algal Culturing Techniques* (R.A. Andersen, Ed.), pp. 287–326. Elsevier, Amsterdam.

Macka, W., Wihlidal, H., Stehlik, G., Washuttl, J., and Bancher, E. (1978). Metabolic studies of Hg-203 on *Chlamydomonas reinhardi*. *Experientia* **34**, 602–603.

Maddelein, M.L., Libessart, N., Bellanger, F., Delrue, B., D'Hulst, C., van den Koornhuyse, N., Fontaine, T., Wieruszeski, J.M., Decq, A., and Ball, S. (1994). Toward an understanding of the biogenesis of the starch granule. Determination of granule-bound and soluble starch synthase functions in amylopectin synthesis. *J. Biol. Chem.* **269**, 25150–25157.

Maeda, I., Hikawa, H., Miyashiro, M., Yagi, K., Miura, Y., Miyasaka, H., Akano, T., Kiyohara, M., Matsumoto, H., and Ikuta, Y. (1994). Enhancement of starch degradation by CO_2 in a marine green alga *Chlamydomonas* sp. MGA161. *J. Ferment. Bioeng.* **78**, 383–385.

Maeda, I., Seto, Y., Ueda, S., Chen, Y., Hari, J., Kawase, M., Miyasaka, H., and Yagii, K. (2006). Simultaneous control of turbidity and dilution rate through adjustment of medium composition in semi-continuous *Chlamydomonas* cultures. *Biotechnol. Bioeng.* **94**, 722–729.

Mages, W., Heinrich, O., Treuner, G., Vlček, D., Daubnerova, I., and Slaninová, M. (2007). Complementation of the *Chlamydomonas reinhardtii arg7-8* (*arg2*) point mutation by recombination with a truncated nonfunctional *ARG7* gene. *Protist* **158**, 435–446.

Maguire, M.P. (1976). Mitotic and meiotic behavior of the chromosomes of the octet strain of *Chlamydomonas reinhardtii*. *Genetica* **46**, 479–502.

Mahan, K.M., Odom, O.W., and Herrin, D.L. (2005). Controlling fungal contamination in *Chlamydomonas reinhardtii* cultures. *BioTechniques* **39**, 457–458.

Mahanty, H.K. and Gresshoff, P.M. (1978). Influence of polychlorinated biphenyls (PCBs) on growth of freshwater algae. *Bot. Gaz.* **139**, 202–206.

Mahjoub, M.R., Montpetit, B., Zhao, L.F., Finst, R.J., Goh, B., Kim, A.C., and Quarmby, L.M. (2002). The *FA2* gene of *Chlamydomonas* encodes a NIMA family kinase with roles in cell cycle progression and microtubule severing during deflagellation. *J. Cell Sci.* **115**, 1759–1768.

Mahjoub, M.R., Qasim Rasi, M., and Quarmby, L.M. (2004). A NIMA-related kinase, Fa2p, localizes to a novel site in the proximal cilia of *Chlamydomonas* and mouse kidney cells. *Mol. Biol. Cell* **15**, 5172–5186.

Mai, J.C. and Coleman, A.W. (1997). The internal transcribed spacer 2 exhibits a common secondary structure in green algae and flowering plants. *J. Mol. Evol.* **44**, 258–271.

Maione, T.E. and Gibbs, M. (1986a). Hydrogenase-mediated activities in isolated chloroplasts of *Chlamydomonas reinhardii*. *Plant Physiol.* **80**, 360–363.

Maione, T.E. and Gibbs, M. (1986b). Association of the chloroplastic respiratory and photosynthetic electron transport chains of *Chlamydomonas reinhardii* with photoreduction and the oxyhydrogen reaction. *Plant Physiol.* **80**, 364–368.

Maitz, G., Haas, E.M., and Castric, P.A. (1982). Purification and properties of the allophanate hydrolase from *Chlamydomonas reinhardii*. *Biochim. Biophys. Acta* **714**, 486–491.

Majeran, W., Olive, J., Drapier, D., Vallon, O., and Wollman, F.A. (2001). The light sensitivity of ATP synthase mutants of *Chlamydomonas reinhardtii*. *Plant Physiol.* **126**, 421–433.

Malmberg, A.E. and VanWinkle-Swift, K.P. (2001). Zygospore germination in *Chlamydomonas monoica* (Chlorophyta): timing and pattern of secondary zygospore wall degradaton in relation to cytoplasmic events. *J. Phycol.* **37**, 86–94.

Mamedov, T.G., Suzuki, K., Miura, K., Kucho, K., and Fukuzawa, H. (2001). Characteristics and sequence of phosphoglycolate phosphatase from a eukaryotic green alga *Chlamydomonas reinhardtii*. *J. Biol. Chem.* **276**, 45573–45579.

Manton, I. (1965). Some phyletic implications of flagellar structure in plants. *Adv. Bot. Res.* **2**, 1–34.

Manuell, A.L., Yamaguchi, K., Haynes, P.A., Milligan, R.A., and Mayfield, S.P. (2005). Composition and structure of the 80S ribosome from the green alga *Chlamydomonas reinhardtii*: 80S ribosomes are conserved in plants and animals. *J. Mol. Biol.* **351**, 266–279.

Manuell, A.L., Beligni, M.V., Elder, J.H., Siefker, D.T., Tran, M., Weber, A., McDonald, T.L., and Mayfield, S.P. (2007). Robust expression of a bioactive mammalian protein in *Chlamydomonas* chloroplast. *Plant Biotechnol. J.* **5**, 402–412.

Marchant, H.J. (1982). Snow algae from the Australian Snowy Mountains. *Phycologia* **21**, 178–184.

Marco, Y. and Rochaix, J.-D. (1980). Organization of the nuclear ribosomal DNA of *Chlamydomonas reinhardii*. *Mol. Gen. Genet.* **177**, 715–723.

Marek, L.F. and Spalding, M.H. (1991). Changes in photorespiratory enzyme activity in response to limiting CO_2 in *Chlamydomonas reinhardtii*. *Plant Physiol.* **97**, 420–425.

Marie, D., Simon, N., and Vaulot, D. (2005). Phytoplankton cell counting by flow cytometry. In: *Algal Culturing Techniques* (R.A. Andersen, Ed.), pp. 253–267. Elsevier, Amsterdam.

Markelova, A.G., Vladimirova, M.G., and Semenenko, V.E. (1990). Ultrastructural localization of ribulose bisphosphate carboxylase in algal cells. *Fiziol. Rast.* **37**, 907–911.

Marshall, P. and Lemieux, C. (1991). Cleavage pattern of the homing endonuclease encoded by the fifth intron in the chloroplast large subunit rRNA-encoding gene of *Chlamydomonas eugametos*. *Gene* **104**, 241–245.

Marshall, W.F. and Rosenbaum, J.L. (2001). Intraflagellar transport balances continuous turnover of outer doublet microtubules: implications for flagellar length control. *J. Cell Biol.* **155**, 405–414.

Marshall, W.F., Qin, H., Rodrigo Brenni, M., and Rosenbaum, J.L. (2005). Flagellar length control system: testing a simple model based on intraflagellar transport and turnover. *Mol. Biol. Cell* **16**, 270–278.

Martin, N.C. and Goodenough, U.W. (1975). Gametic differentiation in *Chlamydomonas reinhardtii*. I. Production of gametes and their fine structure. *J. Cell Biol.* **67**, 587–605.

Martin, N.C., Chiang, K.S., Goodenough, U.W., and Chiang, K.-S. (1976). Turnover of chloroplast and cytoplasmic ribosomes during gametogenesis in *Chlamydomonas reinhardi*. *Dev. Biol.* **51**, 190–201.

Martinez-Rivas, J.M., Vega, J.M., and Marquez, A.J. (1991). Differential regulation of the nitrate-reducing and ammonium-assimilatory systems in synchronous cultures of *Chlamydomonas reinhardtii*. *FEMS Microbiol. Lett.* **78**, 85–88.

Mason, C.B., Bricker, T.M., and Moroney, J.V. (2006). A rapid method for chloroplast isolation from the green alga *Chlamydomonas reinhardtii*. *Nat. Protoc.* **1**, 2227–2230.

Mast, S.O. (1916). The process of orientation in the colonial organism, *Gonium pectorale*, and a study of the structure and function of the eye-spot. *J. Exp. Zool.* **20**, 1–17.

Mast, S.O. (1928). Structure and function of the eye-spot in unicellular and colonial organisms. *Arch. Protistenk.* **60**, 197–220.

Matagne, R.F. (1981). Transmission of chloroplast alleles in somatic fusion products obtained from vegetative cells and/or 'gametes' of *Chlamydomonas reinhardi*. *Curr. Genet.* **3**, 31–36.

Matagne, R.F. and Beckers, M.-C. (1987). Isolation and characterization of biochemical and morphological mutants in *Chlamydomonas smithii*. *Plant Sci.* **49**, 85–88.

Matagne, R.F. and Hermesse, M.-P. (1980a). Chloroplast gene inheritance studied by somatic fusion in *Chlamydomonas reinhardtii*. *Curr. Genet.* **1**, 127–131.

Matagne, R.F. and Hermesse, M.-P. (1980b). Further analyses of somatic fusion products in *Chlamydomonas reinhardii*. *Arch. Int. Physiol. Biochim.* **88**, B291–B292.

Matagne, R.F. and Hermesse, M.-P. (1981). Modification of chloroplast gene transmission in somatic fusion products and vegetative zygotes of *Chlamydomonas reinhardi* by 5-fluorodeoxyuridine. *Genetics* **99**, 371–381.

Matagne, R.F. and Loppes, R. (1975). Isolation and study of mutants lacking a derepressible phosphatase in *Chlamydomonas reinhardi*. *Genetics* **80**, 239–250.

Matagne, R.F. and Mathieu, D. (1983). Transmission of chloroplast genes in triploid and tetraploid zygospores of *Chlamydomonas reinhardtii*: roles of mating-type gene dosage and gametic chloroplast DNA content. *Proc. Natl. Acad. Sci. U. S. A.* **80**, 4780–4783.

Matagne, R.F. and Orbans, A. (1980). Somatic segregation in diploid *Chlamydomonas reinhardii*. *J. Gen. Microbiol.* **119**, 71–77.

Matagne, R.F. and Schaus, M. (1985). About biased and non-biased transmission of chloroplast genes following artificial fusion of gametes in *Chlamydomonas reinhardtii*. *Curr. Genet.* **10**, 81–85.

Matagne, R.F., Loppes, R., and Deltour, R. (1976a). Phosphatases of *Chlamydomonas reinhardi*: biochemical and cytochemical approach with specific mutants. *J. Bacteriol.* **126**, 937–950.

Matagne, R.F., Loppes, R., and Deltour, R. (1976b). Biochemical and cytochemical study of the derepressible phosphatase in *Chlamydomonas reinhardi*. *Arch. Int. Physiol. Biochim.* **84**, 173–174.

Matagne, R.F., Deltour, R., and Ledoux, L. (1979). Somatic fusion between cell wall mutants of *Chlamydomonas reinhardi*. *Nature* **278**, 344–346.

Matagne, R.F., Michel-Wolwertz, M.-R., Munaut, C., Duyckaerts, C., and Sluse, F. (1989). Induction and characterization of mitochondrial DNA mutants in *Chlamydomonas reinhardtii*. *J. Cell Biol.* **108**, 1221–1226.

Matagne, R.F., Remacle, C., and Dinant, M. (1991). Cytoduction in *Chlamydomonas reinhardtii*. *Proc. Natl. Acad. Sci. U. S. A.* **88**, 7447–7450.

Matsuda, A., Yoshimura, K., Sineshchekov, O.A., Hirono, M., and Kamiya, R. (1998). Isolation and characterization of novel *Chlamydomonas* mutants that display phototaxis but not photophobic response. *Cell Motil. Cytoskeleton* **41**, 353–362.

Matsuda, Y. (1980). Occurrence of wall-less cells during synchronous gametogenesis in *Chlamydomonas reinhardtii*. *Plant Cell Physiol.* **21**, 1339–1342.

Matsuda, Y. and Surzycki, S.J. (1980). Chloroplast gene expression in *Chlamydomonas reinhardi. Mol. Gen. Genet.* **180**, 463–474.

Matsuda, Y., Tamaki, S., and Tsubo, Y. (1978). Mating type specific induction of cell wall lytic factor by agglutination of gametes in *Chlamydomonas reinhardtii. Plant Cell Physiol.* **19**, 1253–1261.

Matsuda, Y., Sakamoto, K., Mizuochi, T., Kobata, A., Tamura, G., and Tsubo, Y. (1981). Mating type specific inhibition of gametic differentiation of *Chlamydomonas reinhardtii* by tunicamycin. *Plant Cell Physiol.* **22**, 1607–1611.

Matsuda, Y., Sakamoto, K., and Tsubo, Y. (1983). Biased and non-biased transmission of chloroplast genes in somatic fusion products of *Chlamydomonas reinhardtii. Curr. Genet.* **7**, 339–345.

Matsuda, Y., Saito, T., Yamaguchi, T., and Kawase, H. (1985a). Cell wall lytic enzyme released by mating gametes of *Chlamydomonas reinhardtii* is a metalloprotease and digests the sodium perchlorate-insoluble component of cell wall. *J. Biol. Chem.* **260**, 6373–6377.

Matsuda, Y., Okabayashi, K., and Kitazume, Y. (1985b). Chloroplast development in the chloroplast ribosome deficient mutant (*y-1 ac-20*) of *Chlamydomonas reinhardtii. Plant Cell Physiol.* **26**, 647–653.

Matsuda, Y., Saito, T., Yamaguchi, T., Koseki, M., and Hayashi, K. (1987). Topography of cell wall lytic enzyme in *Chlamydomonas reinhardtii*: form and location of the stored enzyme in vegetative cell and gamete. *J. Cell Biol.* **104**, 321–329.

Matsuda, Y., Saito, T., Umemoto, T., and Tsubo, T. (1988). Transmission patterns of chloroplast genes after polyethylene glycol-induced fusion of gametes in non-mating mutants of *Chlamydomonas reinhardtii. Curr. Genet.* **14**, 53–58.

Matsuda, Y., Koseki, M., Shimada, T., and Saito, T. (1995). Purification and characterization of a vegetative lytic enzyme responsible for liberation of daughter cells during the proliferation of *Chlamydomonas reinhardtii. Plant Cell Physiol.* **36**, 681–689.

Matsumoto, T., Matsuo, M., and Matsuda, Y. (1991). Structural analysis and expression during dark-light transitions of a gene for cytochrome *f* in *Chlamydomonas reinhardtii. Plant Cell Physiol.* **32**, 863–872.

Matsumura, K., Yagi, T., and Yasuda, K. (2003). Role of timer and sizer in regulation of *Chlamydomonas* cell cycle. *Biochem. Biophys. Res. Commun.* **306**, 1042–1049.

Matsuo, T., Onai, K., Okamoto, K., Minagawa, J., and Ishiura, M. (2006). Real-time monitoring of chloroplast gene expression by a luciferase reporter: evidence for nuclear regulation of chloroplast circadian period. *Mol. Cell. Biol.* **26**, 863–870.

Matsuura, K., Lefebvre, P.A., Kamiya, R., and Hirono, M. (2004). Bld10p, a novel protein essential for basal body assembly in *Chlamydomonas*: localization to the cartwheel, the first ninefold symmetrical structure appearing during assembly. *J. Cell Biol.* **165**, 663–671.

Matters, G.L. and Beale, S.I. (1995). Blue-light-regulated expression of genes for two early steps of chlorophyll biosynthesis in *Chlamydomonas reinhardtii. Plant Physiol.* **109**, 471–479.

Mattox, K.R. and Stewart, K.D. (1977). Cell division in the scaly green flagellate *Heteromastix angulata* and its bearing on the origin of the Chlorophyceae. *Am. J. Bot.* **64**, 931–945.

Mattox, K.R. and Stewart, K.D. (1984). Classification of the green algae: a concept based on comparative cytology. In: *Systematics of the Green Algae* (D.E.G. Irvine and D.M. John, Eds.), pp. 29–72. Academic Press, New York.

Maul, J.E., Lilly, J.W., Cui, L.Y., DePamphilis, C.W., Miller, W., Harris, E.H., and Stern, D.B. (2002). The *Chlamydomonas reinhardtii* plastid chromosome: islands of genes in a sea of repeats. *Plant Cell* **14**, 2659–2679.

Mayer, A.M. (1968). *Chlamydomonas*: adaptation phenomena in phototaxis. *Nature* **217**, 875–876.

Mayer, A.M. and Poljakoff-Mayber, A. (1959). The phototactic behavior of *Chlamydomonas snowiae*. *Physiol. Plant.* **12**, 8–14.

Mayfield, S.P. (1991). Over-expression of the oxygen-evolving enhancer 1 protein and its consequences on photosystem II accumulation. *Planta* **185**, 105–110.

Mayfield, S.P. and Kindle, K.L. (1990). Stable nuclear transformation of *Chlamydomonas reinhardtii* by using a *C. reinhardtii* gene as the selectable marker. *Proc. Natl. Acad. Sci. U. S. A.* **87**, 2087–2091.

Mayfield, S.P. and Schultz, J. (2004). Development of a luciferase reporter gene, *luxCt*, for *Chlamydomonas reinhardtii* chloroplast. *Plant J.* **37**, 449–458.

Mayfield, S.P., Bennoun, P., and Rochaix, J.-D. (1987a). Expression of the nuclear encoded OEE1 protein is required for oxygen evolution and stability of photosystem II particles in *Chlamydomonas reinhardtii*. *EMBO J.* **6**, 313–318.

Mayfield, S.P., Rahire, M., Frank, G., Zuber, H., and Rochaix, J.-D. (1987b). Expression of the nuclear gene encoding oxygen-evolving enhancer protein 2 is required for high levels of photosynthetic oxygen evolution in *Chlamydomonas reinhardtii*. *Proc. Natl. Acad. Sci. U. S. A.* **84**, 749–753.

Mayfield, S.P., Schirmer-Rahire, M., Frank, G., Zuber, H., and Rochaix, J.-D. (1989). Analysis of the genes of the OEE1 and OEE3 proteins of the photosystem II complex from *Chlamydomonas reinhardtii*. *Plant Mol. Biol.* **12**, 683–693.

Mayfield, S.P., Cohen, A., Danon, A., and Yohn, C.B. (1994). Translation of the *psbA* mRNA of *Chlamydomonas reinhardtii* requires a structured RNA element contained within the 5′ untranslated region. *J. Cell Biol.* **127**, 1537–1546.

Mayfield, S.P., Franklin, S.E., and Lerner, R.A. (2003). Expression and assembly of a fully active antibody in algae. *Proc. Natl. Acad. Sci. U. S. A.* **100**, 438–442.

Mayfield, S.P., Manuell, A.L., Chen, S., Wu, J., Tran, M., Siefker, D., Muto, M., and Marin-Navarro, J. (2007). *Chlamydomonas reinhardtii* chloroplasts as protein factories. *Curr. Opin. Biotechnol.* **18**, 126–133.

McAteer, M., Donnan, L., and John, P.C.L. (1985). The timing of division in *Chlamydomonas*. *New Phytol.* **99**, 41–56.

McBride, A.C. and Gowans, C.S. (1970). Comparative sensitivity of three species of *Chlamydomonas* to analogs of metabolites. *J. Phycol.* **6**, 54–56.

McBride, A.C. and McBride, J.C. (1975). Uniparental inheritance in *Chlamydomonas eugametos* (Chlorophyceae). *J. Phycol.* **11**, 343–344.

McCarthy, S.S., Kobayashi, M.C., and Niyogi, K.K. (2004). White mutants of *Chlamydomonas reinhardtii* are defective in phytoene synthase. *Genetics* **168**, 1249–1257.

McCourt, R.M. (1995). Green algal phylogeny. *Trends Ecol. Evol.* **10**, 159–163.

McElwain, K.B., Boynton, J.E., and Gillham, N.W. (1993). A nuclear mutation conferring thiostrepton resistance in *Chlamydomonas reinhardtii* affects a chloroplast ribosomal protein related to *Escherichia coli* ribosomal protein L11. *Mol. Gen. Genet.* **241**, 564–572.

McFadden, G.I. and Melkonian, M. (1986). Use of Hepes buffer for microalgal culture media and fixation for electron microscopy. *Phycologia* **25**, 551–557.

McGrath, M.S. and Daggett, P.-M. (1977). Cryopreservation of flagellar mutants of *Chlamydomonas reinhardtii*. *Can. J. Bot.* **55**, 1794–1796.

McGrath, M.S., Daggett, P.-M., and Dilworth, S. (1978). Freeze-drying of algae: Chlorophyta and Chrysophyta. *J. Phycol.* **14**, 521–525.

McHugh, J.P. and Spanier, J.G. (1994). Isolation of cadmium sensitive mutants in *Chlamydomonas reinhardtii* by transformation/insertional mutagenesis. *FEMS Microbiol. Lett.* **124**, 239–244.

McKay, R.M.L. and Gibbs, S.P. (1991a). Composition and function of pyrenoids: cytochemical and immunocytochemical approaches. *Can. J. Bot.* **69**, 1040–1052.

McKay, R.M.L. and Gibbs, S.P. (1991b). Immunocytochemical localization of phosphoribulokinase in microalgae. *Bot. Acta* **104**, 367–373.

McKay, R.M.L., Gibbs, S.P., and Vaughn, K.C. (1991). RuBisCo activase is present in the pyrenoid of green algae. *Protoplasma* **162**, 38–45.

McLean, R.J. and Brown, R.M. (1974). Cell surface differentiation of *Chlamydomonas* during gametogenesis. I. Mating and concanavalin Λ agglutinability. *Dev. Biol.* **36**, 279–285.

McLean, R.J., Katz, K.R., Sedita, N.J., Menoff, A.L., Laurendi, C.J., and Brown, R.M. (1981). Dynamics of concanavalin A binding sites on *Chlamydomonas moewusii* flagellar membranes. *Ber. Deutsch. Bot. Ges.* **94**, 387–400.

McMahon, D. (1971). The isolation of mutants conditionally defective in protein synthesis in *Chlamydomonas reinhardi*. *Mol. Gen. Genet.* **112**, 80–86.

McVittie, A. and Davies, D.R. (1971). The location of the Mendelian linkage groups in *Chlamydomonas reinhardii*. *Mol. Gen. Genet.* **112**, 225–228.

Melis, A. (2007). Photosynthetic H_2 metabolism in *Chlamydomonas reinhardtii* (unicellular green algae). *Planta* **226**, 1075–1086.

Melis, A. and Happe, T. (2004). Trails of green alga hydrogen research – From Hans Gaffron to new frontiers. *Photosynth. Res.* **80**, 401–409.

Melis, A., Zhang, L.P., Forestier, M., Ghirardi, M.L., and Seibert, M. (2000). Sustained photobiological hydrogen gas production upon reversible inactivation of oxygen evolution in the green alga *Chlamydomonas reinhardtii*. *Plant Physiol.* **122**, 127–135.

Melis, A., Seibert, M., and Ghirardi, M.L. (2007). Hydrogen fuel production by transgenic microalgae. *Adv. Exp. Med. Biol.* **616**, 110–121.

Melkonian, M. (1977). The flagellar root system of zoospores of the green alga Chlorosarcinopsis (Chlorosarcinales) as compared with *Chlamydomonas* (Volvocales). *Plant Syst. Evol.* **128**, 79–88.

Melkonian, M. (1982). The functional analysis of the flagellar apparatus in green algae. *Symp. Soc. Exp. Biol.* **35**, 589–606.

Melkonian, M. (1984). Flagellar apparatus ultrastructure in relation to green algal classification. In: *Systematics of the Green Algae* (D.E.G. Irvine and D.M. John, Eds.), pp. 73–120. Academic Press, New York.

Melkonian, M. and Robenek, H. (1979). The eyespot of the flagellate *Tetraselmis cordiformis* Stein (Chlorophyceae): structural specialization of the outer chloroplast membrane and its possible significance in phototaxis of green algae. *Protoplasma* **100**, 183–197.

Melkonian, M. and Robenek, H. (1980). Eyespot membranes of *Chlamydomonas reinhardii*: a freeze-fracture study. *J. Ultrastruct. Res.* **72**, 90–102.

Melkonian, M. and Robenek, H. (1984). The eyespot apparatus of flagellated green algae: a critical review. *Prog. Phycol. Res.* **3**, 193–268.

Melkonian, M., Robenek, H., and Steup, M. (1981). Occurrence and distribution of filipin-sterol complexes in chloroplast envelope membranes of algae and higher plants as visualized by freeze fracture. *Protoplasma* **109**, 349–358.

Mendiola-Morgenthaler, L., Lcu, S., Boschetti, A., and Eichenberger, W. (1985a). Isolation of biochemically active chloroplasts from *Chlamydomonas*. *Plant Sci.* **38**, 33–39.

Mendiola-Morgenthaler, L., Leu, S., Boschetti, A., and Eichenberger, W. (1985b). Isolation of chloroplast envelopes from *Chlamydomonas*. Lipid and polypeptide composition. *Plant Sci.* **41**, 97–104.

Merchán, F., van den Ende, H., Fernández, E., and Beck, C.F. (2001). Low-expression genes induced by nitrogen starvation and subsequent sexual differentiation in *Chlamydomonas reinhardtii*, isolated by the differential display technique. *Planta* **213**, 309–317.

Merchant, S. and Bogorad, L. (1987a). The Cu(II)-repressible plastidic cytochrome *c*: cloning and sequence of a complementary DNA for the pre-apoprotein. *J. Biol. Chem.* **262**, 9062–9067.

Merchant, S. and Bogorad, L. (1987b). Metal ion regulated gene expression: use of a plastocyanin-less mutant of *Chlamydomonas reinhardtii* to study the Cu(II)-dependent expression of cytochrome *c*-552. *EMBO J.* **6**, 2531–2535.

Merchant, S.S., Allen, M.D., Kropat, J., Moseley, J.L., Long, J.C., Tottey, S., and Terauchi, A.M. (2006). Between a rock and a hard place: trace element nutrition in *Chlamydomonas*. *Biochim. Biophys. Acta* **1763**, 578–594.

Merchant, S.S., Prochnik, S.E., Vallon, O., Harris, E.H., Karpowicz, S.J., Witman, G.B., Terry, A., Salamov, A., Fritz-Laylin, L.K., Maréchal-Drouard, L., Marshall, W.F., Qu, L.H., Nelson, D.R., Sanderfoot, A.A., Spalding, M.H., Kapitonov, V.V., Ren, Q., Ferris, P., Lindquist, E., Shapiro, H., Lucas, S.M., Grimwood, J., Schmutz, J., Cardol, P., Cerutti, H., Chanfreau, G., Chen, C.L., Cognat, V., Croft, M.T., Dent, R., Dutcher, S., Fernández, E., Fukuzawa, H., González-Ballester, D., González-Halphen, D., Hallmann, A., Hanikenne, M., Hippler, M., Inwood, W., Jabbari, K., Kalanon, M., Kuras, R., Lefebvre, P.A., Lemaire, S.D., Lobanov, A.V., Lohr, M., Manuell, A., Meier, I., Mets, L., Mittag, M., Mittelmeier, T., Moroney, J.V., Moseley, J., Napoli, C., Nedelcu, A.M., Niyogi, K., Novoselov, S.V., Paulsen, I.T., Pazour, G., Purton, S., Ral, J.P., Riaño-Pachón, D.M., Riekhof, W., Rymarquis, L., Schroda, M., Stern, D., Umen, J., Willows, R., Wilson, N., Zimmer, S.L., Allmer, J., Balk, J., Bisova, K., Chen, C.J., Elias, M., Gendler, K., Hauser, C., Lamb, M.R., Ledford, H., Long, J.C., Minagawa, J., Page, M.D., Pan, J., Pootakham, W., Roje, S., Rose, A., Stahlberg, E., Terauchi, A.M., Yang, P., Ball, S., Bowler, C., Dieckmann, C.L., Gladyshev, V.N., Green, P., Jorgensen, R., Mayfield, S., Mueller-Roeber, B., Rajamani, S., Sayre, R.T., Brokstein, P., Dubchak, I., Goodstein, D., Hornick, L., Huang, Y.W., Jhaveri, J., Luo, Y., Martinez, D., Ngau, W.C., Otillar, B., Poliakov, A., Porter, A., Szajkowski, L., Werner, G., Zhou, K., Grigoriev, I.V., Rokhsar, D.S., and Grossman, A.R. (2007). The *Chlamydomonas* genome reveals the evolution of key animal and plant functions. *Science* **318**, 245–250.

Merendino, L., Falciatore, A., and Rochaix, J.D. (2006). A novel multifunctional factor involved in *trans*-splicing of chloroplast introns in *Chlamydomonas*. *Nucleic Acids Res.* **34**, 262–274.

Mergenhagen, D. (1980). Die Kinetik der Zoosporenfreizetzung bei einem Mutantenstamm von *Chlamydomonas reinhardii*. *Mitt. Inst. Allg. Bot. Hamb.* **17**, 19–26.

Mergenhagen, D. (1984). Circadian clock: genetic characterization of a short period mutant of *Chlamydomonas reinhardii*. *Eur. J. Cell Biol.* **33**, 13–18.

Mergenhagen, D. and Mergenhagen, E. (1987). The biological clock of *Chlamydomonas reinhardii* in space. *Eur. J. Cell Biol.* **43**, 203–207.

Mergenhagen, D. and Mergenhagen, E. (1989). The expression of a circadian rhythm in two strains of *Chlamydomonas reinhardii* in space. *Adv. Space Res.* **9**, 261–270.

Meselson, M. and Stahl, F.W. (1958). The replication of DNA in *Escherichia coli*. *Proc. Natl. Acad. Sci. U. S. A.* **44**, 671–682.

Mesland, D.A. (1976). Mating in *Chlamydomonas eugametos*. A scanning electron microscopical study. *Arch. Microbiol.* **109**, 31–35.

Mesland, D.A., Hoffman, J.L., Caligor, E., and Goodenough, U.W. (1980). Flagellar tip activation stimulated by membrane adhesions in *Chlamydomonas* gametes. *J. Cell Biol.* **84**, 599–617.

Messerli, M.A., Amaral-Zettler, L.A., Zettler, E., Jung, S.K., Smith, P.J., and Sogin, M.L. (2005). Life at acidic pH imposes an increased energetic cost for a eukaryotic acidophile. *J. Exp. Biol.* **208**, 2569–2579.

Mets, L. (1980). Uniparental inheritance of chloroplast DNA sequences in interspecific hybrids of *Chlamydomonas*. *Curr. Genet.* **2**, 131–138.

Mets, L.J. and Bogorad, L. (1974). Two-dimensional polyacrylamide gel electrophoresis: an improved method for ribosomal proteins. *Anal. Biochem.* **57**, 200–210.

Mets, L.J. and Geist, L.J. (1983). Linkage of a known chloroplast gene mutation to the uniparental genome of *Chlamydomonas reinhardii*. *Genetics* **105**, 559–579.

Metting, B. (1986). Population dynamics of *Chlamydomonas sajao* and its influence on soil aggregate stabilization in the field. *Appl. Environ. Microbiol.* **51**, 1161–1164.

Metting, B. (1987). Dynamics of wet and dry aggregate stability from a three-year microalgal soil conditioning experiment in the field. *Soil Sci.* **143**, 139–143.

Metting, B. and Rayburn, W.R. (1983). The influence of a microalgal conditioner on selected Washington soils: an empirical study. *Soil Sci. Soc. Am. J.* **47**, 682–685.

Miadoková, E., Podstavková, S., Vlček, D., and Simonová, M. (1991). Characterization of photoreactivation-deficient mutants of *Chlamydomonas reinhardtii*. *Arch. Protistenk.* **139**, 207–212.

Miadoková, E., Podstavková, S., Červenák, Z., and Vlček, D. (1994). Different responses of repair-deficient strains of *Chlamydomonas reinhardtii* to UV and MNNG treatments. *Biologia* **49**, 633–637.

Michaelis, G., Vahrenholz, C., and Pratje, E. (1990). Mitochondrial DNA of *Chlamydomonas reinhardtii*: the gene for apocytochrome *b* and the complete functional map of the 15.8 kb DNA. *Mol. Gen. Genet.* **223**, 211–216.

Miller, C.C.J., Duckett, J.G., Downes, M.J., Cowell, I., Dowding, A.J., Virtanen, I., and Anderton, B.H. (1985). Plant cytoskeletons contain intermediate filament-related proteins. *Biochem. Soc. Trans.* **13**, 960–961.

Miller, D.H., Lamport, D.T.A., and Miller, M. (1972). Hydroxyproline hetero-oligosaccharides in *Chlamydomonas*. *Science* **176**, 918–920.

Miller, D.H., Mellman, I.S., Lamport, D.T.A., and Miller, M. (1974). The chemical composition of the cell wall of *Chlamydomonas gymnogama* and the concept of a plant cell wall protein. *J. Cell Biol.* **63**, 420–429.

Miller, M.S., Esparza, J.M., Lippa, A.M., Lux, F.G., III, Cole, D.G., and Dutcher, S.K. (2005). Mutant kinesin-2 motor subunits increase chromosome loss. *Mol. Biol. Cell* **16**, 3810–3820.

Millikin, B.E. and Weiss, R.L. (1984a). Distribution of concanavalin A binding carbohydrates during mating in *Chlamydomonas*. *J. Cell Sci.* **66**, 223–239.

Millikin, B.E. and Weiss, R.L. (1984b). Localization of concanavalin A binding carbohydrate in *Chlamydomonas* flagella. *J. Cell Sci.* **68**, 211–226.

Milner, H.W., Lawrence, N.S., and French, C.S. (1950). Colloidal disperson of chloroplast material. *Science* **111**, 633–634.

Minagawa, J. and Crofts, A.R. (1994). A robust protocol for site-directed mutagenesis of the D1 protein in *Chlamydomonas reinhardtii*: a PCR-spliced *psbA* gene in a plasmid conferring spectinomycin resistance was introduced into a *psbA* deletion strain. *Photosynth. Res.* **42**, 121–131.

Minami, S.A. and Goodenough, U.W. (1978). Novel glycopolypeptide synthesis induced by gametic cell fusion in *Chlamydomonas reinhardtii*. *J. Cell Biol.* **77**, 165–181.

Minko, I., Holloway, S.P., Nikaido, S., Carter, M., Odom, O.W., Johnson, C.H., and Herrin, D.L. (1999). *Renilla* luciferase as a vital reporter for chloroplast gene expression in *Chlamydomonas*. *Mol. Gen. Genet.* **262**, 421–425.

Mirnaya, O.N., Fomina-Eshchenko, Yu.G., and Chunayev, A.S. (1990). Localization of the *cbn-1* mutation in the linkage group I of nuclear genes in *Chlamydomonas reinhardtii*. *Genetika* **26**, 958–960.

Misamore, M.J., Gupta, S., and Snell, W.J. (2003). The *Chlamydomonas* Fus1 protein is present on the mating type *plus* fusion organelle and required for a critical membrane adhesion event during fusion with *minus* gametes. *Mol. Biol. Cell* **14**, 2530–2542.

Misquitta, R.W. and Herrin, D.L. (2005). Circadian regulation of chloroplast gene transcription: a review. *Plant Tissue Cult.* **15**, 83–101.

Misumi, O., Yoshida, Y., Nishida, K., Fujiwara, T., Sakajiri, T., Hirooka, S., Nishimura, Y., and Kuroiwa, T. (2007). Genome analysis and its significance in four unicellular algae, *Cyanidioshyzon merolae*, *Ostreococcus tauri*, *Chlamydomonas reinhardtii*, and *Thalassiosira pseudonana*. *J. Plant Res.* **121**, 3–17.

Mitchell, B.F., and Graziano, M.R. (2006). From organelle to protein gel: a 6-wk laboratory project on flagellar proteins. *CBE Life Sci. Educ.* **5**, 239–246.

Mitchell, B.F., Grulich, L.E., and Mader, M.M. (2004). Flagellar quiescence in *Chlamydomonas*: characterization and defective quiescence in cells carrying *sup-pf-1* and *sup-pf-2* outer dynein arm mutations. *Cell Motil. Cytoskeleton* **57**, 186–196.

Mitchell, D.R. (2003). Orientation of the central pair complex during flagellar bend formation in *Chlamydomonas*. *Cell Motil. Cytoskeleton* **56**, 120–129.

Mitchell, D.R. and Brown, K.S. (1994). Sequence analysis of the *Chlamydomonas* alpha and beta dynein heavy chain genes. *J. Cell Sci.* **107**, 635–644.

Mitchell, D.R. and Kang, Y. (1991). Identification of *oda6* as a *Chlamydomonas* dynein mutant by rescue with the wild-type gene. *J. Cell Biol.* **113**, 835–842.

Mitchell, D.R. and Nakatsugawa, M. (2004). Bend propagation drives central pair rotation in *Chlamydomonas reinhardtii* flagella. *J. Cell Biol.* **166**, 709–715.

Mitchell, D.R. and Rosenbaum, J.L. (1985). A motile *Chlamydomonas* flagellar mutant that lacks outer dynein arms. *J. Cell Biol.* **100**, 1228–1234.

Mitchell, D.R. and Sale, W.S. (1999). Characterization of a *Chlamydomonas* insertional mutant that disrupts flagellar central pair microtubule-associated structures. *J. Cell Biol.* **144**, 293–304.

Mitra, A.K. (1950). A peculiar method of sexual reproduction in certain new members of the Chlamydomonadaceae. *Hydrobiologia* **2**, 209–216.

Mitra, M., Mason, C.B., Xiao, Y., Ynalvez, R.A., Lato, S.M., and Moroney, J.V. (2005). The carbonic anhydrase gene families of *Chlamydomonas reinhardtii*. *Can. J. Bot.* **83**, 780–795.

Mittag, M. (1996). Conserved circadian elements in phylogenetically diverse algae. *Proc. Natl. Acad. Sci. U. S. A.* **93**, 14401–14404.

Mittag, M. (2001). Circadian rhythms in microalgae. *Int. Rev. Cytol.* **206**, 213–247.

Mittag, M. (2003). The function of circadian RNA-binding proteins and their cis-acting elements in microalgae. *Chronobiol. Int.* **20**, 529–541.

Mittag, M., Kiaulehn, S., and Johnson, C.H. (2005). The circadian clock in *Chlamydomonas reinhardtii*. What is it for? What is it similar to? *Plant Physiol.* **137**, 399–409.

Miura, Y., Saitoh, C., Matsuoka, S., and Miyamoto, K. (1992). Stably sustained hydrogen production with high molar yield through a combination of a marine green alga and a photosynthetic bacterium. *Biosci. Biotechnol. Biochem.* **56**, 751–754.

Miura, K., Yamano, T., Yoshioka, S., Kohinata, T., Inoue, Y., Taniguchi, F., Asamizu, E., Nakamura, Y., Tabata, S., Yamato, K.T., Ohyama, K., and Fukuzawa, H. (2004). Expression profiling-based identification of CO_2-responsive genes regulated by *CCM1* controlling a carbon-concentrating mechanism in *Chlamydomonas reinhardtii*. *Plant Physiol.* **135**, 1595–1607.

Miyagishima, S.Y., Nishida, K., and Kuroiwa, T. (2003). An evolutionary puzzle: chloroplast and mitochondrial division rings. *Trends Plant Sci.* **8**, 432–438.

Miyasaka, H. and Ikeda, K. (1997). Osmoregulating mechanism of the halotolerant green alga *Chlamydomonas*, strain HS-5. *Plant Sci.* **127**, 91–96.

Miyasaka, H., Ohnishi, Y., Akano, T., Fukatsu, K., Mizoguchi, T., Yagi, K., Maeda, I., Ikuta, Y., Matsumoto, H., Shioji, N., and Miura, Y. (1998). Excretion of glycerol by the marine *Chlamydomonas sp.* strain W-80 in high CO_2 cultures. *J. Ferment. Bioeng.* **85**, 123–125.

Miyasaka, H., Kanaboshi, H., and Ikeda, K. (2000). Isolation of several anti-stress genes from the halotolerant green alga *Chlamydomonas* by simple functional expression screening with *Escherichia coli*. *World J. Microbiol. Biotechnol.* **16**, 23–29.

Moestrup, Ø. (1978). On the phylogenetic validity of the flagellar apparatus in green algae and other chlorophyll A and B containing plants. *BioSystems* **10**, 117–144.

Moharikar, S., D'Souza, J.S., Kulkarni, A.B., and Rao, B.J. (2006). Apoptotic-like cell death pathway is induced in unicellular chlorophyte *Chlamydomonas reinhardtii* (Chlorophyceae) cells following UV irradiation: detection and functional analyses. *J. Phycol.* **42**, 423–433.

Moharikar, S., D'Souza, J.S., and Rao, B.J. (2007). A homologue of the defender against the apoptotic death gene (*dad1*) in UV-exposed *Chlamydomonas* cells is downregulated with the onset of programmed cell death. *J. Biosci.* **32**, 261–270.

Molen, T.A., Rosso, D., Piercy, S., and Maxwell, D.P. (2006). Characterization of the alternative oxidase of *Chlamydomonas reinhardtii* in response to oxidative stress and a shift in nitrogen source. *Physiol. Plant.* **127**, 74–86.

Molnár, A., Schwach, F., Studholme, D.J., Thuenemann, E.C., and Baulcombe, D.C. (2007). miRNAs control gene expression in the single-cell alga *Chlamydomonas*. *Nature* **447**, 1126–1129.

Monk, B.C., Adair, W.S., Cohen, R.A., and Goodenough, U.W. (1983). Topography of *Chlamydomonas*: fine structure and polypeptide components of the gametic flagellar membrane surface and the cell wall. *Planta* **158**, 517–533.

Morel-Laurens, N.M.L. and Bird, D.J. (1984). Effects of cell division on the stigma of wild-type and an "eyeless" mutant of *Chlamydomonas*. *J. Ultrastruct. Res.* **87**, 46–61.

Morel-Laurens, N.M.L. and Feinleib, M.E. (1983). Photomovement in an "eyeless" mutant of *Chlamydomonas*. *Photochem. Photobiol.* **37**, 189–194.

Moreno-Risueno, M.A., Martínez, M., Vicente-Carbajosa, J., and Carbonero, P. (2007). The family of DOF transcription factors: from green unicellular algae to vascular plants. *Mol. Genet. Genomics* **277**, 379–390.

Morgan, R.M., Ivanov, A.G., Priscu, J.C., Maxwell, D.P., and Huner, N.P.A. (1998). Structure and composition of the photochemical apparatus of the Antarctic green alga, *Chlamydomonas subcaudata*. *Photosynth. Res.* **56**, 303–314.

Morgan-Kiss, R., Ivanov, A.G., Williams, J., Khan, M., and Huner, N.P.A. (2002a). Differential thermal effects on the energy distribution between photosystem II and photosystem I in thylakoid membranes of a psychrophilic and a mesophilic alga. *Biochim. Biophys. Acta* **1561**, 251–265.

Morgan-Kiss, R.M., Ivanov, A.G., and Huner, N.P.A. (2002b). The Antarctic psychrophile, *Chlamydomonas subcaudata*, is deficient in state I state II transitions. *Planta* **214**, 435–445.

Morgan-Kiss, R.M., Ivanov, A.G., Pocock, T., Krol, M., Gudynaite-Savitch, L., and Huner, N.P.A. (2005). The Antarctic psychrophile, *Chlamydomonas raudensis* Ettl (UWO241) (Chlorophyceae, Chlorophyta), exhibits a limited capacity to photoacclimate to red light. *J. Phycol.* **41**, 791–800.

Mori, T., Kuroiwa, H., Higashiyama, T., and Kuroiwa, T. (2006). GENERATIVE CELL SPECIFIC 1 is essential for angiosperm fertilization. *Nat. Cell Biol.* **8**, 64–71.

Morita, E., Kuroiwa, H., Kuroiwa, T., and Nozaki, H. (1997). High localization of ribulose-1,5-bisphosphate carboxylase/oxygenase in the pyrenoids of *Chlamydomonas reinhardtii* (Chlorophyta), as revealed by cryofixation and immunogold electron microscopy. *J. Phycol.* **33**, 68–72.

Morita, E., Abe, T., Tsuzuki, M., Fujiwara, S., Sato, N., Hirata, A., Sonoike, K., and Nozaki, H. (1998). Presence of the CO_2-concentrating mechanism in some species of the pyrenoid-less free-living algal genus *Chloromonas* (Volvocales, Chlorophyta). *Planta* **204**, 269–276.

Morita, E., Abe, T., Tsuzuki, M., Fujiwara, S., Sato, N., Hirata, A., Sonoike, K., and Nozaki, H. (1999). Role of pyrenoids in the CO_2-concentrating mechanism: comparative morphology, physiology and molecular phylogenetic analysis of closely related strains of *Chlamydomonas* and *Chloromonas* (Volvocales). *Planta* **208**, 365–372.

Morlon, H., Fortin, C., Adam, C., and Garnier Laplace, J. (2006). Selenite transport and its inhibition in the unicellular green alga *Chlamydomonas reinhardtii*. *Environ. Toxicol. Chem.* **25**, 1408–1417.

Moroney, J.V. and Ynalvez, R.A. (2007). A proposed carbon dioxide concentrating mechanism in *Chlamydomonas reinhardtii*. *Eukaryotic Cell* **6**, 1251–1259.

Moroney, J.V., Wilson, B.J., and Tolbert, N.E. (1986a). Glycolate metabolism and excretion by *Chlamydomonas reinhardtii*. *Plant Physiol.* **82**, 821–826.

Moroney, J.V., Tolbert, N.E., and Sears, B.B. (1986b). Complementation analysis of the inorganic carbon concentrating mechanism of *Chlamydomonas reinhardtii*. *Mol. Gen. Genet.* **204**, 199–203.

Moroney, J.V., Husic, H.D., Tolbert, N.E., Kitayama, M., Manuel, L.J., and Togasaki, R.K. (1989). Isolation and characterization of a mutant of *Chlamydomonas reinhardtii* deficient in the CO_2 concentrating mechanism. *Plant Physiol.* **89**, 897–903.

Morris, G.J., Coulson, G., and Clarke, A. (1979). The cryopreservation of *Chlamydomonas*. *Cryobiology* **16**, 401–410.

Morris, G.J., Coulson, G.E., Clarke, K.J., Grout, B.W.W., and Clarke, A. (1981). Freezing injury in *Chlamydomonas*: a synoptic approach. In: *The Effects of Low Temperature on Biological Systems* (G.J. Morris and A. Clarke, Eds.), pp. 285–306. Academic Press, London.

Morris, R.L., Salinger, A.P., and Rizzo, P.J. (1999). Analysis of lysine-rich histones from the unicellular green alga *Chlamydomonas reinhardtii*. *J. Euk. Microbiol.* **46**, 648–654.

Morton, B.R. (1998). Selection on the codon bias of chloroplast and cyanelle genes in different plant and algal lineages. *J. Mol. Evol.* **46**, 449–459.

Moseley, J., Quinn, J., Eriksson, M., and Merchant, S. (2000). The *Crd1* gene encodes a putative di-iron enzyme required for photosystem I accumulation in copper deficiency and hypoxia in *Chlamydomonas reinhardtii*. *EMBO J.* **19**, 2139–2151.

Moseley, J.L., Chang, C., and Grossman, A.R. (2006). Genome-based approaches to understanding phosphorus deprivation responses and PSR1 control in *Chlamydomonas reinhardtii*. *Eukaryotic Cell* **5**, 26–44.

Moss, A.G., Pazour, G.J., and Witman, G.B. (1995). Assay of *Chlamydomonas* phototaxis. *Methods Cell Biol.* **47**, 281–287.

Mosser, J.L., Mosser, A.G., and Brock, T.D. (1977). Photosynthesis in the snow: the alga *Chlamydomonas nivalis* (Chlorophyceae). *J. Phycol.* **13**, 22–27.

Mottley, J. and Griffiths, D.E. (1977). Minimum inhibitory concentrations of a broad range of inhibitors for the unicellular alga *Chlamydomonas reinhardi* Dangeard. *J. Gen. Microbiol.* **102**, 431–434.

Mouille, G., Maddelein, M.L., Libessart, N., Talaga, P., Decq, A., Delrue, B., and Ball, S. (1996). Preamylopectin processing: a mandatory step for starch biosynthesis in plants. *Plant Cell* **8**, 1353–1366.

Moyano, E., Cárdenas, J., and Muñoz-Blanco, J. (1995). Involvement of $NAD(P)^+$-glutamate dehydrogenase isoenzymes in carbon and nitrogen metabolism in *Chlamydomonas reinhardtii*. *Physiol. Plant.* **94**, 553–559.

Müller, F.W., Igloi, G.L., and Beck, C.F. (1992). Structure of a gene encoding heat-shock protein HSP70 from the unicellular alga *Chlamydomonas reinhardtii*. *Gene* **111**, 165–173.

Mueller, J., Perrone, C.A., Bower, R., Cole, D.G., and Porter, M.E. (2005). The FLA3 KAP subunit is required for localization of kinesin-2 to the site of flagellar assembly and processive anterograde intraflagellar transport. *Mol. Biol. Cell* **16**, 1341–1354.

Münzner, P. and Voigt, J. (1992). Blue light regulation of cell division in *Chlamydomonas reinhardtii*. *Plant Physiol.* **99**, 1370–1375.

Munce, D.B., Cox, J.L., Small, G.D., Vlček, D., Podstavková, S., and Miadoková, E. (1993). Genetic and biochemical analysis of photolyase mutants of *Chlamydomonas reinhardtii*. *Folia Microbiol.* **38**, 435–440.

Muñoz-Blanco, J., Moyano, E., and Cárdenas, J. (1989). Glutamate dehydrogenase isozymes of *Chlamydomonas reinhardtii*. *FEMS Microbiol. Lett.* **61**, 315–318.

Muñoz-Blanco, J., Hidalgo-Martínez, J., and Cárdenas, J. (1990). Extracellular deamination of L-amino acids by *Chlamydomonas reinhardtii* cells. *Planta* **182**, 194–198.

Murakami, S., Kuehnle, K., and Stern, D.B. (2005). A spontaneous tRNA suppressor of a mutation in the *Chlamydomonas reinhardtii* nuclear *MCD1* gene required for stability of the chloroplast *petD* mRNA. *Nucleic Acids Res.* **33**, 3372–3380.

Mus, F., Dubini, A., Seibert, M., Posewitz, M.C., and Grossman, A.R. (2007). Anaerobic acclimation in *Chlamydomonas reinhardtii*: anoxic gene expression, hydrogenase induction and metabolic pathways. *J. Biol. Chem.* **282**, 25475–25486.

Musgrave, A. (1987). Sexual agglutination in *Chlamydomonas eugametos*. In: *Molecular and Cellular Aspects of Algal Development* (W. Weissner and D.G. Robinson, Eds.), pp. 83–89. Springer-Verlag, Berlin.

Musgrave, A., van der Steuyt, P., and Ero, L. (1979). Concanavalin A binding to *Chlamydomonas eugametos* flagellar proteins and its effect on sexual reproduction. *Planta* **147**, 51–56.

Musgrave, A., de Wildt, P., Broekman, R., and van den Ende, H. (1983). The cell wall of *Chlamydomonas eugametos*. Immunological aspects. *Planta* **158**, 82–89.

Musgrave, A., de Wildt, P., Schuring, F., Crabbendam, K., and van den Ende, H. (1985). Sexual agglutination in *Chlamydomonas eugametos* before and after cell fusion. *Planta* **166**, 234–243.

Myers, A.M., Grant, D.M., Rabert, D.K., Harris, E.H., Boynton, J.E., and Gillham, N.W. (1982). Mutants of *Chlamydomonas reinhardtii* with physical alterations in their chloroplast DNA. *Plasmid* **7**, 133–151.

Myroniuk, V.I. and Kurinna, S.M. (2000). Influence of azide and salicylhydroxamate on the respiration of *Chlamydomonas snowiae* Printz and *Dunaliella salina* Teod. *Ukr. Bot. Zh.* **57**, 302–305. [in Ukranian with English abstract]

Myster, S.H., Knott, J.A., O'Toole, E., and Porter, M.E. (1997). The *Chlamydomonas Dhc1* gene encodes a dynein heavy chain subunit required for assembly of the I1 inner arm complex. *Mol. Biol. Cell* **8**, 607–620.

Myster, S.H., Knott, J.A., Wysocki, K.M., O'Toole, E., and Porter, M.E. (1999). Domains in the 1alpha dynein heavy chain required for inner arm assembly and flagellar motility in *Chlamydomonas*. *J. Cell Biol.* **146**, 801–818.

Nagel, K. and Voigt, J. (1989). *In vitro* evolution and preliminary characterization of a cadmium-resistant population of *Chlamydomonas reinhardtii*. *Appl. Environ. Microbiol.* **55**, 526–528.

Nagel, K., Adelmeier, U., and Voigt, J. (1996). Subcellular distribution of cadmium in the unicellular green alga *Chlamydomonas reinhardtii*. *J. Plant Physiol.* **149**, 86–90.

Nagel, G., Ollig, D., Fuhrmann, M., Katcriya, S., Musti, A.M., Bambcrg, E., and Hegemann, P. (2002). Channelrhodopsin-1: a light-gated proton channel in green algae. *Science* **296**, 2395–2398.

Nagel, G., Szellas, T., Huhn, W., Kateriya, S., Adeishvili, N., Berthold, P., Ollig, D., Hegemann, P., and Bamberg, E. (2003). Channelrhodopsin-2, a directly light-gated cation-selective membrane channel. *Proc. Natl. Acad. Sci. U. S. A.* **100**, 13940–13945.

Nagy, A.H., Erdos, G., Beliaeva, N.N., and Gyurjan, I. (1981). Acid phosphatase isoenzymes of *Chlamydomonas reinhardii*. *Mol. Gen. Genet.* **184**, 314–317.

Nagyová, B., Slaninová, M., Galová, E., and Vlček, D. (2003). Cell cycle checkpoints: from yeast to algae. *Biologia (Bratisl.)* **58**, 617–626.

Nakamura, K. and Gowans, C.S. (1964). Nicotinic-acid excreting mutants in *Chlamydomonas*. *Nature* **202**, 826–827.

Nakamura, K. and Gowans, C.S. (1972). A modifier of a *nic* gene in *Chlamydomonas*. *Can. J. Genet. Cytol.* **14**, 733.

Nakamura, K. and Lepard, S.L. (1983). Effect of near-ultraviolet and visible light on the methionine sensitivity of *Chlamydomonas reinhardtii* cultures. *Environ. Exp. Bot.* **23**, 331–338.

Nakamura, K., Bray, D.F., Costerton, J.W., and Wagenaar, E.B. (1973). The eyespot of *Chlamydomonas eugametos*: a freeze-etch study. *Can. J. Bot.* **51**, 817–819.

Nakamura, K., Hepler, W.L., Shaskin, E.J., and Brooks, D.W. (1974). Antialgal activity of chloroplatinic acid on *Chlamydomonas eugametos*. *Can. J. Bot.* **52**, 715–717.

Nakamura, K., Bray, D.F., and Wagenaar, E.B. (1975). Ultrastructure of *Chlamydomonas eugametos* palmelloids induced by chloroplatinic acid treatment. *J. Bacteriol.* **121**, 338–343.

Nakamura, K., Sakon, M., and Hatanaka, M.K. (1976). Chemical factors affecting palmelloid-forming activity of chloroplatinic acid on *Chlamydomonas eugametos*. *Physiol. Plant.* **36**, 293–296.

Nakamura, K., Bray, D.F., and Wagenaar, E.B. (1978). Ultrastructure of a palmelloid-forming strain of *Chlamydomonas eugametos*. *Can. J. Bot.* **56**, 2348–2356.

Nakamura, K., Kozloff, R.A., and Gowans, C.S. (1979). Isolation of tryptophan sensitive strains of the green alga *Chlamydomonas*. *Environ. Exp. Bot.* **19**, 311–314.

Nakamura, K., Lepard, S.L., and MacDonald, S.J. (1981). Is it possible to isolate methionine auxotrophs in *Chlamydomonas reinhardtii*? Consideration of photodynamic action of the amino acid. *Mol. Gen. Genet.* **181**, 292–295.

Nakamura, K., Landry, C.F., Goertzen, C.A., and Ikebuchi, N.W. (1985). Limited uptake as a mechanism for the nonrecoverability of arginine auxotrophs in *Chlamydomonas eugametos*. *Can. J. Bot.* **63**, 909–915.

Nakamura, S. and Ikehara, T. (2000). Mutant male *Chlamydomonas reinhardtii* cells with few chloroplast nucleoids and exceptional chloroplast gene transmission. *J. Plant Res.* **113**, 165–170.

Nakamura, S., Itoh, S., and Kuroiwa, T. (1986). Behavior of chloroplast nucleus during chloroplast development and degeneration in *Chlamydomonas reinhardii*. *Plant Cell Physiol.* **27**, 775–784.

Nakamura, S., Tanaka, G., Maeda, T., Kamiya, R., Matsunaga, T., and Nikaido, O. (1996). Assembly and function of *Chlamydomonas* flagellar mastigonemes as probed with a monoclonal antibody. *J. Cell Sci.* **109**, 57–62.

Nakamura, S., Ogihara, H., Jinbo, K., Tateishi, M., Takahashi, T., Yoshimura, K., Kubota, M., Watanabe, M., and Nakamura, S. (2001). *Chlamydomonas reinhardtii* Dangeard (Chlamydomonadales, Chlorophyceae) mutant with multiple eyespots. *Phycol. Res.* **49**, 115–122.

Nakamura, S., Aoyama, H., and Van Woesik, R. (2003). Strict paternal transmission of mitochondrial DNA of *Chlamydomonas* species is explained by selection against maternal nucleoids. *Protoplasma* **221**, 205–210.

Nakamura, S., Misumi, O., Aoyama, H., Van Woesik, R., and Kuroiwa, T. (2004). Monokaryotic chloroplast mutation has no effect on non-Mendelian transmission of chloroplast and mitochondrial DNA in *Chlamydomonas* species. *Protoplasma* **224**, 107–112.

Nakamura, Y., Kanakagiri, S., Van, K., He, W., and Spalding, M.H. (2005). Disruption of the glycolate dehydrogenase gene in the high-CO_2-requiring mutant *HCR89* of *Chlamydomonas reinhardtii*. *Can. J. Bot.* **83**, 820–833.

Nakayama, T., Watanabe, S., Mitsui, K., Uchida, H., and Inouye, I. (1996). The phylogenetic relationship between the Chlamydomonadales and Chlorococcales inferred from 18SrDNA sequence data. *Phycol. Res.* **44**, 47–55.

Nakazato, E., Fukuzawa, H., Tabata, S., Takahashi, H., and Tanaka, K. (2003). Identification and expression analysis of cDNA encoding a chloroplast recombination protein REC1, the chloroplast RecA homologue in *Chlamydomonas reinhardtii*. *Biosci. Biotechnol. Biochem.* **67**, 2608–2613.

Naumann, B., Stauber, E.J., Busch, A., Sommer, F., and Hippler, M. (2005). N-terminal processing of Lhca3 Is a key step in remodeling of the Photosystem I-light-harvesting complex under iron deficiency in *Chlamydomonas reinhardtii*. *J. Biol. Chem.* **280**, 20431–20441.

Naumann, B., Busch, A., Allmer, J., Ostendorf, E., Zeller, M., Kirchhoff, H., and Hippler, M. (2007). Comparative quantitative proteomics to investigate the remodeling of bioenergetic pathways under iron deficiency in *Chlamydomonas reinhardtii*. *Proteomics* **7**, 3964–3979.

Navarro, M.T., Guerra, E., Fernández, E., and Galván, A. (2000). Nitrite reductase mutants as an approach to understanding nitrate assimilation in *Chlamydomonas reinhardtii*. *Plant Physiol.* **122**, 283–289.

Navarro, M.T., Mariscal, V., Macias, M.I., Fernández, E., and Galván, A. (2005). *Chlamydomonas reinhardtii* strains expressing nitrate reductase under control of the cabII-1 promoter: isolation of chlorate resistant mutants and identification of new loci for nitrate assimilation. *Photosynth. Res.* **82**, 151–161.

Naya, H., Romero, H., Carels, N., Zavala, A., and Musto, H. (2001). Translational selection shapes codon usage in the GC-rich genome of *Chlamydomonas reinhardtii*. *FEBS Lett.* **501**, 127–130.

Nečas, J. (1981). Dependence of the gametogenesis induction, zygote formation and their germination on the culture density of the homothallic alga *Chlamydomonas geitleri* Ettl. *Biol. Plant.* **23**, 278–284.

Nečas, J. (1982a). Regulation of gametogenesis, formation of the zygotes and their maturation and germination by light in *Chlamydomonas geitleri* Ettl. *Arch. Protistenk.* **126**, 229–239.

Nečas, J. (1982b). Comparison of dependence of growth and sexual reproduction of *Chlamydomonas geitleri* on temperature and irradiance. *Biol. Plant.* **24**, 311–313.

Nečas, J. (1983a). Chloroplast inheritance in *Chlamydomonas reinhardii*. *Biol. Listy* **48**, 196–215.

Nečas, J. (1983b). Some peculiarities in gametogenesis, zygote formation, maturation and germination induced by phosphorus limitation of the homothallic alga *Chlamydomonas geitleri*. *Arch. Hydrobiol. Suppl.* **63**, 419–432.

Nečas, J. (1984). Do Volvocal algae form their cultures as autonomous systems? *Biol. Plant.* **26**, 189–196.

Nečas, J. and Pavingerova, D. (1979). Mutagenic effects of short term action of *N*-methyl-*N*′-nitro-*N*-nitrosoguanidine on *Chlamydomonas geitleri*. *Biol. Plant.* **21**, 383–389.

Nečas, J. and Pavingerova, D. (1980). Life cycle of *Chlamydomonas geitleri* Ettl. *Arch. Hydrobiol. Suppl.* **60**, 63–79.

Nečas, J. and Tetík, K. (1981). Some effects of the induction of gametogenesis in the populations of the homothallic alga *Chlamydomonas geitleri* Ettl. *Biol. Plant.* **23**, 270–277.

Nečas, J. and Tetík, K. (1984). The use of synchronization for higher yields of zygotes in the culture of the homothallic alga *Chlamydomonas geitleri*. *Arch. Protistenk.* **128**, 55–68.

Nečas, J. and Tetík, K. (1985). Homeostasis in the culture of *Chlamydomonas geitleri* at nitrogen limitation. *Arch. Protistenk.* **130**, 113–118.

Nečas, J., Sulek, J., and Tetík, K. (1983). A spontaneous *pf*- mutation in *Chlamydomonas geitleri*. *Biochem. Physiol. Pflanz.* **178**, 547–561.

Nečas, J., Tetík, K., and Sulek, J. (1985). Mutation process induced by MNNG in different phases of the cell cycle in *Chlamydomonas geitleri*. I. Some characteristics of the slow cell cycles. *Arch. Hydrobiol. Suppl.* **71**, 621–635.

Nečas, J., Tetík, K., and Sulek, J. (1986a). Mutation process induced by MNNG in different phases of the cell cycle in *Chlamydomonas geitleri*. II. Physiological characteristics of the mutagen effect. *Arch. Hydrobiol. Suppl.* **73**, 129–145.

Nečas, J., Tetík, K., and Sulek, J. (1986b). Mutation process induced by MNNG in different phases of the cell cycle in *Chlamydomonas geitleri*. III. Quantitative alteration of the mutation spectrum during cell cycle course. *Arch. Hydrobiol. Suppl.* **73**, 297–314.

Nečas, J., Tetík, K., and Sulek, J. (1986c). Mutation process induced by MNNG in different phases of the cell cycle in *Chlamydomonas geitleri*. IV. Dependence of the induction of mutagenesis on the mutagen dose in the course of the cell cycle. *Arch. Hydrobiol. Suppl.* **73**, 393–403.

Nedelcu, A.M. (1997). Fragmented and scrambled mitochondrial ribosomal RNA coding regions among green algae: a model for their origin and evolution. *Mol. Biol. Evol.* **14**, 506–517.

Neilson, A.H., Holm-Hansen, O., and Lewin, R.A. (1972). An obligately autotrophic mutant of *Chlamydomonas dysosmos*: a biochemical elucidation. *J. Gen. Microbiol.* **71**, 141–148.

Nelson, J.A.E. and Lefebvre, P.A. (1995). Targeted disruption of the *NIT8* gene in *Chlamydomonas reinhardtii*. *Mol. Cell. Biol.* **15**, 5762–5769.

Nelson, J., Savereide, P.B., and Lefebvre, P.A. (1994). The *CRY1* gene in *Chlamydomonas reinhardtii*: structure and use as a dominant selectable marker for nuclear transformation. *Mol. Cell. Biol.* **14**, 4011–4019.

Netrawali, M.S. and Gandhi, S.R. (1990). Mechanism of cell destructive action of organophosphorus insecticide phosalone in *Chlamydomonas reinhardtii* algal cells. *Bull. Environ. Contam. Toxicol.* **44**, 819–825.

Netrawali, M.S., Gandhi, S.R., and Pednekar, M.D. (1986). Effect of endosulfan, malathion, and permethrin on sexual life cycle of *Chlamydomonas reinhardtii*. *Bull. Environ. Contam. Toxicol.* **36**, 412–420.

Newman, S.M., Boynton, J.E., Gillham, N.W., Randolph-Anderson, B.L., Johnson, A.M., and Harris, E.H. (1990). Transformation of chloroplast ribosomal RNA genes in *Chlamydomonas*. Molecular and genetic characterization of integration events. *Genetics* **126**, 875–888.

Newman, S.M., Gillham, N.W., Harris, E.H., Johnson, A.M., and Boynton, J.E. (1991). Targeted disruption of chloroplast genes in *Chlamydomonas reinhardtii*. *Mol. Gen. Genet.* **230**, 65–74.

Newman, S.M., Harris, E.H., Johnson, A.M., Boynton, J.E., and Gillham, N.W. (1992). Nonrandom distribution of chloroplast recombination events in *Chlamydomonas reinhardtii*: evidence for a hotspot and an adjacent cold region. *Genetics* **132**, 413–429.

Nghia, N.H., Gyurjan, I., Stefanovits, P., Paless, Gy., and Turtoczky, I. (1986). Uptake of Azotobacters by somatic fusion of cell-wall mutants of *Chlamydomonas reinhardii*. *Biochem. Physiol. Pflanzen* **181**, 347–357.

Nguyen, R.L., Tam, L.-W., and Lefebvre, P.A. (2005). The *LF1* gene of *Chlamydomonas reinhardtii* encodes a novel protein required for flagellar length control. *Genetics* **169**, 1415–1424.

Nicastro, D., Schwartz, C., Pierson, J., Gaudette, R., Porter, M.E., and McIntosh, J.R. (2006). The molecular architecture of axonemes revealed by cryoelectron tomography. *Science* **313**, 944–948.

Nicholl, D.S.T., Schloss, J.A., and John, P.C.L. (1988). Tubulin gene expression in the *Chlamydomonas reinhardtii* cell cycle: elimination of environmentally induced artifacts and the measurement of tubulin mRNA levels. *J. Cell Sci.* **89**, 397–403.

Nichols, G.L. and Syrett, P.J. (1978). Nitrate reductase deficient mutants of *Chlamydomonas reinhardii*. Isolation and genetics. *J. Gen. Microbiol.* **108**, 71–77.

Nickelsen, J. (2000). Mutations at three different nuclear loci of *Chlamydomonas* suppress a defect in chloroplast *psbD* mRNA accumulation. *Curr. Genet.* **37**, 136–142.

Nickelsen, J., Fleischmann, M., Boudreau, E., Rahire, M., and Rochaix, J.D. (1999). Identification of *cis*-acting RNA leader elements required for chloroplast *psbD* gene expression in *Chlamydomonas*. *Plant Cell* **11**, 957–970.

Nie, Z.Q. and Wu, M. (1999). The functional role of a DNA primase in chloroplast DNA replication in *Chlamydomonas reinhardtii*. *Arch. Biochem. Biophys.* **369**, 174–180.

Nikaido, S.S. and Johnson, C.H. (2000). Daily and circadian variation in survival from ultraviolet radiation in *Chlamydomonas reinhardtii*. *Photochem. Photobiol.* **71**, 758–765.

Nishikawa, K. and Tominaga, N. (2001). Isolation, growth, ultrastructure, and metal tolerance of the green alga, *Chlamydomonas acidophila* (Chlorophyta). *Biosci. Biotechnol. Biochem.* **65**, 2650–2656.

Nishikawa, K., Yamakoshi, Y., Uemura, I., and Tominaga, N. (2003). Ultrastructural changes in *Chlamydomonas acidophila* (Chlorophyta) induced by heavy metals and polyphosphate metabolism. *FEMS Microbiol. Ecol.* **44**, 253–259.

Nishikawa, K., Onodera, A., and Tominaga, N. (2006a). Phytochelatins do not correlate with the level of Cd accumulation in *Chlamydomonas spp*. *Chemosphere* **63**, 1553–1559.

Nishikawa, K., Machida, H., Yamakoshi, Y., Ohtomo, R., Saito, K., Saito, M., and Tominaga, N. (2006b). Polyphosphate metabolism in an acidophilic alga *Chlamydomonas acidophila* KT-1 (Chlorophyta) under phosphate stress. *Plant Sci.* **170**, 307–313.

Nishimura, Y., Higashiyama, T., Suzuki, L., Misumi, O., and Kuroiwa, T. (1998). The biparental transmission of the mitochondrial genome in *Chlamydomonas reinhardtii* visualized in living cells. *Eur. J. Cell Biol.* **77**, 124–133.

Nishimura, Y., Misumi, O., Matsunaga, S., Higashiyama, T., Yokota, A., and Kuroiwa, T. (1999). The active digestion of uniparental chloroplast DNA in a single zygote of *Chlamydomonas reinhardtii* is revealed by using the optical tweezer. *Proc. Natl. Acad. Sci. U. S. A.* **96**, 12577–12582.

Nishimura, Y., Misumi, O., Kato, K., Inada, N., Higashiyama, T., Momoyama, Y., and Kuroiwa, T. (2002). An mt^+ gamete-specific nuclease that targets mt^- chloroplasts during sexual reproduction in *C. reinhardtii*. *Genes Dev.* **16**, 1116–1128.

Nishiyama, R., Ito, M., Yamaguchi, Y., Koizumi, N., and Sano, H. (2002). A chloroplast-resident DNA methyltransferase is responsible for hypermethylation of chloroplast genes in *Chlamydomonas* maternal gametes. *Proc. Natl. Acad. Sci. U. S. A.* **99**, 5925–5930.

Nishiyama, R., Wada, Y., Mibu, M., Yamaguchi, Y., Shimogawara, K., and Sano, H. (2004). Role of a nonselective *de novo* DNA methyltransferase in maternal inheritance of chloroplast genes in the green alga, *Chlamydomonas reinhardtii*. *Genetics* **168**, 809–816.

Nogales, J., Guijo, M.I., Quesada, A., and Merchán, F. (2004). Functional analysis and regulation of the malate synthase from *Chlamydomonas reinhardtii*. *Planta* **219**, 325–331.

Novis, P.M. (2002). New records of snow algae for New Zealand, from Mt. Philistine, Arthur's Pass National Park. *N.Z. J. Bot.* **40**, 297–312.

Novoselov, S.V. and Gladyshev, V.N. (2003). Non-animal origin of animal thioredoxin reductases: implications for selenocysteine evolution and evolution of protein function through carboxy-terminal extensions. *Protein Sci.* **12**, 372–378.

Novoselov, S.V., Rao, M., Onoshko, N.V., Zhi, H.J., Kryukov, G.V., Xiang, Y.B., Weeks, D.P., Hatfield, D.L., and Gladyshev, V.N. (2002). Selenoproteins and selenocysteine insertion system in the model plant cell system, *Chlamydomonas reinhardtii*. *EMBO J.* **21**, 3681–3693.

Nozaki, H., Itoh, M., Sano, R., Uchida, H., Watanabe, M.M., and Kuroiwa, T. (1995). Phylogenetic relationships within the colonial Volvocales (Chlorophyta) inferred from *rbcL* gene sequence data. *J. Phycol.* **31**, 970–979.

Nozaki, H., Ito, M., Watanabe, M.M., Takano, H., and Kuroiwa, T. (1997). Phylogenetic analysis of morphological species of *Carteria* (Volvocales, Chlorophyta) based on *rbcL* gene sequences. *J. Phycol.* **33**, 864–867.

Nozaki, H., Ohta, N., Takano, H., and Watanabe, M.M. (1999). Re-examination of phylogenetic relationships within the colonial Volvocales (Chlorophyta): an analysis of *atpB* and *rbcL* gene sequences. *J. Phycol.* **35**, 104–112.

Nozaki, H., Misawa, K., Kajita, T., Kato, M., Nohara, S., and Watanabe, M.M. (2000). Origin and evolution of the colonial Volvocales (Chlorophyceae) as inferred from multiple, chloroplast gene sequences. *Mol. Phylogenet. Evol.* **17**, 256–268.

Nozaki, H., Onishi, K., and Morita, E. (2002). Differences in pyrenoid morphology are correlated with differences in the *rbcL* genes of members of the *Chloromonas* lineage (Volvocales, Chlorophyceae). *J. Mol. Evol.* **55**, 414–430.

Nozaki, H., Misumi, O., and Kuroiwa, T. (2003). Phylogeny of the quadriflagellate Volvocales (Chlorophyceae) based on chloroplast multigene sequences. *Mol. Phylogenet. Evol.* **29**, 58–66.

Nozaki, H., Mori, T., Misumi, O., Matsunaga, S., and Kuroiwa, T. (2006). Males evolved from the dominant isogametic mating type. *Curr. Biol.* **16**, R1018–R1020.

Nultsch, W. (1977). Effect of external factors on phototaxis of *Chlamydomonas reinhardtii*. II. Carbon dioxide, oxygen and pH. *Arch. Microbiol.* **112**, 179–185.

Nultsch, W. (1979). Effect of external factors on phototaxis of *Chlamydomonas reinhardtii*. III. Cations. *Arch. Microbiol.* **123**, 93–99.

Nultsch, W. (1983). The photocontrol of movement of *Chlamydomonas*. *Symp. Soc. Exp. Biol.* **36**, 521–539.

Nultsch, W. and Throm, G. (1975). Effect of external factors on phototaxis of *Chlamydomonas reinhardtii*. I. Light. *Arch. Microbiol.* **103**, 175–179.

Nultsch, W., Pfau, J., and Dolle, R. (1986). Effects of calcium channel blockers on phototaxis and motility of *Chlamydomonas reinhardtii*. *Arch. Microbiol.* **144**, 393–397.

Nurani, G. and Franzén, L.G. (1996). Isolation and characterization of the mitochondrial ATP synthase from *Chlamydomonas reinhardtii*. cDNA sequence and deduced protein sequence of the alpha subunit. *Plant Mol. Biol.* **31**, 1105–1116.

Nurani, G., Eriksson, M., Knorpp, C., Glaser, E., and Franzén, L.G. (1997). Homologous and heterologous protein import into mitochondria isolated from the green alga *Chlamydomonas reinhardtii*. *Plant Mol. Biol.* **35**, 973–980.

O'Kelly, C.J. and Floyd, G.L. (1983). Flagellar apparatus absolute orientations and the phylogeny of the green algae. *BioSystems* **16**, 227–251.

O'Neill, M.A. and Roberts, K. (1981). Methylation analysis of cell wall glycoproteins and glycopeptides from *Chlamydomonas reinhardii*. *Phytochemistry* **20**, 25–28.

O'Toole, E.T., Giddings, T.H., McIntosh, J.R., and Dutcher, S.K. (2003). Three-dimensional organization of basal bodies from wild-type and delta-tubulin deletion strains of *Chlamydomonas reinhardtii*. *Mol. Biol. Cell* **14**, 2999–3012.

Oda, T., Hirokawa, N., and Kikkawa, M. (2007). Three-dimensional structures of the flagellar dynein–microtubule complex by cryoelectron microscopy. *J. Cell Biol.* **177**, 243–252.

Odom, O.W., Holloway, S.P., Deshpande, N.N., Lee, J., and Herrin, D.L. (2001). Mobile self-splicing group I introns from the *psbA* gene of *Chlamydomonas reinhardtii*: highly efficient homing of an exogenous intron containing its own promoter. *Mol. Cell. Biol.* **21**, 3472–3481.

Odom, O.W., Baek, K.H., Dani, R.N., and Herrin, D.L. (2008). *Chlamydomonas* chloroplasts can use short dispersed repeats (SDRs) and multiple pathways to repair a double-strand break in the genome. *Plant J.* **53**, 842–853.

Ohad, I., Siekevitz, P., and Palade, G.E. (1967a). Biogenesis of chloroplast membranes. I. Plastid dedifferentiation in a dark-grown algal mutant (*Chlamydomonas reinhardi*). *J. Cell Biol.* **35**, 553–584.

Ohad, I., Siekevitz, P., and Palade, G.E. (1967b). Biogenesis of chloroplast membranes. II. Plastid differentiation during greening of a dark-grown algal mutant (*Chlamydomonas reinhardi*). *J. Cell Biol.* **35**, 553–584.

Ohad, I., Kyle, D.J., and Arntzen, C.J. (1984). Membrane protein damage and repair: removal and replacement of inactivated 32-kilodalton polypeptides in chloroplast membranes. *J. Cell Biol.* **99**, 481–485.

Ohresser, M., Matagne, R.F., and Loppes, R. (1997). Expression of the arylsulphatase reporter gene under the control of the *nit1* promoter in *Chlamydomonas reinhardtii*. *Curr. Genet.* **31**, 264–271.

Ojakian, G.K. and Katz, D.F. (1973). A simple technique for the measurement of swimming speed of *Chlamydomonas*. *Exp. Cell Res.* **81**, 487–491.

Ojakian, G.K. and Satir, P. (1974). Particle movements in chloroplast membranes: quantitative measurements of membrane fluidity by the freeze-fracture technique. *Proc. Natl. Acad. Sci. U. S. A.* **71**, 2052–2056.

Okita, N., Isogai, N., Hirono, M., Kamiya, R., and Yoshimura, K. (2005). Phototactic activity in *Chlamydomonas* "non-phototactic" mutants deficient in Ca^{2+}-dependent control of flagellar dominance or in inner-arm dynein. *J. Cell Sci.* **118**, 529–537.

Okpodu, C.M., Robertson, D., Boss, W.F., Togasaki, R.K., and Surzycki, S.J. (1994). Rapid isolation of nuclei from carrot suspension culture cells using a BioNebulizer. *BioTechniques* **16**, 154–159.

Oldenhof, H., Bisova, K., van den Ende, H., and Zachleder, V. (2004a). Effect of red and blue light on the timing of cyclin-dependent kinase activity and the timing of cell division in *Chlamydomonas reinhardtii*. *Plant Physiol. Biochem.* **42**, 341–348.

Oldenhof, H., Bisova, K., van den Ende, H., and Zachleder, V. (2004b). Blue light delays commitment to cell division in *Chlamydomonas reinhardtii*. *Plant Biol.* **6**, 689–695.

Oldenhof, H., Zachleder, V., and van den Ende, H. (2007). The cell cycle of *Chlamydomonas reinhardtii*: the role of the commitment point. *Folia Microbiol.* **52**, 53–60.

Olive, J., Wollman, F.-A., Bennoun, P., and Recouvreur, M. (1981). Ultrastructure of thylakoid membranes in *Chlamydomonas reinhardtii*: evidence for variations in the partition coefficient of the light-harvesting complex-containing particles upon membrane fracture. *Arch. Biochem. Biophys.* **208**, 456–467.

Olmsted, J.B., Witman, G.B., Carlson, K., and Rosenbaum, J.L. (1971). Comparison of the microtubule proteins of neuroblastoma cells, brain, and *Chlamydomonas* flagella. *Proc. Natl. Acad. Sci. U. S. A.* **68**, 2273–2277.

Olsen, Y., Knutsen, G., and Lien, T. (1983). Characteristics of phosphorus limitation in *Chlamydomonas reinhardtii* (Chlorophyceae) and its palmelloids. *J. Phycol.* **19**, 313–319.

Omoto, C.K., Gibbons, I.R., Kamiya, R., Shingyoji, C., Takahashi, K., and Witman, G.B. (1999). Rotation of the central pair microtubules in eukaryotic flagella. *Mol. Biol. Cell* **10**, 1–4.

Ondarroa, M., Zito, F., Finazzi, G., Joliot, P., Wollman, F.A., and Rich, P.R. (1996). Characterization and electron transfer kinetics of wild type and a mutant *bf* complex in *Chlamydomonas reinhardtii*. *Biochem. Soc. Trans.* **24**, 398S–398S.

Oren, A. (2005). A hundred years of *Dunaliella* research: 1905–2005. *Saline Syst.* **1**, 2.

Osafune, T., Mihara, S., Hase, E., and Ohkuro, I. (1972a). Electron microscope studies on the vegetative life cycle of *Chlamydomonas reinhardi* Dangeard in synchronous culture. I. Some characteristics of changes in subcellular structures during the cell cycle, especially in formation of giant mitochondria. *Plant Cell Physiol.* **13**, 211–227.

Osafune, T., Mihara, S., Hase, E., and Ohkuro, I. (1972b). Electron microscope studies of the vegetative life cycle of *Chlamydomonas reinhardi* Dangeard in synchronous culture. II. Association of mitochondria and the chloroplast at an early developmental stage. *Plant Cell Physiol.* **13**, 981–989.

Osafune, T., Mihara, S., Hase, E., and Ohkuro, I. (1975). Electron microscope studies of the vegetative cellular life cycle of *Chlamydomonas reinhardi* Dangeard in synchronous culture. III. Three-dimensional structures of mitochondria in the cells at intermediate stages of the growth phase of the life cycle. *J. Electron Microsc.* **24**, 247–252.

Osafune, T., Mihara, S., Hase, E., and Ohkuro, I. (1976). Electron microscope studies of the vegetative cellular life cycle of *Chlamydomonas reinhardi* Dangeard in synchronous culture. IV. Mitochondria in dividing cells. *J. Electron Microsc.* **25**, 261–269.

Oshio, H., Shibata, H., Mito, H.N., Yamamoto, M., Harris, E.H., Gillham, N.W., Boynton, J.E., and Sato, R. (1993). Isolation and characterization of a *Chlamydomonas reinhardtii* mutant resistant to photobleaching herbicides. *Z. Naturforsch.* **48c**, 339–344.

Ossenbühl, F. and Nickelsen, J. (2000). *cis-* and *trans-*acting determinants for translation of *psbD* mRNA in *Chlamydomonas reinhardtii*. *Mol. Cell. Biol.* **20**, 8134–8142.

Pajuelo, E., Pajuelo, P., Clemente, M.T., and Marquez, A.J. (1995). Regulation of the expression of ferredoxin-nitrite reductase in synchronous cultures of *Chlamydomonas reinhardtii*. *Biochim. Biophys. Acta* **1249**, 72–78.

Palmer, J.D., Boynton, J.E., Gillham, N.W., and Harris, E.H. (1985). Evolution and recombination of the large inverted repeat in *Chlamydomonas* chloroplast DNA. In: *The Molecular Biology of the Photosynthetic Apparatus* (K.E. Steinback, S. Bonitz, C.J. Arntzen, and L. Bogorad, Eds.), pp. 269–278. Cold Spring Harbor Laboratory, Cold Spring Harbor, New York.

Palombella, A.L. and Dutcher, S.K. (1998). Identification of the gene encoding the tryptophan synthase beta-subunit from *Chlamydomonas reinhardtii*. *Plant Physiol.* **117**, 455–464.

Pan, J. and Snell, W.J. (2002). Kinesin-II is required for flagellar sensory transduction during fertilization in *Chlamydomonas*. *Mol. Biol. Cell* **13**, 1417–1426.

Pan, J.M., Haring, M.A., and Beck, C.F. (1996). Dissection of the blue-light-dependent signal-transduction pathway involved in gametic differentiation of *Chlamydomonas reinhardtii*. *Plant Physiol.* **112**, 303–309.

Pan, J.M., Haring, M.A., and Beck, C.F. (1997). Characterization of blue light signal transduction chains that control development and maintenance of sexual competence in *Chlamydomonas reinhardtii*. *Plant Physiol.* **115**, 1241–1249.

Park, H., Eggink, L.L., Roberson, R.W., and Hoober, J.K. (1999). Transfer of proteins from the chloroplast to vacuoles in *Chlamydomonas reinhardtii* (Chlorophyta): a pathway for degradation. *J. Phycol.* **35**, 528–538.

Parke, J.M., Miller, C.C., Cowell, I., Dodson, A., Dowding, A., Downes, M., Duckett, J.G., and Anderton, B.J. (1987). Monoclonal antibodies against plant proteins recognise animal intermediate filaments. *Cell Motil. Cytoskeleton* **8**, 312–323.

Parker, B.C. (1971). On the evolution of isogamy to oogamy. In: *Contributions in Phycology* (B. Parker and R.M. Brown Jr., Eds.), pp. 47–51. Allen Press, Lawrence KS.

Parker, J.D. and Quarmby, L.M. (2003). *Chlamydomonas fla* mutants reveal a link between deflagellation and intraflagellar transport. *BMC Cell Biol.* **4**, 11.

Pascher, A. (1918). Über die Beziehung der Reduktionsteilung zur Mendelschen Spaltung. *Ber. Deutsch. Bot. Ges.* **36**, 163–168. [English translation by R.E. Reichle in "Genetics of Algae" (R.A. Lewin, Ed.), pp. 302–306. University of California Press, Berkeley, 1976.]

Pascher, A. (1927). *Die Süsswasser-flora Deutschlands, Österreichs und der Schweiz. 4: Volvocales-Phytomonadinae.* Gustav Fischer, Jena. 506 pp.

Pascher, A. (1932). Über Gruppenbildung und "Geshlectswechsel" bei den Gameten einer Chlamydomonadine. *Jahrb. Wiss. Bot.* **57**, 551–580.

Pasquale, S.M. and Goodenough, U.W. (1987). Cyclic AMP functions as a primary sexual signal in gametes of *Chlamydomonas reinhardtii. J. Cell Biol.* **105**, 2279–2292.

Patau, K. (1941). Eine statistiche Bemerkung zu Moewus' Arbeit "Die Analyse von 42 Erblichen Eigenschaften der *Chlamydomonas eugametos* Gruppe". *Z. Abst. Vererbs.* **79**, 317–319.

Patel-King, R.S., Benashski, S.E., Harrison, A., and King, S.M. (1997). A *Chlamydomonas* homologue of the putative murine t complex distorter Tctex-2 is an outer arm dynein light chain. *J. Cell Biol.* **137**, 1081–1090.

Patni, N.J., Dhawale, S.W., and Aaronson, S. (1977). Extracellular phosphatases of *Chlamydomonas reinhardi* and their regulation. *J. Bacteriol.* **130**, 205–211.

Patterson, G.W. (1974). Sterols of some green algae. *Comp. Biochem. Physiol.* **47B**, 453–457.

Paul, J.H. and Cooksey, K.E. (1979). Asparagine metabolism and asparaginase activity in a euryhaline *Chlamydomonas* species. *Can. J. Microbiol.* **25**, 1443–1451.

Pawlowski, J., Holzmann, M., Fahrni, J.F., and Hallock, P. (2001). Molecular identification of algal endosymbionts in large miliolid foraminifera: 1. Chlorophytes. *J. Euk. Microbiol.* **48**, 362–367.

Pazour, G.J. (2004). Comparative genomics: prediction of the ciliary and basal body proteome. *Curr. Biol.* **14**, R575–R577.

Pazour, G.J. and Witman, G.B. (2000). Forward and reverse genetic analysis of microtubule motors in *Chlamydomonas. Methods* **22**, 285–298.

Pazour, G.J., Sineshchekov, O.A., and Witman, G.B. (1995). Mutational analysis of the phototransduction pathway of *Chlamydomonas reinhardtii. J. Cell Biol.* **131**, 427–440.

Pazour, G.J., Wilkerson, C.G., and Witman, G.B. (1998). A dynein light chain is essential for the retrograde particle movement of intraflagellar transport (IFT). *J. Cell Biol.* **141**, 979–992.

Pazour, G.J., Dickert, B.L., Vucica, Y., Seeley, E.S., Rosenbaum, J.L., Witman, G.B., and Cole, D.G. (2000). *Chlamydomonas* IFT88 and its mouse homologue, polycystic kidney disease gene Tg737, are required for assembly of cilia and flagella. *J. Cell Biol.* **151**, 709–718.

Pazour, G.J., Agrin, N., Leszyk, J., and Witman, G.B. (2005). Proteomic analysis of a eukaryotic cilium. *J. Cell Biol.* **170**, 103–113.

Pedersen, L.B., Miller, M.S., Geimer, S., Leitch, J.M., Rosenbaum, J.L., and Cole, D.G. (2005). *Chlamydomonas* IFT172 is encoded by *FLA11*, interacts with CrEB1, and regulates IFT at the flagellar tip. *Curr. Biol.* **15**, 262–266.

Pedley, T.J. and Kessler, J.O. (1990). A new continuum model for suspensions of gyrotactic micro-organisms. *J. Fluid Mech.* **212**, 155–182.

Pedley, T.J., Hill, N.A., and Kessler, J.O. (1988). The growth of bioconvection patterns in a uniform suspension of gyrotactic micro-organisms. *J. Fluid Mech.* **195**, 223–237.

Peltier, G. and Cournac, L. (2002). Chlororespiration. *Annu. Rev. Plant Physiol. Plant Mol. Biol.* **53**, 523–550.

Pérez-Alegre, M., Dubus, A., and Fernández, E. (2005). *REM1*, a new type of long terminal repeat retrotransposon in *Chlamydomonas reinhardtii*. *Mol. Cell. Biol.* **25**, 10628–10638.

Pérez-Vicente, R., Pineda, M., and Cárdenas, J. (1988). Isolation and characterization of xanthine dehydrogenase from *Chlamydomonas reinhardtii*. *Physiol. Plant.* **72**, 101–107.

Pérez-Vicente, R., Cárdenas, J., and Pineda, M. (1991). Distinction between hypoxanthine and xanthine transport in *Chlamydomonas reinhardtii*. *Plant Physiol.* **95**, 126–130.

Pérez-Vicente, R., Alamillo, J.M., Cárdenas, J., and Pineda, M. (1992). Purification and substrate inactivation of xanthine dehydrogenase from *Chlamydomonas reinhardtii*. *Biochim. Biophys. Acta* **1117**, 159–166.

Periz, G., Dharia, D., Miller, S.H., and Keller, L.R. (2007). Flagellar elongation and gene expression in *Chlamydomonas reinhardtii*. *Eukaryotic Cell* **6**, 1411–1420.

Perkins, D.D. (1949). Biochemical mutants in the smut fungus *Ustilago maydis*. *Genetics* **34**, 607–626.

Perkins, D.D. (1953). The detection of linkage in tetrad analysis. *Genetics* **38**, 187–197.

Perkins, D.D. (1962). Crossing-over and interference in a multiply marked chromosome arm of *Neurospora*. *Genetics* **47**, 1253–1274.

Perrone, C.A., Yang, P.F., O'Toole, E., Sale, W.S., and Porter, M.E. (1998). The *Chlamydomonas IDA7* locus encodes a 140-kDa dynein intermediate chain required to assemble the I1 inner arm complex. *Mol. Biol. Cell* **9**, 3351–3365.

Perrone, C.A., Myster, S.H., Bower, R., O'Toole, E.T., and Porter, M.E. (2000). Insights into the structural organization of the I1 inner arm dynein from a domain analysis of the 1b dynein heavy chain. *Mol. Biol. Cell* **11**, 2297–2313.

Petersen, J.L. and Small, G.D. (2001). A gene required for the novel activation of a class II DNA photolyase in *Chlamydomonas*. *Nucleic Acids Res.* **29**, 4472–4481.

Petersen, J.L., Lang, D.W., and Small, G.D. (1999). Cloning and characterization of a class II DNA photolyase from *Chlamydomonas*. *Plant Mol. Biol.* **40**, 1063–1071.

Petridou, S., Foster, K., and Kindle, K. (1997). Light induces accumulation of isocitrate lyase mRNA in a carotenoid-deficient mutant of *Chlamydomonas reinhardtii*. *Plant Mol. Biol.* **33**, 381–392.

Pfannenschmid, F., Wimmer, V.C., Rios, R.M., Geimer, S., Kröckel, U., Leiherer, A., Haller, K., Nemcová, Y., and Mages, W. (2003). *Chlamydomonas* DIP13 and human NA14: a new class of proteins associated with microtubule structures is involved in cell division. *J. Cell Sci.* **116**, 1449–1462.

Pfau, J., Nultsch, W., and Rüffer, U. (1983). A fully automated and computerized system for simultaneous measurements of motility and phototaxis in *Chlamydomonas*. *Arch. Microbiol.* **135**, 259–264.

Pickett-Heaps, J.D. (1975). *Green Algae — Structure, Reproduction and Evolution in Selected Genera*. Sinauer Associates, Sunderland MA. 606 pp.

Piedras, P., Pineda, M., Munoz, J., and Cárdenas, J. (1992). Purification and characterization of an l-amino-acid oxidase from *Chlamydomonas reinhardtii*. *Planta* **188**, 13–18.

Piedras, P., Aguilar, M., and Pineda, M. (1998). Uptake and metabolism of allantoin and allantoate by cells of *Chlamydomonas reinhardtii* (Chlorophyceae). *Eur. J. Phycol.* **33**, 57–64.

Pijst, H.L.A., Ossendorp, F.A., van Egmond, P., Kamps, A.M.I.E., Musgrave, A., and van den Ende, H. (1984a). Sex-specific binding and inactivation of agglutination factor in *Chlamydomonas eugametos*. *Planta* **160**, 529–535.

Pijst, H.L.A., Ossendorp, F.A., van Egmond, P., Kamps, A.M.I.E., Musgrave, A., and van den Ende, H. (1984b). Cyclic AMP is involved in sexual reproduction of *Chlamydomonas eugametos*. *FEBS Lett.* **174**, 132–136.

Pimenova, M.N. and Kondrat'eva, T.F. (1965). Some data on the use of acetate by *Chlamydomonas globosa*. *Mikrobiologiya* **34**, 230–235.

Pineda, M. and Cárdenas, J. (1996). Transport and assimilation of purines in *Chlamydomonas reinhardtii*. *Sci. Mar.* **60**, 195–201.

Pineda, M., Fernández, E., and Cárdenas, J. (1984). Urate oxidase of *Chlamydomonas reinhardii*. *Physiol. Plant.* **62**, 453–457.

Pineda, M., Cabello, P., and Cárdenas, J. (1987). Ammonium regulation of urate uptake in *Chlamydomonas reinhardtii*. *Planta* **171**, 496–500.

Piperno, G. and Luck, D.J.L. (1977). Microtubular proteins of *Chlamydomonas reinhardtii*. An immunochemical study based on the use of an antibody specific for the beta-tubulin subunit. *J. Biol. Chem.* **252**, 383–391.

Piperno, G. and Luck, D.J.L. (1979a). An actin-like protein is a component of axonemes from *Chlamydomonas* flagella. *J. Biol. Chem.* **254**, 2187–2190.

Piperno, G. and Luck, D.J.L. (1979b). Axonemal adenosine triphosphatases from flagella of *Chlamydomonas reinhardtii*. Purification of two dyneins. *J. Biol. Chem.* **254**, 3084–3090.

Piperno, G., Huang, B., and Luck, D.J.L. (1977). Two-dimensional analysis of flagellar proteins from wild-type and paralyzed mutants of *Chlamydomonas reinhardtii*. *Proc. Natl. Acad. Sci. U. S. A.* **74**, 1600–1604.

Piperno, G., Mead, K., and Shestak, W. (1992). The inner dynein arms I2 interact with a "dynein regulatory complex" in *Chlamydomonas* flagella. *J. Cell Biol.* **118**, 1455–1463.

Piperno, G., Mead, K., LeDizet, M., and Moscatelli, A. (1994). Mutations in the "dynein regulatory complex" alter the ATP-insensitive binding sites for inner arm dyneins in *Chlamydomonas* axonemes. *J. Cell Biol.* **125**, 1109–1117.

Piperno, G., Mead, K., and Henderson, S. (1996). Inner dynein arms but not outer dynein arms require the activity of kinesin homologue protein KHp1FLA10 to reach the distal part of flagella in *Chlamydomonas*. *J. Cell Biol.* **133**, 371–379.

Piperno, G., Siuda, E., Henderson, S., Segil, M., Vaananen, H., and Sassaroli, M. (1998). Distinct mutants of retrograde intraflagellar transport (IFT) share similar morphological and molecular defects. *J. Cell Biol.* **143**, 1591–1601.

Planas, D. and Healey, F.P. (1978). Effects of arsenate on growth and phosphorus metabolism of phytoplantkton. *J. Phycol.* **14**, 337–341.

Pocock, T., Lachance, M.A., Pröschold, T., Priscu, J.C., Kim, S.S., and Huner, N.P.A. (2004). Identification of a psychrophilic green alga from Lake Bonney Antarctica: *Chlamydomonas raudensis* Ettl. (UWO 241) Chlorophyceae. *J. Phycol.* **40**, 1138–1148.

Podstavková, S., Vlček, D., and Miadoková, E. (1994). Repair genes of *Chlamydomonas reinhardtii*. *Biologia* **49**, 629–631.

Polle, J.E.W., Benemann, J.R., Tanaka, A., and Melis, A. (2000). Photosynthetic apparatus organization and function in the wild type and a chlorophyll *b*-less mutant of *Chlamydomonas reinhardtii*. Dependence on carbon source. *Planta* **211**, 335–344.

Pollio, A., Cennamo, P., Ciniglia, C., de Stefano, M., Pinto, G., and Huss, A.R.V. (2005). *Chlamydomonas pitschmannii* Ettl, a little known species from thermoacidic environments. *Protist* **156**, 287–302.

Pollock, S.V., Colombo, S.L., Prout, D.L., Jr., Godfrey, A.C., and Moroney, J.V. (2003). Rubisco activase is required for optimal photosynthesis in the green alga *Chlamydomonas reinhardtii* in a low-CO_2 atmosphere. *Plant Physiol.* **133**, 1854–1861.

Pollock, S.V., Pootakham, W., Shibagaki, N., Moseley, J.L., and Grossman, A.R. (2005). Insights into the acclimation of *Chlamydomonas reinhardtii* to sulfur deprivation. *Photosynth. Res.* **86**, 475–489.

Ponamarev, M.V. and Cramer, W.A. (1998). Perturbation of the internal water chain in cytochrome *f* of oxygenic photosynthesis: loss of the concerted reduction of cytochromes *f* and b_6. *Biochemistry* **37**, 17199–17208.

Popescu, C.E., Borza, T., Bielawski, J.P., and Lee, R.W. (2006). Evolutionary rates and expression level in *Chlamydomonas*. *Genetics* **172**, 1567–1576.

Porter, M.E., Knott, J.A., Gardner, L.C., Mitchell, D.R., and Dutcher, S.K. (1994). Mutations in the *SUP-PF-1* locus of *Chlamydomonas reinhardtii* identify a regulatory domain in the beta-dynein heavy chain. *J. Cell Biol.* **126**, 1495–1507.

Portney, M.D. and Rosen, H. (1980). The effect of caffeine on repair in *Chlamydomonas reinhardtii*. II. Interaction of repair systems. *Mutat. Res.* **70**, 311–321.

Posewitz, M.C., Smolinski, S.L., Kanakagiri, S., Melis, A., Seibert, M., and Ghirardi, M.L. (2004). Hydrogen photoproduction is attenuated by disruption of an isoamylase gene in *Chlamydomonas reinhardtii*. *Plant Cell* **16**, 2151–2163.

Posewitz, M.C., Mulder, D.W., and Peters, J.W. (2008). New frontiers in hydrogenase structure and biosynthesis. *Curr. Chem. Biol.* **2**, 178–199.

Pottier, L., Pruvost, J., Deremetz, J., Cornet, J.F., Legrand, J., and Dussap, C.G. (2005). A fully predictive model for one-dimensional light attenuation by *Chlamydomonas reinhardtii* in a torus photobioreactor. *Biotechnol. Bioeng.* **91**, 569–582.

Pouneva, I. (1997). Evaluation of algal culture viability and physiological state by fluorescent microscopic methods. *Bulgarian J. Plant Physiol.* **23**, 67–76.

Prasad, M.N., Drej, K., Skawinska, A., and Strzalka, K. (1998). Toxicity of cadmium and copper in *Chlamydomonas reinhardtii* wild-type (WT 2137) and cell wall deficient mutant strain (*CW 15*). *Bull. Environ. Contam. Toxicol.* **60**, 306–311.

Pratje, E., Schnierer, S., and Dujon, B. (1984). Mitochondrial DNA of *Chlamydomonas reinhardtii*: the DNA sequence of a region showing homology with mammalian URF2. *Curr. Genet.* **9**, 75–82.

Pratje, E., Vahrenholz, C., Buehler, S., and Michaelis, G. (1989). Mitochondrial DNA of *Chlamydomonas reinhardtii*: the *ND4* gene encoding a subunit of NADH dehydrogenase. *Curr. Genet.* **16**, 61–64.

Preble, A.M., Giddings, T.H., and Dutcher, S.K. (2001). Extragenic bypass suppressors of mutations in the essential gene *BLD2* promote assembly of basal bodies with abnormal microtubules in *Chlamydomonas reinhardtii*. *Genetics* **157**, 163–181.

Preuss, D. and Mets, L. (2002). Plant centromere functions defined by tetrad analysis and artificial chromosomes. *Plant Physiol.* **129**, 421–422.

Prieto, R. and Fernández, E. (1993). Toxicity of and mutagenesis by chlorate are independent of nitrate reductase activity in *Chlamydomonas reinhardtii*. *Mol. Gen. Genet.* **237**, 429–438.

Prieto, R., Dubus, A., Galván, A., and Fernández, E. (1996a). Isolation and characterization of two new negative regulatory mutants for nitrate assimilation in *Chlamydomonas reinhardtii* obtained by insertional mutagenesis. *Mol. Gen. Genet.* **251**, 461–471.

Prieto, R., Pardo, J.M., Niu, X.M., Bressan, R.A., and Hasegawa, P.M. (1996b). Salt-sensitive mutants of *Chlamydomonas reinhardtii* isolated after insertional tagging. *Plant Physiol.* **112**, 99–104.

Pringsheim, E.G. (1930). Neue Chlamydomonadaceen, welche in Reinkultur gewonnen wurden. *Arch. Protistenk.* **69**, 95–102.

Pringsheim, E.G. (1937). Beiträge zur Physiologie saprotropher Algen und Flagellaten. 3. Mitteilung: Die Stellung der Azetatflagellaten in einem physiologischen Ernährungssystem. *Planta* **27**, 61–72.

Pringsheim, E.G. (1946a). *Pure Cultures of Algae. Their Preparation and Maintenance*. Cambridge University Press, Cambridge. 119 pp.

Pringsheim, E.G. (1946b). The biphasic or soil–water culture method for growing algae and flagellata. *J. Ecol.* **33**, 193–204.

Pringsheim, E.G. (1954). *Algenreinkulturen, ihre Herstellung und Erhaltung.* Jena. 109 pp.

Pringsheim, E.G. (1962). *Chlamydomonas pallens,* a new organism proposed for assays of vitamin B12. *Nature* **195**, 604.

Pringsheim, E.G. (1963). Chlorophyllarme Algen. I. *Chlamydomonas pallens nov. spec. Arch. Mikrobiol.* **45**, 136–144.

Pröschold, T., Marin, B., Schlösser, U.G., and Melkonian, M. (2001). Molecular phylogeny and taxonomic revision of *Chlamydomonas* (Chlorophyta). I. Emendation of *Chlamydomonas* Ehrenberg and *Chloromonas* Gobi, and description of *Oogamochlamys gen. nov.* and *Lobochlamys gen. nov. Protist* **152**, 265–300.

Pröschold, T. and Silva, P.C. (2007). Proposal to change the listed type of *Chlamydomonas* Ehrenb., *nom. cons* (Chlorophyta). *Taxon* **56**, 595–596.

Pröschold, T., Harris, E.H., and Coleman, A. (2005). Portrait of a species: *Chlamydomonas reinhardtii. Genetics* **170**, 1601–1610.

Provasoli, L. and Carlucci, A.F. (1974). Vitamins and growth regulators. In: *Algal Physiology and Biochemistry* (W.D.P. Stewart, Ed.), pp. 741–787. Blackwell, Oxford.

Pucheu, N., Oettmeier, W., Heisterkamp, U., Masson, K., and Wildner, G.F. (1984). Metribuzin-resistant mutants of *Chlamydomonas reinhardii. Z. Naturforsch.* **39c**, 437–439.

Puck, T.T. and Kao, F. (1967). Genetics of somatic mammalian cells, V. Treatment with 5-bromodeoxyuridine and visible light for isolation of nutritionally deficient mutants. *Proc. Natl. Acad. Sci. U. S. A.* **58**, 1227–1234.

Purton, S. (2007). Tools and techniques for chloroplast transformation of *Chlamydomonas. Adv. Exp. Med. Biol.* **616**, 34–45.

Purton, S. and Rochaix, J.-D. (1994). Complementation of a *Chlamydomonas reinhardtii* mutant using a genomic cosmid library. *Plant Mol. Biol.* **24**, 533–537.

Purton, S. and Rochaix, J.-D. (1995). Characterization of the *ARG7* gene of *Chlamydomonas reinhardtii* and its application to nuclear transformation. *Eur. J. Phycol.* **30**, 141–148.

Quader, H. and Filner, P. (1980). The action of antimitotic herbicides on flagellar regeneration in *Chlamydomonas reinhardtii*: a comparison with the action of colchicine. *Eur. J. Cell Biol.* **21**, 301–304.

Quader, H., Cherniack, J., and Filner, P. (1978). Participation of calcium in flagellar shortening and regeneration in *Chlamydomonas reinhardii. Exp. Cell Res.* **113**, 295–301.

Quarmby, L.M. (1996). Ca^{2+} influx activated by low pH in *Chlamydomonas. J. Gen. Physiol.* **108**, 351–361.

Quarmby, L.M. and Mahjoub, M.R. (2005). Caught Nek-ing: cilia and centrioles. *J. Cell Sci.* **118**, 5161–5169.

Quarmby, L.M. and Parker, J.D.K. (2005). Cilia and the cell cycle? *J. Cell Biol.* **169**, 707–710.

Quesada, A. and Fernàndez, E. (1994). Expression of nitrate assimilation related genes in *Chlamydomonas reinhardtii. Plant Mol. Biol.* **24**, 185–194.

Quesada, A., Galván, A., Schnell, R.A., Lefebvre, P.A., and Fernández, E. (1993). Five nitrate assimilation-related loci are clustered in *Chlamydomonas reinhardtii. Mol. Gen. Genet.* **240**, 387–394.

Quinn, J.M. and Merchant, S. (1998). Copper-responsive gene expression during adaptation to copper deficiency. *Methods Enzymol.* **297**, 263–279.

Quinn, J., Li, H.H., Singer, J., Morimoto, B., Mets, L., Kindle, K., and Merchant, S. (1993). The plastocyanin-deficient phenotype of *Chlamydomonas reinhardtii Ac-208* results from a frame-shift mutation in the nuclear gene encoding preapoplastocyanin. *J. Biol. Chem.* **268**, 7832–7841.

Quinn, J.M., Eriksson, M., Moseley, J.L., and Merchant, S. (2002). Oxygen deficiency responsive gene expression in *Chlamydomonas reinhardtii* through a copper-sensing signal transduction pathway. *Plant Physiol.* **128**, 463–471.

Quinones, M.A., Galvan, A., Fernández, E., and Aparicio, P.J. (1999). Blue-light requirement for the biosynthesis of an NO_2^- transport system in the *Chlamydomonas reinhardtii* nitrate transport mutant S10. *Plant Cell Environ.* **22**, 1169–1175.

Quisel, J.D., Wykoff, D.D., and Grossman, A.R. (1996). Biochemical characterization of the extracellular phosphatases produced by phosphorus-deprived *Chlamydomonas reinhardtii*. *Plant Physiol.* **111**, 839–848.

Racey, T.J. and Hallett, F.R. (1983a). A low angle quasi-elastic light scattering investigation of *Chlamydomonas reinhardtii*. *J. Muscle Res. Cell Motil.* **4**, 321–331.

Racey, T.J. and Hallett, F.R. (1983b). A quasi-elastic light scattering and cinematographical comparison of three strains of motile *Chlamydomonas reinhardtii*: a wild type strain, a colchicine resistant mutant and a backward swimming mutant. *J. Muscle Res. Cell Motil.* **4**, 333–351.

Raja, R., Hemaiswarya, S., and Rengasamy, R. (2007). Exploitation of *Dunaliella* for beta-carotene production. *Appl. Microbiol. Biotechnol.* **74**, 517–523.

Rajam, M.V. and Kumar, S.V. (2006). Green alga (*Chlamydomonas reinhardtii*). *Methods Mol. Biol.* **344**, 421–434.

Rajamani, S., Siripornadulsil, S., Falcao, V., Torres, M., Colepicolo, P., and Sayre, R. (2007). Phycoremediation of heavy metals using transgenic microalgae. *Adv. Exp. Med. Biol.* **616**, 99–109.

Ral, J.P., Colleoni, C., Wattebled, F., Dauvillee, D., Nempont, C., Deschamps, P., Li, Z., Morell, M.K., Chibbar, R., Purton, S., D'Hulst, C., and Ball, S.G. (2006). Circadian clock regulation of starch metabolism establishes GBSSI as a major contributor to amylopectin synthesis in *Chlamydomonas reinhardtii*. *Plant Physiol.* **142**, 305–317.

Ramanis, Z. and Luck, D.J.L. (1986). Loci affecting flagellar assembly and function map to an unusual linkage group in *Chlamydomonas reinhardtii*. *Proc. Natl. Acad. Sci. U. S. A.* **83**, 423–426.

Ramazanov, Z., Rawat, M., Henk, M.C., Mason, C.B., Matthews, S.W., and Moroney, J.V. (1994). The induction of the CO_2-concentrating mechanism is correlated with the formation of the starch sheath around the pyrenoid of *Chlamydomonas reinhardtii*. *Planta* **195**, 210–216.

Ramesh, V.M., Bingham, S.E., and Webber, A.N. (2004). A simple method for chloroplast transformation in *Chlamydomonas reinhardtii*. *Methods Mol. Biol.* **274**, 301–307.

Randall, J., Cavalier-Smith, T., McVittie, A., Warr, J.R., and Hopkins, J.M. (1967). Developmental and control processes in the basal bodies and flagella of *Chlamydomonas reinhardii*. *Dev. Biol. Suppl.* **1**, 43–83.

Randolph-Anderson, B.L., Boynton, J.E., Gillham, N.W., Harris, E.H., Johnson, A.M., Dorthu, M.-P., and Matagne, R.F. (1993). Further characterization of the respiratory deficient *dum-1* mutation of *Chlamydomonas reinhardtii* and its use as a recipient for mitochondrial transformation. *Mol. Gen. Genet.* **236**, 235–244.

Randolph-Anderson, B.L., Boynton, J.E., Gillham, N.W., Huang, C., and Liu, X.-Q. (1995a). The chloroplast gene encoding ribosomal protein S4 in *Chlamydomonas reinhardtii* spans an inverted repeat-unique sequence juntion and can be mutated to suppress a streptomycin dependence mutation in ribosomal protein S12. *Mol. Gen. Genet.* **247**, 295–305.

Randolph-Anderson, B., Boynton, J.E., Dawson, J., Dunder, E., Eskes, R., Gillham, N.W., Johnson, A., Perlman, P.S., Suttie, J., and Heiser, W.C. (1995b). Sub-micron gold particles are superior to larger particles for efficient biolistic transformation of organelles and some cell types. *BioRad Technical Bulletin no. 2015*.

[http://www.bio-medicine.org/biology-technology/Sub-Micron-Gold-Particles-Are-Superior-to-Larger-Particles-for-Efficient-Biolistic-Transformation-of-Organelles-and-Some-Cells-1201-1/]

Randolph-Anderson, B.L., Sato, R., Johnson, A.M., Harris, E.H., Hauser, C.R., Oeda, K., Ishige, F., Nishio, S., Gillham, N.W., and Boynton, J.E. (1998). Isolation and characterization of a mutant protoporphyrinogen oxidase gene from *Chlamydomonas reinhardtii* conferring resistance to porphyric herbicides. *Plant Mol. Biol.* **38**, 839–859.

Ranum, L.P.W., Thompson, M.D., Schloss, J.A., Lefebvre, P.A., and Silflow, C.D. (1988). Mapping flagellar genes in *Chlamydomonas* using restriction fragment length polymorphisms. *Genetics* **120**, 109–122.

Rao, M., Carlson, B.A., Novoselov, S.V., Weeks, D.P., Gladyshev, V.N., and Hatfield, D.L. (2003). *Chlamydomonas reinhardtii* selenocysteine tRNA[Ser]Sec. *RNA* **9**, 923–930.

Raper, J.R. (1952). Chemical regulation of sexual processes in the Thallophytes. *Bot. Rev.* **18**, 447–545.

Ravina, C.G., Chang, C.I., Tsakraklides, G.P., McDermott, J.P., Vega, J.M., Leustek, T., Gotor, C., and Davies, J.P. (2002). The *sac* mutants of *Chlamydomonas reinhardtii* reveal transcriptional and posttranscriptional control of cysteine biosynthesis. *Plant Physiol.* **130**, 2076–2084.

Rawat, M., Henk, M.C., Lavigne, L.L., and Moroney, J.V. (1996). *Chlamydomonas reinhardtii* mutants without ribulose-1,5-bisphosphate carboxylase-oxygenase lack a detectable pyrenoid. *Planta* **198**, 263–270.

Ray, D.A., Solter, K.M., and Gibor, A. (1978). Flagellar surface differentiation. Evidence for multiple sites involved in mating of *Chlamydomonas reinhardi*. *Exp. Cell Res.* **114**, 185–189.

Raynaud, C., Loiselay, C., Wostrikoff, K., Kuras, R., Girard-Bascou, J., Wollman, F.A., and Choquet, Y. (2007). Evidence for regulatory function of nucleus-encoded factors on mRNA stabilization and translation in the chloroplast. *Proc. Natl. Acad. Sci. U. S. A.* **104**, 9093–9098.

Reboud, X. (2002). Response of *Chlamydomonas reinhardtii* to herbicides: negative relationship between toxicity and water solubility across several herbicide families. *Bull. Environ. Contam. Toxicol.* **69**, 554–561.

Reboud, X., Majerus, N., Gasquez, J., and Powles, S. (2007). *Chlamydomonas reinhardtii* as a model system for pro-active herbicide resistance evolution research. *Biol. J. Linn. Soc.* **91**, 257–266.

Reed, R.B., Simione, F.P., and McGrath, M.S. (1976). Preservation of encysted *Polytomella*. *J. Gen. Microbiol.* **97**, 29–34.

Reinhard, L.W. (1922). [obituary]. *Bot. Not.* **1922**, 287.

Reisdorph, N.A. and Small, G.D. (2004). The *CPH1* gene of *Chlamydomonas reinhardtii* encodes two forms of cryptochrome whose levels are controlled by light-induced proteolysis. *Plant Physiol.* **134**, 1546–1554.

Remacle, C., Baurain, D., Cardol, P., and Matagne, R.F. (2001a). Mutants of *Chlamydomonas reinhardtii* deficient in mitochondrial complex I: characterization of two mutations affecting the *nd1* coding sequence. *Genetics* **158**, 1051–1060.

Remacle, C., Duby, F., Cardol, P., and Matagne, R.F. (2001b). Mutations inactivating mitochondrial genes in *Chlamydomonas reinhardtii*. *Biochem. Soc. Trans.* **29**, 442–446.

Remacle, C., Cardol, P., Coosemans, N., Gaisne, M., and Bonnefoy, N. (2006). High-efficiency biolistic transformation of *Chlamydomonas* mitochondria can be used to insert mutations in complex I genes. *Proc. Natl. Acad. Sci. U. S. A.* **103**, 4771–4776.

Remias, D., Lütz-Meindl, U., and Lütz, C. (2005). Photosynthesis, pigments and ultrastructure of the alpine snow alga *Chlamydomonas nivalis Eur. J. Phycol.* **40**, 259–268.

Remillard, S.P. and Witman, G.B. (1982). Synthesis, transport, and utilization of specific flagellar proteins during flagellar regeneration in *Chlamydomonas*. *J. Cell Biol.* **93**, 615–631.

Renaut, S., Replansky, T., Heppleston, A., and Bell, G. (2006). The ecology and genetics of fitness in *Chlamydomonas*. XIII. Fitness of long-term sexual and asexual populations in benign environments. *Evolution Int. J. Org. Evolution* **60**, 2272–2279.

Renner, O. (1958). Auch etwas über F. Moewus, *Forsythia* und *Chlamydomonas*. *Z. Naturforsch.* **13b**, 399–403.

Renner, T. and Waters, E.R. (2007). Comparative genomic analysis of the Hsp70s from five diverse photosynthetic eukaryotes. *Cell Stress Chaperones* **12**, 172–185.

Renninger, S., Dieckmann, C.L., and Kreimer, G. (2006). Toward a protein map of the green algal eyespot: analysis of eyespot globule-associated proteins. *Phycologia* **45**, 199–212.

Rexach, J., Montero, B., Fernández, E., and Galván, A. (1999). Differential regulation of the high affinity nitrite transport systems III and IV in *Chlamydomonas reinhardtii*. *J. Biol. Chem.* **274**, 27801–27806.

Rexach, J., Llamas, A., Fernández, E., and Galván, A. (2002). The activity of the high-affinity nitrate transport system I (NRT2;1, NAR2) is responsible for the efficient signalling of nitrate assimilation genes in *Chlamydomonas reinhardtii*. *Planta* **215**, 606–611.

Rexroth, S., Meyer zu Tittingdorf, J.M., Krause, F., Dencher, N.A., and Seelert, H. (2003). Thylakoid membrane at altered metabolic state: challenging the forgotten realms of the proteome. *Electrophoresis* **24**, 2814–2823.

Reyes-Prieto, A., El-Hafidi, M., Moreno-Sánchez, R., and González-Halphen, D. (2002). Characterization of oxidative phosphorylation in the colorless chlorophyte *Polytomella sp.* Its mitochondrial respiratory chain lacks a plant-like alternative oxidase. *Biochim. Biophys. Acta* **1554**, 170–179.

Reynoso, G.T. and de Gamboa, B.A. (1982). Salt tolerance in the fresh water algae *Chlamydomonas reinhardii*: effect of proline and taurine. *Comp. Biochem. Physiol.* **73A**, 95–99.

Rhodes, R.G. (1981). Heterothallism in *Chlamydomonas acidophila* Negoro isolated from acidic strip-mine ponds. *Phycologia* **20**, 81–82.

Riano-Pachon, D.M., Ruzicic, S., Dreyer, I., and Mueller-Roeber, B. (2007). PlnTFDB: an integrative plant transcription factor database. *BMC Bioinformatics* **8**, 42.

Riekhof, W.R., Ruckle, M.E., Lydic, T.A., Sears, B.B., and Benning, C. (2003). The sulfolipids 2′-*O*-acyl-sulfoquinovosyldiacylglycerol and sulfoquinovosyldiacylglycerol are absent from a *Chlamydomonas reinhardtii* mutant deleted in SQD1. *Plant Physiol.* **133**, 864–874.

Ringo, D.L. (1967a). Flagellar motion and fine structure of the flagellar apparatus in *Chlamydomonas*. *J. Cell Biol.* **33**, 543–571.

Ringo, D.L. (1967b). The arrangement of subunits in flagellar fibers. *J. Ultrastruct. Res.* **17**, 266–277.

Rioboo, C., Franqueira, D., Canle, M.L., Herrero, C., and Cid, A. (2001). Microalgal bioassays as a test of pesticide photodegradation efficiency in water. *Bull. Environ. Contam. Toxicol.* **67**, 233–238.

Ris, H. and Plaut, W. (1962). Ultrastructure of DNA-containing areas in the chloroplast of *Chlamydomonas*. *J. Cell Biol.* **13**, 383–391.

Rizzo, P.J. (1985). Histones in protistan evolution. *BioSystems* **18**, 249–262.

Roberts, K. (1974). Crystalline glycoprotein cell walls of algae: their structure, composition and assembly. *Philos. Trans. R. Soc. Lond., B, Biol. Sci.* **268**, 129–146.

Roberts, A.M. (2006). Mechanisms of gravitaxis in *Chlamydomonas*. *Biol. Bull.* **210**, 78–80.

Roberts, D.G.W., Lamb, M.R., and Dieckmann, C.L. (2001). Characterization of the *EYE2* gene required for eyespot assembly in *Chlamydomonas reinhardtii*. *Genetics* **158**, 1037–1049.

Roberts, K. (1979). Hydroxyproline: its asymmetric distribution in a cell wall glyco-protein. *Planta* **146**, 275–279.

Roberts, K., Gurney-Smith, M., and Hills, G.J. (1972). Structure, composition and morphogenesis of the cell wall of *Chlamydomonas reinhardi*. I. Ultrastructure and preliminary chemical analysis. *J. Ultrastruct. Res.* **40**, 599–613.

Roberts, K., Gay, M.R., and Hills, G.J. (1980). Cell wall glycoproteins from *Chlamy-domonas reinhardii* are sulphated. *Physiol. Plant.* **49**, 421–424.

Roberts, K., Shaw, P.J., and Hills, G.J. (1981). High-resolution electron microscopy of glycoproteins: the crystalline cell wall of *Lobomonas*. *J. Cell Sci.* **51**, 295–313.

Roberts, K., Hills, G.J., and Shaw, P. (1982). Structure of algal cell walls. In: *Electron Microscopy of Proteins* (J.R. Harris, Ed.), Vol. **3**, pp. 1–40. Academic Press, London.

Roberts, K., Grief, C., Hills, G.J., and Shaw, P.J. (1985a). Cell wall glycoproteins: structure and function. *J. Cell Sci.* **2**, 105–127.

Roberts, K., Phillips, J., Shaw, P., Grief, C., and Smith, E. (1985b). An immunological approach to the plant cell wall. In: *Biochemistry of Plant Cell Walls* (C.T. Brett and J.R. Hillman, Eds.), pp. 125–154. Cambridge University Press, Cambridge.

Robertson, D., Woessner, J.P., Boynton, J.E., and Gillham, N.W. (1989). Molecular characterization of two point mutants in the chloroplast *atpB* gene of the green alga *Chlamydomonas reinhardtii* defective in assembly of the ATP synthase complex. *J. Biol. Chem.* **264**, 2331–2337.

Robertson, D., Boynton, J.E., and Gillham, N.W. (1990). Cotranscription of the wild type chloroplast *atpE* gene encoding the CF_1/CF_0 epsilon subunit with the 3′ half of the *rps7* gene in *Chlamydomonas reinhardtii* and characterization of frame-shift mutations in *atpE*. *Mol. Gen. Genet.* **221**, 155–163.

Robinson, D.G. and Schlösser, U.G. (1978). Cell wall regeneration by protoplasts of *Chlamydomonas*. *Planta* **141**, 83–92.

Rochaix, J.-D. (1978). Restriction endonuclease map of the chloroplast DNA of *Chlamydomonas reinhardii*. *J. Mol. Biol.* **126**, 597–617.

Rochaix, J.-D. and Darlix, J.-L. (1982). Composite structure of the chloroplast 23S ribosomal RNA genes of *Chlamydomonas reinhardii*. Evolutionary and function-al implications. *J. Mol. Biol.* **159**, 383–395.

Rochaix, J.-D. and van Dillewijn, J. (1982). Transformation of the green alga *Chlamy-domonas reinhardii* with yeast DNA. *Nature* **296**, 70–72.

Rochaix, J.-D., van Dillewijn, J., and Rahire, M. (1984). Construction and charac-terization of autonomously replicating plasmids in the green unicellular alga *Chlamydomonas reinhardii*. *Cell* **36**, 925–931.

Rochaix, J.-D., Erickson, J., Goldschmidt-Clermont, M., Herz, M., Spreitzer, R., and Vallet, J.-M. (1985). Strategy, progress and prospects of transformation in *Chlamy-domonas reinhardii*. In: *Molecular Form and Function of the Plant Genome* (L. van-Vloten-Doting, G.S.P. Groot, and T.C. Hall, Eds.), pp. 579–592. *NATO Adv. Study Inst., Ser. A* **83**. Plenum, New York.

Rochaix, J.-D., Erickson, J., Goldschmidt-Clermont, M., Herz, M., Spreitzer, R.J., and Vallet, J.-M. (1986). Molecular genetics of photosynthesis and transforma-tion in *Chlamydomonas reinhardii*. *Symp. Soc. Dev. Biol.* **43**, 27–43.

Rochaix, J.-D., Kuchka, M., Mayfield, S., Schirmer-Rahire, M., Girard-Bascou, J., and amd Bennoun, P. (1989). Nuclear and chloroplast mutations affect the synthesis or stability of the chloroplast *psbC* gene product in *Chlamydomonas reinhardtii*. *EMBO J.* **8**, 1013–1021.

Rochaix, J.-D., Goldschmidt-Clermont, M., and Merchant, S. (Eds.) (1998). *The Molecular Biology of Chloroplasts and Mitochondria in Chlamydomonas*. *Advances in Photosynthesis*, Vol. 7. Kluwer Academic, Dordrecht. 733 pp.

Rodríguez, H., Haring, M.A., and Beck, C.F. (1999). Molecular characterization of two light-induced, gamete-specific genes from *Chlamydomonas reinhardtii* that encode hydroxyproline-rich proteins. *Mol. Gen. Genet.* **261**, 267–274.

Rodríguez, M.C., Barsanti, L., Passarelli, V., Evangelista, V., Conforti, V., and Gualtieri, P. (2007). Effects of chromium on photosynthetic and photoreceptive apparatus of the alga *Chlamydomonas reinhardtii*. *Environ. Res.* **105**, 234–239.

Roesler, K.R., Marcotte, B.L., and Ogren, W.L. (1992). Functional importance of arginine 64 in *Chlamydomonas reinhardtii* phosphoribulokinase. *Plant Physiol.* **98**, 1285–1289.

Roffey, R.A., Golbeck, J.H., Hille, C.R., and Sayre, R.T. (1991). Photosynthetic electron transport in genetically altered photosystem II reaction centers of chloroplasts. *Proc. Natl. Acad. Sci. U. S. A.* **88**, 9122–9126.

Rohr, J., Sarkar, N., Balenger, S., Jeong, B.R., and Cerutti, H. (2004). Tandem inverted repeat system for selection of effective transgenic RNAi strains in *Chlamydomonas*. *Plant J.* **40**, 611–621.

Rolland, N., Dorne, A.J., Amoroso, G., Sültemeyer, D.F., Joyard, J., and Rochaix, J.D. (1997). Disruption of the plastid *ycf10* open reading frame affects uptake of inorganic carbon in the chloroplast of *Chlamydomonas*. *EMBO J.* **16**, 6713–6726.

Rolland, N., Ferro, M., Seigneurin-Berny, D., Garin, J., Douce, R., and Joyard, J. (2003). Proteomics of chloroplast envelope membranes. *Photosynth. Res.* **78**, 205–230.

Rollins, M.J., Harper, J.D.I., and John, P.C.L. (1983). Synthesis of individual proteins, including tubulins and chloroplast membrane proteins, in synchronous cultures of the eukaryote *Chlamydomonas reinhardtii*. Elimination of periodic changes in protein synthesis and enzyme activity under constant environmental conditions. *J. Gen. Microbiol.* **129**, 1899–1919.

Rooney, D.W., Yen, B.-C., and Mikita, D.J. (1971). Synchronization of *Chlamydomonas* division with intermittent hypothermia. *Exp. Cell Res.* **65**, 94–98.

Rosen, H. and Ebersold, W.T. (1972). Recombination in relation to ultraviolet sensitivity in *Chlamydomonas reinhardi*. *Genetics* **71**, 247–253.

Rosen, H., Rehn, M.M., and Johnson, B.A. (1980). The effect of caffeine on repair in *Chlamydomonas reinhardtii*. I. Enhancement of recombination repair. *Mutat. Res.* **70**, 301–309.

Rosenbaum, J.L., Moulder, J.E., and Ringo, D.L. (1969). Flagellar elongation and shortening in *Chlamydomonas*. The use of cycloheximide and colchicine to study the synthesis and assembly of flagellar proteins. *J. Cell Biol.* **41**, 600–619.

Rosenbaum, J.L., Binder, L.I., Granett, S., Dentler, W.L., Snell, W., Sloboda, R., and Haimo, L. (1975). Directionality and rate of assembly of chick brain tubulin onto pieces of neurotubules, flagellar axonemes, and basal bodies. *Ann. N. Y. Acad. Sci.* **253**, 147–177.

Rosowski, J.R. and Hoshaw, R.W. (1970). Staining algal pyrenoids with carmine after fixation in an acidified hypochlorite solution. *Stain Technol.* **45**, 293–298.

Rosowski, J.R. and Hoshaw, R.W. (1988). Advanced anisogamy in *Chlamydomonas monadina* (Chlorophyceae) with special reference to vacuolar activity during sexuality. *Phycologia* **27**, 494–504.

Rott, R., Drager, R.G., Stern, D.B., and Schuster, G. (1996). The 3' untranslated regions of chloroplast genes in *Chlamydomonas reinhardtii* do not serve as efficient transcriptional terminators. *Mol. Gen. Genet.* **252**, 676–683.

Rott, R., Levy, H., Drager, R.G., Stern, D.B., and Schuster, G. (1998a). 3'-Processed mRNA is preferentially translated in *Chlamydomonas reinhardtii* chloroplasts. *Mol. Cell. Biol.* **18**, 4605–4611.

Rott, R., Liveanu, V., Drager, R.G., Stern, D.B., and Schuster, G. (1998b). The sequence and structure of the 3'-untranslated regions of chloroplast transcripts are important determinants of mRNA accumulation and stability. *Plant Mol. Biol.* **36**, 307–314.

Rova, M., Franzén, L.-G., Fredriksson, P.-O., and Styring, S. (1994). Photosystem II in a mutant of *Chlamydomonas reinhardtii* lacking the 23 kDa psbP protein shows increased sensitivity to photoinhibition in the absence of chloride. *Photosynth. Res.* **39**, 75–83.

Rova, E.M., McEwen, B., Fredriksson, P.O., and Styring, S. (1996). Photoactivation and photoinhibition are competing in a mutant of *Chlamydomonas reinhardtii* lacking the 23-kDa extrinsic subunit of photosystem II. *J. Biol. Chem.* **271**, 28918–28924.

Royer, H.-D. and Sager, R. (1979). Methylation of chloroplast DNAs in the life cycle of *Chlamydomonas*. *Proc. Natl. Acad. Sci. U. S. A.* **76**, 5794–5798.

Rubin, R.W. and Filner, P. (1973). Adenosine 3′,5′-cyclic monophosphate in *Chlamydomonas reinhardtii*. Influence on flagellar function and regneration. *J. Cell Biol.* **56**, 628–635.

Rubinelli, P., Siripornadulsil, S., Gao-Rubinelli, F., and Sayre, R.T. (2002). Cadmium- and iron-stress-inducible gene expression in the green alga *Chlamydomonas reinhardtii*: evidence for H43 protein function in iron assimilation. *Planta* **215**, 1–13.

Rubio, V., Linhares, F., Solano, R., Martin, A.C., Iglesias, J., Leyva, A., and Paz-Ares, J. (2001). A conserved MYB transcription factor involved in phosphate starvation signaling both in vascular plants and in unicellular algae. *Genes Dev.* **15**, 2122–2133.

Rüfenacht, A. and Boschetti, A. (2000). Chloroplasts of the green alga *Chlamydomonas reinhardtii* possess at least four distinct stromal processing proteases. *Photosynth. Res.* **63**, 249–258.

Rüffer, U. and Nultsch, W. (1985). High-speed cinematographic analysis of the movement of *Chlamydomonas*. *Cell Motil.* **5**, 251–263.

Rüffer, U. and Nultsch, W. (1987). Comparison of the beating of *cis*- and *trans*-flagella of *Chlamydomonas* cells held on micropipettes. *Cell Motil. Cytoskeleton* **7**, 87–93.

Rüffer, U. and Nultsch, W. (1990). Flagellar photoresponses of *Chlamydomonas* cells held on micropipettes. I. Change in flagellar beat frequency. *Cell Motil. Cytoskeleton* **15**, 162–167.

Rüffer, U. and Nultsch, W. (1991). Flagellar photoresponses of *Chlamydomonas* cells held on micropipettes. II. Change in flagellar beat pattern. *Cell Motil. Cytoskeleton* **18**, 269–278.

Rüffer, U. and Nultsch, W. (1995). Flagellar photoresponses of *Chlamydomonas* cells held on micropipettes. *Bot. Acta* **108**, 255–265.

Rüffer, U. and Nultsch, W. (1997). Flagellar photoresponses of *ptx1*, a nonphototactic mutant of *Chlamydomonas*. *Cell Motil. Cytoskeleton* **37**, 111–119.

Rüffer, U. and Nultsch, W. (1998). Flagellar coordination in *Chlamydomonas* cells held on micropipettes. *Cell Motil. Cytoskeleton* **41**, 297–307.

Ruiz, F.A., Marchesini, N., Seufferheld, M., Govindjee, and Docampo, R. (2001). The polyphosphate bodies of *Chlamydomonas reinhardtii* possess a proton-pumping pyrophosphatase and are similar to acidocalcisomes. *J. Biol. Chem.* **276**, 46196–46203.

Ruiz-Binder, N.E., Geimer, S., and Melkonian, M. (2002). *In vivo* localization of centrin in the green alga *Chlamydomonas reinhardtii*. *Cell Motil. Cytoskeleton* **52**, 43–55.

Rumpf, R., Vernon, D., Schreiber, D., and Birky, C.W., Jr. (1996). Evolutionary consequences of the loss of photosynthesis in Chlamydomonadaceae: phylogenetic analysis of Rrn18 (18S rDNA) in 13 *Polytoma* strains (Chlorophyta). *J. Phycol.* **32**, 119–126.

Rupp, G. and Porter, M.E. (2003). A subunit of the dynein regulatory complex in *Chlamydomonas* is a homologue of a growth arrest-specific gene product. *J. Cell Biol.* **162**, 47–57.

Rupp, G., O'Toole, E., Gardner, L.C., Mitchell, B.F., and Porter, M.E. (1996). The *suppf-2* mutations of *Chlamydomonas* alter the activity of the outer dynein arms by modification of the gamma-dynein heavy chain. *J. Cell Biol.* **135**, 1853–1865.

Rupp, G., O'Toole, E., and Porter, M.E. (2001). The *Chlamydomonas PF6* locus encodes a large alanine/proline-rich polypeptide that is required for assembly of a central pair projection and regulates flagellar motility. *Mol. Biol. Cell* **12**, 739–751.

Rupprecht, J., Hankamer, B., Mussgnug, J.H., Ananyev, G., Dismukes, C., and Kruse, O. (2006). Perspectives and advances of biological H_2 production in microorganisms. *Appl. Microbiol. Biotechnol.* **72**, 442–449.

Ryan, F.J. (1955). Attempt to reproduce some of Moewus' experiments on *Chlamydomonas* and *Polytoma*. *Science* **122**, 470.

Ryan, R., Grant, D., Chiang, K.S., and Swift, H. (1978). Isolation and characterization of mitochondrial DNA from *Chlamydomonas reinhardtii*. *Proc. Natl. Acad. Sci. U. S. A.* **75**, 3268–3272.

Rymarquis, L.A., Handley, J.M., Thomas, M., and Stern, D.B. (2005). Beyond complementation. Map-based cloning in *Chlamydomonas reinhardtii*. *Plant Physiol.* **137**, 557–566.

Rymarquis, L.A., Higgs, D.C., and Stern, D.B. (2006). Nuclear suppressors define three factors that participate in both 5′ and 3′ end processing of mRNAs in *Chlamydomonas* chloroplasts. *Plant J.* **46**, 448–461.

Sack, L., Zeyl, C., Bell, G., Sharbel, T., Reboud, X., Bernhardt, T., and Koelewyn, H. (1994). Isolation of four new strains of *Chlamydomonas reinhardtii* (Chlorophyta) from soil samples. *J. Phycol.* **30**, 770–773.

Sager, R. (1954). Mendelian and non-Mendelian inheritance of streptomycin resistance in *Chlamydomonas reinhardi*. *Proc. Natl. Acad. Sci. U. S. A.* **40**, 356–363.

Sager, R. (1955). Inheritance in the green alga *Chlamydomonas reinhardi*. *Genetics* **40**, 476–489.

Sager, R. (1972). *Cytoplasmic Genes and Organelles*. Academic Press, New York. 405 pp.

Sager, R. (1977). Genetic analysis of chloroplast DNA in *Chlamydomonas*. *Adv. Genet.* **19**, 287–340.

Sager, R. and Grabowy, C. (1983). Differential methylation of chloroplast DNA regulates maternal inheritance in a methylated mutant of *Chlamydomonas*. *Proc. Natl. Acad. Sci. U. S. A.* **80**, 3025–3029.

Sager, R. and Grabowy, C. (1985). Sex in *Chlamydomonas*: sex and the single chloroplast. In: *The Origin and Evolution of Sex* (H.O. Halvorson and A. Monroy, Eds.), pp. 113–121. MBL Lectures in Biology, Vol. 7. Alan R. Liss, New York.

Sager, R. and Granick, S. (1953). Nutritional studies with *Chlamydomonas reinhardi*. *Ann. N. Y. Acad. Sci.* **56**, 831–838.

Sager, R. and Granick, S. (1954). Nutritional control of sexuality in *Chlamydomonas reinhardi*. *J. Gen. Physiol.* **37**, 729–742.

Sager, R. and Ishida, M.R. (1963). Chloroplast DNA in *Chlamydomonas*. *Proc. Natl. Acad. Sci. U. S. A.* **50**, 725–730.

Sager, R. and Kitchin, R. (1975). Selective silencing of eukaryotic DNA. *Science* **189**, 426–433.

Sager, R. and Lane, D. (1972). Molecular basis of maternal inheritance. *Proc. Natl. Acad. Sci. U. S. A.* **69**, 2410–2413.

Sager, R. and Palade, G.E. (1954). Chloroplast structure in green and yellow strains of *Chlamydomonas*. *Exp. Cell Res.* **7**, 584–588.

Sager, R. and Palade, G.E. (1957). Structure and development of the chloroplast in *Chlamydomonas*. I. The normal green cell. *J. Biophys. Biochem. Cytol.* **3**, 463–487.

Sager, R. and Ramanis, Z. (1963). The particulate nature of nonchromosomal genes in *Chlamydomonas*. *Proc. Natl. Acad. Sci. U. S. A.* **50**, 260–268.

Sager, R. and Ramanis, Z. (1965). Recombination of nonchromosomal genes in *Chlamydomonas*. *Proc. Natl. Acad. Sci. U. S. A.* **53**, 1053–1061.

Sager, R. and Ramanis, Z. (1967). Biparental inheritance of nonchromosomal genes induced by ultraviolet irradiation. *Proc. Natl. Acad. Sci. U. S. A.* **58**, 931–935.

Sager, R. and Ramanis, Z. (1968). The pattern of segregation of cytoplasmic genes in *Chlamydomonas*. *Proc. Natl. Acad. Sci. U. S. A.* **61**, 324–331.

Sager, R. and Ramanis, Z. (1970). A genetic map of non-Mendelian genes in *Chlamydomonas*. *Proc. Natl. Acad. Sci. U. S. A.* **65**, 593–600.

Sager, R. and Ramanis, Z. (1973). The mechanism of maternal inheritance in *Chlamydomonas*: biochemical and genetic studies. *Theor. Appl. Genet.* **43**, 101–108.

Sager, R. and Ramanis, Z. (1974). Mutations that alter the transmission of chloroplast genes in *Chlamydomonas*. *Proc. Natl. Acad. Sci. U. S. A.* **71**, 4698–4702.

Sager, R. and Ramanis, Z. (1976a). Chloroplast genetics of *Chlamydomonas*. I. Allelic segregation ratios. *Genetics* **83**, 303–321.

Sager, R. and Ramanis, Z. (1976b). Chloroplast genetics of *Chlamydomonas*. II. Mapping by cosegregation frequency analysis. *Genetics* **83**, 323–340.

Sager, R., Grabowy, C., and Sano, H. (1981). The *mat-1* gene in *Chlamydomonas* regulates DNA methylation during gametogenesis. *Cell* **24**, 41–47.

Saito, T. and Matsuda, Y. (1984). Sexual agglutinin of mating-type minus gametes in *Chlamydomonas reinhardii*. I. Loss and recovery of agglutinability of gametes treated with EDTA. *Exp. Cell Res.* **152**, 322–330.

Saito, T. and Matsuda, Y. (1991). Isolation and characterization of *Chlamydomonas* temperature-sensitive mutants affecting gametic differentiation under nitrogen-starved conditions. *Curr. Genet.* **19**, 65–71.

Saito, T., Tsubo, Y., and Matsuda, Y. (1988). A new assay system to classify non-mating mutants and to distinguish between vegetative cell and gamete in *Chlamydomonas reinhardtii*. *Curr. Genet.* **14**, 59–63.

Saito, T., Inoue, M., Yamada, M., and Matsuda, Y. (1998). Control of gametic differentiation and activity by light in *Chlamydomonas reinhardtii*. *Plant Cell Physiol.* **39**, 8–15.

Sakamoto, W., Kindle, K.L., and Stern, D.B. (1993). In vivo analysis of *Chlamydomonas* chloroplast *petD* gene expression using stable transformation of beta-glucuronidase translational fusions. *Proc. Natl. Acad. Sci. U. S. A.* **90**, 497–501.

Sakamoto, W., Sturm, N.R., Kindle, K.L., Stern, D.B., and Chen, X. (1994). Function of the *Chlamydomonas reinhardtii petD* 5′ untranslated region in regulating the accumulation of subunit IV of the cytochrome b_6/f complex. *Plant J.* **6**, 503–512.

Saks, N.M. (1982). Primary production and release of assimilated carbon by *Chlamydomonas provasolii* in culture. *Mar. Biol.* **70**, 205–208.

Salisbury, J.L. (1988). The lost neuromotor apparatus of *Chlamydomonas* rediscovered. *J. Protozool.* **35**, 574–577.

Salisbury, J.L. (1989). Centrin and the algal flagellar apparatus. *J. Phycol.* **25**, 201–206.

Salisbury, J.L. (1995). Centrin, centrosomes, and mitotic spindle poles. *Curr. Opin. Cell Biol.* **7**, 39–45.

Salisbury, J.L., Baron, A.T., and Sanders, M.A. (1988). The centrin-based cytoskeleton of *Chlamydomonas reinhardtii*: distribution in interphase and mitotic cells. *J. Cell Biol.* **107**, 635–641.

Salvador, M.L., Klein, U., and Bogorad, L. (1993a). 5′ Sequences are important positive and negative determinants of the longevity of *Chlamydomonas* chloroplast gene transcripts. *Proc. Natl. Acad. Sci. U. S. A.* **90**, 1556–1560.

Salvador, M.L., Klein, U., and Bogorad, L. (1993b). Light-regulated and endogenous fluctuations of chloroplast transcript levels in *Chlamydomonas*. Regulation by transcription and RNA degradation. *Plant J.* **3**, 213–219.

Samson, M.R., Klis, F.M., Crabbendam, K.J., van Egmond, P., and van den Ende, H. (1987a). Purification, visualization and characterization of the sexual agglutinins of the green alga *Chlamydomonas moewusii yapensis*. *J. Gen. Microbiol.* **133**, 3183–3191.

Samson, M.R., Klis, F.M., Homan, W.L., van Egmond, P., Musgrave, A., and van den Ende, H. (1987b). Composition and properties of the sexual agglutinins of the flagellated green alga *Chlamydomonas eugametos*. *Planta* **170**, 314–321.

Sanders, M.A. and Salisbury, J.L. (1989). Centrin-mediated microtubule severing during flagellar excision in *Chlamydomonas reinhardtii*. *J. Cell Biol.* **108**, 1751–1760.

Sanders, M.A. and Salisbury, J.L. (1994). Centrin plays an essential role in microtubule severing during flagellar excision in *Chlamydomonas reinhardtii*. *J. Cell Biol.* **124**, 795–805.

Sanders, M.A. and Salisbury, J.L. (1995). Immunofluorescence microscopy of cilia and flagella. *Methods Cell Biol.* **47**, 163–169.

Sanford, J.C. (1990). Biolistic plant transformation. *Physiol. Plant.* **79**, 206–209.

Sanford, J.C., Klein, T.M., Wolf, E.D., and Allen, N. (1987). Delivery of substances into cells and tissues using a particle bombardment process. *Part. Sci. Technol.* **5**, 27–37.

Sano, H., Royer, H.D., and Sager, R. (1980). Identification of 5-methylcytosine in DNA fragments immobilized on nitrocellulose paper. *Proc. Natl. Acad. Sci. U. S. A.* **77**, 3581–3585.

Sano, H., Grabowy, C., and Sager, R. (1981). Differential activity of DNA methyltransferase in the life cycle of *Chlamydomonas reinhardi*. *Proc. Natl. Acad. Sci. U. S. A.* **78**, 3118–3122.

Sano, H., Grabowy, C., and Sager, R. (1984). Loss of chloroplast DNA methylation during dedifferentiation of *Chlamydomonas reinhardi* gametes. *Mol. Cell. Biol.* **4**, 2103–2108.

Sapp, J. (1987). What counts as evidence or who was Franz Moewus and why was everybody saying such terrible things about him? *Hist. Philos. Life Sci.* **9**, 277–308.

Sapp, J. (1990). *Where the Truth Lies. Franz Moewus and the Origins of Molecular Biology*. Cambridge University Press, Cambridge. 340 pp.

Sarkar, N., Lemaire, S., Wu-Scharf, D., Issakidis-Bourguet, E., and Cerutti, H. (2005). Functional specialization of *Chlamydomonas reinhardtii* cytosolic thioredoxin h1 in the response to alkylation-induced DNA damage. *Eukaryotic Cell* **4**, 262–273.

Sasa, T., Suda, S., Watanabe, M.M., and Takaichi, S. (1992). A yellow marine *Chlamydomonas*. Morphology and pigment composition. *Plant Cell Physiol.* **33**, 527–534.

Sato, C. (1976). A conditional cell division mutant of *Chlamydomonas reinhardii* having an increased level of colchicine resistance. *Exp. Cell Res.* **101**, 251–259.

Sato, N., Sugimoto, K., Meguro, A., and Tsuzuki, M. (2003). Identification of a gene for UDP-sulfoquinovose synthase of a green alga, *Chlamydomonas reinhardtii*, and its phylogeny. *DNA Res.* **10**, 229–237.

Schaechter, M. and DeLamater, E.D. (1955). Mitosis of *Chlamydomonas*. *Am. J. Bot.* **42**, 417–422.

Schaechter, M. and DeLamater, E.D. (1956). Studies on meiosis in *Chlamydomonas*. *J. Elisha Mitchell Soc.* **72**, 73–80.

Schaller, K. and Uhl, R. (1997). A microspectrophotometric study of the shielding properties of eyespot and cell body in *Chlamydomonas*. *Biophys. J.* **73**, 1573–1578.

Schimmer, O. (1983). Effect of re-irradiation with UV-A on inactivation and mutation induction in *arg⁻* cells of *Chlamydomonas reinhardii* pretreated with furocoumarins plus UV-A. *Mutat. Res.* **109**, 195–205.

Schimmer, O. and Abel, G. (1986). Mutagenicity of a furocoumarin epoxide, heraclenin, in *Chlamydomonas reinhardii*. *Mutat. Res.* **169**, 47–50.

Schimmer, O. and Arnold, C.G. (1969). Untersuchungen zur Lokalisation eines ausserkaryotischen Gens bei *Chlamydomonas reinhardi*. *Arch. Mikrobiol.* **66**, 199–202.

Schimmer, O. and Arnold, C.G. (1970a). Untersuchungen über Reversions- und Segregationsverhalten eines ausserkaryotischen Gens von *Chlamydomonas reinhardii* zur Bestimmung des Erbträgers. *Mol. Gen. Genet.* **107**, 281–290.

Schimmer, O. and Arnold, C.G. (1970b). Über die Zahl der Kopien eines ausserkaryotischen Gens bei *Chlamydomonas reinhardii*. *Mol. Gen. Genet.* **107**, 366–371.

Schimmer, O. and Arnold, C.G. (1970c). Hin- und Rücksegregation eines ausserkaryotischen Gens bei *Chlamydomonas reinhardii*. *Mol. Gen. Genet.* **108**, 33–40.

Schimmer, O. and Arnold, C.G. (1970d). Die Suppression der ausserkaryotisch bedingten Streptomycin-Abhängigkeit bei *Chlamydomonas reinhardii*. *Arch. Mikrobiol.* **73**, 195–200.

Schimmer, O. and Kühne, K. (1990). Mutagenic compounds in an extract from Rutae Herba (*Ruta graveolens* L.). II. UV-A mediated mutagenicity in the green alga *Chlamydomonas reinhardtii* by furoquinoline alkaloids and furocoumarins present in a commercial tincture from Rutae Herba. *Mutat. Res.* **243**, 57–62.

Schlapfer, P. and Eichenberger, W. (1983). Evidence for the involvement of diacylglyceryl (*N,N,N*-trimethyl)homoserine in the desaturation of oleic and linoleic acids in *Chlamydomonas reinhardi* (Chlorophyceae). *Plant Sci. Lett.* **32**, 243–252.

Schloss, J. (1990). A *Chlamydomonas* gene encodes a G protein beta subunit-like polypeptide. *Mol. Gen. Genet.* **221**, 443–452.

Schloss, J.A. and Croom, H.B. (1991). Normal *Chlamydomonas* nuclear gene structure on linkage group XIX. *J. Cell Sci.* **100**, 877–881.

Schlösser, U.G. (1966). Enzymatisch gesteuerte Freisetzung von Zoosporen bei *Chlamydomonas reinhardii* Dangeard in Synchronkultur. *Arch. Mikrobiol.* **54**, 129–159.

Schlösser, U.G. (1976). Entwicklungsstadien- und sippenspezifische Zellwand-Autolysine bei der Freisetzung von Fortpflanzungszellen in der Gattung *Chlamydomonas*. *Ber. Deutsch. Bot. Ges.* **89**, 1–56.

Schlösser, U.G. (1981). Release of reproduction cells by action of cell wall autolytic factors in *Chlamydomonas* and *Geminella*. *Ber. Deutsch. Bot. Ges.* **94**, 373–374.

Schlösser, U.G. (1982). Sammlung von Algenkulturen, Pflanzenphysiologisches Institut der Universität Göttingen. *Ber. Deutsch. Bot. Ges.* **95**, 181–276.

Schlösser, U.G. (1984). Species-specific sporangium autolysins (cell-wall-dissolving enzymes) in the genus *Chlamydomonas*. In: *Systematics of the Green Algae* (D.E.G. Irvine and D.M. John, Eds.), pp. 409–418. Academic Press, New York.

Schlösser, U.G., Sachs, H., and Robinson, D.G. (1976). Isolation of protoplasts by means of a 'species-specific' autolysine in *Chlamydomonas*. *Protoplasma* **88**, 51–64.

Schmeisser, E.T., Baumgartel, D.M., and Howell, S.H. (1973). Gametic differentiation in *Chlamydomonas reinhardi*: cell cycle dependency and rates in attainment of mating competency. *Dev. Biol.* **31**, 31–37.

Schmidt, G.W., Matlin, K.S., and Chua, N.-H. (1977). A rapid procedure for selective enrichment of photosynthetic electron transport mutants. *Proc. Natl. Acad. Sci. U. S. A.* **74**, 610–614.

Schmidt, J.A. and Eckert, R. (1976). Calcium couples flagellar reversal to photo-stimulation in *Chlamydomonas reinhardtii*. *Nature* **262**, 713–715.

Schmidt, M., Gessner, G., Luff, M., Heiland, I., Wagner, V., Kaminski, M., Geimer, S., Eitzinger, N., Reissenweber, T., Voytsekh, O., Fiedler, M., Mittag, M., and Kreimer, G. (2006). Proteomic analysis of the eyespot of *Chlamydomonas reinhardtii* provides novel insights into its components and tactic movements. *Plant Cell* **18**, 1908–1930.

Schmidt, M., Luff, M., Mollwo, A., Kaminski, M., Mittag, M., and Kreimer, G. (2007). Evidence for a specialized localization of the chloroplast ATP-synthase subunits a, b, and g in the eyespot apparatus of *Chlamydomonas reinhardtii* (Chlorophyceae). *J. Phycol.* **43**, 284–294.

Schmidt, R.J., Richardson, C.B., Gillham, N.W., and Boynton, J.E. (1983). Sites of synthesis of chloroplast ribosomal proteins in *Chlamydomonas*. *J. Cell Biol.* **96**, 1451–1463.

Schmidt, R.J., Myers, A.M., Gillham, N.W., and Boynton, J.E. (1984). Chloroplast ribosomal proteins of *Chlamydomonas* synthesized in the cytoplasm are made as precursors. *J. Cell Biol.* **98**, 2011–2018.

Schneider, M. and Rochaix, J.-D. (1986). Sequence organization of the chloroplast ribosomal spacer of *Chlamydomonas reinhardii*: uninterrupted *tRNAile* and *tRNAala* genes and extensive secondary structure. *Plant Mol. Biol.* **6**, 265–270.

Schneider, M., Darlix, J.-L., Erickson, J., and Rochaix, J.-D. (1985). Sequence organization of repetitive elements in the flanking regions of the chloroplast ribosomal unit of *Chlamydomonas reinhardii*. *Nucleic Acids Res.* **13**, 8531–8541.

Schnell, R.A. and Lefebvre, P.A. (1993). Isolation of the *Chlamydomonas* regulatory gene *NIT2* by transposon tagging. *Genetics* **134**, 737–747.

Schneyour, A. and Avron, M. (1975). A method for producing, selecting and isolating photosynthetic mutants of *Euglena gracilis*. *Plant Physiol.* **55**, 142–144.

Schönfeld, C., Wobbe, L., Borgstädt, R., Kienast, A., Nixon, P.J., and Kruse, O. (2004). The nucleus-encoded protein MOC1 is essential for mitochondrial light acclimation in *Chlamydomonas reinhardtii*. *J. Biol. Chem.* **279**, 50366–50374.

Schoppmeier, J., Mages, W., and Lechtreck, K.F. (2005). GFP as a tool for the analysis of proteins in the flagellar basal apparatus of *Chlamydomonas*. *Cell Motil. Cytoskeleton* **61**, 189–200.

Schötz, G. (1972). Dreidimensionale, maßstabgetreue Rekonstruktion einer grünen Flagellatenzelle nach Elektronenmikroskopie von Serienschnitten. *Planta* **102**, 152–159.

Schötz, F., Bathelt, H., Arnold, C.-G., and Schimmer, O. (1972). Die Architektur und Organisation der *Chlamydomonas*-Zelle. Ergebnisse der Elektronenmikroskopie von Serienschnitten und der daraus resultierenden dreidimensionalen Rekonstruktion. *Protoplasma* **75**, 229–254.

Schroda, M. (2004). The *Chlamydomonas* genome reveals its secrets: chaperone genes and the potential roles of their gene products in the chloroplast. *Photosynth. Res.* **82**, 221–240.

Schroda, M. (2006). RNA silencing in *Chlamydomonas*: mechanisms and tools. *Curr. Genet.* **49**, 69–84.

Schroda, M., Vallon, O., Wollman, F.A., and Beck, C.F. (1999). A chloroplast-targeted heat shock protein 70 (HSP70) contributes to the photoprotection and repair of photosystem II during and after photoinhibition. *Plant Cell* **11**, 1165–1178.

Schroda, M., Blöcker, D., and Beck, C.F. (2000). The *HSP70A* promoter as a tool for the improved expression of transgenes in *Chlamydomonas*. *Plant J.* **21**, 121–131.

Schroda, M., Beck, C.F., and Vallon, O. (2002). Sequence elements within an *HSP70* promoter counteract transcriptional transgene silencing in *Chlamydomonas*. *Plant J.* **31**, 445–455.

Schulz-Baldes, M. and Lewin, R.A. (1975). Manganese encrustation of zygospores of a *Chlamydomonas* (Chlorophyta: Volvocales). *Science* **188**, 1119–1120.

Schulz-Raffelt, M., Lodha, M., and Schroda, M. (2007). Heat shock factor 1 is a key regulator of the stress response in *Chlamydomonas*. *Plant J.* **52**, 286–295.

Schwarz, C., Elles, I., Kortmann, J., Piotrowski, M., and Nickelsen, J. (2007). Synthesis of the D2 protein of photosystem II in *Chlamydomonas* is controlled by a high molecular mass complex containing the RNA stabilization factor Nac2 and the translational activator RBP40. *Plant Cell* **19**, 3627–3639.

Schwinck, I. (1960). Isolation of deoxyribonucleic acid from *Chlamydomonas reinhardi*. *J. Protozool.* **7**, 294–297.

Scudo, F.M. (1967). The adaptive value of sexual dimorphism: I. Anisogamy. *Evolution* **21**, 285–291.

Sears, B.B. (1980a). Changes in chloroplast genome composition and recombination during the maturation of zygospores of *Chlamydomonas reinhardtii*. *Curr. Genet.* **2**, 1–8.

Sears, B.B. (1980b). Disappearance of the heteroplasmic state for chloroplast markers in zygospores of *Chlamydomonas reinhardtii*. *Plasmid* **3**, 18–34.

Sears, B.B. (1980c). Elimination of plastids during spermatogenesis and fertilization in the plant kingdom. *Plasmid* **4**, 233–255.

Sears, B.B. and VanWinkle-Swift, K. (1994). The salvage/turnover/repair (STOR) model for uniparental inheritance in *Chlamydomonas*: DNA as a source of sustenance. *J. Hered.* **85**, 366–376.

Sears, B.B., Boynton, J.E., and Gillham, N.W. (1980). The effect of gametogenesis regimes on the chloroplast genetic system of *Chlamydomonas reinhardtii*. *Genetics* **96**, 95–114.

Sedova, T.V. (1998a). Comparative cytological investigations of unicellular green algae. VIII. Some peculiarities of mitosis in *Chlamydomonas* of the Amphichloris-group (Chlamydomonadaceae, Chlorophyta). *Bot. Zh. St. Petersburg* **83**, 32–35.

Sedova, T.V. (1998b). Comparative cytological investigation of unicellular green algae. VII. Some peculiarities of mitosis in *Chlamydomonas* of the Pleiochloris-group (Chlamydomonadaceae, Chlorophyta). *Bot. Zh. St. Petersburg* **83**, 43–46.

Sedova, T.V. (1998c). Comparative cytological investigation of unicellular green algae. VI. Some peculiarities of mitosis in *Chlamydomonas yellowstonensis* of the Sphaerella-group (Chlamydomonadales, Chlorophyta). *Bot. Zh. St. Petersburg* **83**, 50–53.

Sedova, T.V. (1999). Comparative cytological investigation of unicellular green algae. IX. Some peculiarities of mitosis in *Chlamydomonas* of the Agloë-group (Chlamydomonadales, Chlorophyta). *Bot. Zh. St. Petersburg* **84**, 60–63.

Sedova, T.V. (2001a). Comparative cytological study of unicellular green algae. XI. Some peculiarities of mitosis in *Chlamydomonas*, group of Chlorogoniella (Chlamydomonadaceae, Chlorophyta). *Bot. Zh. St. Petersburg* **86**, 84–88.

Sedova, T.V. (2001b). Comparative cytological investigation of unicellular green algae. XII. Some peculiarities of mitosis in *Chlamydomonas* from the group Chlamydella (Chlamydomonadales, Chlorophyta). *Bot. Zh. St. Petersburg* **86**, 56–64.

Sedova, T.V. (2002). Comparative cytological investigation of unicellular green algae. XIII. Some peculiarities of mitosis in *Chlamydomonas* of the Euchlamydomonas-group (Chlamydomonadales, Chlorophyta). *Bot. Zh. St. Petersburg* **87**, 36–46.

Segal, R.A. and Luck, D.J.L. (1985). Phosphorylation in isolated *Chlamydomonas* axonemes: a phosphoprotein may mediate the Ca^{2+}-dependent photophobic response. *J. Cell Biol.* **101**, 1702–1712.

Semin, B.K., Davletshina, L.N., Novakova, A.A., Kiseleva, T.Y., Lanchinskaya, V.Y., Aleksandrov, A.Y., Seifulina, N., Ivanov, I.I., Seibert, M., and Rubin, A.B. (2003). Accumulation of ferrous iron in *Chlamydomonas reinhardtii*. Influence of CO_2 and anaerobic induction of the reversible hydrogenase. *Plant Physiol.* **131**, 1756–1764.

Shalguev, V.I., Kaboev, O.K., Sizova, I.A., Hegemann, P., and Lanzov, V.A. (2005). Identification of *Chlamydomonas reinhardtii* Rad51C: recombinational characteristics. *Mol. Biol. (Moscow)* **39**, 98–104.

Sharaf, M.A. and Rooney, D.W. (1982). Changes in cyclic nucleotide levels correlated with growth, division, and morphology in *Chlamydomonas* chemostat culture. *Biochem. Biophys. Res. Commun.* **105**, 1461–1465.

Shaw, P.J. and Hills, G.J. (1984). The three-dimensional structure of the cell wall glycoprotein of *Chlorogonium elongatum*. *J. Cell Sci.* **68**, 271–284.

Shepherd, H.S., Boynton, J.E., and Gillham, N.W. (1979). Mutations in nine chloroplast loci of *Chlamydomonas* affecting different photosynthetic functions. *Proc. Natl. Acad. Sci. U. S. A.* **76**, 1353–1357.

Shigyo, M., Tabei, N., Yoneyama, T., and Yanagisawa, S. (2007). Evolutionary processes during the formation of the plant-specific Dof transcription factor family. *Plant Cell Physiol.* **48**, 179–185.

Shimogawara, K., Fujiwara, S., Grossman, A., and Usuda, H. (1998). High-efficiency transformation of *Chlamydomonas reinhardtii* by electroporation. *Genetics* **148**, 1821–1828.

Shimogawara, K., Wykoff, D.D., Usuda, H., and Grossman, A.R. (1999). *Chlamydomonas reinhardtii* mutants abnormal in their responses to phosphorus deprivation. *Plant Physiol.* **120**, 685–693.

Shinozaki, K., Ohme, M., Tanaka, M., Wakasugi, T., Hayashida, N., Matsubayashi, T., Zaita, N., Chunwongse, J., Obokata, J., Yamaguchi-Shinozaki, K., Ohto, C., Torazawa, K., Meng, B.Y., Sugita, M., Deno, H., Kamogashira, T., Yamada, K., Kusuda, J., Takaiwa, F., Kato, A., Tohdoh, N., Shimada, H., and Sugiura, M. (1986). The complete nucleotide sequence of the tobacco chloroplast genome: its gene organization and expression. *EMBO J.* **5**, 2043–2049.

Short, M.B., Solari, C.A., Ganguly, S., Powers, T.R., Kessler, J.O., and Goldstein, R.E. (2006). Flows driven by flagella of multicellular organisms enhance long-range molecular transport. *Proc. Natl. Acad. Sci. U. S. A.* **103**, 8315–8319.

Shrager, J., Hauser, C., Chang, C.W., Harris, E.H., Davies, J., McDermott, J., Tamse, R., Zhang, Z.D., and Grossman, A.R. (2003). *Chlamydomonas reinhardtii* genome project. A guide to the generation and use of the cDNA information. *Plant Physiol.* **131**, 401–408.

Siderius, M., Musgrave, A., van den Ende, H., Koerten, H., Cambier, P., and van der Meer, P. (1996). *Chlamydomonas eugametos* (Chlorophyta) stores phosphate in polyphosphate bodies together with calcium. *J. Phycol.* **32**, 402–409.

Siersma, P.W. and Chiang, K.-S. (1971). Conservation and degradation of chloroplast and cytoplasmic ribosomes in *Chlamydomonas reinhardtii*. *J. Mol. Biol.* **58**, 167–185.

Silflow, C.D., Lefebvre, P.A., McKeithan, T.W., Schloss, J.A., Keller, L.R., and Rosenbaum, J.L. (1981). Expression of flagellar protein genes during flagellar regeneration in *Chlamydomonas*. *Cold Spring Harb. Symp. Quant. Biol.* **46**, 157–169.

Silflow, C.D., Liu, B., LaVoie, M., Richardson, E.A., and Palevitz, B.A. (1999). Gamma-tubulin in *Chlamydomonas*: characterization of the gene and localization of the gene product in cells. *Cell Motil. Cytoskeleton* **42**, 285–297.

Silflow, C.D., LaVoie, M., Tam, L.W., Tousey, S., Sanders, M., Wu, W.C., Borodovsky, M., and Lefebvre, P.A. (2001). The Vfl1 protein in *Chlamydomonas* localizes in a rotationally asymmetric pattern at the distal ends of the basal bodies. *J. Cell Biol.* **153**, 63–74.

Silk, G.W., Dela Cruz, F., and Wu, M. (1990). Nucleotide sequence of the chloroplast gene for the 4 kD K polypeptide of photosystem II (*psbK*) and the *psbK-tufA* intergenic region of *Chlamydomonas reinhardtii*. *Nucleic Acids Res.* **18**, 4930.

Silverberg, B.A. (1974). The presence of unusual microtubular structures in senescent cells of *Chlamydomonas dysosmos*. *Arch. Microbiol.* **98**, 199–206.

Silverberg, B.A. (1975). An ultrastructural and cytochemical characterization of microbodies in the green algae. *Protoplasma* **83**, 269–925.

Silverberg, B.A. and Sawa, T. (1984). Cytochemical localization of oxidase activities with diaminobenzidine in the green alga *Chlamydomonas dysosmos*. *Protoplasma* **81**, 177–188.

Simard, C., Lemieux, C., and Bellemare, G. (1988). Cloning and sequencing of a cDNA encoding the small subunit precursor of ribulose-1,5-bisphosphate carboxylase from *Chlamydomonas moewusii*. *Curr. Genet.* **14**, 461–470.

Simon, A., Glöckner, G., Felder, M., Melkonian, M., and Becker, B. (2006). EST analysis of the scaly green flagellate *Mesostigma viride* (Streptophyta): implications for the evolution of green plants (Viridiplantae). *BMC Plant Biol.* **6**, 2.

Simpson, L., Thiemann, O.H., Savill, N.J., Alfonzo, J.D., and Maslov, D.A. (2000). Evolution of RNA editing in trypanosome mitochondria. *Proc. Natl. Acad. Sci. U. S. A.* **97**, 6986–6993.

Sinclair, J.H. (1972). Buoyant density of ribosomal genes in *Chlamydomonas reinhardii*. *Exp. Cell Res.* **74**, 569–571.

Sineshchekov, O.A. and Spudich, J.A. (2005). Sensory rhodopsin signaling in green flagellate algae. In: *Handbook of Photosensory Receptors* (Briggs W.R. and Spudich, J.L. Eds.), pp. 25–42. Wiley-Vch, Weinheim.

Sineshchekov, O., Lebert, M., and Häder, D.P. (2000). Effects of light on gravitaxis and velocity in *Chlamydomonas reinhardtii*. *J. Plant Physiol.* **157**, 247–254.

Sineshchekov, O.A., Jung, K.H., and Spudich, J.L. (2002). Two rhodopsins mediate phototaxis to low- and high-intensity light in *Chlamydomonas reinhardtii*. *Proc. Natl. Acad. Sci. U. S. A.* **99**, 8689–8694.

Singer, B., Sager, R., and Ramanis, Z. (1976). Chloroplast genetics of *Chlamydomonas*. III. Closing the circle. *Genetics* **83**, 341–354.

Siomi, H. and Siomi, M.C. (2007). Expanding RNA physiology: MicroRNAs in a unicellular organism. *Genes Dev.* **21**, 1153–1156.

Siu, C.-H., Swift, H., and Chiang, K.-S. (1976a). Characterization of cytoplasmic and nuclear genomes in the colorless alga *Polytoma*. I. Ultrastructural analysis of organelles. *J. Cell Biol.* **69**, 352–370.

Siu, C.-H., Swift, H., and Chiang, K.-S. (1976b). Characterization of cytoplasmic and nuclear genomes in the colorless alga *Polytoma*. II. General characterization of organelle nucleic acids. *J. Cell Biol.* **69**, 371–382.

Siu, C.-H., Chiang, K.-S., and Swift, H. (1976c). Characterization of cytoplasmic and nuclear genomes in the colorless alga *Polytoma*. III. Ribosomal RNA cistrons of the nucleus and leucoplast. *J. Cell Biol.* **69**, 383–392.

Sizova, I.A., Lapina, T.V., Frolova, O.N., Alexandrova, N.N., Akopiants, K.E., and Danilenko, V.N. (1996). Stable nuclear transformation of *Chlamydomonas reinhardtii* with a *Streptomyces rimosus* gene as the selective marker. *Gene* **181**, 13–18.

Sizova, I., Fuhrmann, M., and Hegemann, P. (2001). A *Streptomyces rimosus aphVIII* gene coding for a new type phosphotransferase provides stable antibiotic resistance to *Chlamydomonas reinhardtii*. *Gene* **277**, 221–229.

Sjoblad, R.D. and Frederikse, P.H. (1981). Chemotactic responses of *Chlamydomonas reinhardtii*. *Mol. Cell. Biol.* **1**, 1057–1060.

Skuja, H. (1949). Drei Fälle von sexueller Reproduktion in der Gattung *Chlamydomonas* Ehrnb. *Svensk. Bot. Tidskr.* **43**, 586–602.

Slaninová, M., Ševčovičová, A., Nagyová, B., Miadoková, E., Vlčková, V., and Vlček, D. (2002). *Chlamydomonas reinhardtii UVS11* gene is required for cell cycle arrest in response to DNA damage. *Arch. Hydrobiol. Suppl.* **145**, 97–107.

Slaninová, M., Nagyová, B., Gálová, E., Hendrychová, J., Bišová, K., Zachleder, V., and Vlček, D. (2003). The alga *Chlamydomonas reinhardtii UVS11* gene is responsible for cell division delay and temporal decrease in histone H1 kinase activity caused by UV irradiation. *DNA Repair* **2**, 737–750.

Sleigh, M.A. (1981). Flagellar beat patterns and their possible evolution. *BioSystems* **14**, 423–431.

Sloboda, R.D. and Howard, L. (2007). Localization of EB1, IFT polypeptides, and kinesin-2 in *Chlamydomonas* flagellar axonemes via immunogold scanning electron microscopy. *Cell Motil. Cytoskeleton* **64**, 446–460.

Small, G.D., Min, B., and Lefebvre, P.A. (1995). Characterization of a *Chlamydomonas reinhardtii* gene encoding a protein of the DNA photolyase/blue light photoreceptor family. *Plant Mol. Biol.* **28**, 443–454.

Smart, E.J. and Selman, B.R. (1993). Complementation of a *Chlamydomonas reinhardtii* mutant defective in the nuclear gene encoding the chloroplast coupling factor 1 (CF$_1$) gamma subunit (atpC). *J. Bioenerg. Biomembr.* **25**, 275–284.

Smith, A.C. and Purton, S. (2002). The transcriptional apparatus of algal plastids. *Eur. J. Phycol.* **37**, 301–311.

Smith, E., Roberts, K., Hutchings, A., and Galfre, G. (1984). Monoclonal antibodies to the major structural glycoprotein of the *Chlamydomonas* cell wall. *Planta* **161**, 330–338.

Smith, E.F. and Lefebvre, P.A. (1996). *PF16* encodes a protein with armadillo repeats and localizes to a single microtubule of the central apparatus in *Chlamydomonas* flagella. *J. Cell Biol.* **132**, 359–370.

Smith, E.F. and Lefebvre, P.A. (1997). *PF20* gene product contains WD repeats and localizes to the intermicrotubule bridges in *Chlamydomonas* flagella. *Mol. Biol. Cell* **8**, 455–467.

Smith, G.M. (1938). *Cryptogamic Botany. Algae and Fungi*. McGraw-Hill, New York. 545 pp.

Smith, G.M. (1946). The nature of sexuality in *Chlamydomonas*. *Am. J. Bot.* **33**, 625–630.

Smith, G.M. (1950). Sexuality, zygote formation, and zygote germination in *Chlamydomonas*. *Proc. Seventh Int. Bot. Cong.*, 836–837.

Smith, G.M. and Regnery, D.C. (1950). Inheritance of sexuality in *Chlamydomonas reinhardi*. *Proc. Natl. Acad. Sci. U. S. A.* **36**, 246–248.

Smith, T.A. and Kohorn, B.D. (1994). Mutations in a signal sequence for the thylakoid membrane identify multiple protein transport pathways and nuclear suppressors. *J. Cell Biol.* **126**, 365–374.

Smyth, R.D. and Berg, H.C. (1982). Change in flagellar beat frequency of *Chlamydomonas* in response to light. *Prog. Clin. Biol. Res.* **80**, 211–215.

Smyth, R.D. and Ebersold, W.T. (1985). Genetic investigation of a negatively phototactic strain of *Chlamydomonas reinhardtii*. *Genet. Res.* **46**, 133–148.

Smyth, R.D., Martinek, G.W., and Ebersold, W.T. (1975). Linkage of six genes in *Chlamydomonas reinhardtii* and the construction of linkage test strains. *J. Bacteriol.* **124**, 1615–1617.

Snell, W.J. (1976). Mating in *Chlamydomonas*: a system for the study of specific cell adhesion. II. A radioactive flagella-binding assay for quantitation of adhesion. *J. Cell Biol.* **68**, 70–79.

Snell, W.J. (1980). Gamete induction and flagellar adhesion in *Chlamydomonas reinhardi*. In: *Handbook of Phycological Methods. Vol. III. Developmental and Cytological Methods* (E. Gantt, Ed.), pp. 37–45. Cambridge University Press, Cambridge.

Snell, W.J. (1983). Characterization of the *Chlamydomonas* flagellar collar. *Cell Motil.* **3**, 273–280.

Snell, W.J., Buchanan, M., and Clausell, A. (1982). Lidocaine reversibly inhibits fertilization in *Chlamydomonas*: a possible role for calcium in sexual signalling. *J. Cell Biol.* **94**, 607–612.

Snell, W.J., Clausell, A., and Moore, W.S. (1983). Flagellar adhesion in *Chlamydomonas* induces synthesis of two high molecular weight cell surface proteins. *J. Cell Biol.* **96**, 589–597.

Snell, W.J., Eskue, W.A., and Buchanan, M.J. (1989). Regulated secretion of a serine protease that activates an extracellular matrix-degrading metalloprotease during fertilization in *Chlamydomonas. J. Cell Biol.* **109**, 1689–1694.

Snow, R. (1966). An enrichment method for auxotrophic yeast mutants using the antibiotic 'Nystatin.' *Nature* **211**, 206–207.

Snow, R. (1979). Maximum likelihood estimation of linkage and interference from tetrad data. *Genetics* **92**, 231–245.

Sodeinde, O.A. and Kindle, K.L. (1993). Homologous recombination in the nuclear genome of *Chlamydomonas reinhardtii. Proc. Natl. Acad. Sci. U. S. A.* **90**, 9199–9203.

Solari, C.A., Ganguly, S., Kessler, J.O., Michod, R.E., and Goldstein, R.E. (2006a). Multicellularity and the functional interdependence of motility and molecular transport. *Proc. Natl. Acad. Sci. U. S. A.* **103**, 1353–1358.

Solari, C.A., Kessler, J.O., and Michod, R.E. (2006b). A hydrodynamics approach to the evolution of multicellularity: flagellar motility and germ-soma differentiation in volvocalean green algae. *Am. Nat.* **167**, 537–554.

Solter, K.M. and Gibor, A. (1977). Evidence for role of flagella as sensory transducers in mating of *Chlamydomonas reinhardi. Nature* **265**, 444–445.

Solter, K.M. and Gibor, A. (1978). The relationship between tonicity and flagellar length. *Nature* **275**, 651–652.

Somanchi, A., Barnes, D., and Mayfield, S.P. (2005). A nuclear gene of *Chlamydomonas reinhardtii, Tba1,* encodes a putative oxidoreductase required for translation of the chloroplast *psbA* mRNA. *Plant J.* **42**, 341–352.

Sommaruga, R., Obernosterer, I., Herndl, G.J., and Psenner, R. (1997). Inhibitory effect of solar radiation on thymidine and leucine incorporation by freshwater and marine bacterioplankton. *Appl. Environ. Microbiol.* **63**, 4178–4184.

Song, L. and Dentler, W.L. (2001). Flagellar protein dynamics in *Chlamydomonas. J. Biol. Chem.* **276**, 29754–29763.

Sonneborn, T.M. (1951). Some current problems of genetics in the light of investigations on *Chlamydomonas* and *Paramecium. Cold Spring Harb. Symp. Quant. Biol.* **16**, 483–503.

Sosa, F.M., Ortega, T., and Barea, J.L. (1978). Mutants from *Chlamydomonas reinhardii* affected in their nitrate assimilation capability. *Plant Sci. Lett.* **11**, 51–58.

Spalding, M.H., Spreitzer, R.J., and Ogren, W.L. (1983). Carbonic anhydrase-deficient mutant of *Chlamydomonas reinhardii* requires elevated carbon dioxide concentration for photoautotrophic growth. *Plant Physiol.* **73**, 268–272.

Spalding, M.H., Winder, T.L., Anderson, J.C., Geraghty, A.M., and Marek, L.F. (1991). Changes in protein and gene expression during induction of the CO_2-concentrating mechanism in wild-type and mutant *Chlamydomonas. Can. J. Bot.* **69**, 1008–1016.

Spalding, M.H., Van, K., Wang, Y., and Nakamura, Y. (2002). Acclimation of *Chlamydomonas* to changing carbon availability. *Funct. Plant Biol.* **29**, 221–230.

Spanier, J.G., Graham, J.E., and Jarvik, J.W. (1992). Isolation and preliminary characterisation of three *Chlamydomonas* strains interfertile with *Chlamydomonas reinhardtii* (Chlorophyta). *J. Phycol.* **28**, 822–828.

Spessert, R. and Waffenschmidt, S. (1990). Studies on the vegetative autolysin during the vegetative life cycle in *Chlamydomonas reinhardtii. Eur. J. Cell Biol.* **51**, 17–22.

Spiegel, P.C., Chevalier, B., Sussman, D., Turmel, M., Lemieux, C., and Stoddard, B.L. (2006). The structure of I-CeuI homing endonuclease: evolving asymmetric DNA recognition from a symmetric protein scaffold. *Structure* **14**, 869–880.

Spijkerman, E. (2005). Inorganic carbon acquisition by *Chlamydomonas acidophila* across a pH range. *Can. J. Bot.* **83**, 872–878.

Spijkerman, E., Barua, D., Gerloff-Elias, A., Kern, J., Gaedke, U., and Heckathorn, S.A. (2007). Stress responses and metal tolerance of *Chlamydomonas acidophila* in metal-enriched lake water and artificial medium. *Extremophiles* **11**, 551–562.

Spreitzer, R.J. (1980). A uniparental mutant of *Chlamydomonas reinhardii* with altered ribulose-1,5-bisphosphate carboxylase. Ph.D. dissertation, Case Western Reserve University, Cleveland, Ohio.

Spreitzer, R.J. and Chastain, C.J. (1987). Heteroplasmic suppression of an amber mutation in the *Chlamydomonas* chloroplast gene that encodes the large subunit of ribulosebisphosphate carboxylase/oxygenase. *Curr. Genet.* **11**, 611–616.

Spreitzer, R.J. and Mets, L. (1980). Non-mendelian mutation affecting ribulose-1,5-bisphosphate carboxylase structure and activity. *Nature* **285**, 114–115.

Spreitzer, R.J. and Mets, L. (1981). Photosynthesis-deficient mutants of *Chlamydomonas reinhardii* with associated light-sensitive phenotypes. *Plant Physiol.* **67**, 565–569.

Spreitzer, R.J. and Mets, L. (1982). An assessment of arsenate selection as a method for obtaining nonphotosynthetic mutants of *Chlamydomonas*. *Genetics* **100**, 417–425.

Spreitzer, R.J. and Ogren, W.L. (1983a). Nuclear suppressors of the photosensitivity associated with defective photosynthesis in *Chlamydomonas reinhardii*. *Plant Physiol.* **71**, 35–39.

Spreitzer, R.J. and Ogren, W.L. (1983b). Rapid recovery of chloroplast mutations affecting ribulosebisphosphate carboxylase/oxygenase in *Chlamydomonas reinhardtii*. *Proc. Natl. Acad. Sci. U. S. A.* **80**, 6293–6297.

Spreitzer, R.J., Chastain, C.J., and Ogren, W.L. (1984). Chloroplast gene suppression of defective ribulosebisphosphate carboxylase/oxygenase in *Chlamydomonas reinhardii*: evidence for stable heteroplasmic genes. *Curr. Genet.* **9**, 83–89.

Spreitzer, R.J., Goldschmidt-Clermont, M., Rahire, M., and Rochaix, J.-D. (1985). Nonsense mutations in the *Chlamydomonas* chloroplast gene that codes for the large subunit of ribulosebisphosphate carboxylase/oxygenase. *Proc. Natl. Acad. Sci. U. S. A.* **82**, 5460–5464.

Spudich, J.L. and Sager, R. (1980). Regulation of the *Chlamydomonas* cell cycle by light and dark. *J. Cell Biol.* **85**, 136–145.

Stabenau, H. (1974). Verteilung von Microbody-Enzymen aus *Chlamydomonas* in Dichtegradienten. *Planta* **118**, 35–42.

Stabenau, H., Winkler, U., and Säftel, W. (1993). Localization of glycolate dehydrogenase in two species of *Dunaliella*. *Planta* **191**, 362–364.

Stahl, N. and Mayer, A.M. (1963). Experimental differentiation between phototaxis and motility in *Chlamydomonas snowiae*. *Science* **141**, 1282–1284.

Stamm, A. (1980). Der Einfluss von Sulfit auf das Wachstum und die CO_2-Fixierung einzelliger Grünalgen. *Environ. Pollution (Series A)* **22**, 91–99.

Starling, D. and Randall, J. (1971). The flagella of temporary dikaryons of *Chlamydomonas reinhardii*. *Genet. Res.* **18**, 107–113.

Starr, R.C. (1949). A method of effecting zygospore germination in certain Chlorophyceae. *Proc. Natl. Acad. Sci. U. S. A.* **35**, 453–456.

Starr, R.C. (1971). Algal cultures – sources and methods of cultivation. *Methods Enzymol.* **23**, 29–53.

Starr, R.C. (1973). Special methods – Dry soil samples. In: *Handbook of Phycological Methods. Vol. I. Culture Methods and Growth Measurements* (J.R. Stein, Ed.), pp. 159–167. Cambridge University Press, Cambridge.

Starr, R.C., Marner, F.J., and Jaenicke, L. (1995). Chemoattraction of male gametes by a pheromone produced by female gametes of *Chlamydomonas*. *Proc. Natl. Acad. Sci. U. S. A.* **92**, 641–645.

Stauber, E.J. and Hippler, M. (2004). *Chlamydomonas reinhardtii* proteomics. *Plant Physiol. Biochem.* **42**, 989–1001.

Stauber, E.J., Fink, A., Markert, C., Kruse, O., Johanningmeier, U., and Hippler, M. (2003). Proteomics of *Chlamydomonas reinhardtii* light-harvesting proteins. *Eukaryotic Cell* **2**, 978–994.

Stavis, R.L. (1974). The effect of azide on phototaxis in *Chlamydomonas reinhardi*. *Proc. Natl. Acad. Sci. U. S. A.* **71**, 1824–1827.

Stavis, R.L. and Hirschberg, R. (1973). Phototaxis in *Chlamydomonas reinhardtii*. *J. Cell Biol.* **59**, 367–377.

Stein, M., Jacquot, J.-P., and Miginiac-Maslow, M. (1993). A cDNA clone encoding *Chlamydomonas reinhardtii* preferredoxin. *Plant Physiol.* **102**, 1349–1350.

Steinkötter, J., Bhattacharya, D., Semmelroth, I., Bibeau, C., and Melkonian, M. (1994). Prasinophytes form independent lineages within the Chlorophyta: evidence from ribosomal RNA sequence comparisons. *J. Phycol.* **30**, 340–345.

Stevens, D.R. and Purton, S. (1997). Genetic engineering of eukaryotic algae: progress and prospects. *J. Phycol.* **33**, 713–722.

Stevens, D.R., Rochaix, J.-D., and Purton, S. (1996). The bacterial phleomycin resistance gene *ble* as a dominant selectable marker in *Chlamydomonas*. *Mol. Gen. Genet.* **251**, 23–30.

Stewart, K.D. and Mattox, K.R. (1975). Some aspects of mitosis in primitive green algae: phylogeny and function. *BioSystems* **7**, 310–315.

Stewart, K.D. and Mattox, K.R. (1978). Structural evolution in the flagellated cells of green algae and land plants. *BioSystems* **10**, 145–152.

Stokes, K.D. and Osteryoung, K.W. (2003). Early divergence of the FtsZ1 and FtsZ2 plastid division gene families in photosynthetic eukaryotes. *Gene* **320**, 97–108.

Storms, R. and Hastings, P.J. (1975). Definition by electron microscopy of two periods in meiosis in *Chlamydomonas reinhardi* when recombination is susceptible to DNA synthesis inhibition. *Can. J. Genet. Cytol.* **17**, 467–468.

Storms, R. and Hastings, P.J. (1977). A fine structure analysis of meiotic pairing in *Chlamydomonas reinhardi*. *Exp. Cell Res.* **104**, 39–46.

Straley, S.C. and Bruce, V.G. (1979). Stickiness to glass. Circadian changes in the cell surface of *Chlamydomonas reinhardi*. *Plant Physiol.* **63**, 1175–1181.

Strehlow, K. (1929). Über die Sexualität einiger Volvocales. *Z. Bot.* **21**, 625–692.

Stroebel, D., Choquet, Y., Popot, J.-L., Picot, D., and Popot, J.L. (2003). An atypical haem in the cytochrome b_6f complex. *Nature* **426**, 413–418.

Sturm, N.R., Kuras, R., Büschlen, S., Sakamoto, W., Kindle, K.L., Stern, D.B., and Wollman, F.A. (1994). The *petD* gene is transcribed by functionally redundant promoters in *Chlamydomonas reinhardtii* chloroplasts. *Mol. Cell. Biol.* **14**, 6171–6179.

Su, Q. and Boschetti, A. (1994). Substrate- and species-specific processing enzymes for chloroplast precursor proteins. *Biochem. J.* **300**, 787–792.

Su, X., Kaska, D.D., and Gibor, A. (1990). Induction of cytosine-rich poly(A)+ RNAs in *Chlamydomonas reinhardtii* by cell wall removal. *Exp. Cell Res.* **187**, 54–58.

Sueoka, N. (1960). Mitotic replication of deoxyribonucleic acid in *Chlamydomonas reinhardi*. *Proc. Natl. Acad. Sci. U. S. A.* **46**, 83–91.

Sugimoto, I. and Takahashi, Y. (2003). Evidence that the PsbK polypeptide is associated with the photosystem II core antenna complex CP43*. *J. Biol. Chem.* **278**, 45004–45010.

Sumper, M. and Hallmann, A. (1998). Biochemistry of the extracellular matrix of *Volvox*. *Int. Rev. Cytol.* **180**, 51–85.

Sun, M., Qian, K.X., Su, N., Chang, H.Y., Liu, J.X., and Shen, G. (2003). Foot-and-mouth disease virus VP1 protein fused with cholera toxin B subunit expressed in *Chlamydomonas reinhardtii* chloroplast. *Biotechnol. Lett.* **25**, 1087–1092.

Sunda, W.G., Price, N.M., and Morel, F.M.M. (2005). Trace metal ion buffers and their use in culture studies. In: *Algal Culturing Techniques* (R.A. Andersen, Ed.), pp. 35–63. Elsevier, Amsterdam.

Surzycki, R., Cournac, L., Peltier, G., and Rochaix, J.D. (2007). Potential for hydrogen production with inducible chloroplast gene expression in *Chlamydomonas*. *Proc. Natl. Acad. Sci. U. S. A.* **104**, 17548–17553.

Surzycki, S. (1971). Synchronously grown cultures of *Chlamydomonas reinhardi*. *Methods Enzymol.* **23**, 67–73.

Suss, K.H., Prokhorenko, I., and Adler, K. (1995). *In situ* association of Calvin cycle enzymes, ribulose-1,5-bisphosphate carboxylase/oxygenase activase, ferredoxin-NADP+ reductase, and nitrite reductase with thylakoid and pyrenoid membranes of *Chlamydomonas reinhardtii* chloroplasts as revealed by immuno-electron microscopy. *Plant Physiol.* **107**, 1387–1397.

Suzuki, J.Y. and Bauer, C.E. (1992). Light-independent chlorophyll biosynthesis: involvement of the chloroplast gene *chlL* (*frxC*). *Plant Cell* **4**, 929–940.

Suzuki, K. and Spalding, M.H. (1989). Adaptation of *Chlamydomonas reinhardtii* high-CO_2-requiring mutants to limiting CO_2. *Plant Physiol.* **90**, 1195–1200.

Suzuki, L. and Johnson, C.H. (2001). Algae know the time of day: circadian and photoperiodic programs. *J. Phycol.* **37**, 933–942.

Suzuki, L. and Johnson, C.H. (2002). Photoperiodic control of germination in the unicell *Chlamydomonas*. *Naturwissenschaften* **89**, 214–220.

Suzuki, L., Woessner, J.P., Uchida, H., Kuroiwa, H., Yuasa, Y., Waffenschmidt, S., Goodenough, U.W., and Kuroiwa, T. (2000). A zygote-specific protein with hydroxyproline-rich glycoprotein domains and lectin-like domains involved in the assembly of the cell wall of *Chlamydomonas reinhardtii* (Chlorophyta). *J. Phycol.* **36**, 571–583.

Suzuki, T., Yamasaki, K., Fujita, S., Oda, K., Iseki, M., Yoshida, K., Watanabe, M., Daiyasu, H., Toh, H., Asamizu, E., Tabata, S., Miura, K., Fukuzawa, H., Nakamura, S., and Takahashi, T. (2003). Archaeal-type rhodopsins in *Chlamydomonas*: model structure and intracellular localization. *Biochem. Biophys. Res. Commun.* **301**, 711–717.

Sweet, J.M., Carda, B., and Small, G.D. (1981). Repair of 3-methyladenine and 7-methylguanine in nuclear DNA of *Chlamydomonas*: requirement for protein synthesis. *Mutat. Res.* **84**, 73–82.

Swiatek, M., Kuras, R., Sokolenko, A., Higgs, D., Olive, J., Cinque, G., Mueller, B., Eichacker, L.A., Stern, D.B., Bassi, R., Herrmann, R.G., and Wollman, F.A. (2001). The chloroplast gene *ycf9* encodes a photosystem II (PSII) core sub-unit, PsbZ, that participates in PSII supramolecular architecture. *Plant Cell* **13**, 1347–1367.

Syrett, P.J. (1962). Nitrogen assimilation. In: *Physiology and Biochemistry of Algae* (R.A. Lewin, Ed.), pp. 171–188. Academic Press, New York.

Szybalski, W. and Bryson, V. (1952). Genetic studies on microbial cross-resistance to toxic agents. 1. Cross-resistance of *E. coli* to 15 antibiotics. *J. Bacteriol.* **64**, 289–299.

Szyszka, B., Ivanov, A.G., and Huner, N.P. (2007). Psychrophily is associated with differential energy partitioning, photosystem stoichiometry and polypeptide phosphorylation in *Chlamydomonas raudensis*. *Biochim. Biophys. Acta* **1767**, 789–800.

Taillon, B.E., Adler, S.A., Suhan, J.P., and Jarvik, J.W. (1992). Mutational analysis of centrin: an EF-hand protein associated with three distinct contractile fibers in the basal body apparatus of *Chlamydomonas*. *J. Cell Biol.* **119**, 1613–1624.

Takada, S., Wilkerson, C.G., Wakabayashi, K., Kamiya, R., and Witman, G.B. (2002). The outer dynein arm-docking complex: composition and characterization of a subunit (Oda1) necessary for outer arm assembly. *Mol. Biol. Cell* **13**, 1015–1029.

Takahama, U., Egashira, T., and Nakamura, K. (1985). Photoinactivation of a *Chlamydomonas* mutant (*NL-11*) in the presence of methionine: roles of H_2O_2 and O_2^-. *Photochem. Photobiol.* **41**, 149–152.

Takahashi, H., Braby, C.E., and Grossman, A.R. (2001). Sulfur economy and cell wall biosynthesis during sulfur limitation of *Chlamydomonas reinhardtii*. *Plant Physiol.* **127**, 665–673.

Takahashi, Y., Goldschmidt-Clermont, M., Soen, S.-Y., Franzén, L.-G., and Rochaix, J.-D. (1991). Directed chloroplast transformation in *Chlamydomonas reinhardtii*: insertional inactivation of the *psaC* gene encoding the iron sulfur protein destabilizes photosystem I. *EMBO J.* **10**, 2033–2040.

Takahashi, Y., Goldschmidt-Clermont, M., and Rochaix, J. D. (1992). Directed mutagenesis of the chloroplast gene *psaC* in *Chlamydomonas reinhardtii*. In: *Research in Photosynthesis*, Vol. 3 (N. Murata, Ed.), pp. 393–396. Kluwer Academic, Dordrecht.

Takahashi, Y., Matsumoto, H., Goldschmidt-Clermont, M., and Rochaix, J.-D. (1994). Directed disruption of the *Chlamydomonas* chloroplast *psbK* gene destabilizes the photosystem II reaction center complex. *Plant Mol. Biol.* **24**, 779–788.

Takahashi, Y., Rahire, M., Breyton, C., Popot, J.-L., Joliot, P., and Rochaix, J.-D. (1996). The chloroplast *ycf7* (*petL*) open reading frame of *Chlamydomonas reinhardtii* encodes a small functionally important subunit of the cytochrome b_6f complex. *EMBO J.* **15**, 3498–3506.

Takeda, T., Yoshimura, K., Yoshii, M., Kanahoshi, H., Miyasaka, H., and Shigeoka, S. (2000). Molecular characterization and physiological role of ascorbate peroxidase from halotolerant *Chlamydomonas sp.* W80 strain *Arch. Biochem. Biophys.* **376**, 82–90.

Takeda, T., Miyao, K., Tamoi, M., Kanaboshi, H., Miyasaka, H., and Shigeoka, S. (2003). Molecular characterization of glutathione peroxidase-like protein in halotolerant *Chlamydomonas sp.* W80. *Physiol. Plant.* **117**, 467–475.

Tam, L.W. and Lefebvre, P.A. (1993). Cloning of flagellar genes in *Chlamydomonas reinhardtii* by DNA insertional mutagenesis. *Genetics* **135**, 375–384.

Tam, L.W. and Lefebvre, P.A. (1995). Insertional mutagenesis and isolation of tagged genes in *Chlamydomonas*. *Methods Cell Biol.* **47**, 519–523.

Tam, L., Dentler, W.L., and Lefebvre, P.A. (2003). Defective flagellar assembly and length regulation in *LF3* null mutants in *Chlamydomonas*. *J. Cell Biol.* **163**, 597–607.

Tam, L.-W., Wilson, N.F., and Lefebvre, P.A. (2007). A CDK-related kinase regulates the length and assembly of flagella in *Chlamydomonas*. *J. Cell Biol.* **176**, 819–829.

Tamaki, S., Matsuda, Y., and Tsubo, Y. (1981). The isolation and properties of the lytic enzyme of the cell wall released by mating gametes of *Chlamydomonas reinhardtii*. *Plant Cell Physiol.* **22**, 127–133.

Tamoi, M., Kanaboshi, H., Miyasaka, H., and Shigeoka, S. (2001). Molecular mechanisms of the resistance to hydrogen peroxide of enzymes involved in the Calvin cycle from halotolerant *Chlamydomonas sp.* W80. *Arch. Biochem. Biophys.* **390**, 176–185.

Tamoi, M., Nagaoka, M., and Shigeoka, S. (2005). Immunological properties of sedoheptulose 1,7 bisphosphatase from *Chlamydomonas sp.* W80. *Biosci. Biotechnol. Biochem.* **69**, 848–851.

Tan, C.K. and Hastings, P.J. (1977). DNA synthesis during meiosis of eight-spored strains of *Chlamydomonas reinhardi*. *Mol. Gen. Genet.* **152**, 311–317.

Tanaka, S., Ikeda, K., and Miyasaka, H. (2004). Isolation of a new member of group 3 late embryogenesis abundant protein gene from a halotolerant green alga by a functional expression screening with cyanobacterial cells. *FEMS Microbiol. Lett.* **236**, 41–45.

Tang, B., Sitomer, A., and Jackson, T. (1997). Population dynamics and competition in chemostat models with adaptive nutrient uptake. *J. Math. Biol.* **35**, 453–479.

Tang, D.K., Qiao, S.Y., and Wu, M. (1995). Insertion mutagenesis of *Chlamydomonas reinhardtii* by electroporation and heterologous DNA. *Biochem. Mol. Biol. Int.* **36**, 1025–1035.

Tatsuzawa, H., Takizawa, E., Wada, M., and Yamamoto, Y. (1996). Fatty acid and lipid composition of the acidophilic green alga *Chlamydomonas sp.*. *J. Phycol.* **32**, 598–601.

Taub, F.B. and Dollar, A.M. (1968). Improvement of a continuous culture apparatus for long term use. *Appl. Microbiol.* **16**, 232–235.

Telser, A. (1977). The inhibition of flagellar regeneration in *Chlamydomonas reinhardii* by inhalational anesthetic halothane. *Exp. Cell Res.* **107**, 247–252.

Teramoto, H., Ishii, A., Kimura, Y., Hasegawa, K., Nakazawa, S., Nakamura, T., Higashi, S.I., Watanabe, M., and Ono, T.A. (2006). Action spectrum for expression of the high intensity light-inducible Lhc-like gene *Lhl4* in the green alga *Chlamydomonas reinhardtii*. *Plant Cell Physiol.* **47**, 419–425.

Tetali, S.D., Mitra, M., and Melis, A. (2007). Development of the light-harvesting chlorophyll antenna in the green alga *Chlamydomonas reinhardtii* is regulated by the novel *Tla1* gene. *Planta* **225**, 813–829.

Tetík, K. (1976). Study of mechanical destruction of algal cells and the possibility of using it to obtain cellular organelles and isolate nucleic acids. *Sov. Plant Physiol.* **23**, 367–374. [Russian original *Fiziol. Rast.* **23**, 427–235].

Tetík, K. and Nečas, J. (1979). The cell cycles of *Chlamydomonas geitleri* Ettl. *Arch. Protistenk.* **122**, 201–217.

Tetík, K. and Sulek, J. (1994). Problems accompanying the separation of zygospores of the alga *Chlamydomonas geitleri* (Chlorophyta) by density gradient centrifugation. *Biologia* **49**, 639–644.

Tetík, K. and Zadrazil, S. (1982). Characterization of DNA of the alga *Chlamydomonas geitleri* Ettl. *Biol. Plant.* **24**, 202–210.

Thacker, A. and Syrett, P.J. (1972). The assimilation of nitrate and ammonium by *Chlamydomonas reinhardi*. *New Phytol.* **71**, 423–433.

Thomas, W.H. (1972). Observations on snow algae in California. *J. Phycol.* **8**, 1–9.

Thomas, D.J. and Herbert, S.K. (2005). An inexpensive apparatus for growing photosynthetic microorganisms in exotic atmospheres. *Astrobiology* **5**, 75–82.

Thompson, A.J., Yuan, X., Kudlicki, W., and Herrin, D.L. (1992). Cleavage and recognition pattern of a double-strand-specific endonuclease (I-CreI) encoded by the chloroplast 23S rRNA intron of *Chlamydomonas reinhardtii*. *Gene* **119**, 247–251.

Thompson, R.J. and Mosig, G. (1985). An ATP-dependent supercoiling topoisomerase of *Chlamydomonas reinhardtii* affects accumulation of specific chloroplast transcripts. *Nucleic Acids Res.* **13**, 873–891.

Thompson, R.J. and Mosig, G. (1987). Stimulation of a *Chlamydomonas* chloroplast promoter by novobiocin *in situ* and in *E. coli* implies regulation by torsional stress in the chloroplast DNA. *Cell* **48**, 281–287.

Thompson, R.J., Davies, J.P., and Mosig, G. (1985). 'Dark-lethality' of certain *Chlamydomonas reinhardtii* strains is prevented by dim blue light. *Plant Physiol.* **79**, 903–907.

Thyssen, C., Schlichting, R., and Giersch, C. (2001). The CO_2-concentrating mechanism in the physiological context: lowering the CO_2 supply diminishes culture growth and economises starch utilisation in *Chlamydomonas reinhardtii*. *Planta* **213**, 629–639.

Toby, A.L. and Kemp, C.L. (1975). Mutant enrichment in the colonial alga, *Eudorina elegans*. *Genetics* **81**, 243–251.

Togasaki, R.K. and Hudock, M.O. (1972). Effect(s) of inorganic arsenate on the growth of *Chlamydomonas reinhardi*. *Plant Physiol.* **49**, s52.

Toguri, T., Muto, S., Mihara, S., and Miyachi, S. (1989). Synthesis and degradation of carbonic anhydrase in a synchronized culture of *Chlamydomonas reinhardtii*. *Plant Cell Physiol.* **30**, 533–539.

Tolbert, N.E., Harrison, M., and Selph, N. (1983). Aminooxyacetate stimulation of glycolate formation and excretion by *Chlamydomonas*. *Plant Physiol.* **72**, 1075–1083.

Tomson, A.M., Demets, R., Bakker, N.P.M., Stegwee, D., and van den Ende, H. (1985). Gametogenesis in liquid cultures of *Chlamydomonas eugametos*. *J. Gen. Microbiol.* **131**, 1553–1560.

Tomson, A.M., Demets, R., Sigon, C.A.M., Stegwee, D., and van den Ende, H. (1986). Cellular interactions during the mating process in *Chlamydomonas eugametos*. *Plant Physiol.* **81**, 522–526.

Trainor, F.R. (1958). Control of sexuality in *Chlamydomonas chlamydogama*. *Am. J. Bot.* **45**, 621–626.

Trainor, F.R. (1959). A comparative study of sexual reproduction in four species of *Chlamydomonas*. *Am. J. Bot.* **46**, 65–70.

Trainor, F.R. (1960). Mating in *Chlamydomonas chlamydogama* at various temperatures under continuous illumination. *Am. J. Bot.* **47**, 482–484.

Trainor, F.R. (1961). Temperature and sexuality in *Chlamydomonas chlamydogama*. *Can. J. Bot.* **39**, 1273–1280.

Trainor, F.R. (1975). Is a reduced level of nitrogen essential for *Chlamydomonas eugametos* mating in nature? *Phycologia* **14**, 167–170.

Trainor, F.R. (1985a). Survival of algae in a desiccated soil: a 25 year study. *Phycologia* **24**, 79–82.

Trainor, F.R. (1985b). On the restoration of fertility to a sexually inactive, heterothallic species of *Chlamydomonas* (Chlorophyta). *Br. Phycol. J.* **20**, 1–4.

Trainor, F.R. and Cain, J.R. (1986). Famous algal genera. I. *Chlamydomonas*. In: *Progress in Phycological Research*, vol. 4 (F.E. Round and D.J. Chapman, Eds.), pp. 81–126. Biopress Ltd., Bristol.

Trainor, F. and Roskosky, F.G. (1963). The effect of a temperature shift on mating in *Chlamydomonas eugametos*. *Can. J. Bot.* **41**, 673–680.

Trainor, F.R., Wallett, S., and Grochowski, J. (1991). A useful algal growth medium. *J. Phycol.* **27**, 460–461.

Trebst, A. and Depka, B. (1997). Role of carotene in the rapid turnover and assembly of photosystem II in *Chlamydomonas reinhardtii*. *FEBS Lett.* **400**, 359–362.

Trebst, A., Depka, B., Jager, J., and Oettmeier, W. (2004). Reversal of the inhibition of photosynthesis by herbicides affecting hydroxyphenylpyruvate dioxygenase by plastoquinone and tocopheryl derivatives in *Chlamydomonas reinhardtii*. *Pest Manag. Sci.* **60**, 669–674.

Treier, U. and Beck, C.F. (1991). Changes in gene expression patterns during the sexual life cycle of *Chlamydomonas reinhardtii*. *Physiol. Plant.* **83**, 633–639.

Treier, U., Fuchs, S., Weber, M., Wakarchuk, W.W., and Beck, C.F. (1989). Gametic differentiation in *Chlamydomonas reinhardtii*: light dependence and gene expression patterns. *Arch. Microbiol.* **152**, 572–577.

Tremblay, S.D. and Lafontaine, J.G. (1992). Composition of nuclear dense bodies and nucleolus-associated bodies in interphase nuclei of the unicellular green alga *Chlamydomonas reinhardtii*. *Biol. Cell* **76**, 67–72.

Tremblay, S.D., Gugg, S., and Lafontaine, J.G. (1992). Ultrastructural, cytochemical and immunocytochemical investigation of the interphase nucleus in the unicellular green alga *Chlamydomonas reinhardtii*. *Biol. Cell* **76**, 73–86.

Triemer, R.E. and Brown, R.M., Jr. (1974). Cell division in *Chlamydomonas moewusii*. *J. Phycol.* **10**, 419–433.

Triemer, R.E. and Brown, R M , Jr (1975a). The ultrastructure of fertilization in *Chlamydomonas moewusii*. *Protoplasma* **84**, 315–325.

Triemer, R.E. and Brown, R.M., Jr. (1975b). Fertilization in *Chlamydomonas reinhardi*, with special reference to the structure, development, and fate of the choanoid body. *Protoplasma* **85**, 99–107.

Triemer, R.E. and Brown, R.M., Jr. (1976). Ultrastructure of meiosis in *Chlamydomonas reinhardtii*. *Br. Phycol. J.* **12**, 23–44.

Tsubo, Y. (1956). Observations on sexual reproduction in a *Chlamydomonas*. *Bot. Mag. (Tokyo)* **69**, 1–6.

Tsubo, Y. (1957). On the mating reaction of a *Chlamydomonas*, with special references to clumping and chemotaxis. *Bot. Mag. (Tokyo)* **70**, 327–334.

Tsubo, Y. (1961). Chemotaxis and sexual behavior in *Chlamydomonas*. *J. Protozool.* **8**, 114–121.

Tsubo, Y. and Matsuda, Y. (1984). Transmission of chloroplast genes in crosses between *Chlamydomonas reinhardtii* diploids: correlation with chloroplast nucleoid behavior in young zygotes. *Curr. Genet.* **8**, 223–229.

Tsuboi, M. (2002). Raman scattering anisotropy of biological systems. *J. Biomed. Opt.* **7**, 435–441.

Tsuru, S. (1973). Preservation of marine and fresh water algae by means of freezing and freeze-drying. *Cryobiology* **10**, 445–452.

Tugarinov, V.V. and Levchenko, A.B. (1976). Interaction between light-sensitivity and antibiotic resistance mutations in *Chlamydomonas reinhardi*. *Sov. Genet.* **12**, 336–342. [Russian original *Genetika* **12**, 103–110].

Tural, B. and Moroney, J.V. (2005). Regulation of the expression of photorespiratory genes in *Chlamydomonas reinhardtii*. *Can. J. Bot.* **83**, 810–819.

Turkina, M.V., Kargul, J., Blanco-Rivero, A., Villarejo, A., Barber, J., and Vener, A.V. (2006). Environmentally-modulated phosphoproteome of photosynthetic membranes in the green alga *Chlamydomonas reinhardtii*. *Mol. Cell. Proteomics* **5**, 1412–1425.

Turmel, M., Lemieux, C., and Lee, R.W. (1980). Net synthesis of chloroplast DNA throughout the synchronized vegetative cell-cycle of *Chlamydomonas*. *Curr. Genet.* **2**, 229–232.

Turmel, M., Lemieux, C., and Lee, R.W. (1981). Dispersive labelling of *Chlamydomonas* chloroplast DNA in ^{15}N–^{14}N density transfer experiments. *Curr. Genet.* **4**, 91–197.

Turmel, M., Gutell, R.R., Mercier, J.-P., Otis, C., and Lemieux, C. (1993). Analysis of the chloroplast large subunit ribosomal RNA gene from 17 *Chlamydomonas* taxa. Three internal transcribed spacers and 12 group I intron insertion sites. *J. Mol. Biol.* **232**, 446–467.

Turner, M.F. (1979). Nutrition of some marine microalgae with special reference to vitamin requirements and utilization of nitrogen and carbon sources. *J. Mar. Biol. Assoc. U.K.* **59**, 535–552.

Tüzün, I., Bayramoğlu, G., Yalçin, E., Başaran, G., Celik, G., and Arica, M.Y. (2005). Equilibrium and kinetic studies on biosorption of Hg(II), Cd(II) and Pb(II) ions onto microalgae *Chlamydomonas reinhardtii*. *J. Environ. Manage.* **77**, 85–92.

Twiss, M.R. (1990). Copper tolerance of *Chlamydomonas acidophila* (Chlorophyceaaae) isolated from acidic copper-contaminated soils. *J. Phycol.* **26**, 655–659.

Uhl, R. and Hegemann, P. (1990). Adaptation of *Chlamydomonas* phototaxis: I. A light-scattering apparatus for measuring the phototactic rate of microorganisms with high time resolution. *Cell Motil. Cytoskeleton* **15**, 230–244.

Uhlik, D.J. and Bold, H.C. (1970). Two new species of *Chlamydomonas*. *J. Phycol.* **6**, 106–110.

Umen, J.G. and Goodenough, U.W. (2001a). Control of cell division by a retinoblastoma protein homolog in *Chlamydomonas*. *Genes Dev.* **15**, 1652–1661.

Umen, J.G. and Goodenough, U.W. (2001b). Chloroplast DNA methylation and inheritance in *Chlamydomonas*. *Genes Dev.* **15**, 2585–2597.

Uniacke, J. and Zerges, W. (2007). Photosystem II assembly and repair are differentially localized in *Chlamydomonas*. *Plant Cell* **19**, 3640–3654.

Vahrenholz, C., Pratje, E., Michaelis, G., and Dujon, B. (1985). Mitochondrial DNA of *Chlamydomonas reinhardtii*: sequence and arrangement of *URF5* and the gene for cytochrome oxidase subunit I. *Mol. Gen. Genet.* **201**, 213–224.

Vahrenholz, C., Ricmen, G., Pratje, E., Dujon, B., and Michaelis, G. (1993). Mitochondrial DNA of *Chlamydomonas reinhardtii*: the structure of the ends of the linear 15.8-kb genome suggests mechanisms for DNA replication. *Curr. Genet.* **24**, 241–247.

Vaistij, F.E., Boudreau, E., Lemaire, S.D., Goldschmidt-Clermont, M., and Rochaix, J.D. (2000). Characterization of Mbb1, a nucleus-encoded tetratricopeptide-like repeat protein required for expression of the chloroplast *psbB/psbT/psbH* gene cluster in *Chlamydomonas reinhardtii*. *Proc. Natl. Acad. Sci. U. S. A.* **97**, 14813–14818.

Valle, O., Lien, T., and Knutsen, G. (1981). Fluorometric determination of DNA and RNA in *Chlamydomonas* using ethidium bromide. *J. Biochem. Biophys. Methods* **4**, 271–277.

Vallet, J.-M. and Rochaix, J.-D. (1985). Chloroplast origins of DNA replication are distinct from chloroplast ARS sequences in two green algae. *Curr. Genet.* **9**, 321–324.

Vallet, J.-M., Rahire, M., and Rochaix, J.-D. (1984). Localization and sequence analysis of chloroplast DNA sequences of *Chlamydomonas reinhardii* that promote autonomous replication in yeast. *EMBO J.* **3**, 415–421.

Vallon, O. and Wollman, F.-A. (1995). Mutations affecting O-glycosylation in *Chlamydomonas reinhardtii* cause delayed cell wall degradation and sex-limited sterility. *Plant Physiol.* **108**, 703–712.

Vallon, O., Bulté, L., Kuras, R., Olive, J., and Wollman, F.A. (1993). Extensive accumulation of an extracellular l-amino-acid oxidase during gametogenesis of *Chlamydomonas reinhardtii*. *Eur. J. Biochem.* **215**, 351–360.

Van, K. and Spalding, M.H. (1999). Periplasmic carbonic anhydrase structural gene (*Cah1*) mutant in *Chlamydomonas reinhardtii*. *Plant Physiol.* **120**, 757–764.

Van, K.J., Wang, Y.J., Nakamura, Y., and Spalding, M.H. (2001). Insertional mutants of *Chlamydomonas reinhardtii* that require elevated CO_2 for survival. *Plant Physiol.* **127**, 607–614.

van den Ende, H. (1985). Sexual agglutination in Chlamydomonads. *Adv. Microb. Physiol.* **26**, 89–123.

van den Ende, H. (1994). Vegetative and gametic development in the green alga *Chlamydomonas*. *Adv. Bot. Res.* **20**, 125–161.

van den Ende, H. and VanWinkle-Swift, K.P. (1994). Mating-type differentiation and mate selection in the homothallic *Chlamydomonas monoica*. *Curr. Genet.* **25**, 209–216.

van den Koornhuyse, N., Libessart, N., Delrue, B., Zabawinski, C., Decq, A., Iglesias, A., Carton, A., Preiss, J., and Ball, S. (1996). Control of starch composition and structure through substrate supply in the monocellular alga *Chlamydomonas reinhardtii*. *J. Biol. Chem.* **271**, 16281–16287.

van Dijk, K., Marley, K.E., Jeong, B.R., Xu, J., Hesson, J., Cerny, R.L., Waterborg, J.H., and Cerutti, H. (2005). Monomethyl histone H3 lysine 4 as an epigenetic mark for silenced euchromatin in *Chlamydomonas*. *Plant Cell* **17**, 2439–2453.

van Lis, R., Atteia, A., Mendoza-Hernández, G., and González-Halphen, D. (2003). Identification of novel mitochondrial protein components of *Chlamydomonas reinhardtii*. A proteomic approach. *Plant Physiol.* **132**, 318–330.

van Lis, R., González-Halphen, D., and Atteia, A. (2005). Divergence of the mitochondrial electron transport chains from the green alga *Chlamydomonas reinhardtii* and its colorless close relative *Polytomella sp. Biochim. Biophys. Acta* **1708**, 23–34.

van Lis, R., Mendoza-Hernández, G., Croth, G., and Attela, A. (2007). New insights into the unique structure of the F_0F_1-ATP synthase from the Chlamydomonad algae *Polytomella sp.* and *Chlamydomonas reinhardtii. Plant Physiol.* **144**, 1190–1199.

Van Wijk, K.J. (2004). Plastid proteomics. *Plant Physiol. Biochem.* **42**, 963–977.

VanDover, B. (1974). Oogamy in a species of *Chlamydomonas. J. Phycol.* **10 suppl**, 13.

VanWinkle-Swift, K.P. (1977). Maturation of algal zygotes: alternative experimental approaches for *Chlamydomonas reinhardtii* (Chlorophyceae). *J. Phycol.* **13**, 225–231.

VanWinkle-Swift, K.P. (1978). Uniparental inheritance is promoted by delayed division of the zygote in *Chlamydomonas. Nature* **275**, 749–751.

VanWinkle-Swift, K.P. (1980). A model for the rapid vegetative segregation of multiple chloroplast genomes in *Chlamydomonas*: assumptions and predictions of the model. *Curr. Genet.* **1**, 113–125.

VanWinkle-Swift, K.P. and Aubert, B. (1983). Uniparental inheritance in a homothallic alga. *Nature* **303**, 167–169.

VanWinkle-Swift, K.P. and Bauer, J.C. (1982). Self-sterile and maturation-defective mutants of the homothallic alga, *Chlamydomonas monoica* (Chlorophyceae). *J. Phycol.* **18**, 312–317.

VanWinkle-Swift, K.P. and Hahn, J.-H. (1986). The search for mating-type-limited genes in the homothallic alga *Chlamydomonas monoica. Genetics* **113**, 601–619.

VanWinkle-Swift, K.P. and Rickoll, W.L. (1997). The zygospore wall of *Chlamydomonas monoica* (Chlorophyceae): morphogenesis and evidence for the presence of sporopollenin. *J. Phycol.* **33**, 655–665.

VanWinkle-Swift, K.P. and Thuerauf, D.J. (1991). The unusual sexual preferences of a *Chlamydomonas* mutant may provide insight into mating-type evolution. *Genetics* **127**, 103–115.

VanWinkle-Swift, K., Baron, K., McNamara, A., Minke, P., Burrascano, C., and Maddock, J. (1998). The *Chlamydomonas* zygospore: mutant strains of *Chlamydomonas monoica* blocked in zygospore morphogenesis comprise 46 complementation groups. *Genetics* **148**, 131–137.

Vashishtha, M., Walther, Z., and Hall, J.L. (1996a). The kinesin-homologous protein encoded by the *Chlamydomonas FLA10* gene is associated with basal bodies and centrioles. *J. Cell Sci.* **109**, 541–549.

Vashishtha, M., Segil, G., and Hall, J.L. (1996b). Direct complementation of *Chlamydomonas* mutants with amplified YAC DNA. *Genomics* **36**, 459–467.

Vega, J.M., Garbayo, I., Dominguez, M.J., and Vigara, J. (2006). Effect of abiotic stress on photosynthesis and respiration in *Chlamydomonas reinhardtii*-Induction of oxidative stress. *Enzyme Microb. Technol.* **40**, 163–167.

Versluis, M., Schuring, F., Klis, F.M., van Egmond, P., and van den Ende, H. (1993). The sexual agglutinins in *Chlamydomonas eugametos* are sulphated glycoproteins. *J. Gen. Microbiol.* **139**, 763–767.

Viala, G. (1966). L'astaxanthine chez le *Chlamydomonas nivalis* Wille. *C. R. Acad. Sci.* **263**, 1383–1386.

Viala, G. (1967). Recherches sur le *Chlamydomonas nivalis* Wille dans les Pyrénées. *Bull. Soc. Bot. Fr.* **114**, 75–79.

Vigneault, B. and Campbell, P.G.C. (2005). Uptake of cadmium by freshwater green algae: Effects of pH and aquatic humic substances. *J. Phycol.* **41**, 55–61.

Visconti, N. and Delbrück, M. (1953). The mechanism of genetic recombination in phage. *Genetics* **38**, 5–33.

Visviki, I. and Palladino, J. (2001). Growth and cytology of *Chlamydomonas acidophila* under acidic stress *Bull. Environ. Contam. Toxicol.* **66**, 623–630.

Visviki, I. and Rachlin, J.W. (1994a). Acute and chronic exposure of *Dunaliella salina* and *Chlamydomonas bullosa* to copper and cadmium: effects on growth. *Arch. Environ. Contam. Toxicol.* **26**, 149–153.

Visviki, I. and Rachlin, J.W. (1994b). Acute and chronic exposure of *Dunaliella salina* and *Chlamydomonas bullosa* to copper and cadmium: effects of ultrastructure. *Arch. Environ. Contam. Toxicol.* **26**, 154–163.

Visviki, I. and Santikul, D. (2000). The pH tolerance of *Chlamydomonas applanata* (Volvocales, Chlorophyta). *Arch. Environ. Contam. Toxicol.* **38**, 147–151.

Vladimirescu, A. and Romanowska-Duda, Z. (2003). Polyethylene glycol (PEG) induced uptake of bacterial cells into *Chlamydomonas reinhardtii* CW 10 protoplasts: a fluorescence study. *Biologia (Bratisl.)* **58**, 633–635.

Vladimirova, M.G., Markelova, A.G., and Semenenko, B.E. (1982). Use of the cytoimmunofluorescent method to clarify localization of ribulose bisphosphate carboxylase in pyrenoids of unicellular algae. *Sov. Plant Physiol.* **29**, 725–734. [Russian original *Fiziol. Rast.* **29**, 941–950]

Vlček, D., Podstavková, S., and Miadoková, E. (1981). Study of selected characteristics of the *Chlamydomonas reinhardii* strains with altered sensitivity to UV radiation. *Biol. Plant.* **23**, 427–433.

Vlček, D., Podstavková, S., Miadoková, E., Adams, G.M.W., and Small, G.D. (1987). General characteristics, molecular and genetic analysis of two new UV-sensitive mutants of *Chlamydomonas reinhardtii*. *Mutat. Res.* **183**, 169–175.

Vlček, D., Podstavková, S., Miadoková, E., and Vlčková, V. (1991). The repair systems in green algae as compared with the present knowledge in heterotrophic microorganisms. *Arch. Protistenk.* **139**, 193–199.

Vlček, D., Podstavková, S., and Miadoková, E. (1995). Interactions between photolyase and dark repair processes in *Chlamydomonas reinhardtii*. *Mutat. Res.* **336**, 251–256.

Vlček, D., Slivková, A., Podstavková, S., and Miadoková, E. (1997). A *Chlamydomonas reinhardtii* UV-sensitive mutant *uvs15* is impaired in a gene involved in several repair pathways. *Mutat. Res.* **385**, 243–249.

Vlček, D., Ševčovičová, A., Sviežená, B., Galová, E., and Miadoková, E. (2008). *Chlamydomonas reinhardtii*: a convenient model system for the study of DNA repair in photoautotrophic eukaryotes. *Curr. Genet.* **53**, 1–22.

Vogeler, H.-P., Voigt, J., and Koenig, W.A. (1990). Polypeptide pattern of the insoluble wall component of *Chlamydomonas reinhardtii* and its variation during the vegetative cell cycle. *Plant Sci.* **71**, 119–128.

Voigt, J. (1985a). Extraction by lithium chloride of hydroxyproline-rich glycoproteins from intact cells of *Chlamydomonas reinhardii*. *Planta* **164**, 379–389.

Voigt, J. (1985b). Macromolecules released into the culture medium during the vegetative cell cycle of the unicellular green alga *Chlamydomonas reinhardii*. *Biochem. J.* **226**, 259–268.

Voigt, J. (1986). Biosynthesis and turnover of cell wall glycoproteins during the vegetative cell cycle of *Chlamydomonas reinhardii*. *Z. Naturforsch.* **41c**, 885–896.

Voigt, J. and Frank, R. (2003). 14-3-3 proteins are constituents of the insoluble glycoprotein framework of the *Chlamydomonas* cell wall. *Plant Cell* **15**, 1399–1413.

Voigt, J. and Münzner, P. (1994). Blue light-induced lethality of a cell wall-deficient mutant of the unicellular green alga *Chlamydomonas reinhardtii*. *Plant Cell Physiol.* **35**, 99–106.

Voigt, J., Mergenhagen, D., Muenzner, P., Vogeler, H.-P., and Nagel, K. (1989). Effects of light and acetate on the liberation of zoospores by a mutant strain of *Chlamydomonas reinhardtii*. *Planta* **178**, 456–462.

Voigt, J., Mergenhagen, D., Wachholz, I., Manshard, E., and Mix, M. (1990). Cell-wall abnormalities of a *Chlamydomonas reinhardtii* mutant strain under suboptimal growth conditions. *Planta* **183**, 65–68.

Voigt, J., Muenzner, P., and Vogeler, H.-P. (1991). The cell-wall glycoproteins of *Chlamydomonas reinhardtii*: analysis of the *in vitro* translation products. *Plant Sci.* **75**, 129–142.

Voigt, J., Liebich, I., Hinkelmann, B., and Kiess, M. (1996a). Immunological identification of a putative precursor of the insoluble glycoprotein framework of the *Chlamydomonas* cell wall. *Plant Cell Physiol.* **37**, 91–101.

Voigt, J., Hinkelmann, B., Liebich, I., and Mix, M. (1996b). Alteration of the cell surface during the vegetative cell cycle of the unicellular green alga *Chlamydomonas reinhardtii*. *Plant Cell Physiol.* **37**, 726–733.

Voigt, J., Hinkelmann, B., and Harris, E.H. (1997). Production of cell wall polypeptides by different cell wall mutants of the unicellular green alga *Chlamydomonas reinhardtii*. *Microbiol. Res.* **152**, 189–198.

Voigt, J., Nagel, K., and Wrann, D. (1998). A cadmium-tolerant *Chlamydomonas* mutant strain impaired in photosystem II activity. *J. Plant Physiol.* **153**, 566–573.

Voigt, J., Liebich, I., Kiess, M., and Frank, R. (2001). Subcellular distribution of 14-3-3 proteins in the unicellular green alga *Chlamydomonas reinhardtii*. *Eur. J. Biochem.* **268**, 6449–6457.

Voigt, J., Stevanovic, S., Schirle, M., Fausel, M., Maier, J., Adam, K.-H., and Marquardt, O. (2004). A 14-3-3 protein of *Chlamydomonas reinhardtii* associated with the endoplasmic reticulum: nucleotide sequence of the cDNA and the corresponding gene and derived amino acid sequence. *Biochim. Biophys. Acta* **1679**, 180–194.

Voigt, J., Woestemeyer, J., and Frank, R. (2007). The chaotrope-soluble glycoprotein GP2 is a precursor of the insoluble glycoprotein framework of the *Chlamydomonas* cell wall. *J. Biol. Chem.* **282**, 30381–30392.

von Gromoff, E.D. and Beck, C.F. (1993). Genes expressed during sexual differentiation of *Chlamydomonas reinhardtii*. *Mol. Gen. Genet.* **241**, 415–421.

von Gromoff, E.D., Trieier, U., and Beck, C.F. (1989). Three light-inducible heat shock genes of *Chlamydomonas reinhardtii*. *Mol. Cell. Biol.* **9**, 3911–3918.

Vyhnalek, V. (1990). Phosphate-limited growth of *Chlamydomonas geitleri* Ettl Volvocales Chlorophyceae in a cyclostat. *Arch. Hydrobiol. Suppl.* **87**, 43–56.

Waddell, J., Wang, X.-M., and Wu, M. (1984). Electron microscopic localization of the chloroplast DNA replicative origins in *Chlamydomonas reinhardii*. *Nucleic Acids Res.* **12**, 3843–3856.

Waffenschmidt, S., Spessert, R., and Jaenicke, L. (1988). Oligosaccharide side chains of wall molecules are essential for cell-wall lysis in *Chlamydomonas reinhardtii*. *Planta* **175**, 513–519.

Waffenschmidt, S., Woessner, J.P., Beer, K., and Goodenough, U.W. (1993). Isodityrosine cross-linking mediates insolubilization of cell walls in *Chlamydomonas*. *Plant Cell* **5**, 809–820.

Waffenschmidt, S., Kusch, T., and Woessner, J.P. (1999). A transglutaminase immunologically related to tissue transglutaminase catalyzes cross-linking of cell wall proteins in *Chlamydomonas reinhardtii*. *Plant Physiol.* **121**, 1003–1015.

Wagner, V., Fiedler, M., Markert, C., Hippler, M., and Mittag, M. (2004). Functional proteomics of circadian expressed proteins from *Chlamydomonas reinhardtii*. *FEBS Lett.* **559**, 129–135.

Wagner, V., Gessner, G., and Mittag, M. (2005). Functional proteomics: a promising approach to find novel components of the circadian system. *Chronobiol. Int.* **22**, 403–415.

Wagner, V., Gessner, G., Heiland, I., Kaminski, M., Hawat, S., Scheffler, K., and Mittag, M. (2006). Analysis of the phosphoproteome of *Chlamydomonas reinhardtii* provides new insights into various cellular pathways. *Eukaryotic Cell* **5**, 457–468.

Wagner, V., Ullmann, K., Mollwo, A., Kaminski, M., Mittag, M., and Kreimer, G. (2008). The phosphoproteome of a *Chlamydomonas reinhardtii* eyespot fraction includes key proteins of the light signaling pathway. *Plant Physiol.* **146**, 772–788.

Wakasugi, T., Nagai, T., Kapoor, M., Sugita, M., Ito, M., Ito, S., Tsudzuki, J., Nakashima, K., Tsudzuki, T., Suzuki, Y., Hamada, A., Ohta, T., Inamura, A., Yoshinaga, K., and Sugiura, M. (1997). Complete nucleotide sequence of the chloroplast genome from the green alga *Chlorella vulgaris*: the existence of genes possibly involved in chloroplast division. *Proc. Natl. Acad. Sci. U. S. A.* **94**, 5967–5972.

Walker, T.L., Collet, C., and Purton, S. (2005). Algal transgenics in the genomic era. *J. Phycol.* **41**, 1077–1093.

Walne, P.L. (1967). The effects of colchicine on cellular organization in *Chlamydomonas*. II. Ultrastructure. *Am. J. Bot.* **54**, 564–577.

Walne, P.L. and Arnott, H.J. (1967). The comparative ultrastructure and possible function of eyespots: *Euglena granulata* and *Chlamydomonas eugametos*. *Planta* **77**, 325–353.

Waltenberger, H., Schneid, C., Grosch, J.O., Bareiss, A., and Mittag, M. (2001). Identification of target mRNAs for the clock-controlled RNA-binding protein Chlamy 1 from *Chlamydomonas reinhardtii*. *Mol. Gen. Genet.* **265**, 180–188.

Walther, Z., Vashishtha, M., and Hall, J.L. (1994). The *Chlamydomonas FLA10* gene encodes a novel kinesin-homologous protein. *J. Cell Biol.* **126**, 175–188.

Wang, D., Kong, D.D., Wang, Y.D., Hu, Y., He, Y.K., and Sun, J.S. (2003). Isolation of two plastid division *ftsZ* genes from *Chlamydomonas reinhardtii* and its evolutionary implication for the role of FtsZ in plastid division. *J. Exp. Bot.* **54**, 1115–1116.

Wang, L. and Roossinck, M.J. (2006). Comparative analysis of expressed sequences reveals a conserved pattern of optimal codon usage in plants. *Plant Mol. Biol.* **61**, 699–710.

Wang, Q. and Snell, W.J. (2003). Flagellar adhesion between mating type *plus* and mating type *minus* gametes activates a flagellar protein-tyrosine kinase during fertilization in *Chlamydomonas*. *J. Biol. Chem.* **278**, 32936–32942.

Wang, S.B., Hu, Q., Sommerfeld, M., and Chen, F. (2004). Cell wall proteomics of the green alga *Haematococcus pluvialis* (Chlorophyceae). *Proteomics* **4**, 692–708.

Wang, S.B., Chen, F., Sommerfeld, M., and Hu, Q. (2005). Isolation and proteomic analysis of cell wall-deficient *Haematococcus pluvialis* mutants. *Proteomics* **5**, 4839–4851.

Wang, S.C., Schnell, R.A., and Lefebvre, P.A. (1998). Isolation and characterization of a new transposable element in *Chlamydomonas reinhardtii*. *Plant Mol. Biol.* **38**, 681–687.

Wang, S.L. and Liu, X.Q. (1997). Identification of an unusual intein in chloroplast ClpP protease of *Chlamydomonas eugametos*. *J. Biol. Chem.* **272**, 11869–11873.

Wang, W.-Y. (1978). Genetic control of chlorophyll biosynthesis in *Chlamydomonas reinhardtii*. *Int. Rev. Cytol. Suppl.* **8**, 335–364.

Wang, W.-Y., Wang, W.L., Boynton, J.E., and Gillham, N.W. (1974). Genetic control of chlorophyll biosynthesis in *Chlamydomonas*. Analysis of mutants at two loci mediating the conversion of protoporphyrin-IX to magnesium protoprophyrin *J. Cell Biol.* **63**, 806–823.

Wang, W.-Y., Boynton, J.E., Gillham, N.W., and Gough, S. (1975). Genetic control of chlorophyll biosynthesis in *Chlamydomonas*: analysis of a mutant affecting synthesis of delta-aminolevulinic acid. *Cell* **6**, 75–84.

Wang, X.D., Liu, X.J., Yang, S., Li, A.L., and Yang, Y.L. (2007). Removal and toxicological response of triazophos by *Chlamydomonas reinhardtii*. *Bull. Environ. Contam. Toxicol.* **78**, 67–71.

Wang, X.-M., Chang, C.H., Waddell, J., and Wu, M. (1984). Cloning and delimiting one chloroplast DNA replicative origin of *Chlamydomonas*. *Nucleic Acids Res.* **12**, 3857–3872.

Wang, Y. and Spalding, M.H. (2006). An inorganic carbon transport system responsible for acclimation specific to air levels of CO_2 in *Chlamydomonas reinhardtii*. *Proc. Natl. Acad. Sci. U. S. A.* **103**, 10110–10115.

Wang, Z.F., Yang, J., Nie, Z.Q., and Wu, M. (1991). Purification and characterization of a gamma-like DNA polymerase from *Chlamydomonas reinhardtii*. *Biochemistry* **30**, 1127–1131.

Ward, M.A. (1970a). Whole cell and cell-free hydrogenases of algae. *Phytochemistry* **9**, 259–266.

Ward, M.A. (1970b). Adaptation of hydrogenase in cell-free preparations from *Chlamydomonas*. *Phytochemistry* **9**, 267–274.

Wargo, M.J., Dymek, E.E., and Smith, E.F. (2005). Calmodulin and PF6 are components of a complex that localizes to the C1 microtubule of the flagellar central apparatus. *J. Cell Sci.* **118**, 4655–4665.

Warr, J.R. (1968). A mutant of *Chlamydomonas reinhardi* with abnormal cell division. *J. Gen. Microbiol.* **52**, 243–251.

Warr, J.R., McVittie, A., Randall, J., and Hopkins, J.M. (1966). Genetic control of flagellar structure in *Chlamydomonas reinhardii*. *Genet. Res.* **7**, 335–351.

Watanabe, M.M. (2005). Freshwater culture media. In: *Algal Culturing Techniques* (R.A. Andersen, Ed.), pp. 13–200. Elsevier, Amsterdam.

Watanabe, F., Nakano, Y., Tamura, Y., and Yamanaka, H. (1991). Vitamin B12 metabolism in a photosynthesizing green alga, *Chlamydomonas reinhardtii*. *Biochim. Biophys. Acta* **1075**, 36–41.

Waterborg, J.H. (1998). Dynamics of histone acetylation in *Chlamydomonas reinhardtii*. *J. Biol. Chem.* **273**, 27602–27609.

Waterborg, J.H., Robertson, A.J., Tatar, D.L., Borza, C.M., and Davie, J.R. (1995). Histones of *Chlamydomonas reinhardtii* – Synthesis, acetylation, and methylation. *Plant Physiol.* **109**, 393–407.

Wattebled, F., Buleón, A., Bouchet, B., Ral, J.P., Lienard, L., Delvalle, D., Binderup, K., Dauvillée, D., Ball, S., and D'Hulst, C. (2002). Granule-bound starch synthase I – A major enzyme involved in the biogenesis of B-crystallites in starch granules. *Eur. J. Biochem.* **269**, 3810–3820.

Wattebled, F., Ral, J.P., Dauvillée, D., Myers, A.M., James, M.G., Schlichting, R., Giersch, C., Ball, S., and D'Hulst, C. (2003). STA11, a *Chlamydomonas reinhardtii* locus required for normal starch granule biogenesis, encodes disproportionating enzyme. Further evidence for a function of α-1,4 glucanotransferases during starch granule biosynthesis in green algae. *Plant Physiol.* **132**, 137–145.

Weeks, D.P. and Collis, P.S. (1976). Induction of microtubule protein synthesis in *Chlamydomonas reinhardi* during flagellar regeneration. *Cell* **9**, 15–27.

Weeks, D.P. and Collis, P.S. (1979). Induction and synthesis of tubulin during the cell cycle and life cycle of *Chlamydomonas reinhardi*. *Dev. Biol.* **69**, 400–407.

Weeks, D.P., Collis, P., and Gealt, M.A. (1977). Control of induction of tubulin synthesis in *Chlamydomonas reinhardi*. *Nature* **268**, 667–668.

Wegener, D., Treier, U., and Beck, C.F. (1989). Procedures for the generation of mature *Chlamydomonas reinhardtii* zygotes for molecular and biochemical analyses. *Plant Physiol.* **90**, 512–515.

Weger, H.G., Chadderton, A.R., Lin, M., Guy, R.D., and Turpin, D.H. (1990a). Cytochrome and alternative pathway respiration during transient ammonium assimilation by *N*-limited *Chlamydomonas reinhardtii*. *Plant Physiol.* **94**, 1131–1136.

Weger, H.G., Guy, R.D., and Turpin, D.H. (1990b). Cytochrome and alternative pathway respiration in green algae. *Plant Physiol.* **93**, 356–360.

Weinberger, P., De Chacin, C., and Czuba, M. (1987). Effects of nonyl phenol, a pesticide surfactant, on some metabolic processes of *Chlamydomonas segnis*. *Can. J. Bot.* **65**, 696–702.

Weiss, M. and Pick, U. (1996). Primary structure and effect of pH on the expression of the plasma membrane H⁺-ATPase from *Dunaliella acidophila* and *Dunaliella salina*. *Plant Physiol.* **112**, 1693–1702.

Weiss, R.L. (1983a). Coated vesicles in the contractile vacuole-/mating structure region of *Chlamydomonas*. *J. Ultrastruct. Res.* **85**, 33–44.

Weiss, R.L. (1983b). Fine structure of the snow alga (*Chlamydomonas nivalis*) and associated bacteria. *J. Phycol.* **19**, 200–204.

Weiss, R.L. (1984). Ultrastructure of the flagellar roots in *Chlamydomonas* gametes. *J. Cell Sci.* **67**, 133–143.

Weiss, R.L., Goodenough, D.A., and Goodenough, U.W. (1977a). Membrane particle arrays associated with the basal body and with contractile vacuole secretion in *Chlamydomonas*. *J. Cell Biol.* **72**, 133–143.

Weiss, R.L., Goodenough, D.A., and Goodenough, U.W. (1977b). Membrane differentiations at sites specialized for cell fusion. *J. Cell Biol.* **72**, 144–160.

Weiss-Magasic, C., Lustigman, B., and Lee, L.H. (1997). Effect of mercury on the growth of *Chlamydomonas reinhardtii*. *Bull. Environ. Contam. Toxicol.* **59**, 828–833.

Weissig, H. and Beck, C.F. (1991). Action spectrum for the light-dependent step in gametic differentiation of *Chlamydomonas reinhardtii*. *Plant Physiol.* **9**, 118–121.

Wells, R. and Sager, R. (1971). Denaturation and the renaturation kinetics of chloroplast DNA from *Chlamydomonas reinhardi*. *J. Mol. Biol.* **58**, 611–622.

Wemmer, K.A. and Marshall, W.F. (2007). Flagellar length control in *Chlamydomonas* – A paradigm for organelle size regulation. *Int. Rev. Cytol.* **260**, 175–212.

Werner, R. and Mergenhagen, D. (1998). Mating type determination of *Chlamydomonas reinhardtii* by PCR. *Plant Mol. Biol. Rep.* **16**, 295–299.

Werner, T.P., Amrhein, N., and Freimoser, F.M. (2007). Inorganic polyphosphate occurs in the cell wall of *Chlamydomonas reinhardtii* and accumulates during cytokinesis. *BMC Plant Biol.* **7**, 51.

Wetherell, D.F. (1958). Obligate phototrophy in *Chlamydomonas eugametos*. *Physiol. Plant.* **11**, 260–274.

Wetherell, D.F. and Krauss, R.W. (1956). Colchicine-induced polyploidy in *Chlamydomonas*. *Science* **124**, 25–26.

White, R.A., Wolfe, G.R., Komine, Y., and Hoober, J.K. (1996). Localization of light-harvesting complex apoproteins in the chloroplast and cytoplasm during greening of *Chlamydomonas reinhardtii* at 38C. *Photosynth. Res.* **47**, 267–280.

Whitehouse, H.L.K. (1950). Mapping chromosome centromeres by the analysis of unordered tetrads. *Nature* **165**, 893.

Whitehouse, H.L.K. (1957). Mapping chromosome centromeres from tetratype frequencies. *J. Genet.* **55**, 348–360.

Whitelegge, J.P. (2003). Thylakoid membrane proteomics. *Photosynth. Res.* **78**, 265–277.

Whiteway, M.S. and Lee, R.W. (1977). Chloroplast DNA content increases with nuclear ploidy in *Chlamydomonas*. *Mol. Gen. Genet.* **157**, 11–15.

Wiedeman, V.E., Walne, P.L., and Trainor, F.R. (1964). A new technique for obtaining axenic cultures of algae. *Can. J. Bot.* **42**, 958–959.

Wiese, L. (1981). On the evolution of anisogamy from isogamous monoecy and on the origin of sex. *J. Theor. Biol.* **89**, 573–580.

Wiese, L. and Hayward, P.C. (1972). On sexual agglutination and mating-type substances in isogamous dioecious Chlamydomonads III. The sensitivity of sex cell contact to various enzymes.. *Am. J. Bot.* **59**, 530–536.

Wiese, L. and Mayer, R.A. (1982). Unilateral tunicamycin sensitivity of gametogenesis in dioecious isogamous *Chlamydomonas* species. *Gamete Res.* **5**, 1–9.

Wiese, L. and Shoemaker, D.W. (1970). On sexual agglutination and mating type substances (gamones) in isogamous heterothallic Chlamydomonads. II. The effect of concanavalin A upon the mating type reaction. *Biol. Bull.* **138**, 88–95.

Wiese, I. and Wiese, W. (1977). On speciation by evolution of gametic incompatibility: a model case in *Chlamydomonas. Am. Nat.* **111**, 733–742.

Wiese, L., Wiese, W., and Edwards, D.A. (1979). Inducible anisogamy and the evolution of oogamy from isogamy. *Ann. Bot.* **44**, 131–139.

Wiese, L., Williams, L.A., and Baker, D.L. (1983). A general and fundamental molecular bipolarity of the sex cell contact mechanism as revealed by tunicamycin and bacitracin in *Chlamydomonas. Am. Nat.* **122**, 806–816.

Wildner, G.F., Heisterkamp, U., and Trebst, A. (1990). Herbicide cross-resistance and mutations of the *psbA* gene in *Chlamydomonas reinhardtii. Z. Naturforsch.* **45c**, 1142–1150.

Wilkerson, C.G., King, S.M., and Witman, G.B. (1994). Molecular analysis of the gamma heavy chain of *Chlamydomonas* flagellar outer-arm dynein. *J. Cell Sci.* **107**, 497–506.

Wilkerson, C.G., King, S.M., Koutoulis, A., Pazour, G.J., and Witman, G.B. (1995). The 78,000 M(r) intermediate chain of *Chlamydomonas* outer arm dynein is a WD-repeat protein required for arm assembly. *J. Cell Biol.* **129**, 169–178.

Williams, B.D., Velleca, M.A., Curry, A.M., and Rosenbaum, J.L. (1989). Molecular cloning and sequence analysis of the *Chlamydomonas* gene coding for radial spoke protein 3: flagellar mutation *pf-14* is an ochre allele. *J. Cell Biol.* **109**, 235–245.

Williams, W.E., Gorton, H.L., and Vogelmann, T.C. (2003). Surface gas-exchange processes of snow algae. *Proc. Natl. Acad. Sci. U. S. A.* **100**, 562–566.

Willmund, F. and Schroda, M. (2005). Heat shock protein 90C is a bona fide Hsp90 that interacts with plastidic HSP70B in *Chlamydomonas reinhardtii. Plant Physiol.* **138**, 2310–2322.

Wilski, S., Johanningmeier, U., Hertel, S., and Oettmeier, W. (2006). Herbicide binding in various mutants of the photosystem II D1 protein of *Chlamydomonas reinhardtii. Pestic. Biochem. Physiol.* **84**, 157–164.

Wilson, N.F., O'Connell, J.S., Lu, M., and Snell, W.J. (1999). Flagellar adhesion between *mt*(+) and *mt*(−) *Chlamydomonas* gametes regulates phosphorylation of the *mt*(+)-specific homeodomain protein GSP1. *J. Biol. Chem.* **274**, 34383–34388.

Winder, T. and Spalding, M.H. (1988). Imazaquin and chlorsulfuron resistance and cross resistance in mutants of *Chlamydomonas reinhardtii. Mol. Gen. Genet.* **213**, 394–399.

Witman, G.B., Wirschell, M., Pazour, G., Yoda, A., Hirono, M., and Kamiya, R. (2004). Oda5p, a novel axonemal protein required for assembly of the outer dynein arm and an associated adenylate kinase. *Mol. Biol. Cell* **15**, 2729–2741.

Wiseman, A., Gillham, N.W., and Boynton, J.E. (1977). Nuclear mutations affecting mitochondrial structure and function in *Chlamydomonas. J. Cell Biol.* **73**, 56–77.

Witman, G.B. (1975). The site of *in vivo* assembly of flagellar microtubules. *Ann. N. Y. Acad. Sci.* **253**, 178–191.

Witman, G.B. (1986). Isolation of *Chlamydomonas* flagella and flagellar axonemes. *Methods Enzymol.* **134**, 180–190.

Witman, G.B. (1993). *Chlamydomonas* phototaxis. *Trends Cell Biol.* **3**, 403–408.

Witman, G.B., Carlson, K., Berliner, J., and Rosenbaum, J.L. (1972). *Chlamydomonas* flagella. I. Isolation and electrophoretic analysis of microtubules, matrix, membranes, and mastigonemes. *J. Cell Biol.* **54**, 507–539.

Witman, G.B., Plummer, J., and Sander, G. (1978). *Chlamydomonas* flagellar mutants lacking radial spokes and central tubules. Structure, composition and function of specific axonemal components. *J. Cell Biol.* **76**, 729–747.

Wodzinski, R.S. and Alexander, M. (1978). Effect of sulfur dioxide on algae. *J. Environ. Qual.* **7**, 358–360.

Wodzinski, R.S. and Alexander, M. (1980). Effects of nitrogen dioxide on algae. *J. Environ. Qual.* **9**, 34–36.

Woelfle, M.A., Thompson, R.J., and Mosig, G. (1993). Roles of novobiocin-sensitive topoisomerases in chloroplast DNA replication in *Chlamydomonas reinhardtii. Nucleic Acids Res.* **21**, 4231–4238.

Woessner, J.P. and Goodenough, U.W. (1989). Molecular characterization of a zygote wall protein: an extensin-like molecule in *Chlamydomonas reinhardtii. Plant Cell* **1**, 901–911.

Woessner, J.P. and Goodenough, U.W. (1992). Zygote and vegetative cell wall proteins in *Chlamydomonas reinhardtii* share a common epitope, (SerPro)x. *Plant Sci.* **83**, 65–76.

Woessner, J.P. and Goodenough, U.W. (1994). Volvocine cell walls and their constituent glycoproteins: an evolutionary perspective. *Protoplasma* **181**, 245–258.

Woessner, J.P., Masson, A., Harris, E.H., Bennoun, P., Gillham, N.W., and Boynton, J.E. (1984). Molecular and genetic analysis of the chloroplast ATPase of *Chlamydomonas. Plant Mol. Biol.* **3**, 177–190.

Woessner, J.P., Gillham, N.W., and Boynton, J.E. (1986). The sequence of the chloroplast *atpB* gene and its flanking regions in *Chlamydomonas reinhardtii. Gene* **44**, 17–28.

Woessner, J.P., Gillham, N.W., and Boynton, J.E. (1987). Chloroplast genes encoding subunits of the H^+-ATPase complex of *Chlamydomonas reinhardtii* are rearranged compared to higher plants: sequence of the *atpE* gene and location of the *atpF* and *atpI* genes. *Plant Mol. Biol.* **8**, 151–158.

Woessner, J.P., Molendijk, A.J., van Egmond, P., Klis, F.M., Goodenough, U.W., and Haring, M.A. (1994). Domain conservation in several volvocalean cell wall proteins. *Plant Mol. Biol.* **26**, 947–960.

Wolf, K.W. (1995). Centromere structure and chromosome number in mitosis of the colourless phytoflagellate *Polytoma papillatum* (Chlorophyceae, Volvocales, Chlamydomonadaceae). *Genome* **38**, 1249–1254.

Wolfe, G.R., Park, H., Sharp, W.P., and Hoober, J.K. (1997). Light-harvesting complex apoproteins in cytoplasmic vacuoles in *Chlamydomonas reinhardtii* (Chlorophyta). *J. Phycol.* **33**, 377–386.

Wollman, F.-A., Olive, J., Bennoun, P., and Recouvreur, M. (1980). Organization of the photosystem II centers and their associated antennae in the thylakoid membranes: a comparative ultrastructural, biochemical, and biophysical study of *Chlamydomonas* wild type and mutants lacking in photosystem II reaction centers. *J. Cell Biol.* **87**, 728–735.

Wollman, F.A., Minai, L., and Nechushtai, R. (1999). The biogenesis and assembly of photosynthetic proteins in thylakoid membranes. *Biochim. Biophys. Acta* **1411**, 21–85.

Wostrikoff, K., Choquet, Y., Wollman, F.A., and Girard-Bascou, J. (2001). TCA1, a single nuclear-encoded translational activator specific for *petA* mRNA in *Chlamydomonas reinhardtii* chloroplast. *Genetics* **159**, 119–132.

Wright, R.L., Chojnacki, B., and Jarvik, J.W. (1983). Abnormal basal-body number, location, and orientation in a striated fiber-defective mutant of *Chlamydomonas reinhardtii. J. Cell Biol.* **96**, 1697–1707.

Wright, R.L., Salisbury, J., and Jarvik, J.W. (1985). A nucleus-basal body connector in *Chlamydomonas reinhardtii* that may function in basal body localization or segregation. *J. Cell Biol.* **101**, 1903–1912.

Wright, R.L., Adler, S.A., Spanier, J.G., and Jarvik, J.W. (1989). Nucleus-basal body connector in *Chlamydomonas*: evidence for a role in basal body segregation and against essential roles in mitosis or in determining cell polarity. *Cell Motil. Cytoskeleton* **14**, 516–526.

Wu, H.Y. and Kuchka, M.R. (1995). A nuclear suppressor overcomes defects in the synthesis of the chloroplast *psbD* gene product caused by mutations in two distinct nuclear genes of *Chlamydomonas*. *Curr. Genet.* **27**, 263–269.

Wu, M., Lou, J.K., Chang, D.Y., Chang, C.H., and Nie, Z.Q. (1986a). Structure and function of a chloroplast DNA replication origin of *Chlamydomonas reinhardtii*. *Proc. Natl. Acad. Sci. U. S. A.* **83**, 6761–6765.

Wu, M., Kong, X.F., and Kung, S.D. (1986b). Prokaryotic promoters in the chloroplast DNA replication origin of *Chlamydomonas reinhardtii*. *Curr. Genet.* **10**, 819–822.

Wu-Scharf, D., Jeong, B.R., Zhang, C.M., and Cerutti, H. (2000). Transgene and transposon silencing in *Chlamydomonas reinhardtii* by a DEAH-Box RNA helicase. *Science* **290**, 1159–1162.

Wurtz, E.A., Boynton, J.E., and Gillham, N.W. (1977). Perturbation of chloroplast DNA amounts and chloroplast gene transmission in *Chlamydomonas reinhardtii* by 5-fluorodeoxyuridine. *Proc. Natl. Acad. Sci. U. S. A.* **74**, 4552–4556.

Wurtz, E.A., Sears, B.B., Rabert, D.K., Shepherd, H.S., Gillham, N.W., and Boynton, J.E. (1979). A specific increase in chloroplast gene mutations following growth of *Chlamydomonas* in 5-fluorodeoxyuridine. *Mol. Gen. Genet.* **170**, 235–242.

Wykoff, D.D., Davies, J.P., Melis, A., and Grossman, A.R. (1998). The regulation of photosynthetic electron transport during nutrient deprivation in *Chlamydomonas reinhardtii*. *Plant Physiol.* **117**, 129–139.

Xiang, Y., Zhang, J., and Weeks, D.P. (2001). The *Cia5* gene controls formation of the carbon concentrating mechanism in *Chlamydomonas reinhardtii*. *Proc. Natl. Acad. Sci. U. S. A.* **98**, 5341–5346.

Xie, Z.Y. and Merchant, S. (1996). The plastid-encoded *ccsA* gene is required for heme attachment to chloroplast c-type cytochromes. *J. Biol. Chem.* **271**, 4632–4639.

Xie, Z., Dreyfuss, B.W., Kuras, R., Wollman, F.A., Girard-Bascou, J., and Merchant, S. (1998). Genetic analysis of chloroplast *c*-type cytochrome assembly in *Chlamydomonas reinhardtii*: one chloroplast locus and at least four nuclear loci are required for heme attachment. *Genetics* **148**, 681–692.

Xiong, L. and Sayre, R.T. (2004). Engineering the chloroplast encoded proteins of *Chlamydomonas*. *Photosynth. Res.* **80**, 411–419.

Xu, R., Bingham, S.E., and Webber, A.N. (1993). Increased mRNA accumulation in a *psaB* frame-shift mutant of *Chlamydomonas reinhardtii* suggests a role for translation in *psaB* mRNA stability. *Plant Mol. Biol.* **22**, 465–474.

Yamaguchi, K., Prieto, S., Beligni, M.V., Haynes, P.A., McDonald, W.H., Yates, J.R.I., and Mayfield, S.P. (2002). Proteomic characterization of the small subunit of *Chlamydomonas reinhardtii* chloroplast ribosome: identification of a novel S1 domain-containing protein and unusually large orthologs of bacterial S2, S3, and S5. *Plant Cell* **14**, 2957–2974.

Yamaguchi, K., Beligni, M.V., Prieto, S., Haynes, P.A., McDonald, W.H., Yates, J.R., and Mayfield, S.P. (2003). Proteomic characterization of the *Chlamydomonas reinhardtii* chloroplast ribosome: identification of proteins unique to the 70S ribosome. *J. Biol. Chem.* **278**, 33774–33785.

Yang, C. and Yang, P. (2006). The flagellar motility of *Chlamydomonas pf25* mutant lacking an AKAP-binding protein is overtly sensitive to medium conditions. *Mol. Biol. Cell* **17**, 227–238.

Yang, P., Yang, C., Sale, W.S., Yang, W., Moroney, J.V., and Moore, T.S. (2004). Flagellar radial spoke protein 2 is a calmodulin binding protein required for motility in *Chlamydomonas reinhardtii*. *Eukaryotic Cell* **3**, 72–81.

Yang, P., Diener, D.R., Yang, C., Kohno, T., Pazour, G.J., Dienes, J.M., Agrin, N.S., King, S.M., Sale, W.S., Kamiya, R., Rosenbaum, J.L., and Witman, G.B. (2006). Radial spoke proteins of *Chlamydomonas* flagella. *J. Cell Sci.* **119**, 1165–1174.

Yang, S.Y. and Tsuboi, M. (1999). Polarizing microscopy of eyespot of *Chlamydomonas*: *in situ* observation of its location, orientation, and multiplication. *Biospectroscopy* **5**, 93–100.

Yannai, Y., Epel, B.L., and Neumann, J. (1976). Photophosphorylation in stable chloroplast fragments from the alga *Chlamydomonas reinhardi*. *Plant Sci. Lett.* **7**, 295–304.

Yehudai-Resheff, S., Zimmer, S.L., Komine, Y., and Stern, D.B. (2007). Integration of chloroplast nucleic acid metabolism into the phosphate deprivation response in *Chlamydomonas reinhardtii*. *Plant Cell* **19**, 1023–1038.

Yildiz, F.H., Davies, J.P., and Grossman, A.R. (1994). Characterization of sulfate transport in *Chlamydomonas reinhardtii* during sulfur-limited and sulfur-sufficient growth. *Plant Physiol.* **104**, 981–987.

Yildiz, F.H., Davies, J.P., and Grossman, A. (1996). Sulfur availability and the *SAC1* gene control adenosine triphosphate sulfurylase gene expression in *Chlamydomonas reinhardtii*. *Plant Physiol.* **112**, 669–675.

Yohn, C.B., Cohen, A., Danon, A., and Mayfield, S.P. (1996). Altered mRNA binding activity and decreased translation initiation in a nuclear mutant lacking translation of the chloroplast *psbA* mRNA. *Mol. Cell Biol.* **16**, 3560–3566.

Yoshida, K., Igarashi, E., Mukai, M., Hirata, K., and Miyamoto, K. (2003). Induction of tolerance to oxidative stress in the green alga, *Chlamydomonas reinhardtii*, by abscisic acid. *Plant Cell Environ.* **26**, 451–457.

Yoshida, K., Igarashi, E., Wakatsuki, E., Miyamoto, K., and Hirata, K. (2004). Mitigation of osmotic and salt stresses by abscisic acid through reduction of stress-derived oxidative damage in *Chlamydomonas reinhardtii*. *Plant Sci.* **167**, 1335–1341.

Yoshimura, K., Matsuo, Y., and Kamiya, R. (2003). Gravitaxis in *Chlamydomonas reinhardtii* studied with novel mutants. *Plant Cell Physiol.* **44**, 1112–1118.

Yoshimura, S., Ranjbar, R., Inoue, R., Katsuda, T., and Katoh, S. (2006). Effective utilization of transmitted light for astaxanthin production by *Haematococcus pluvialis*. *J. Biosci. Bioeng.* **102**, 97–101.

Yoshioka, S., Taniguchi, F., Miura, K., Inoue, T., Yamano, T., and Fukuzawa, H. (2004). The novel Myb transcription factor LCR1 regulates the CO_2-responsive gene *Cah1*, encoding a periplasmic carbonic anhydrase in *Chlamydomonas reinhardtii*. *Plant Cell* **16**, 1466–1477.

Yu, W., Zhang, D., and Spreitzer, R.J. (1992). Sequences of the *Chlamydomonas reinhardtii* chloroplast genes encoding tRNASer and ribosomal protein L20. *Plant Physiol.* **100**, 1079–1080.

Zabawinski, C., van den Koornhuyse, N., D'Hulst, C., Schlichting, R., Giersch, C., Delrue, B., Lacroix, J.M., Preiss, J., and Ball, S. (2001). Starchless mutants of *Chlamydomonas reinhardtii* lack the small subunit of a heterotetrameric ADP-glucose pyrophosphorylase. *J. Bacteriol.* **183**, 1069–1077.

Zachleder, V. (1984). Optimization of nucleic acids assay in green and blue-green algae: extraction procedures and the light-activated diphenylamine reaction for DNA. *Arch. Hydrobiol. Suppl.* **67**, 313–328.

Zachleder, V. and van den Ende, H. (1992). Cell cycle events in the green alga *Chlamydomonas eugametos* and their control by environmental factors. *J. Cell Sci.* **102**, 469–474.

Zamora, I. and Marshall, W.F. (2005). A mutation in the centriole-associated protein centrin causes genomic instability via increased chromosome loss in *Chlamydomonas reinhardtii*. *BMC Biol.* **3**, 15.

Zamora, I., Feldman, J.L., and Marshall, W.F. (2004). PCR-based assay for mating type and diploidy in *Chlamydomonas*. *BioTechniques* **37**, 534–536.

Žárský, V., Kalina, T., and Sulek, J. (1985). Notes on the sexual reproduction of *Chlamydomonas geitleri* Ettl. *Arch. Protistenk.* **130**, 343–353.

Zerges, W., Girard-Bascou, J., and Rochaix, J.-D. (1997). Translation of the chloroplast *psbC* mRNA is controlled by interactions between its 5′ leader and the nuclear loci *TBC1* and *TBC3* in *Chlamydomonas reinhardtii*. *Mol. Cell. Biol.* **17**, 3440–3448.

Zerges, W., Auchincloss, A.H., and Rochaix, J.D. (2003). Multiple translational control sequences in the 5′ leader of, the chloroplast *psbC* mRNA interact with nuclear gene products in *Chlamydomonas reinhardtii*. *Genetics* **163**, 895–904.

Zeyl, C., Bell, G., and Da Silva, J. (1994). Transposon abundance in sexual and asexual populations of *Chlamydomonas reinhardtii*. *Evolution* **48**, 1406–1409.

Zhang, C.M., Wu-Scharf, D., Jeong, B., and Cerutti, H. (2002). A WD40-repeat containing protein, similar to a fungal co-repressor, is required for transcriptional gene silencing in *Chlamydomonas*. *Plant J.* **31**, 25–36.

Zhang, H. and Mitchell, D.R. (2004). Cpc1, a *Chlamydomonas* central pair protein with an adenylate kinase domain. *J. Cell Sci.* **117**, 4179–4188.

Zhang, H., Herman, P.L., and Weeks, D.P. (1994). Gene isolation through genomic complementation using an indexed library of *Chlamydomonas reinhardtii* DNA. *Plant Mol. Biol.* **24**, 663–672.

Zhang, L.P., Niyogi, K.K., Baroli, I., Nemson, J.A., Grossman, A.R., and Melis, A. (1997). DNA insertional mutagenesis for the elucidation of a Photosystem II repair process in the green alga *Chlamydomonas reinhardtii*. *Photosynth. Res.* **53**, 173–184.

Zhang, Y. and Wang, L. (2005). The WRKY transcription factor superfamily: its origin in eukaryotes and expansion in plants. *BMC Evol. Biol.* **5**, 1.

Zhang, Y. and Wu, M. (1993). Fluorescence microscopy on dynamic changes of frx B distribution in *Chlamydomonas reinhardtii*. *Protoplasma* **172**, 57–63.

Zhang, Y.-H. and Robinson, D.G. (1986a). The endomembranes of *Chlamydomonas reinhardii*: a comparison of the wildtype with the wall mutants *CW 2* and *CW 15*. *Protoplasma* **133**, 186–194.

Zhang, Y.-H. and Robinson, D.G. (1986b). On the fixation of *Chlamydomonas reinhardii*. *Ber. Deutsch. Bot. Ges.* **99**, 179–188.

Zhang, Y.-H. and Robinson, D.G. (1990). Cell-wall synthesis in *Chlamydomonas reinhardtii*: an immunological study on the wild type and wall-less mutants *cw2* and *cw15*. *Planta* **180**, 229–236.

Zhang, Y.-H., Lang, W.C., and Robinson, D.G. (1989). *In vitro* localization of hydroxyproline O-glycosyl transferases in *Chlamydomonas reinhardii*. *Plant Cell Physiol.* **30**, 617–622.

Zhang, Z., Shrager, J., Jain, M., Chang, C.W., Vallon, O., and Grossman, A.R. (2004). Insights into the survival of *Chlamydomonas reinhardtii* during sulfur starvation based on microarray analysis of gene expression. *Eukaryotic Cell* **3**, 1331–1348.

Zhao, B., Schneid, C., Iliev, D., Schmidt, E.-M., Wagner, V., Wollnik, F., and Mittag, M. (2004). The circadian RNA-binding protein CHLAMY 1 represents a novel type heteromer of RNA recognition motif and lysine homology domain-containing subunits. *Eukaryotic Cell* **3**, 815–825.

Zhao, H., Lu, M., Singh, R., and Snell, W.J. (2001). Ectopic expression of a *Chlamydomonas* mt^+-specific homeodomain protein in mt^- gametes initiates zygote development without gamete fusion. *Genes Dev.* **15**, 2767–2777.

Zhao, T., Li, G., Mi, S., Li, S., Hannon, G.J., Wang, X.J., and Qi, Y. (2007). A complex system of small RNAs in the unicellular green alga *Chlamydomonas reinhardtii*. *Genes Dev.* **21**, 1190–1203.

Zicker, A.A., Kadakia, C.S., and Herrin, D.L. (2007). Distinct roles for the 5′ and 3′ untranslated regions in the degradation and accumulation of chloroplast *tufA* mRNA: identification of an early intermediate in the *in vivo* degradation pathway. *Plant Mol. Biol.* **63**, 689–702.

Zito, F., Vinh, J., Popot, J.L., and Finazzi, G. (2002). Chimeric fusions of subunit IV and PctL in the b_6f complex of *Chlamydomonas reinhardtii*. Structural implications and consequences on state transitions. *J. Biol. Chem.* **277**, 12446–12455.

Zorin, B., Hegemann, P., and Sizova, I. (2005). Nuclear-gene targeting by using single-stranded DNA avoids illegitimate DNA integration in *Chlamydomonas reinhardtii*. *Eukaryotic Cell* **4**, 1264–1272.

Index

Page numbers in italics indicate tables or figures.

435

Index to *Chlamydomonas* species other than *C. reinhardtii*

Page numbers in italics indicate tables or figures.

Printed and bound by CPI Group (UK) Ltd, Croydon, CR0 4YY

03/10/2024

01040313-0017

AN
INTRODUCTION
TO THE
MATHEMATICS OF
FINANCIAL
DERIVATIVES